PSA 1994

VOLUME ONE

PSA 1994

PROCEEDINGS OF THE 1994
BIENNIAL MEETING
OF THE
PHILOSOPHY OF SCIENCE
ASSOCIATION

volume one

Symposia and Invited Papers

edited by

DAVID HULL,
MICKY FORBES
&
RICHARD M. BURIAN

1994
Philosophy of Science Association
East Lansing, Michigan

Copyright © 1994 by the
Philosophy of Science Association

All Rights Reserved

No part of this book may be utilized, or reproduced in any form or by any means electronic or mechanical, without written permission from the publishers except in the case of brief quotations embodied in critical articles and reviews.

Library of Congress Catalog Card Number 72-624169

Paper Edition: ISBN 0-917586-35-2
Cloth Edition: ISBN 0-917586-36-0

ISSN: 0270-8647

Manufactured in the United States of America

CONTENTS

Preface ix
PSA 1994 Program xi
Synopsis xxiii

Part I. Philosophy of Biology

1. *The Selection of Alleles and the Additivity of Variance* 3
   Sahotra Sarkar, Dibner Institute MIT and McGill University

2. *Optimization in Evolutionary Ecology* 13
   Robert C. Richardson, University of Cincinnati

3. *The Super Bowl and the Ox-Phos Controversy: "Winner-Take-All" Competition in Philosophy of Science* 22
   Douglas Allchin, Cornell University

4. *Ecological Explanation and the Population-Growth Thesis* 34
   Kristin Shrader-Frechette, University of South Florida

5. *Defending Robustness: the Bacterial Mesosome as a Test Case* 46
   Sylvia Culp, Western Michigan University

Part II. Explanation, Induction, and Linguistic Representation

6. *Explaining Brute Facts* 61
   Eric Barnes, Southern Methodist University

7. *Scientific Explanation: From Covering Law to Covering Theory* 69
   Fritz Rohrlich, Syracuse University

8. *Why There Can't be a Logic of Induction* 78
   Stuart S. Glennan, Butler University

9. *Earman on the Projectibility of Grue* 87
   Marc Lange, University of California, Los Angeles

10. *A Representational Reconstruction of Carnap's Quasianalysis* 96
    Thomas Mormann, Universität München

Part III. Spacetime and Related Matters

11. *Locality/Separability: Is This Necessarily a Useful Distinction?* 107
    James T. Cushing, University of Notre Dame

12. *Spacetime and Holes* 117
    Carolyn Brighouse, Occidental College and University of Southern California

13.	*Non-Turing Computers and Non-Turing Computability* Mark Hogarth, University of Cambridge	126

Part IV. Philosophy of Chemistry

14.	*Spectrometers as Analogues of Nature* Daniel Rothbart, George Mason University	141
15.	*Ideal Reaction Types and the Reactions of Real Alloys* Jeffry L. Ramsey, Oregon State University	149
16.	*Has Chemistry Been at Least Approximately Reduced to Quantum Mechanics?* Eric R. Scerri, London School of Economics	160

Part V. Realism and its Guises

17.	*Could Theoretical Entities Save Realism?* Mohamed Elsamahi, The University of Calgary	173
18.	*Realism, Convergence, and Additivity* Cory Juhl, University of Texas at Austin and Kevin T. Kelly, Carnegie-Mellon University	181
19.	*Austere Realism and the Worldly Assumptions of Inferential Statistics* J.D. Trout, Loyola University of Chicago	190
20.	*Retrieving the Point of the Realism-Instrumentalism Debate: Mach vs. Planck on Science Education Policy* Steve Fuller, University of Durham	200

Part VI. Quantum Mechanics and Cosmology

21.	*On the Paradoxical Aspects of New Quantum Experiments* Lev Vaidman, Tel-Aviv University	211
22.	*The Bohmian Model of Quantum Cosmology* Craig Callender and Robert Weingard, Rutgers University	218
23.	*Should We Believe in the Big Bang?: A Critique of the Integrity of Modern Cosmology* Graeme Rhook and Mark Zangari, La Trobe University	228

Part VII. Statistics and Experimental Reasoning

24.	*On the Nature of Bayesian Convergence* James Hawthorne, University of Oklahoma	241

25.	*The Extent of Dilation of Sets of Probabilities and the Asymptotics of Robust Bayesian Inference* Timothy Herron, Teddy Seidenfeld, and Larry Wasserman, Carnegie-Mellon University	250
26.	*In Search of a Pointless Decision Principle* Prasanta S. Bandyopadhayay, University of Rochester	260
27.	*The New Experimentalism, Topical Hypotheses, and Learning from Error* Deborah G. Mayo, Virginia Polytechnic Institute and State University	270
28.	*Of Nulls and Norms* Peter Godfrey-Smith, Stanford University	280

Part VIII. Historical Case Studies and Methodology

29.	*Experiment, Speculation and Law: Faraday's Analysis of Arago's Wheel* Friedrich Steinle, Georg-August-Universität, Göttingen	293
30.	*Scientists' Responses to Anomalous Data: Evidence from Psychology, History, and Philosophy of Science* William F. Brewer and Clark A. Chinn, University of Illinois at Urbana-Champaign	304
31.	*Methodology, Epistemology and Conventions: Popper's Bad Start* John Preston, University of Reading	314
32.	*Sherlock Holmes, Galileo, and the Missing History of Science* Neil Thomason, University of Melbourne	323
33.	*How to Remain (Reasonably) Optimistic: Scientific Realism and the "Luminiferous Ether"* John Worrall, The London School of Economics	334

Part IX. Quantum Mechanics: Decoherence and Related Matters

34.	*Making Sense of Approximate Decoherence* Guido Bacciagaluppi and Meir Hemmo, University of Cambridge	345
35.	*The 'Decoherence' Approach to the Measurement Problem in Quantum Mechanics* Andrew Elby, University of California at Berkeley	355
36.	*Wavefunction Tails in the Modal Interpretation* Michael Dickson, University of Notre Dame	366

Part X. Games, Explanations, Authority, and Justification

37. *The Microeconomic Interpretation of Games* — 379
    Chantale LaCasse and Don Ross, University of Ottawa

38. *Ceteris Paribus Laws and Psychological Explanations* — 388
    Charles Wallis, University of Rochester

39. *The Epistemic Authority of Expertise* — 398
    Robert Pierson, York University

40. *Circular Justifications* — 406
    Harold I. Brown, Northern Illinois University

Part XI. Philosophy of Psychology and Perception

41. *Is Cognitive Neuropsychology Possible?* — 417
    Jeffrey Bub, University of Maryland

42. *The Scope of Psychology* — 428
    Keith Butler, Washington University

43. *Perception and Proper Explanatory Width* — 437
    Mark Rollins, Washington University

44. *Is Seeing Believing?* — 446
    David Hilbert, California Institute of Technology

45. *Simplicity, Cognition and Adaptation: Some Remarks on Marr's Theory of Vision* — 454
    Daniel Gilman, The Pennsylvania State University

PREFACE

This volume contains the contributed papers for the 1994 Biennial meeting of the Philosophy of Science Association held in New Orleans, Louisiana. It is published before the meeting, with a second volume, containing the symposia papers to be published afterward. The organization of the papers records that of the program. Northwestern University provides generous support for the preparation of the volumes.

The 45 contributed papers contained in this volume were selected from a field of 139 submissions under very tight time restrictions. Because of the limited time available for presentation of papers at PSA 1994 and because of the need to represent a broad sampling of topics, a number of excellent papers submitted for the meeting could not be included on the program. The heroic job of refereeing was performed by the program committee, consisting of Richard M. Burian (chair), David Gooding, Gary Hatfield, Don Howard, Helen Longino, Miriam Solomon, and James Woodward. Because of a high concentration of papers in the philosophy of physics, we also called on the services of RIG Hughes (University of South Carolina), who graciously agreed to referee a batch of contributed papers. We are especially grateful to him for his help. Richard Burian would like to express special appreciation to Carolyn Furrow, who undertook a very large secretarial burden for the Program Committee in addition to her enormous work load as sole secretary of the Center for the Study of Science in Society at Virginia Tech. As always, she did the entire job with grace and efficiency. The Center for the Study of Science in Society and the College of Arts and Sciences at Virginia Tech also provided generous support to the Program Committee, for which we are grateful. Micky Forbes supervised the editing and processing of the papers to produce uniform camera-ready copy. The PSA Business Office saw the copy through to publication.

We owe thanks to the members of the Program Committee and RIG Hughes for the considerable time and energy they contributed on behalf of the PSA and to our authors for their cooperation with the imperatives of a tight publication schedule.

We are grateful to Northwestern University for financial support and its Academic Computing Center for the use of their microcomputing facilities. Special thanks are due Wendy Ward whose expertise and imagination in utilizing these facilities shows up on nearly every page of the volume.

Richard M. Burian	David Hull and
Center for the Study of Science in Society	Micky Forbes
Virginia Polytechnic Institute and	Department of Philosophy
State University	Northwestern University

Program

* Indicates the Organizer(s) of a Symposium

Thursday, October 13, 3:00 - 10:00 p.m.

PSA Governing Board

Friday, October 14, 9:00 - 11:45 a.m.

Philosophy of Psychology as Philosophy of Science
Symposium

    Chair: Gary Hatfield, University of Pennsylvania

    Speakers: *Introduction*
                           Gary Hatfield, University of Pennsylvania

                           Psychology and the Semantic View
                         Valerie Hardcastle, Virginia Polytechnic Institute
                             and State University

                         The Accuracy Problem in Social Perception Research
                         Barbara Von Eckardt, University of
                            Nebraska—Lincoln

                         What is Psychophysics?
                         Lawrence A. Shapiro*, University of Wisconsin

Fields, Particles, and Quantum Theories
Symposium

    Chair: Linda Wessels, Indiana University

    Speakers: *The History and Philosophy of Quantum Field*
                                Theory
                           Don Robinson*, University of Illinois at Chicago

                         What are Quanta and What Does it Matter ?
                         Nick Huggett*, Rutgers University

                         More Ado About Nothing
                         Michael Redhead, Cambridge University

                         Remarks on the 'Interpretation' of Quantum Field
                                Theory
                         Simon Saunders, Harvard University

Philosophy of Biology
Contributed Papers

    Chair: Elizabeth Lloyd, University of California—Berkeley

    Papers: *The Selection of Alleles and the Additivity of Variance*
                            Sahotra Sarkar, Dibner Institute, MIT
                              and McGill University

                            Optimization in Evolutionary Ecology
                            Robert C. Richardson, University of Cincinnati

                            The Super Bowl and the Ox-Phos Controversy:
                              "Winner-Take-All" Competition in Philosophy of
                            Science
                            Douglas Allchin, Cornell University

                            Ecological Explanation and the Population-Growth
                              Thesis
                            Kristin Shrader-Frechette, University of South Florida

                            Defending Robustness: The Bacterial Mesosome as
                              a Test Case
                            Sylvia Culp, Western Michigan University

Explanation, Induction, and Linguistic Representation
Contributed Papers

    Chair: Jane Duran, University of California—Santa Barbara

    Papers: *Explaining Brute Facts*
                            Eric Barnes, Southern Methodist University

                            Scientific Explanation: From Covering Law
                              to Covering Theory
                            Fritz Rohrlich, Syracuse University

                            Why There Can't be a Logic of Induction
                            Stuart S. Glennan, Butler University

                            Earman on the Projectibility of Grue
                            Marc Lange, University of California at Los Angeles

                            A Representational Reconstruction of Carnap's
                              Quasianalysis
                            Thomas Mormann, Universität München

Friday, October 14, 1:15 - 3:00 p.m.

Spacetime and Related Matters
Contributed Papers

  Chair: Don Howard, University of Kentucky

  Papers: *Locality/Separability: Is This Necessarily a Useful Distinction?*
          James T. Cushing, University of Notre Dame

          Spacetime and Holes
          Carolyn Brighouse, Occidental College and University of Southern California

          Non-Turing Computers and Non-Turing Computability
          Mark Hogarth, University of Cambridge

Philosophy of Chemistry
Contributed Papers

  Chair: Clark Glymour, Carnegie-Mellon University

  Papers: *Spectrometers as Analogues of Nature*
          Daniel Rothbart, George Mason University

          Ideal Reaction Types and the Reactions of Real Alloys
          Jeffry L. Ramsey, Oregon State University

          Has Chemistry Been at Least Approximately Reduced to Quantum Mechanics?
          Eric R. Scerri, London School of Economics

Friday, October 14, 3:15 - 5:20 p.m.

Symposium in Honor of Robert S. Cohen: Perspectives on the Social Relations of History and Philosophy of Science
With financial support from the International Union of History and Philosophy of Science

  Chair: Richard Burian*, Virginia Polytechnic Institute and State University

  Speakers: Dominique Lecourt, Université de Paris VII—Denis Diderot
          Kostas Gavroglu, National Technical University, Athens
          Marx Wartofsky*, Bernard Baruch College, City University of New York

  Response: Robert S. Cohen, Boston University

Feminist Perspectives on Special Sciences
Symposium

    Chair: Vivian Weil, Illinois Institute of Technology

    Speakers: *Descriptive and Normative Aspects of Philosophy of Science: Methodological Issues in Feminist Philosophy of Science*
Elizabeth Potter, Mills College

                       Methodological Issues in the Construction of Gender as a Meaningful Variable in Scientific Studies of Cognition
Phyllis Rooney, Oakland University

                       Evidence in and for Research into Sex Differences in Cognitive Abilities in Behavioral Endocrinology
Lynn Hankinson Nelson*, Rowan College of New Jersey and Jack Nelson*, Temple University

    Commentator: Peter Machamer, University of Pittsburgh

Do Explanations or Predictions (or Neither) Provide More Evidential support for Scientific Theories?
Symposium

    Chair: Helen Longino, Rice University and University of Minnesota

    Speakers: *Dynamics of Theory Change: The Role of Prediction*
Stephen Brush, University of Maryland

                       Empirical and Rational Components in Scientific Confirmation
Abner Shimony, Boston University

                       The Historical Character of Evidence
Peter Achinstein*, Johns Hopkins University

Realism and its Guises
Contributed Papers

    Chair: Arthur Fine, Northwestern University

    Papers: *Could Theoretical Entities Save Realism?*
Mohamed Elsamahi, The University of Calgary

                       Realism, Convergence, and Additivity
Cory Juhl, University of Texas at Austin
and Kevin T. Kelly, Carnegie-Mellon University

Austere Realism and the Worldly Assumptions of Inferential Statistics
J.D. Trout, Loyola University of Chicago

Retrieving the Point of Realism-Instrumentalism Debate: Mach vs. Planck on Science Education Policy
Steve Fuller, University of Durham

Quantum Mechanics and Cosmology
Contributed Papers

Chair: George Gale, University of Missouri, Kansas City

Papers: *On the Paradoxical Aspects of New Quantum Experiments*
Lev Vaidman, Tel-Aviv University

The Bohmian Model of Quantum Cosmology
Craig Callender and Robert Weingard, Rutgers University

Should We Believe in the Big Bang?: A Critique of the Integrity of Modern Cosmology
Graeme Rhook and Mark Zangari, La Trobe University

Search Heuristics, Experimentation, and Technology in Molecular and Cell Biology
Symposium

Chair: Robert Richardson, University of Cincinnati

Speakers: *Deciding on the Data: Epistemological Problems Surrounding Instruments and Research Techniques in Cell Biology*
William Bechtel, Georgia State University

Reasoning Strategies in Molecular Biology
Lindley Darden, University of Maryland at College Park

Interactions Among Theory, Experiment, and Technology in Molecular Biology: The Solution of the Antibody Diversity Problem
Kenneth F. Schaffner*, George Washington University

Friday, October 14, 5:45 - 7:00 p.m.

Presidential Address

    Introduction: Bas van Fraassen, Princeton University

                              Viewing Science
                              Ronald Giere, University of Minnesota

Saturday, October 15, 9:00 - 11:45 a.m.

Foundational Projects in Mathematics at the Beginning of the 20th Century in Their Systematic and Historical Contexts
Symposium

    Chair: Alan W. Richardson*, University of California at San Diego

    Speakers: *The Contemporary Interest of an Old Doctrine*
                              William Demopoulos, University of Western Ontario

                              The Role of Intuition in Poincaré's Philosophy of Mathematics
                              Janet Folina, Macalester College

                              Hilbert and the Axiomatization of Physics
                              Peter Clark, University of St. Andrews

                              Hilbert and the Axiomatization of Physics
                              Michael Hallett, McGill University

Science and Philosophy in the Classic Texts
Symposium

    Chair: Daniel Jones, National Endowment for the Humanities

    Speakers: *Putting Philosophy of Science to the Test: The Case of Aristotle's Biology*
                              James G. Lennox, University of Pittsburgh

                              Methodological Judgment and Critical Thinking in Galileo's Dialogue
                              Maurice A. Finocchiaro*, University of Nevada at Las Vegas

                              Newton's Opticks as Classic: On Teaching the Texture of Science
                              Dennis L. Sepper, University of Dallas

    Commentator: Ernan McMullin, University of Notre Dame

Integrating Cognitive and Social Models of Science
Symposium

    Chair: Martin Leckey, Monash University

    Speakers: *Cognitive and Social Dimensions of Persuasion in Science*
                          Alvin Goldman, University of Arizona

                          Multivariate Models of Scientific Change
                          Miriam Solomon, Temple University

                          Why Science? Nonreductionist Integrations of Cognitive and Social Explanations
                          Paul Thagard*, University of Waterloo

    Commentator: Peter Taylor, Cornell University

Statistics and Experimental Reasoning
Contributed Papers

    Chair: John Earman, University of Pittsburgh

    Papers: *On the Nature of Bayesian Convergence*
                          James Hawthorne, University of Oklahoma

                          The Extent of Dilation of Sets of Probabilities and the Asymptotics of Robust Bayesian Inference
                          Timothy Herron, Teddy Seidenfeld, and Larry Wasserman, Carnegie-Mellon University

                          In Search of a Pointless Decision Principle
                          Prasanta S. Bandyopadhayay, University of Rochester

                          The New Experimentalism, Topical Hypotheses, and Learning from Error
                          Deborah G. Mayo, Virginia Polytechnic Institute and State University

                          Of Nulls and Norms
                          Peter Godfrey-Smith, Stanford University

Historical Case Studies and Methodology
Contributed Papers

    Chair: David Gooding, University of Bath

    Papers: *Experiment, Speculation and Law: Faraday's Analysis of Arago's Wheel*
                          Friedrich Steinle, Georg-August-Universität, Göttingen

Scientists' Responses to Anomalous Data: Evidence from Psychology, History, and Philosophy of Science
William F. Brewer and Clark A. Chinn, University of Illinois at Urbana-Champaign

Methodology, Epistemology and Conventions: Popper's Bad Start
John Preston, University of Reading

Sherlock Holmes, Galileo, and the Missing History of Science
Neil Thomason, University of Melbourne

How to Remain (Reasonably) Optimistic: Scientific Realism and the "Luminiferous Ether"
John Worrall, The London School of Economics

Saturday, October 15, 12:00 - 1:00 p.m.

PSA Business Meeting

Saturday, October 15, 1:15 - 3:00 p.m.

Quantum Mechanics: Decoherence and Related Matters
Contributed Papers

    Chair: R.I.G. Hughes, University of South Carolina

    Papers: *Making Sense of Approximate Decoherence*
                           Guido Bacciagaluppi and Meir Hemmo, University of Cambridge

                         The 'Decoherence' Approach to the Measurement Problem in Quantum Mechanics
                         Andrew Elby, University of California at Berkeley

                         Wavefunction Tails in the Modal Interpretation
                         Michael Dickson, University of Notre Dame

Games, Explanations, Authority, and Justification
Contributed Papers

    Chair: Cristina Bicchieri, Carnegie-Mellon University

    Papers: *The Microeconomic Interpretation of Games*
                           Chantale LaCasse and Don Ross, University of Ottawa

                         Ceteris Paribus Laws and Psychological Explanations
                         Charles Wallis, University of Rochester

The Epistemic Authority of Expertise
Robert Pierson, York University

Circular Justifications
Harold I. Brown, Northern Illinois University

Saturday, October 15, 3:15 - 5:20 p.m.

Policy Issues in Human Genetics
Symposium

    Chair: Elliott Sober*, University of Wisconsin

    Speakers: *Implications of the Human Genome Project*
                           Philip Kitcher, University of California at San Diego

                           'Informed Consent' in Mass Genetic Screening
                           Diane Paul, University of Massachusetts—Boston

                           The Political Economy of the Human Genome Project
                           Alexander Rosenberg, University of California at
                              Riverside

Probability and the Art of Judgment
Symposium

    Chair: Maria Carla Galavotti, Universitá di Bologna

    Speakers: *Prevision: Its Logical Limits and its Subjective Span*
                           Bas van Fraassen, Princeton University

                           Learning from Uncertain Evidence
                           Brian Skyrms*, University of California at Irvine

                           Probability and Confidence
                           S.L. Zabell, Northwestern University

Sunday, October 16, 9:00 - 11:45 a.m.

Unity and Disunity in Physics and Biology
Symposium

    Chair: James Woodward, California Institute of Technology

    Speakers: *The Dappled World*
                           Nancy Cartwright, London School of Economics

Unified Theories, Disparate Things
Margaret Morrison*, Trinity College

The Limits of Scientific Theory
John Dupré, Stanford University

Pluralism, Normative Naturalism, and Biological Taxonomy
Marc Ereshefsky, The University of Calgary

Discourse, Practice, Context: From HPS to Interdisciplinary Science Studies
Symposium

    Chair: Alison Wylie*, University of Western Ontario

    Speakers: *Cultural Studies as a Model for Science Studies*
                                    Joseph Rouse, Wesleyan University

                                Contextualizing Science: From Science Studies to Cultural Studies
                                Vassiliki Betty Smocovitis, University of Florida

                                After Representation: Science Studies in the Performative Idiom
                                Andrew Pickering, University of Illinois, Urbana-Champaign

                                Scientific Practice and Normativity
                                Brian S. Baigrie, University of Toronto

The Technological Infrastructure of Science
Symposium

    Chair: Joseph C. Pitt*, Virginia Polytechnic Institute and State University

    Speakers: *Between the Magnet and the Marketplace: Nuclear Physics at the Department of Terrestrial Magnetism During the 1930s*
                                Michael Aaron Dennis, Cornell University

                                Scientific Objects versus Technological Artifacts
                                Peter Kroes, Eindhoven Technological University

                                Meaning in a Material Medium
                                Davis Baird, University of South Carolina

    Commentator: Allan Franklin, University of Colorado

Philosophy of Psychology and Perception
Contributed Papers

    Chair: Nancy Nersessian, Georgia Institute of Technology

    Papers: *Is Cognitive Neuropsychology Possible?*
                           Jeffrey Bub, University of Maryland

                           The Scope of Psychology
                           Keith Butler, Washington University

                         Perception and Proper Explanatory Width
                         Mark Rollins, Washington University

                         Is Seeing Believing?
                         David Hilbert, California Institute of Technology

                         Simplicity, Cognition and Adaptation: Some
                            Remarks on Marr's Theory of Vision
                         Daniel Gilman, The Pennsylvania State University

Notes: In addition to the sessions listed above, the PSA will co-sponsor various sessions initiated by HSS and 4S. To avoid conflicts with sessions planned by the other societies it may be necessary to move some of the PSA sessions. There will also be some interest group meetings sponsored by members of the three societies. Two that are initiated by members of the PSA are known as of this date. These are:

Interest group in the Philosophy of the Social Sciences, organized by Professor Harold Kincaid (University of Alabama, Birmingham). The interest group will have a lunch meeting at the noon hour, Friday, October 14.

International History, Philosophy, and Science Teaching Group, meeting organized by Professor Michael Matthews (University of New South Wales). The interest group will meet 7:30 - 9:30 p.m. on Friday, October 14.

Synopsis

The following brief summaries, arranged here alphabetically by author, provide an introduction to each of the papers in this volume.

1. *The Super Bowl and the Ox-Phos Controversy: "Winner-take-all" Competition in Philosophy of Science.* **Douglas Allchin.** Several diagrams and tables from review articles during the Ox-Phos Controversy serve as an occasion to assess the nature of competition in models of theory choice in science. Many models follow "Super-Bowl" principles of polar, either-or, winner-take-all competition. A significant alternative highlighted by this episode, however, is the differentiation of domains. Incommensurability and the partial divergence of overlapping domains serve both as signals and context for shifting frameworks of competition. Appropriate strategies may thus help researchers diagnose the status of competition and shape their research accordingly.

2. *Making Sense of Approximate Decoherence.* **Guido Bacciagaluppi and Meir Hemmo.** In realistic situations where a macroscopic system interacts with an external environment, decoherence of the quantum state, as derived in the decoherence approach, is only approximate. We argue that this can still give rise to facts, provided that during the decoherence process states that are, respectively, always close to eigenvectors of pointer position and record observable are correlated. We show in a model that this is always the case.

3. *In Search of a Pointless Decision Principle.* **Prasanta S. Bandyopadhayay.** I advance a decision principle called the "weak dominance principle"(WDP) based on the interval notion of probability to deal with the Ellsberg type paradox(ETP). Given ETP, I explain three things: (i) Why WDP is a better principle than many principles e.g. Kyburg's principle and Gardenfors and Sahlin's principle, (ii) Why one should not, contrary to many principles, expect a unique solution in ETP, and (iii) What is the relationship between WDP and the principles mentioned above. I prove also that WDP induces a strict partial ordering on the intervals to which it is applied.

4. *Explaining Brute Facts.* **Eric Barnes.** I aim to show that one way of testing the mettle of a theory of scientific explanation is to inquire what that theory entails about the status of brute facts. Here I consider the nature of brute facts, and survey several contemporary accounts of explanation vis a vis this subject (the Friedman-Kitcher theory of explanatory unification, Humphreys' causal theory of explanation, and Lipton's notion of 'explanatory loveliness'). One problem with these accounts is that they seem to entail that brute facts represent a gap in scientific understanding. I argue that brute facts are non-mysterious and indeed are even explainable by the lights of Salmon's ontic conception of explanation (which I endorse here). The plausibility of various models of explanation, I suggest, depends to some extent on the tendency of their proponents to focus on certain examples of explananda - I ponder brute facts qua explananda here as a way of helping us to recognize this dependency.

5. *Scientists' Responses to Anomalous Data: Evidence from Psychology, History, and Philosophy of Science.* **William F. Brewer and Clark A. Chinn.** This paper presents an analysis of the forms of response that scientists make when confronted with anomalous data. We postulate that there are seven ways in which an individual who currently holds a theory can respond to anomalous data: (1) ignore the data; (2) reject the data; (3) exclude the data from the domain of the current theory; (4) hold the data in abeyance; (5) reinterpret the data; (6) make peripheral changes to the current theory; or (7) change the theory. We analyze psychological experiments and

cases from the history of science to support this proposal. Implications for the philosophy of science are discussed.

6. *Spacetime and Holes.* **Carolyn Brighouse.** John Earman and John Norton have argued that substantivalism leads to a radical form of indeterminism within local spacetime theories. I compare their argument to more traditional arguments typical in the Relationist/Substantivalist dispute and show that they all fail for the same reason. All these arguments ascribe to the substantivalist a particular way of talking about possibility. I argue that the substantivalist is not committed to the modal claims required for the arguments to have any force, and show that this naturally leads to an alteration in the way determinism is characterized for local spacetime theories.

7. *Circular Justifications.* **Harold I. Brown.** The thesis of this paper is that philosophers are often too hasty in rejecting justifications because the argument that yields the justification is circular. Circularity is distinguished from vicious circularity and several examples are examined in which a proposed justification is circular in a precise sense, but not viciously circular. These include an observational procedure which could yield a velocity in excess of the velocity of light even though the impossibility of such velocities is assumed at a key step in analyzing the data, and an argument that uses a specific argument form to show that that form is invalid.

8. *Is Cognitive Neuropsychology Possible?* **Jeffrey Bub.** The aim of cognitive neuropsychology is to articulate the functional architecture underlying normal cognition, on the basis of cognitive performance data involving brain-damaged subjects. Glymour (forthcoming) formulates a discovery problem for cognitive neuropsychology, in the sense of formal learning theory, concerning the existence of a reliable methodology, and argues that the problem is insoluble: granted certain apparently plausible assumptions about the form of neuropsychological theories and the nature of the available evidence, a reliable methodology does not exist! I argue for a reformulation of the discovery problem in terms of an alternative characterization of relevant evidence in neuropsychology.

9. *The Scope of Psychology.* **Keith Butler.** Descartes' conception of the mind as a private entity, separable (in various ways) from the body and the world around it, has come under increasingly vigorous attack in recent years. A new and very different sort of expansion of the scope of psychology has recently been advanced by John Haugeland, who argues quite ingeniously that the Cartesian divisions between mind, body, and world are psychologically otiose. I demur, citing several traditional individuative criteria that are immune to Haugeland's case.

10. *The Bohmian Model of Quantum Cosmology.* **Craig Callender and Robert Weingard.** A realist causal model of quantum cosmology (QC) is developed. By applying the de Broglie-Bohm interpretation of quantum mechanics to QC, we resolve the notorious 'problem of time' in QC, and derive exact equations of motion for cosmological dynamical variables. Due to this success, it is argued that if the situation in QC is used as a yardstick by which other interpretations are measured, the de Broglie-Bohm theory seems uniquely fit as an interpretation of quantum mechanics.

11. *Defending Robustness: The Bacterial Mesosome as a Test Case.* **Sylvia Culp.** Rasmussen (1993) argues that, because electron microscopists did not use robustness and would not have been warranted in using it as a criterion for the reality or the artifactuality of mesosomes, the bacterial mesosome serves as a test case for robustness that it fails. I respond by arguing that a more complete reading of the research literature on the mesosome shows that ultimately the more robust body of

data did not support the mesosome and that electron microscopists used and were warranted in using robustness as a criterion for the artifactuality of mesosomes.

12. *Locality/Separability: Is This Necessarily a Useful Distinction?* **James T. Cushing.** In the philosophy of science, we are to assess critically and on their intrinsic merits various proposals for a consistent interpretation of quantum mechanics, including resolutions of the measurement problem and accounts of the long-range Bell correlations. In this paper I suggest that the terms of debate may have been so severely and unduly constrained by the reigning orthodoxy that we labor unproductively with an unhelpful vocabulary and set of definitions and distinctions. I present an alternative conceptual framework, free of many of the standard conundrums.

13. *Wavefunction Tails in the Modal Interpretation.* **Michael Dickson.** I review the modal interpretation of quantum mechanics, some versions of which rely on the biorthonormal decomposition of a statevector to determine which properties are physically possessed. Some have suggested that these versions fail in the case of inaccurate measurements, i.e., when one takes tails of the wavefunction into account. I show that these versions of the modal interpretation are satisfactory in such cases. I further suggest that a more general result is possible, namely, that these versions of the modal interpretation never encounter the sort of trouble that has been claimed to arise in the case of inaccurate measurement.

14. *The 'Decoherence' Approach to the Measurement Problem in Quantum Mechanics.* **Andrew Elby.** Decoherence results from the dissipative interaction between a quantum system and its environment. As the system and environment become entangled, the reduced density operator describing the system "decoheres" into a mixture (with the interference terms damped out). This formal result prompts some to exclaim that the measurement problem is solved. I will scrutinize this claim by examining how modal and relative-state interpretations can use decoherence. Although decoherence cannot rescue these interpretations from general metaphysical difficulties, decoherence may help these interpretations to pick out a preferred basis. I will explore whether decoherence solves nagging technical problems associated with selecting a preferred basis.

15. *Could Theoretical Entities Save Realism?* **Mohamed Elsamahi.** Hacking and other entity realists suggest a strategy to build scientific realism on a stronger foundation than inference to the best explanation. They argue that if beliefs in the existence of theoretical entities are derived from experimentation rather than theories, they can escape the antirealist's criticism and provide a stronger ground for realism. In this paper, an outline and a critique of entity realism are presented. It will be argued that entity realism cannot stand as a separate position from classical realism. Thus, entity realism cannot avoid the problems facing classical realism..

16. *Retrieving the Point of the Realism-Instrumentalism Debate: Mach vs. Planck on Science Education Policy.* **Steve Fuller.** I aim to recover some of the original cultural significance that was attached to the realism-instrumentalism debate (RID) when it was hotly contested by professional scientists in the decades before World War I. Focusing on the highly visible Mach-Planck exchange of 1908-13, I show that arguments about the nature of scientific progress were used to justify alternative visions of science education. Among the many issues revealed in the exchange are realist worries that instrumentalism would subserve science entirely to human interests, as well as instrumentalist worries that realism could become the basis of a science-based religion. I conclude by addressing some issues relating to RID that are now occluded because of Planck's triumph over Mach.

17. *Simplicity, Cognition and Adaptation: Some Remarks on Marr's Theory of Vision.* **Daniel Gilman.** A large body of research in computational vision science stems from the pioneering work of David Marr. Recently, Patricia Kitcher and others have criticized this work as depending upon optimizing assumptions, assumptions which are held to be inappropriate for evolved cognitive mechanisms just as anti-adaptationists (e.g., Lewontin and Gould) have argued they are inappropriate for other evolved physiological mechanisms. The paper discusses the criticism and suggests that it is, in part, misdirected. It is further suggested that the criticism leads to interesting questions about how one formulates constraints--across "levels of organization" and disciplinary boundaries--on one's models of complex systems, such as human vision.

18. *Why There Can't be a Logic of Induction.* **Stuart S. Glennan.** In this paper I offer a criticism of Carnap's inductive logic which also applies to other formal methods of inductive inference. Criticisms of Carnap's views have typically centered upon the justification of his particular choice of inductive method. I argue that the real problem is not that there is an agreed upon method for which no justification can be found, but that different methods are justified in different circumstances.

19. *Of Nulls and Norms.* **Peter Godfrey-Smith.** Neyman-Pearson methods in statistics distinguish between Type I and Type II errors. Through rigid control of Type I error, the "null" hypothesis typically receives the benefit of the doubt. I compare philosophers' interpretations of this feature of Neyman-Pearson tests with interpretations given in statistics textbooks. The pragmatic view of the tests advocated by Neyman, largely rejected by philosophers, lives on in many textbooks. Birnbaum thought the pragmatic view has a useful "heuristic" role in understanding testing. I suggest that it may have the opposite effect.

20. *On the Nature of Bayesian Convergence.* **James Hawthorne.** The objectivity of Bayesian induction relies on the ability of evidence to produce a *convergence to agreement* among agents who initially disagree about the plausibilities of hypotheses. I will describe three sorts of Bayesian convergence. The first reduces the objectivity of inductions about simple "occurrent events" to the objectivity of posterior probabilities for theoretical hypotheses. The second reveals that evidence will generally induce *converge to agreement* among agents on the posterior probabilities of theories *only if* the convergence is 0 or 1. The third establishes conditions under which evidence will *very probably* compel posterior probabilities of theories to converge to 0 or 1.

21. *The Extent of Dilation of Sets of Probabilities and the Asymptotics of Robust Bayesian Inference.* **Timothy Herron, Teddy Seidenfeld, and Larry Wasserman.** We report two issues concerning diverging sets of Bayesian (conditional) probabilities—divergence of "posteriors"—that can result with increasing evidence. Consider a set $\mathcal{P}$ of probabilities typically, but not always, based on a set of Bayesian "priors." Fix E, an event of interest, and X, a random variable to be observed. With respect to $\mathcal{P}$, when the set of conditional probabilities for E, given X, strictly contains the set of unconditional probabilities for E, for each possible outcome $X = x$, call this phenomenon dilation of the set of probabilities (Seidenfeld and Wasserman 1993). Thus, *dilation* contrasts with the asymptotic merging of posterior probabilities reported by Savage (1954) and by Blackwell and Dubins (1962).

(1) In a wide variety of models for Robust Bayesian inference the extent to which X dilates E is related to a model specific index of how far key elements of $\mathcal{P}$ are from a distribution that makes X and E independent.

(2) At a fixed confidence level, $(1-\alpha)$, Classical interval estimates A_n for, e.g., a Normal mean θ have length $O(n^{-1/2})$ (for sample size n). Of course, the confidence

level correctly reports the (prior) probability that $\theta \in A_n$, $P(A_n) = 1-\alpha$, independent of the prior for θ. However, as shown by Pericchi and Walley (1991), if an ε-contamination class is used for the prior on the parameter θ, there is asymptotic (posterior) dilation for the A_n, given the data. If, however, the intervals A'_n are chosen with length $O(\sqrt{\log(n)}/n)$, then there is no asymptotic dilation. We discuss the asymptotic rates of dilation for ClassClassical and Bayesian interval estimates and relate these to Bayesian hypothesis testing.

22. *Is Seeing Believing?* **David Hilbert.** One of the traditional problems of philosophy is the nature of the connection between perceptual experience and empirical knowledge. That there is an intimate connection between the two is rarely doubted. Three case studies of visual deficits due to brain damage are used to motivate the claim that perceptual experience is neither necessary nor sufficient for perceptual knowledge. Acceptance of this claim leaves a mystery as to the epistemic role, if any, of perceptual experience. It is argued that one function of perceptual experience is to provide information about the sources of beliefs, both as to which perceptual modality and within a given modality. This information is useful in assessing the reliability of perceptual beliefs.

23. *Non-Turing Computers and Non-Turing Computability.* **Mark Hogarth.** A true Turing machine (TM) requires an infinitely long paper tape. Thus a TM can be housed in the infinite world of Newtonian spacetime (the spacetime of common sense), but not necessarily in our world, because our world—at least according to our best spacetime theory, general relativity—may be finite. All the same, one can argue for the "existence" of a TM on the basis that there is no such housing problem in some other relativistic worlds that are similar ("close") to our world. But curiously enough—and this is the main point of this paper—some of these close worlds have a special spacetime structure that allows TMs to perform certain Turing unsolvable tasks. For example, in one kind of spacetime a TM can be used to solve first-order predicate logic and the halting problem. And in a more complicated spacetime, TMs can be used to decide arithmetic. These new computers serve to show that Church's thesis is a thoroughly contingent claim. Moreover, since these new computers share the fundamental properties of a TM in ordinary operation (e.g. intuitive, finitely programmed, limited in computational capability), a computability theory based on these non-Turing computers is no less worthy of investigation than orthodox computability theory. Some ideas about this new mathematical theory are given.

24. *Realism, Convergence, and Additivity.* **Cory Juhl and Kevin T. Kelly.** In this paper, we argue for the centrality of countable additivity to realist claims about the convergence of science to the truth. In particular, we show how classical sceptical arguments can be revived when countable additivity is dropped.

25. *The Microeconomic Interpretation of Games.* **Chantale LaCasse and Don Ross.** This paper is part of a larger project defending of the foundations of microeconomics against recent criticisms by philosophers. Here, we undermine one source of these criticisms, arising from philosophers' disappointment with the performance of microeconomic tools, in particular game theory, when these are applied to normative decision theory. Hollis and Sugden have recently articulated such disappointment in a sophisticated way, and have argued on the basis of it that the economic conception of rationality is inadequate. We argue, however, that their claim rests upon a misunderstanding of the concept of a game as it is used in microeconomics.

26. *Earman on the Projectibility of Grue.* **Marc Lange.** In *Bayes or Bust?*, John Earman attempts to express in Bayesian terms a sense of "projectibility" in which it is

logically impossible for "All emeralds are green" and "All emeralds are grue" simultaneously to be projectible. I argue that Earman overlooks an important sense in which these two hypotheses cannot both be projectible. This sense is important because it allows projectibility to be connected to lawlikeness, as Goodman intended. Whether this connection suggests a way to resolve Goodman's famous riddle remains unsettled, awaiting an account of lawlikeness. I explore one line of thought that might prove illuminating.

27. *The New Experimentalism, Topical Hypotheses, and Learning from Error.* **Deborah G. Mayo.** An important theme to have emerged from the new experimentalist movement is that much of actual scientific practice deals not with appraising full-blown theories but with the manifold local tasks required to arrive at data, distinguish fact from artifact, and estimate backgrounds. Still, no program for working out a philosophy of experiment based on this recognition has been demarcated. I suggest why the new experimentalism has come up short, and propose a remedy appealing to the practice of standard error statistics. I illustrate a portion of my proposal using Galison's (1987) experimental narrative on neutral currents.

28. *A Representational Reconstruction of Carnap's Quasianalysis.* **Thomas Mormann.** According to general wisdom, Carnap's quasianalysis is an ingenious but definitively flawed approach to epistemology and philosophy of science. I argue that this assessment is mistaken. Rather, Carnapian quasianalysis can be reconstructed as a special case of a general theory of structural representation. This enables us to exploit some interesting analogies of quasianalysis with the representational theory of measurement. It is shown how Goodman's well-known objections against the quasianalytical approach may be defused in the new framework. As an application, I sketch how the thesis of empirical underdetermination of theories may be elucidated in the framework of quasianalysis.

29. *The Epistemic Authority of Expertise.* **Robert Pierson.** When is it more rational to think for oneself or to defer to the relevant expert? Expertise is either closed-system oriented and lay-person oriented. The first sort is concerned primarily with controlling and manipulating a discipline's defining set of variables as a closed or relatively closed system. The second sort is simply in the business of "advising" clients. I argue that when expert claims are of the first sort, the layperson must defer to the experts; but when experts either extrapolate from their closed-systems, or if they are of the second sort, then the layperson should think for herself.

30. *Methodology, Epistemology and Conventions: Popper's Bad Start.* **John Preston.** Popper's conception of methodology and its relationship to epistemology is examined, and found wanting. Popper argues that positivist criteria of demarcation fail because they are attempts to discover a difference in the natures of empirical science and metaphysics. His alternative to naturalism is that a plausible criterion of demarcation is a proposal for an agreement, or convention. But this conventionalism about methodology is misplaced. Methodological rules are conventions, but which methodological rules are followed by scientists it is not itself a matter of convention. This casts doubt upon the status of Popper's famous criterion of demarcation.

31. *Ideal Reaction Types and the Reactions of Real Alloys.* **Jeffry L. Ramsey.** Research on the oxidation of alloys supports the claim that natural scientists can and do use ideal type concepts when confronted with analytical or computational intractability. In opposition to those who collapse ideal types into 'standard' theoretical concepts, I argue ideal types possess a unique structure, function and axiology. In

phenomenologically complex situations, scientists use these features to articulate experiment with theory generally and in particular to discover new boundary conditions. This conceptual articulation is achieved using models rather than objective perceptual attributes alone. The analysis supports a claim of local rather than global identities of methodology.

32. *Should we Believe in the Big Bang?: A Critique of the Integrity of Modern Cosmology.* **Graeme Rhook and Mark Zangari.** We analyse aspects of the Big Bang program in modern cosmology, with special focus on the strategies employed by its adherents both in defending the theory against anomalous data and in dismissing rival accounts. We illustrate this by critically examining four aspects of Big Bang cosmology: the interpretation of the cosmic red-shift, the explanation of the cosmic background radiation, the inflation hypothesis and the search for dark matter. We conclude that the Big Bang's dominance of contemporary cosmology is not justified by the degree of experimental support it receives relative to rival theories.

33. *Optimization in Evolutionary Ecology.* **Robert C. Richardson.** Optimization models treat natural selection as a process tending to produce maximal adaptedness to the environment, measured on some "criterion scale" defining the optimal phenotype. These models are descriptively adequate if they describe the outcomes of evolutionary processes. They are dynamically adequate if the variables which describe the outcomes also are responsible for those evolutionary outcomes. Optimality models can be descriptively adequate, but dynamically unrealistic. Relying on cases from evolutionary ecology, I provide reasons to question the dynamic adequacy of optimality models, and offer reasons for distinguishing, at least at a theoretical level, between satsificing and optimizing.

34. *Scientific Explanation: From Covering Law to Covering Theory.* **Fritz Rohrlich.** A new model of scientific explanation is proposed: the covering *theory* model. Its goal is understanding. One chooses the appropriate scientific theory and a model within it. From these follows the functioning of the explanandum, i.e. the way in which the model portrays it on one particular cognitive level. It requires an ontology and knowledge of the causal processes, probabilities, or potentialities (propensities) according to which it functions. This knowledge yields understanding. Explanations across cognitive levels demand pluralistic ontologies. An explanation is believed or only accepted depending on the credibility of the theory and the idealizations in the model.

35. *Perception and Proper Explanatory Width.* **Mark Rollins.** Marr's theory of vision is often said to exemplify wide psychology. The claim rests primarily on Marr's appeal to a high level theory of computational functions. I agree that Marr's theory embodies an exemplary form of wide psychology; what is exemplary about it is the appeal to perceptual tasks. But I argue that the result of invoking task considerations is that we should not adhere to Marr's own conception of proper explanatory width. There is no one conception of width that has a priviledged place in explanation.

36. *Spectrometers as Analogues of Nature.* **Daniel Rothbart.** The success of chemistry is directly credited to the capacity of instruments to provide human contact to the structures of physical reality. Empiricist philosophers have given scant attention to instruments as a separate topic of inquiry on the grounds that reliability of instruments is reducible to the epistemology of common sense experience. I argue that the reliability of many modern instruments is based on their design as analogical replication of natural systems. Scientists designed absorption spectrometers as artifi-

cial technological replicas of familiar physical systems. Such designs are generated by analogical projections of theoretical insights from known physical systems to unknown terrain. Instrumentation enables scientists to extend theoretical understanding to previously hidden domains. After exploring this analogical function of instruments, the nature of instrumental data is discussed, followed by an explicit rejection of both skepticism and naive realism. In the end I argue for an experimental realism which lacks any theory-neutral access to the fundamental analogies of nature.

37. *The Selection of Alleles and the Additivity of Variance.* **Sahotra Sarkar.** It is shown that, for technical reasons, the additivity of variance criterion employed by Lloyd (1988) to define a unit of selection is, in almost all models of selection, inconsistent with the possibility that genes are sometimes not the unit of selection. A case when the latter view is particularly attractive is that of heterosis, and the additivity criterion is inadequate in even such an extreme case. The connection between that criterion and the so-called "fundamental theorem of natural selection" is briefly explored. Skepticism is expressed about the value of measures such as variance in efforts to resolve any of the disputes about the "units of selection."

38. *Has Chemistry Been at Least Approximately Reduced to Quantum Mechanics?* **Eric R. Scerri.** Differing views on reduction are briefly reviewed and a suggestion is made for a working definition of 'approximate reduction'. Ab initio studies in quantum chemistry are then considered, including the issues of convergence and error bounds. This includes an examination of the classic studies on CH2 and the recent work on the Si2C molecule. I conclude that chemistry has not even been approximately reduced.

39. *Ecological Explanation and the Population-Growth Thesis.* **Kristin Shrader-Frechette.** Many ecologists have dismissed alleged ecological laws as tautological or trivial. This essay investigates the epistemological status of one prominent such "law," the population-growth thesis, and argues for 4 claims: (1) Once interpreted, the thesis cannot be denied the status of empirical law on the grounds that it is always and everywhere untestable. (2) Contrary to Peters' (1991) claim, some interpretations of the thesis have significant heuristic power. (3) One can use the reasoning of Brandon (1990), Lloyd (1987), and Sober (1984) to show that some interpretations of the thesis are not *a priori*. (4) Even if the thesis is *a priori*, it has explanatory power as a "schematic law."

40. *Experiment, Speculation and Law: Faraday's Analysis of Arago's Wheel.* **Friedrich Steinle.** Faraday's view of the mutual relation of speculative theories and laws of nature implies that there should be a procedure, leading from speculative considerations to a system of facts and laws in which theories do no longer play any role. In order to make out the degree in which Faraday's claims correspond to his practice, the way in which he gains an explanation of Arago's effect is analyzed. The thesis is proposed that he indeed has a procedure of leaving theories aside. It is intimately connected with certain methodological guidelines of his experimentation.

41. *Sherlock Holmes, Galileo, and the Missing History of Science.* **Neil Thomason.** There is a common (although not universal) claim among historians and philosophers that Copernican theory predicted the phases of Venus. This claim ignores a prominant feature of the writings of, among others, Copernicus, Galileo and Kepler—the possibility that Venus might be self illuminating or translucent. I propose that such over-simplifications of the history of science emerges from "psychological predictivism", the tendency to infer from "E is good evidence for H" to "H

predicts E." If this explanation is correct, then in cases where evidence is less blatant the history of science (and philosophies of science that rely on it) has probably been seriously distorted in a predictivist direction.

42. *Austere Realism and the Worldly Assumptions of Inferential Statistics.* **J.D. Trout.** Inferential statistical tests—such as analysis of variance, t-tests, chi-square and Wilcoxin signed ranks—now constitute a principal class of methods for the testing of scientific hypotheses. In this paper I will consider the role of one statistical concept (statistical power) and two statistical principles or assumptions (homogeneity of variance and the independence of random error), in the reliable application of selected statistical methods. I defend a tacit but widely-deployed naturalistic principle of explanation (E): Philosophers should not treat as inexplicable or basic those correlational facts that scientists themselves do not treat as irreducible. In light of (E), I contend that the conformity of epistemically reliable statistical tests to these concepts and assumptions entails at least the following modest or austere realist commitment: (C) The populations under study have a stable theoretical or unobserved structure that metaphysically grounds the observed values; the objects therefore have a fixed value independent of our efforts to measure them. (C) provides the best explanation for the correlation between the joint use of statistical assumptions and statistical tests, on the one hand, and methodological success on the other.

43. *On the Paradoxical Aspects of New Quantum Experiments.* **Lev Vaidman.** Two recently proposed quantum experiments are analyzed. The first allows to find an object without "touching" it. The second allows to teleport quantum states, transmitting a very small amount of information. It is shown that in the standard approach these experiments are in conflict with the intuitive notions of causality and locality. It is argued that the situation is less paradoxical in the framework of the many-worlds interpretation of quantum theory.

44. *Ceteris Paribus Laws and Psychological Explanations.* **Charles Wallis.** I argue that Fodor's (1991) analysis of ceteris paribus laws fails to underwrite his appeal to such laws in his sufficient conditions for representation. It also renders his appeal to ceteris paribus laws impotent against the major problem for his theory of representation. Finally, Fodor's analysis fails to provide useful solutions to the traditional problems associated with a thoroughgoing understanding of ceteris paribus clauses. The analysis, therefore, fails to bolster Fodor's (1975, 1990) position that special science laws are of necessity ceteris paribus laws and that one must recognize them as scientifically legitimate.

45. *How to Remain (Reasonably) Optimistic: Scientific Realism and the "Luminiferous Ether".* **John Worrall.** Fresnel's theory of light was (a) impressively predictively successful yet (b) was based on an "entity" (the elastic-solid ether) that we now "know" does not exist. Does this case "confute" scientific realism as Laudan suggested? Previous attempts (by Hardin and Rosenberg and by Kitcher) to defuse the episode's anti-realist impact. The strongest form of realism compatible with this case of theory-rejection is in fact *structural* realism. This view was developed by Poincaré who also provided reasons to think that it is the only realist view of theories that really makes sense.

Part I

PHILOSOPHY OF BIOLOGY

The Selection of Alleles and the Additivity of Variance[1]

Sahotra Sarkar

Dibner Institute, MIT and McGill University

1. Introduction

Since the mid-1970s, when philosophical scrutiny of biology began to be focused on evolutionary theory, the center of philosophical attention in that field has been what is called the "units of selection" controversy.[2] In particular, two problems have attracted philosophical and, on occasion, biological attention: (i) should natural selection be regarded as capable of operating on groups of individuals as distinct units, or should such a process only be regarded as a special type of selection on the constituent individuals of such groups even when every member of the group, by virtue of its membership of that group, is affected by selection in an identical manner?; and (ii) similarly, should selection be regarded as "ultimately" operating only on individual alleles since, even when the genotype seems to feel the action of selection more directly, any change in the genotypic composition of a population also consists of a change in its allelic composition?

The first of these problems has sometimes been regarded as an empirical one because the question whether individuals are being identically selected by virtue of their membership of (an adequately defined) group can be given a fecund experimentally determinate interpretation (see Wade 1978). However, it has usually been regarded as an interpretive question since, even when all experimental issues are resolved, any positive answer to the question posed in the last sentence requires a construal of "by virtue of". Its experimental interpretation is hardly straightforward. The second problem has almost always been regarded as an interpretive one since, here, the experimental situation is usually uncontroversial: no proponent of universal genic selectionism would probably dispute a claim that the fitness of an allele may depend on the genotypic composition of a population.

Many proposals have been mooted to resolve these interpretive questions by suggesting criteria to identify the appropriate unit (or units) of selection in a given context. The most influential of the ones that have recently been proposed, and one that is superficially very attractive because it apparently captures some types of experimental practice, has been due to Lloyd (1988). Drawing on earlier insights of Wimsatt (1980), she has argued for the following definition:

> A unit of selection is any entity-type for which there is an additive component of variance for some specific component of fitness, F*, among all entities within a system at that level which does *not appear* as an additive component of variance in (some decomposition of) F* among all entities at any lower level (1988, p. 70; italics in the original).

Though this definition is stated clearly enough, Lloyd's discussion of it less than clear because two different qualifications are introduced: (i) it is explicitly required that "additivity" should be read as "transformable into additivity" though no definite constraint is put on the set of allowable transformations; (ii) implicitly, it often appears as if the criterion should be read as requiring that there be *only* an additive component of variance (see, e.g., 1988, p. 71). In its unadulterated form, following Lloyd, this criterion will be referred to as the "additivity criterion."

In one sense this criterion allows for pluralism because it permits a multiplicity of units on which selection may act: different components of fitness may pick units at different levels. In another, and more interesting sense, while the definition as stated should allow a multiplicity of units even for a *single* component of fitness, Lloyd's discussion usually denies pluralism and implies that only one unit is preferred. This certainly seems to be the case in her discussion of genic selectionism where—in yet another far from clear discussion—she seems to endorse the possibility that the genotype, *rather than* the gene, can sometimes be the unit of selection.

Restricting attention to the problem of genotypic versus genic selection, the primary purpose of this paper is to show that Lloyd's definition, if taken literally (or if taken just without the second qualification which, in any case, is never explicitly stated), always admits the gene as a unit of selection. This is inconsistent with the possibility that the genotype *rather than* the gene is even *sometimes* the unit of selection.[3] The basis for this conclusion is largely technical. In Section 2 a simple model of selection is analyzed in enough detail to show that there is always an additive genetic component of fitness even if selection is manifestly operating only on the genotype. In Section 3 the extent to which this conclusion can be carried over to more complex situations is discussed. These considerations make Lloyd's first (and only explicit) qualification of her criterion largely irrelevant. However, the discussion of Section 3 also shows that the second qualification—whether or not Lloyd intended it—in no way mitigates these conclusions. In fact, explicit endorsement of the second qualification (along with the additivity criterion), when genes are involved, amounts to a conflation of genic selectionism with an endorsement of what Fisher (1930, and until his death) called "the fundamental theorem of natural selection." Finally, Section 4 notes the limitations of the analysis being attempted here and makes some suggestions about ways of handling the "units of selection". However, none of these considerations are, in any strict sense, consequences of the analysis of Section 2.

2. Natural Selection and Additive Genetic Variance

The techniques used in this section are not original. They are based on Kingman (1961), Li (1969) and are very slight refinements of the textbook treatment of Nagylaki (1992). The strategy used in this section is a generalization of one suggested by Lewontin (1991). The basic idea is to construct explicit models of genotypic selection in which the alleles still provide an additive component of variance to the fitness. The strategy followed is the one that is standard in attempts to explore the applicability of the so-called fundamental theorem of natural selection: the change in mean fitness of a population over one generation is computed and the resulting expression is inspected for an additive component of genetic variance.

The model analyzed in this section (often called the "standard selection model") is the simplest one of interest. Nevertheless, it suffices for the argument because it subsumes a very simple example which would separate genotypic from genic selectionists. This is the case of heterosis (heterozygote superiority in fitness over both homozygotes) at one locus with two alleles as seen, for instance, in the case of one of the loci for hemoglobin in several West African populations (see, e. g., Edelstein 1986). Because of the malarial resistance that sickle cell hemoglobin (HbS) provides, the heterozygote with both sickle cell (HbS) and normal (HbA) hemoglobin alleles has slightly higher fitness than the HbA homozygote and much higher fitness than the HbS homozygote. It will be shown below that, even in such cases, the alleles provide an additive component of variance to the mean fitness. For those who are prone to admit at least occasional genotypic, rather than genic, selection but do not regard heterosis as a critical example, the range of the applicability of the conclusions obtained here—it puts no restrictions on genotypic fitnesses even when many loci are involved (as will be discussed in detail in the next section)—should remove any lingering doubts about the impotence of the additivity criterion.

What is known as the "standard selection model" (e. g., Nagylaki 1992, p. 51) makes the following thirteen assumptions: (i) an infinite population; (ii) full diploidy (that is, the locus being considered is not on a special sex-determining chromosome); (iii) two sexes; (iv) dioecy (which means that each individual will be of exactly one of the two sexes); (v) initial genotypic frequencies are the same for both sexes; (vi) random mating; (vii) segregation of alleles in gametes; (viii) independent assortment of alleles; (ix) discrete generations; (x) non-overlapping generations; (xi) selection operating on exactly one locus; (xii) frequency-independent genotypic fitnesses; and (xiii) time-independent genotypic fitnesses. The analysis below does not make assumption (xiii). In general, each of these assumptions can be relaxed to generate models of biological interest, and the effects of relaxing them will be discussed in Section 3.

Notation:[4] In a given generation, let n be the number of alleles at the fitness-determining locus with A_i ($i=1, ..., n$) being the i-th allele; let A_iA_j be the genotype with the i-th and the j-th allele in that order. Let P_{ij} ($i, j=1, ..., n$) be the frequency of A_iA_j, let w_{ij} be its fitness; let p_i ($i=1, ..., n$) be the frequency of A_i; and let $\overline{w}$ be the mean fitness of the population. In the next generation, let P_{ij}', w_{ij}', p_i', and $\overline{w}'$ be the corresponding quantities.

Definitions: The allelic fitness, w_i of the allele A_i ($i = 1, ..., n$) is defined as:

(i) $\quad w_i = \dfrac{1}{p_i} \sum_{j=1}^{n} P_{ij} w_{ij}$.

The change in the frequency, Δp_i, of the allele, A_i ($i = 1, ..., n$), over one generation is defined as:

(ii) $\quad \Delta p_i = p_i' - p_i$.

Similarly, the change in the genotypic fitness, Δw_{ij} ($i, j = 1, ..., n$), over one generation is defined as:

(iii) $\quad \Delta w_{ij} = w_{ij}' - w_{ij}$.

The change in mean fitness, $\Delta \overline{w}$, over one generation is defined as:

(iv) $\quad \Delta \overline{w} = \overline{w}' - \overline{w}$.

Finally, the mean of the fitness changes over one generation, $\overline{\Delta w}$, is defined as:

(v) $\quad \overline{\Delta w} = \sum_{i=1}^{n}\sum_{j=1}^{n} P_{ij}\Delta w_{ij}$.

Analysis:[5] The following eight claims are trivial consequences of the assumptions and definitions:

(vi) $\quad P_{ij} = P_{ji}$;

(vii) $\quad p_i = \sum_{j=1}^{n} P_{ij}$;

(viii) $\quad p_i' = \sum_{j=1}^{n} P_{ij}'$;

(ix) $\quad P_{ij} = p_i p_j$;

(x) $\quad P_{ij}' = p_i' p_j'$;

(xi) $\quad \overline{w} = \sum_{i=1}^{n}\sum_{j=1}^{n} w_{ij} P_{ij}$;

(xii) $\quad \overline{w}' = \sum_{i=1}^{n}\sum_{j=1}^{n} w_{ij}' P_{ij}'$;

(xiii) $\quad P_{ij}' = \dfrac{w_{ij} P_{ij}}{\overline{w}}$.

From (i) and (ix), it follows that:

(xiv) $\quad w_i = \sum_{j=1}^{n} w_{ij} p_j$.

From (xiii):

(xv) $\quad \sum_{j=1}^{n} P_{ij}' = \sum_{j=1}^{n} \dfrac{w_{ij} P_{ij}}{\overline{w}}$.

Using (viii) and (x) for the first step, and (xiv) for the second, from (xv):

(xvi) $\quad p_i' = p_i \dfrac{\sum_{j=1}^{n} w_{ij} p_j}{\overline{w}} = p_i \dfrac{w_i}{\overline{w}}$.

Therefore, using (ii),

(xvii) $\quad \Delta p_i = p_i \dfrac{w_i - \overline{w}}{\overline{w}}$.

It is now easy to calculate the change in mean fitness over one generation. Using (iv), (ix), (x), (xi) and (xii):

(xviii) $\quad \Delta \overline{w} = \sum_{i=1}^{n}\sum_{j=1}^{n}(p_i{}' p_j{}' w_{ij}{}' - p_i p_j w_{ij})$.

Using (ii) and (v), from (xviii), after algebraic manipulation:

(xix) $\quad \Delta \overline{w} = \overline{\Delta w} + \sum_{i=1}^{n}\sum_{j=1}^{n}(2 p_j \Delta p_i + \Delta p_i \Delta p_j) w_{ij}$.

Now, using (xiv) in the second step, and (xvii) in the third,

(xx) $\quad \sum_{i=1}^{n}\sum_{j=1}^{n} p_j \Delta p_i w_{ij} = \sum_{i=1}^{n} \Delta p_i \sum_{j=1}^{n} p_j w_{ij} = \sum_{i=1}^{n} w_i \Delta p_i = \frac{1}{\overline{w}}\sum_{i=1}^{n} p_i w_i (w_i - \overline{w})$.

Algebraic manipulation suffices to show that:

(xxi) $\quad \sum_{i=1}^{n} p_i w_i (w_i - \overline{w}) = \sum_{i=1}^{n} p_i (w_i - \overline{w})^2$.

But $\quad \sum_{i=1}^{n} p_i (w_i - \overline{w})^2$

is half the genic—or additive genetic—variance, V_g, in fitness due to the allelic heterogeneity in the population. Therefore, using (xv) and (xxi), (xix) can be written in the form:

(xxii) $\quad \Delta \overline{w} = \overline{\Delta w} + \frac{V_g}{\overline{w}} + \sum_{i=1}^{n}\sum_{j=1}^{n} \Delta p_i \Delta p_j w_{ij}$.

This is the equation on which the claims of this paper are based.

What equation (xxii) shows is that there is an additive component of variance, that can be attributed to the individual alleles, in the mean fitness of the population.[6] Not only that, the fitness increment due to this additive component is 0 if and only if equilibrium is reached in the allelic composition of the population.[7] Moreover, it is obvious that this component cannot be relegated to some lower level of organization—there are no lower levels in the genetic hierarchy. Therefore, according to Lloyd's additivity criterion, the allele (or, speaking a little loosely, the gene) is a unit of selection. Now note that the model analyzed above made no assumption about the possible interactions between the alleles. In particular, if $n = 2$, it can incorporate cases of heterosis such as that in the case of sickle cell hemoglobin. Moreover, none of the terms on the R.H.S. of equation (xxii) can be interpreted as being proportional to an additive genotypic component of variance, that is, to

$$\sum_{i=1}^{n}\sum_{j=1}^{n} P_{ij}(w_{ij} - \overline{w})^2.$$

So there seems to be no prospect for ever finding genotypic selection in this model using the additivity criterion. The conclusion to be drawn is that the additivity criterion is simply inconsistent with the view that there are circumstances in which genotypes, *rather than* individual alleles, are the units of selection.

3. Technical Comments.[8]

The model above was a particularly simple one, and it did not take into account the qualifications that Lloyd makes. The first two of the following sets of remarks show that the conclusion reached above remains valid in a more general setting. The last explores the relation between the additivity criterion and the so-called fundamental theorem of natural selection. Throughout it will be clear that Lloyd's qualifications do not, in any way, rescue the additivity criterion:

(i) Of the thirteen assumptions that are made in the standard selection model, it is trivial to qualify the last six and still obtain an equation corresponding to (xxii) with one term being proportional to the genic variance. This has been shown by a wide variety of methods, and with different degrees of rigor, in attempts to explore the range of applicability of the so-called fundamental theorem of natural selection which, taken strictly, demands that there only be a genic variance term on the R.H.S. of these equations. While no systematic fully general treatment seems to exist, the one-locus cases are all found in Nagylaki (1992) or can be very easily proved using the methods there. That the multiple locus cases would also give rise to the same result is seen most easily from the exploration of Ewens (1989) of the general range of applicability of that "theorem." The genic variance term is ubiquitous. The sixth assumption (random mating) is almost never required to prove these results. Removing assumptions (ii) - (v) should make no difference, though an explicit proof of that does not seem to exist. The other three assumptions are those of an infinite population, and Mendel's laws (limiting independent assortment to alleles at the same locus). Moreover, the assumptions (ii) - (v) are about either diploidy, two sexes, or sex-linkage. Almost all disputants over the units of selection would be happy to live with these assumptions. In any case, there is little reason to suspect that a biologically plausible violation of any of these assumptions would lead to a disappearance of an additive genetic component of variance in fitness. The possible complexity of other models, therefore, does not detract from the criticism of the additivity criterion that is being offered here;

(ii) The genic variance term in equation (xxii) is already additive. Therefore the qualification "transformable into additivity" is irrelevant to this discussion. The other possible qualification, that of requiring *only* an additive component of variance to the fitness, is more interesting. This will be called the "modified Lloyd criterion." However, even it turns out to be unacceptable for three reasons. First, and most important, even if genotypic selection is all that is occurring, the R.H.S. of equation (xxii) is simply not proportional to the genotypic variance,

$$\sum_{i=1}^{n}\sum_{j=1}^{n}P_{ij}(w_{ij}-\overline{w})^2.$$

Second, the third term in the R.H.S. of equation (xxii) is equal to 0 if and only if there is no dominance, that is, for $i, j = 1, ..., n$, $|w_{ii} - w_{ij}| = |w_{jj} - w_{ij}|$

(Li 1969). Few genotypic selectionists would demand, and almost no genic selectionist would admit, that the slightest bit of dominance dooms the case for alleles being the units of selection.[9] Third, even if this extreme view is adopted, i. e., it is required that there be no dominance at all for genic selection to take place, this modified proposal would still fall afoul of $\overline{\Delta w}$. Suppose that there is no dominance, but the fitnesses change from generation to generation such that the allelic fitnesses change at a constant (or transparently rule-governed) rate for each allele. Surely, this would be a prime candidate for genic selection. Yet, $\overline{\Delta w} \neq 0$ and the R.H.S. of equation (xxii) would not reduce to only a term proportional to the genic variance. The situation is the same with more complex models;

(iii) Fisher's (1930) "fundamental theorem of natural selection," if taken literally, requires that the *only* term on the R.H.S. of equation (xxii), and in the corresponding equations in more complex models, be a term that is proportional to the genic variance. Thus, the gene would be a unit of selection by the modified Lloyd criterion if and only if Fisher's theorem could be derived in a model. This would, indeed, provide a strikingly clear explication of genic selectionism. Few genic selectionists, however, would probably accept it. The reason for this is simple: Fisher's theorem, in this unadulterated form, holds for a vanishingly few number of cases. It does not even hold in the standard selection model.[10]

4. Conclusions

It is worth emphasis that the analysis given above, and the appended technical comments, only directly argue for a rather narrow negative conclusion: the dispute between genotypic and genic selectionism cannot be resolved on the basis of the additivity criterion unless one is always willing to countenance the gene as a (if not as the only) unit of selection. Going slightly beyond what is strictly implied by the technical discussion, however, this conclusion should generate some general wariness about the use of variances or similar statistical measures to identify those biological units which, in some sense, interact most systematically with their environment in such ways that it results in natural selection.[11] This sort of skepticism, if sustained, would affect other analyses of the units of selection, such as those of Wimsatt (1980), which also refer to variances but which are not directly affected by the negative result given above. At the very least, the analysis shows that the experimental identification of a genic component of variance of fitness does not suffice to demonstrate that the allele is a unit of selection.

Finally, lest the actual conclusions of this paper be over-interpreted, the following cautionary remarks are in order:

(i) This analysis does not, in any way, resolve the dispute between genic and genotypic selectionism. In particular, even though V_g is not equal to 0 under almost all circumstances, this does not indicate a victory for genic selectionism. In general, genic selectionists do not endorse the additivity criterion. Even if they now adopt that criterion with the intent of claiming victory over their detractors, they would be faced with the unsettling problem that just as $V_g \neq 0$ in almost all circumstances, there are many other terms that are also not equal to 0 in almost all circumstances;

(ii) This analysis does not have any direct implications for other units-of-selection disputes, in particular that between individual and group selection. The reason for this is that there are two different hierarchies: (i) that of allele → locus → gene complex → genotype, etc.; and (ii) that of molecule → organelle (including chromosome) → cell → tissue → organism → group, etc., that are distinct but have normally been conflated in discussions of the units of selection (see, however, Falk and Sarkar (1992)), even though this distinction is sometimes implicitly recognized.[12] For instance, in Lewontin's (1970) delineation of the units of selection, which is a particularly careful explication of the second hierarchy, there is no mention of genes. The probable reason for the usual conflation is a tendency to conflate an individual organism with a genotype.[13] The existence of these two separate hierarchies raises the possibility that the same form of selection can potentially simultaneously pick units at widely disparate levels in the two hierarchies. This point, too, has sometimes been recognized. Williams (1966) has argued that selection almost always acts on an individual organism *and* on a single allele at the same time. Another consequence of the existence of two separate hierarchies is that results obtained for one may not be pertinent to the other. An example of this is that the analysis given here cannot be directly re-interpreted in the context of group and individual selection though a parallel development is probably possible.

Notes

[1]The influence of Richard Lewontin on this paper should be obvious. My skepticism about the value of variance-based analyses in theoretical population genetics arose from conversation with John Maynard Smith. Discussions with J. F. Crow, T. Nagylaki and, in particular, W.C. Wimsatt have been useful. This analysis is part of a larger project of attempting to give a comprehensive account of the structure of contemporary evolutionary theory, with full attention to the technical details and complexities of the models and strategies employed in that field. I apologize for several references to that incomplete work but these are needed to situate the arguments being offered here in the context of past work on the units of selection. Please note that the basic—and technical—argument of this paper is fully elaborated here and does not rely on that work.

[2]A useful collection of the early papers is found in Brandon and Burian (1984). The most comprehensive philosophical discussion that is available of the later work is in Lloyd (1988).

[3]For some special one-locus, two-allele cases of genotypic selection, Godfrey-Smith (1992) has reached a similar conclusion though the arguments he offers are largely intuitive, rather than formal, and not particularly compelling. The analytic treatment given here, in Section 2, is fully general and provides the needed rigorous basis for Godfrey-Smith's claims. Besides the technical rigor and the level of generality being offered here, there is an important philosophical difference between the position advocated by Godfrey-Smith and the one defended here. Godfrey-Smith starts with an assumption that there is some method for identifying the "real" unit of selection. No such claim is made in this paper.

[4]This follows Nagylaki (1992).

[5]Throughout, whenever i and j are not summed over, it will be assumed that they range over $1, ..., n$.

[6]Directly, what this equation shows is that there is such an additive component in the fitness increment over one generation. Nevertheless, since the current fitness is a *sum* of these increments, the equation can be interpreted in the way it is in the text.

[7] If, for $i = 1, ..., n$, $w_i = \overline{w}$, then V_g is obviously 0. However, for V_g ever to be equal to 0, the same relations must hold since every term in

$$\sum_{i=1}^{n} p_i(w_i - \overline{w})^2 \text{ is non-negative.}$$

[8] The discussions in this section and the next consist of summaries of what will be found in full detail in Sarkar (forthcoming).

[9] However, even if they did endorse so extreme a claim, the first objection raised here would still prevent genotypic selectionists from obtaining any solace.

[10] This is exactly why those who see some value in that theorem choose to interpret it more loosely in a variety of ways. Once again, no comprehensive review of these developments seems to exist, but a full discussion is being attempted in Sarkar (forthcoming).

[11] This is probably yet another example of the old methodological point that the analysis of variance does not, *by itself*, provide any indication of what the underlying "causes" may be (see, especially, Lewontin 1974).

[12] The relations between these distinctions and roughly similar ones proposed by Hull (1989) and Brandon (1990) will also be found in Sarkar (forthcoming).

[13] This conflation probably results from the extension of intuitions generated by the properties of relatively large animals to all organisms. It is illegitimate: trees can form a mosaic of genotypes; in the case of Dictyostelium, free-living and reproducing cells with different genotypes can come together to produce a single slug which behaves as an individual organism.

References

Brandon, R. (1990), *Adaptation and Environment*. Princeton: Princeton University Press.

Brandon, R. and Burian, B., (eds.). (1984), *Genes, Organisms and Populations: Controversies over the Units of Selection*. Cambridge, MA: MIT Press.

Edelstein, S.J. (1986), *The Sickled Cell: From Myths to Molecules*. Cambridge, MA: Harvard University Press.

Ewens, W.J. (1989), "An Interpretation and Proof of the Fundamental Theorem of Natural Selection." *Theoretical Population Biology* 36: 167-180.

Falk, R. and Sarkar, S. (1992), "Harmony from Discord." *Biology and Philosophy* 7: 463-472.

Godfrey-Smith, P. (1992), "Additivity and the Units of Selection." In Hull, D., Forbes, M. and Okruhlik, K. (eds.). *PSA 1992*, Vol I. East Lansing: Philosophy of Science Association, pp. 315-328.

Hull, D.L. (1989), *The Metaphysics of Evolution*. Stony Brook: SUNY Press.

Kingman, J.F.C. (1961), "A Mathematical Problem in Population Genetics." *Proceedings of the Cambridge Philosophical Society* 57: 574-582.

Lewontin, R.C. (1970), "The Units of Selection." *Annual Review of Ecology and Systematics* 1: 1-18.

_____. (1974), "The Analysis of Variance and the Analysis of Causes." *American Journal of Human Genetics* 26: 400-411.

_____. (1991), "The Structure and Confirmation of Evolutionary Theory." *Biology and Philosophy* 6: 461-466.

Li, C.C. (1969), "Increments of Average Fitness for Multiple Alleles." *Proceedings of the National Academy of Sciences (USA)* 62: 395-398.

Lloyd, E.A. (1988), *The Structure and Confirmation of Evolutionary Theory*. Westport: Greenwood Press.

Nagylaki, T. (1992), *An Introduction to Theoretical Population Genetics*. Berlin: Springer-Verlag.

Sarkar, S. (Forthcoming), *Drift, Selection, and Fitness: The Conceptual Framework of Evolutionary Theory*.

Wade, M.J. (1978), "A Critical Review of the Models of Group Selection." *Quarterly Review of Biology* 53: 101-114.

Williams, G.C. (1966), *Adaptation and Natural Selection*. Princeton: Princeton University Press.

Wimsatt, W.C. (1980), "Reductionist Research Strategies and Their Biases in the Units of Selection Controversy." In Nickles, T., (ed.). *Scientific Discovery: Case Studies*. Dordrecht: Reidel, pp. 213-259.

Optimization in Evolutionary Ecology[1]

Robert C. Richardson

University of Cincinnati

1. Optimality and Adaptation

   Within evolutionary ecology, there is a substantial tradition treating natural selection as an optimizing agent tending to produce maximal adaptedness to the environment. This includes the work of Sewall Wright and R.A. Fisher, as well as a number of influential modern defenders such as G.C. Williams, J.R. Krebs, Richard Levins, Robert MacArthur, E.O. Wilson, and John Maynard Smith. In one widely deployed class of models, the phenotype or behavior of organisms is predicted on the basis of optimization principles, assuming "... that the organisms' preferences are related to evolutionary fitness and that the options more preferred must lead to greater survival and reproduction" (Real and Caraco 1986, 371-2). Optimality is measured by some "currency," or what Richard Lewontin (1987) calls a "criterion scale," providing an independent operational measure of fitness.

   Evolution is conceived in these models of as a process analogous to economic decision-making. Two common variants concern foraging (Stephens and Krebs 1986) and "life history strategies" (Stearns 1992) which are analogues of consumer choice and investment, respectively. As consumer choice involves allocation of resources, in models of optimal foraging, a predator allocates energy in different ways, with a family of indifference curves defining equal fitness values; the optimal solution is the distribution with the highest fitness value. We can determine the optimal allocation of resources for foraging, MacArthur tells us, if we know the structure of the environment, the "functional morphology" of the species, and have a "knowledge of its economic goals" (1972, 59). Birds encounter a variety of potential prey items. If these differ in nutritional value, we should expect to see systematic preferences for one type of prey item over others. The classical model of prey choice (MacArthur and Pianka 1966), assumes a predator encountering two prey types giving expected yields E_1 and E_2, with probabilities p_1 and p_2. If the "profit" from consuming one of these types is the ratio of yield to their respective "handling times" h_1 and h_2, then a predator confronted with these two types should specialize on one if

$$(p_1E_1)/(1 + p_1h_1) > (p_1E_1 + p_2E_2)/(1 + p_1h_1 + p_2h_2).$$

A predator should specialize if the "profit" from specialization is greater than the "profit" from exploiting both prey types. A model of the optimum phenotype predicts the pattern of preferences as a function of parameters for abundance of prey types, probability of capture, and yield. As investment concerns the optimal allocation of resources, organisms may invest more or less in reproduction. Lack claims that "clutch-size has been evolved through natural selection to correspond with the largest number of young for which the parents can on average find food" (1954, 22). Assuming that selection favors a maximum of young fledged, Lack assumed that maximizing fitness gain per clutch, measured by the number of young fledged, is equivalent to maximizing lifetime fitness.

In the simplest cases, the evolutionary problems are deterministic, but in more realistic cases they take on the complexity associated with decisions under risk or uncertainty. Thus, in Charnov's version of the marginal value theorem, the goal of maximizing caloric intake is used to explain breadth of diet (Charnov and Orians 1973; Charnov 1976a). Prey types are ranked in terms of the ratio of energy gained to handling time, and the marginal value theorem tells us that whether a prey type of a given rank is taken depends on the abundance of prey of higher rank: a prey type will be taken if it increases the rate of caloric intake. Analogous results hold for choosing a "patch" in which to forage when resources are unevenly distributed: an organism should move to a new patch when the rate of return drops below the average rate for the habitat. Under risk, decisions are based on known probability distributions. The choice is between remaining within a patch or shifting to a new patch, with an array of patches varying in quality in a predictable way. Under uncertainty, the problem concerns decisions based on unknown probability distributions. The problem, thus, also requires determining the character of the probabilistic array. These are important differences, but do not affect the morals suggested here.

Two caveats are needed. First, I will focus on evolutionary questions, rather than behavioral ones: the problem concerns evolved patterns of response, rather than the regulation of behavior. Hymenoptera may produce an optimum number of females for every male, but they make no optimal choices. Analogously, there may be an optimal height for a plant in a given environment, and though growth may be subject to evolutionary pressures which optimize height, the plant does not calculate the optimal height. Natural selection, rather than the organism, is the optimizing agent. Second, the question is not whether *evolution* is an optimizing process. The question, rather, is whether selection is an optimizing process; more properly, where fitness is understood as a propensity, the question is whether *selection* is a component of evolutionary models which will tend to increase fitness so long as there is significant and heritable variation (cf. Mills and Beatty 1979; Brandon 1990; Richardson and Burian 1992). At most, natural selection, constrained by history, genetics, and competing goals, will tend to maximize fitness (cf. Maynard Smith 1978; Parker and Maynard Smith 1990). Optimality models provide a surrogate for fitness which selection should tend to optimize. The actual evolutionary result may fall far short of the optimal one.

2. Static and Dynamic Adequacy

Two distinct requirements may be imposed on optimization models. An optimality analysis must be *descriptively adequate*. This is a matter of *static* adequacy: a model must describe a result, and show it is optimal. This is independent of *how* that solution was produced. It is silent on the evolutionary trajectories and on the factors shaping the evolutionary outcome. Additionally, we might require that an optimality analysis be *dynamically adequate*; that is, that it not only describe the results of selection, but also reflect causally significant processes involved in producing the result.

This requires more than simple descriptive adequacy. Insofar as we are interested not only in projecting outcomes, but also in explaining their evolution, it is natural that a dynamic condition be met. If foraging strategies and clutch size are adaptations, then explaining their presence depends on knowing their evolutionary ancestry (cf. Brandon 1990, ch. 1). MacArthur's (1957, 1960) early work on the abundance of species indicated that the existing data on relative abundance of species fit better with a model based on non-overlapping niches. He went on to claim that this would be expected if niche breadth is a consequence of competition. These data, MacArthur concluded, fit best with an analysis assuming competitive exclusion. The analysis of pattern is a vehicle for understanding evolutionary history.

There is a qualitative fit between the predicted distribution of species and a competitive model. The fit is by no means perfect: common species are more abundant than the models would predict, and rare species are less abundant than predicted. MacArthur pointed out that if the environment is heterogeneous, then it is possible to improve the fit between data and model: "The divergence from the ideal curve may, in fact, be regarded on this hypothesis as a measure of heterogeneity. Experimentally, for bird communities, this appears to explain most of the 'steep' curves" (MacArthur 1957, 293). This is a standard strategy in handling a mismatch between model and prediction, in order to explain apparent lack of optimal design. A deviation from initial predictions is explained by superimposing a second application of the same model, taking up the slack in the fit by *assuming* optimal design. In a similar way, foraging models also get at best an imperfect fit between behavior and theoretical models. In experimental work with great tits, Krebs and his collaborators provided meal worms of different sizes and manipulated the frequency of encounter. Though the models predict a discrete threshold, the actual preference function is a sigmoid one (Krebs, *et. al.* 1977). The general fit is impressive nonetheless, and there are a variety of ways to explain the deviation from the theoretical prediction: there may be incomplete information concerning the value of prey or alternatives, or undetected differences in prey quality. Again, the slack between data and the optimal model is taken up by assuming foraging is optimal. This kind of retrofitting of model to domain is not inherently objectionable, though it does virtually insure that *some* optimality analysis can be fit to the domain being modelled (cf. Gould & Lewontin 1979). As Charnov says, "... the failure of an 'efficiency' model to account for a behavior may well lead to insights as to how other ultimate factors affect behavior" (1976b, 150). Under the assumption of optimality, this retrofitting is important and useful.

Optimality models are often equilibrium models. It is not clear how realistic this assumption is. There is evidence supporting Lack's view that clutches tend toward the maximum, and when there are deviations, the tendency has been to produce smaller than optimal clutch sizes; however, neither the assumption that the number of surviving offspring is optimal nor the assumption of equilibrium is unproblematic. In work on clutch size in the blue tit (*Parus caeruleus*), Nadev Nur provides some reason to be skeptical about Lack's claim that the average number of surviving offspring is optimized. Nur shows that there is considerable variation in the number of eggs laid, both within years and from one year to the next, ranging from 3 to 15, and with a mean of 9.2. When measured in terms of the number of surviving offpring, the optimal clutch sizes also varied, but differed from the modal value of 9.0. Optimal clutch sizes and actual average brood sizes are different. Nur's tentative conclusion is that the population is subject to fluctuating selection pressures and that, as a consequence, the population is not in equilibrium (1987, 75). Whether Nur is right or not in appealing to environmental fluctuations to explain the patterns in the blue tit, optimality models which describe the response at equilibrium will be inapplicable to these natural populations.

There are, thus, reasons to doubt the descriptive adequacy of these models. The classical line, as we have seen, is to claim that the results *are* optimal given other limitations which constrain the problem. Though not optimal in their simplest form, results may nonetheless turn out to be optimal once other constraints are taken into account. Thus, there is a cost to reproduction, and since what must finally be optimized is lifetime fitness, this too can be used to explain the tendency for clutch sizes to be smaller than the optimal. Life history models have relied centrally on accepting tradeoffs between immediate reproductive output and other factors such as survivorship of offspring, parental survival and subsequent reproduction. We must distinguish the immediate effects of a "choice" from its costs in terms of long term fitness. Similarly, though Charnov's marginal value theorem (1976b) fits at least qualitatively with observations of mantid behavior (*Hierodulla crassa*), the model assumes that foraging does not increase risk of mortality. This is not true in general. If a result is not optimal given one description of the problem, the problem may be reformulated, incorporating other parameters in such a way that it becomes optimal.

Similar skepticism is appropriate concerning optimization models applied to human decision making; it was reflection on the factors which make optimality unrealistic that led Herbert Simon (1983) to defend a "satisficing" rule for human decision-making. Satisficing substitutes a stopping rule for one that would maximize utility. Imagine searching through alternatives until finding some choice above a specified threshold. That option is then embraced *without any further search or evaluation*.[2] This was meant to yield an improved and more realistic decision procedure within the constraints of what Simon (1981, 1983) terms "bounded rationality." We are relieved of an exhaustive search through the available options. We simply look at enough options to find one which has consequences which are minimally acceptable. A satisficing rule also relieves us from the need for a utility function strong enough to allow for systematic comparisons and rank orderings of outcomes; systematic comparisons are unnecessary. Analogous observations apply within evolutionary theory. Natural selection is not presented simultaneously with all the alternative forms possible (Oster and Wilson 1978); indeed, the mere fact that the number of possible genotypes far exceeds the number of actual genotypes insures that natural selection is not generally an optimizing procedure and that the results of natural selection will not be optimal. Additionally, selection provides an imperfect mechanism for achieving optimal solutions. Simple evolutionary models do predict that mean fitness will increase each generation. In reality evolution will tend at most to reach local optima. There is no guarantee that the result will be one that maximizes fitness.

The parallels suggest that there may be advantages to an analysis in terms of satisfaction rather than optimization; however, it has become common to maintain that satisficing is merely a case of optimizing under side constraints. Krebs and McCleery point out that animals may only use "rules of thumb" which approximate optimal decisions, but that these may in fact *be* optimal when additional constraints are taken into account (1984, 118-9). In a similar vein, Dennett writes, in defense of optimization:

> ... poor old Mother Nature makes do, opportunistically and short-sightedly exploiting whatever is at hand—until we add: she isn't perfect, but *she does the best she can.* Satisficing itself can often be shown to be the *optimal* strategy when "costs of searching" are added as a constraint (1983, 75-76).

It is not clear whether Dennett thinks evolution, or natural selection, optimizes. It is clear that he does not think satisfaction provides a significant alternative to optimization. Once we take into account the costs of additional search, the probability of finding preferable alternatives, or even the costs of assessing alternatives, we may find that the solutions arrived at are optimal. This is true in a limited sense. Given an evolutionary outcome, we can describe a problem, with sufficiently elaborate constraints, for which it is optimal. Without independent motivation for introducing the constraints and without independent measures of their significance, this provides little comfort and amounts to merely a *post hoc* reconstruction. These are not grounds for a blanket condemnation of optimality models, or of the various strategies of model-building which they deploy. As Stephens and Krebs insist, there is a difference between such *post hoc* modifications, and the "refinement" of hypotheses (1986, 207). As a grounds for defending optimization, the mere possiblity of introducing some unspecified constraints after the fact should leave us unmoved.

3. Satisfaction and Optimization

There are principled reasons for maintaining that optimization and satisfaction are not generally equivalent. The critical question for models in evolutionary ecology is *not* merely whether a result is optimal given *some* description of the problem, but whether there is a dymanically adequate model under which the result is optimal. Foraging models standardly assume that energy intake provides a useful measure of optimality. Natural selection then should optimize energy intake insofar as this optimizes fitness; in the face of a failure to optimize energy intake, the response is to search for other dimensions which are to be simultaneously optimized. In accounts of life histories such trade-offs are ubiquitous, including trade-offs between current reproduction and survival, current and future reproduction, reproduction and growth, and number and size of offspring. Such modifications of the initial, and rather austere, optimizing model can maintain descriptive adequacy, but amount to a restructuring of the problem. In place of a simple problem of optimization, we substitute a problem which involves simultaneous optimization in several dimensions. This exacts a cost. First of all, problems involving simultaneous optimization are substantially more complex than the initial simple problem. If these factors are not independent–and they are not if there are evolutionary "trade-offs" involved–then there will be an array of "solutions" each of which defines a local optimum. Secondly, the methodological problem of showing that a model is dynamically adequate becomes increasingly overwhelming. We need independent evidence for the significance of each of the constraints, and for their interaction. We become increasingly hard pressed to show that the model is realistic, and we are increasingly unable to explain the outcome using the criterion scale adopted.

These reservations can be reinforced. There are two different dimensions to the distinction between satisficing and optimizing. One concerns whether the constraints defining an optimal solution are dynamically significant. The second concerns the "structure of the environment" (Simon 1956). In many cases of human decision making, alternatives are presented sequentially rather than simultaneously, and must be accepted or rejected as presented. History becomes critical in explaining a decision. Evolution also presents variations in specific contexts, and sequentially rather than simultaneously. History matters. Factors other than selection affect evolution and lead to systematic deviations from optimal results, including not only mutation, but the amount of variation, the kind of variation, and the order in which variation is presented. Populations can be subjected to essentially the same selective forces and evolve in radically divergent ways. This is nicely illustrated by a series of experiments with geographically mixed populations of *Drosophila* by Dobzhansky (1957). Under similar selection pressures, the evolutionary responses were "erratic." In the early phases of the experiment the responses of the populations were similar. In subsequent stages, they diverged considerably. While Dobzhansky allows that such divergence "may conceivably be due to undetected differences between the environments in the replicate experiments" (389), but considers this unlikely because similar experiments with uniform populations yielded convergent evolutionary trajectories. He explains it this way:

> The geographically mixed populations do not have, at least at the start, a stabilized genotype. ... These genes segregate in the experimental populations, and the chromosomes with the different and microscopically recognizable gene arrangements are thus placed on a highly variable genetic background. Natural selection may, then, perpetuate whatever gene combinations confer a high fitness on their carriers in the experimental environments. If differential gene constellations are thus adaptively valuable, natural selection will tend to establish the ones which happen to appear first in a given population; since natural selection lacks foresight, the genotype favored need not even be the fittest of all possible ones in a given set of hybrids. Different genotypes may appear and be established in replicate experiments (391).

The indeterminacy in the process apparently arises in part because the number of possible genotypes is much larger than the number of flies in the experiments, so that the actual range of genotypic variations is a small sample from the possible. In heterogeneous populations, there is sufficient variation that the initial populations lie within reach of more than one adaptive peak; as a consequence, the final states will vary. We could, of course, incorporate additional constraints, including the array of available genotypes and their order of introduction, as further constraints on the structure of the optimization problem. However, as we continue embellishing on the constraints, optimization *per se* does increasingly little explanatory work. The explanatory burden is carried by the constraints. The initial simplicity which makes optimization attractive vanishes under the diversity of constraints.

4. Satisfaction and Adaptation

Satisfaction and optimization are thus dynamically distinct. Satisfaction requires only meeting some minimal threshold, rather than finding even a local maximum; and it depends critically on the extent and order of presentation. The next question is whether satisficing presents a *significant* alternative to optimizing; that is, in the context of evolutionary ecology, whether the differences between satisfaction and optimization would lead to importantly different results, or whether we would do better to look for more elaborate constraints in embellishing our models.[2] I doubt that a shift to satisficing will resolve the central methodological problems confronting optimality models. In the case of human judgment, satisficing does provide systematic advantages, in part because it does not require more than an ordinal ranking of choices, and does not require an exhaustive ordering of alternatives. Since we do require more than an ordinal ranking for evolutionary purposes, there is no corresponding gain here. The appeal to satisfaction does, however, have some systematic theoretical advantages insofar as it rests on a relative rather than an absolute measure of fitness (cf. Fetzer 1992); it does not presuppose an exhaustive ordering of alternatives; and it can at least reduce the complexity of the problems to be dealt with in operationalizing an account of fitness.

The most I can offer here is suggestive of the shape of analyses of adaptation in terms of satisfaction. One dimension of satisficing depends on threshold effects, on which I will concentrate. I have no view as to their relative importance. Gillespie and Caraco (1987) investigated differences in the foraging strategy of orb-weaving spiders in the wild. They found some showed risk aversion and some risk proneness, depending on the average prey availability. This shows up in terms of whether they moved the sites of their webs regularly or stared relatively stable; that, in turn, depended on the specific habitat, whether lake or creek. All that appears to be required in order to shift strategies is for average prey availability to exceed a threshold for successful reproduction. Fitness is treated as a step function dependent on prey consumption; the determining factor in foraging strategy is the probability that the actual prey consumed will exceed some threshold over a season. Across a wide range of differences in "profit," there is no tendency to optimize. There is no tendency to "sample" the average quality of the environment in order to achieve even a local optimum. It may be that any differences within the two regimes are uncorrelated with fitness, though there is no reason to think this is so; nonetheless, in terms of the criterion scale on which fitness is measured, the differences make no difference.

A parallel can be drawn with the distinction between what Bruce Wallace (1968, 1975) calls "hard" and "soft" selection. Hard selection requires a cutoff point with an invariant fitness scale, both density and frequency independent; the result would be that as average fitness levels vary within the population, the absolute number of individuals leaving progeny varies as well. Soft selection, by contrast, is both frequency and density dependent, with no fixed threshold. Fitness in this case is a function of the competitive environment; accordingly, the absolute number of individuals leaving progeny would be fixed (assuming limiting and fixed resources), though relative numbers could fluctuate. Under hard selection, both population size and composition are affected by, say, the range of habitat available or changes in the habitat structure, whereas only genetic composition is affected under soft selection. In the latter case, the representation in the adult population changes as a function of relative fitness levels. Soft selection is in a regime governed by competition, and, for that reason, might be expected to maximize fitness. Hard selection is a satisficing option. Selection depends only on meeting some minimum threshold.

Notes

[1] This work was supported by the National Science Foundation (DIR-8921837), and the Taft Faculty Committee of the University of Cincinnati. I am indebted to many people for discussion of this work at one time or another, including John Beatty, Robert Brandon, Richard Burian, W. R. Carter, Daniel Dennett, Christopher Gauker, Donald Gustafson, Lawrence Jost, W. E. Morris, Elliott Sober, Miriam Solomon and William Wimsatt.

[2] The procedure can be refined to allow for adjustment of the threshold value either up or down, depending on the ease with which we find options at or above the threshold value, thus allowing for the satisficing rule to adjust sufficiently to reach a decision even in an environment which is fundamentally inhospitable.

[3] Some biologists have taken this prospect seriously, including Stearns (1976 and 1982), Myers (1983), Krebs and McCleery (1984), and Stephens and Krebs (1986).

References

Brandon, R.N. (1990), *Adaptation and Environment*. Princeton, NJ: Princeton University Press.

Charnov, E.L. (1976a), "Optimal Foraging: the Marginal Value Theorem," *Theoretical Population Biology* 9: 129-36.

_____. (1976b), "Optimal Foraging: Attack Strategy of a Mantid," *American Naturalist* 110: 141-51.

Charnov, E.L. and Orians, G.H. (1973), "Optimal Foraging: Some Theoretical Explorations," unpublished manuscript.

Dennett, D.C. (1983), "Intentional Systems in Cognitive Ethology: The 'Panglossian Paradigm' Defended," *The Behavioral and Brain Sciences* 6: 75-76.

Dobzhansky, T. (1957), "Mendelian Populations as Genetic Systems,"*Cold Spring Harbor Symposium in Quantitative Biology. Population Studies: Animal Ecology and Demography* 22: 385-93.

Fetzer, J.H. (1992), "Is Evolution an Optimizing Process?" draft version.

Gillespie, R.G., and Caraco, T. 1987. "Risk Sensitive Foraging Strategies in Two Spider Populations," *Ecology* 68: 887-99.

Gould, S.J. and Lewontin, R.C. (1979), "The Spandrels of San Marco and the Panglossian Paradigm," *Proceedings of the Royal Society of London*, B, 205: 581-598.

Krebs, J.R., and McCleery, R.H. (1984), "Optimization in Behavioral Ecology," in J. R. Krebs and N. D. Davies (eds.), *Behavioral Ecology: An Evolutionary Approach*. Sunderland: Sinauer, pp. 91-121.

Krebs, J.R., Erichsen, J.T., Webber, M.I., and Charnov, E.L. (1977), "Optimal Prey Selection in the Great Tit (*Parus major*)," *Animal Behavior* 25: 30-38.

Lack , D. (1954), *The Natural Regulation of Animal Numbers*. Oxford: The Clarendon Press.

Lewontin, R.C. (1987), "The Shape of Optimality," in J. Dupré, J. (ed.), *The Latest on the Best: Essays on Evolution and Optimality*. Cambridge: M. I. T. Press & Bradford Books. pp. 151-160.

MacArthur, R.H. (1957), "On the Relative Abundance of Bird Species," *Proceedings of the National Academy of Sciences* 43: 293-5.

_____. (1960), "On the Relative Abundance of Species," *American Naturalist* 94:25-36.

_____. (1972), *Geographical Ecology: Patterns in the Distribution of Species*. New York: Harper & Row.

MacArthur, R.H., and Pianca, E.R. (1966), "On Optimal Use of a Patchy Environment," *American Naturalist* 100: 60-9.

Maynard Smith, J. (1978), "Optimization Theory in Evolution," *Annual Review of Ecology and Systematics* 9: 31-56.

Mills, S.K., and Beatty, J.H. (1979), "The Propensity Interpretation of Fitness," *Philosophy of Science* 46: 263-286.

Myers, J.P. (1983), "Commentary," in A. H. Brush and G. A. Clark (eds.), *Perspectives in Ornithology*. Cambridge: Cambridge University Press. pp. 216-21.

Nur, N. (1987), "Alternative Reproductive Tactics in Birds: Individual Variation in Clutch Size," in P.P.G. Bateson aand P.H. Klopfer (eds.), *Perspectives in Ethology*. New York: Plenum Press. pp. 49-77.

Oster, G.F., and Wilson, E.O. (1978), *Caste and Ecology in the Social Insects*. Princeton: Princeton University Press.

Parker, G.A., and Smith, J.M. (1990), "Optimality Theory in Evolutionary Biology," *Nature* 348: 27-33.

Perrins, C.M. (1965), "Population Fluctuations and Clutch-Size in the Great Tit, *Parus major* L." *Journal of Animal Ecology* 34: 601-47

Real, L.A., and Caraco, T. (1986), "Risk and Foraging in Stochastic Environments," *Annual Review of Ecology and Systematics* 17: 371-390.

Richardson, R.C., and Burian, R.M. (1992), "A Defence of Propensty Interpretations of Fitness," in D. Hull and M. Forbes (eds.), *PSA 1992*, vol. 1. East Lansing: Philosophy of Science Association. pp. 349-62.

Simon, H. (1956), "Rational Choice and the Structure of the Environment", reprinted in Simon, *Models of Thought*. New Haven: Yale University Press. chapter 1.2.

_ _ _ _ _ . (1981), *The Sciences of the Artificial* , Second Edition. Cambridge: M.I.T. Press.

_ _ _ _ _ . (1983), *Reason in Human Affairs* Stanford: Stanford University Press.

Stearns, S.C. (1976), "Life-History Tactics: A Review of the Ideas," *The Quarterley Review of Biology* 51: 3-47

_ _ _ _ _ _ _. (1982), "On Fitness," in D. Mossakowski and G. Roth (eds.), *Environmental Adaptation and Evolution*. Stuttgart: Gustav Fischer. pp. 3-17.

_ _ _ _ _ _ _. (1992), *The Evolution of Life Histories*. New York: Oxford University Press.

Stephens, D.W., and Krebs, J.R. (1986), *Foraging Theory*. Princeton: Princeton University Press.

Wallace, B. (1968), "Polymorphism, Population Size, and Genetic Load," in R.C. Lewontin (ed.), *Populat6ion Biology and its Evolution* pp. pp. 87-108.

_ _ _ _ _ _. (1975), "Hard and Soft Selection Revisited," *Evolution* 29: 465-74.

The Super Bowl and the Ox-Phos Controversy: "Winner-take-all" Competition in Philosophy of Science

Douglas Allchin

Cornell University

1. Introduction

Imagine two theories in a scientific controversy cast as competing teams in the Super Bowl, and you may get a "scoreboard of experimental evidence" such as the following, published in a review article in 1970 (after Racker 1970, 135):

SCOREBOARD OF EXPERIMENTAL EVIDENCE FOR THE CHEMICAL AND CHEMIOSMOTIC HYPOTHESIS OF ENERGY GENERATION DURING OXIDATIVE PHOSPHORYLATION AND PHOTOPHOSPHORYLATION

	Chemical	Chemiosmotic
Role of the membrane	−	+
Ion transport	+	×
Action of uncoupling agents	−	+
Isolation of high-energy intermediates	−	±
^{32}Pi-ATP exchange ADP-ATP exchange $H_2{}^{18}O$ exchanges	+	−

This figure compares two hypotheses in a debate in bioenergetics in the 1960s and 70s known as the Ox-Phos Controversy (Rowen 1986; Allchin 1990; Weber 1991). But its format, suggested by its title, bears a striking resemblance to the half-time recaps in televised football games: parallel assessments in several categories ask us to compare, say, how many yards rushed, number of first downs, passes completed, evidence for the role of the membrane, or evidence for ion transport, etc. (see also Sindermann 1982). Why did the review author—Efraim Racker, a research biochemist—borrow the scoreboard framework from sports to convey his assessment in science? Should one—can one—evaluate the performance of each scientific "team," infer a probable winner and loser from the plus-minus ratings in each column, and decide which hypothesis we should bet on or, given the final "score," which we should rationally support? Indeed, current biology textbooks would lead one to believe that the Chemiosmotic Hypothesis triumphed when its originator, Peter Mitchell, "won" the Nobel Prize in Chemistry in 1978. For those who view theory-choice in science as a matter of theory competition, the scoreboard may be a quite natural expression for assessing alternative hypotheses, construed (like athletic adversaries) as "rivals." Here, I explore Racker's figure, along with the Super Bowl metaphor as a model of competition, to consider more fully the nature of competition in philosophical conceptions of science.

In what follows, I analyze Racker's figure (§2) and contrast it, first, to three other comparative diagrams and tables published around the same time (§3), and then to a later "Revised Scoreboard" (§4). The most salient feature of Racker's scoreboard is, perhaps, that it frames the debate in polar, either-or, winner-take-all terms, though the outcome of the controversy suggests a pattern of differentiation or partitioning of domains among the hypotheses (§2). Incompatible or incommensurable theories, one finds, may not necessarily be mutually exclusive. There are two frameworks for interpreting competition, each applicable in separate contexts (§3). These observations further suggest strategies for scientists—to analyze their discourse in cases of disagreement and shape further research, and to bridge contexts of discovery and justification (§4).

2. A Scoreboard of Experimental Evidence?

Though Racker was a biochemist, not a philosopher, his "scoreboard of experimental evidence" above epitomized a notion of competition that has been fundamental to philosophy of science for at least the past three decades. As we came to understand the significance of alternative theories (logically, historically), we focused on ways to discriminate between them—and justification became linked to theory choice. Approaches to theory choice, however, have consistently drawn on competition as an underlying theme. Kuhn, for example, referred repeatedly to conflicting paradigms as "competitors" (1962, 147-50, 154-55) and he has suggested how the process of science may fit in a Darwinian framework (pp. 171-73; 1990 PSA Presidential address; 1992). Laudan, likewise, consistently portrays theories as rivals, scored on a scale of progressiveness (1977; 1992). For Laudan (1977), rationality itself emerges from comparing competing theories. Bayesian approaches inscribe competition in quantitative comparisons and sustain the notion of a crucial experiment as decisive between competing theories (Howson and Urbach 1989, 91-92). Even those who reorient their focus away from theory and more towards experiment (e.g., Galison 1987 and Franklin 1986) often build their accounts on episodes of competition. Finally, Hull's more social, evolutionary model adopts an explicit Darwinian metaphor with a vengeance: science is propelled by curiosity and credit and is regulated by "the visible hand" of competition (1988, Chap. 10; note also the dust jacket image alluding to "science red in tooth and claw"). For Hull, competition is a critical feature of the social structure of justification when we view "science as a process." Competition is a widespread—and often explicit—metaphor in philosophy of science, bridging Bayesian, historical, experimental, and even more sociological perspectives.

Racker's scoreboard allows one to notice a particular theme present, but not made explicit, in many of the philosophical models. Most saliently, perhaps, the scoreboard frames the scientific debate in polar, either-or, winner-take-all terms. I call this (somewhat archly) the "Super-Bowl model" of competition. The Super Bowl is *not* science, of course. But it does vividly exemplify certain *features* of competition also present in many philosophies of science (on exemplification, see Goodman 1976, 52-60; 1978, 63-65, 133-37). In the Super Bowl, as in Racker's assessment:

(1) competitors are assumed to be in the same category (are functionally equivalent or intersubstitutable) (principle of functional symmetry);

(2) competition is limited to two contenders (principle of bipolarity);

(3) only one can win (principle of either-or); and

(4) the winner wins exclusively and absolutely (principle of winner-take-all).

The winner-take-all principle (based on the other three), in particular, characterizes the rhetorical essence of the Super Bowl: the championship, the title, the best, being #1; and it may remind us of philosophical efforts to articulate a method by which we may select the single "most rational" theory (e.g., Laudan, Laudan and Donovan 1988; Niiniluoto 1992; Kukla 1992). A winner-take-all principle implies that only one theory is "right" or "completely right," and all others—even those we may call "half-right" or "partly right"—are ultimately "wrong" (see, e.g., Popper 1975). Again, while science is hardly the highly conventionalized practice of sports, when we view science through the Super-Bowl principles, we get Racker's scoreboard: either the Chemical Hypothesis is justified or the Chemi-Osmotic Hypothesis is justified, not parts of both, nor both partly (nor even some third alternative). The two solutions are exhaustive and mutually exclusive.

Participants in the Ox-Phos Controversy tended to interpret their disagreement in these "Super-Bowl" terms. In disagreeing about oxidative phosphorylation (or ox-phos)—how ATP is produced in the cell—they implicitly assumed only one theory could be correct. In the text which accompanied his table, for instance, Racker wryly characterized the then-raging debate: "In reading discussions of the proponents of the two hypotheses," he noted, "one gains the impression that the evidence against the formulations of the opponent is overwhelming." Indeed, the structure (and title) of the scoreboard implicitly invites us to compare the two columns to determine the all-inclusive "winner." The two vertical columns allow us to tally the various positive ('+') and negative ('−') scores for each hypothesis, and balance them with other scores (marked by '×') that represented "serious discrepanices" that still "need to be answered by decisive experiments" (pp. 132-37). Although the weight of the evidence might strike the casual observer as favoring the Chemi-Osmotic Hypothesis, Racker himself reached the opposite conclusion. He acknowledged that the novel chemiosmotic hypothesis was an "ingenious scheme" and evidence for it was "mounting"; nonetheless, he regarded some its assumptions as "formidable and controversial" (p. 132). He admitted, finally: "having been raised by the music of substrate-level phosphorylation [the basis of the more conventional approach] my own prejudices induce me to lean toward some aspects of the chemical hypothesis" (p. 137). Racker's assessment was governed by assumptions about dichotomous choice among mutually exclusive alternatives.

Ironically, perhaps, Racker's scoreboard also introduced information in a way that suggests another approach to competition. That is, when one views the table in terms of its horizontal rows, rather than its vertical columns, one finds the evidence divided or partitioned into separate categories. Racker identified specific sets of experiments and their corresponding phenomena: domains or perhaps sub-domains (*sensu* Shapere 1984), or data-domains (Ackerman 1986). When one sorts the observations in this way—by domain—the distribution of evidence becomes more clearly articulated. Both hypotheses claimed to describe the critical intermediate energy state in the transfer of energy to ATP and thus their domains overlapped significantly. However, the evidence for their related claims was not uniformly distributed.

The sorting of evidence is even more striking when one notes that the negative ('−') scores did not document direct counter-evidence or anomalous mismatches between predictions and observations. Rather, they indicated results that were "difficult to explain" or "more distressing to" each hypothesis. That is, there was lack of evidence or, more properly perhaps, lack of a theoretical concept through which one could even situate or address the evidence (see also Laudan, 1977, on non-refuting anomalies). The Chemical Hypothesis, for example, did not concern itself with the membrane, though data seemed to indicate that the presence of an intact, closed membrane was essential to the phosphorylation process. The Chemi-Osmotic Hypothesis, on the other hand, was ill-equipped conceptually to explain how certain atoms were transferred during the reactions (the 'exchange reactions' in the table). Plus and minus ratings were thus awkward parallels, not representing opposite evaluations of "right" and "wrong"—or even "right" versus "more right." Instead, domains were deemed more or less relevant, and thus reflected more or less favorably on each hypothesis. In an either-or, win-lose approach, of course, one disregards precisely this cross-characterization of the evidence. However, one can sort evidence or observations, not just theories. The scoreboard (as its label suggests) characterized the status of the "experimental evidence" more than of the two hypotheses.

The alternative claims about ox-phos, presumed to be incompatible, were thus found to be compatible by articulating new sub-domains and understanding how they related to each other. In a sense, differentiation dissolved the competition. But to suggest that the theories never competed at all would betray the history, here. The competition resulted precisely from differing perceptions about how one could generalize each hypothesis across the various domains (another variant of Goodman's, 1963, problem of defining induction classes; also the problem of "rightness of categorization," 1978, 127 and Chap. 7; 1976, 169-73).

An alternative to winner-take-all competition, then, is differentiation. That is, one may differentiate the "competing" theories by sorting or partitioning the domains appropriate to each. Indeed, in retrospect, we can say that Racker foreshadowed—though surely without prescience—how the conflict or competition between the two hypotheses would eventually be resolved. When the Ox-Phos Controversy finally subsided, both hypotheses remained, though they explained different, intersecting (or adjacent) domains, as suggested in Racker's "scoreboard of experimental evidence."[1]

3. Incommensurability and the "Winner-Take-All" Principle

Racker's analysis suggests that philosophers must qualify or revise substantially many models of science and include differentiation as a possible outcome of theory competition or conceptual disagreement. But it also reminds us of the context in which scientists themselves must make these assessments. How would one know in the midst of this debate whether a winner-take-all framework was appropriate or not? The challenge introduced by Racker's analysis is to articulate the different contexts in which each framework of competition may properly apply. That is, in what particular types of occasions does each model of competition function?

One must examine the resources available in the context of the controversy. As noted earlier, participants in the ox-phos episode tended to cast the debate themselves according to implicit "Super Bowl" principles. Nowhere is this orientation more evident than in another, extraordinary review of the same two hypotheses by another prominent biochemist, E.C. Slater, in the year following Racker's (Slater 1971). In a dominant and explicitly parallel structure, each concept for one hypothesis is presented and compared against a concept for the other hypothesis. Every diagram depicting relationships for one hypothesis is carefully paired with a corresponding diagram for the rival hypothesis. The survey is systematic and thorough. Yet in terms of the outcome of the controversy, Slater's analysis was less effectively framed than Racker's: why? Here, the ineffective strategy is far more telling philosophically than the successful one. When viewed in more detail, Slater's review reveals how incommensurability itself may indicate where differentiation may be appropriate.

Consider, for example, a pair of diagrams representing alternative versions of the "Sequence of components of [the] respiratory chain" (Figure 2a). The respiratory chain is a series of proteins and other molecules where electrons (labeled "2e") cascade down energy levels to oxygen (this is precisely where the oxygen we breathe is ultimately used). Even without knowing any biochemistry, one can easily recognize the FMN or Q or c_1 in both columns and note the clearly differing sequences and varying positions of the supplementary arrows into and out of the chain. For someone familiar with the two hypotheses, however, the similarity in the diagrams disguises a fair amount of conceptual shoehorning. Proponents of the chemical (or "C") hypothesis (depicted on the left) viewed the energy conversion as a stepwise release of energy from molecule to molecule. For them, the sequence was a familiar image of energy flow, detailing each intermediate step. For the chemiosmotic (labeled "C-O") hypothesis, however, the function of each component was coupled to its position in the mitochondrial membrane, essential for understanding how an energized gradient across the membrane could be generated. The order of the components was largely incidental to how the electrons moved through space. Information about sequence alone was inadequate and perhaps peripheral or misleading. Additional domain items needed to be considered at the same time. For the chemical hypothesis, then, the sequence diagram was central to the answer about energy transfer; for the chemiosmotic hypothesis, the diagram described a state of affairs, but was far from "the" answer—and did not even

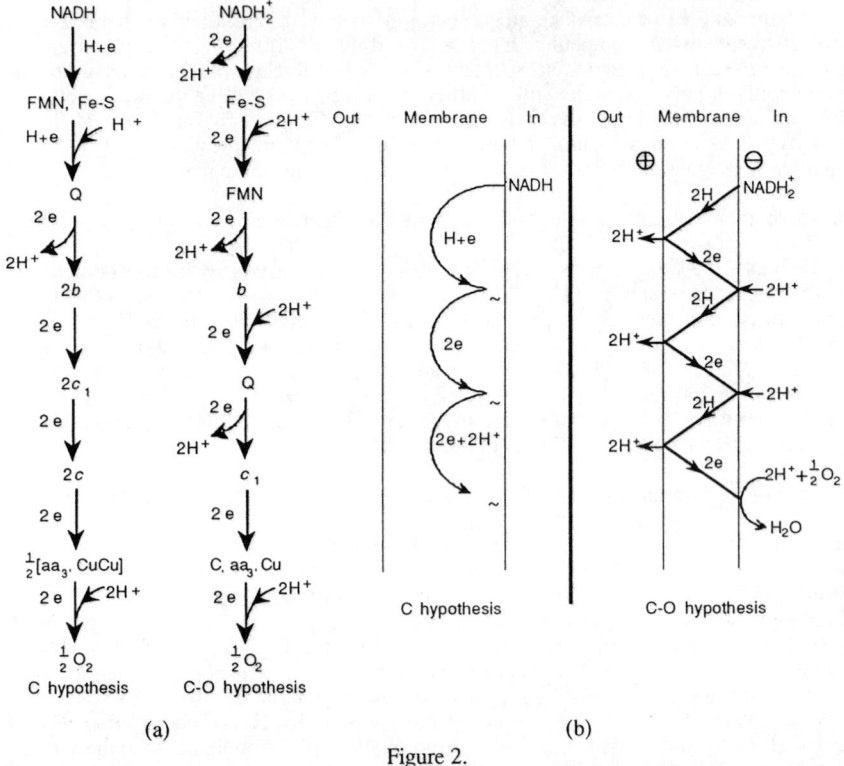

Figure 2.
Comparisons of chemical (C) and chemiosmotic (C-O) hypotheses after Slater (1971, 44-45)

address the significant causal questions. The sequence diagram thus represented an effort to make chemiosmotic claims conform to a conventional chemical framework—ostensibly so that the two could be compared in parallel.

Given what we know about theoretical bias, one might suspect that the (mis)interpretation was based on a singular, inflexible perspective; however, the shoehorning occurred in reverse in the very next pair of diagrams (Figure 2b). Here, the chemiosmotic hypothesis was represented in a drawing taken directly from one of its original documents (Mitchell's 1966 monograph). The diagram conveyed the fundamental chemiosmotic claim that oxidative phosphorylation was a "vectorial" process (having a spatial dimension as well as scalar magnitude). As electrons shifted to lower energy levels within the membrane (diagrammatically, downwards), hydrogen ions moved critically across the membrane ("$2H^+$," from right to left). The chemical hypothesis, on the other hand, made no essential claims about the position of ox-phos components or the physical pathway of the reactions. It assumed they were irrelevant. As a result, the diagram could depict no more than an arbitrary scalloped pathway punctuated by squiggle symbols ('~') representing the high-energy bonds of proposed intermediate molecules. Even though the membrane was where the reactions undeniably took place, there was no meaningful physical correlate to the semi-circular pathways. As in Racker's scoreboard, there was a causal category for one hypothesis with no corresponding category for the other. In this case, the claims of the chemical hypothesis were shoehorned to fit into a chemiosmotic framework. The shoehorning reflected a view that the hypotheses were funcitonally commensurable and thus could be evaluated as mutually exclusive by either-or rules.

The parallel assessment continued into a table of objections (Figure 3), another "scoreboard" of sorts. Here, the awkward matching of categories was even more striking. Slater considered eight objections to each hypothesis, some overlapping with those mentioned by Racker (C#1, C-O#4, C#6). But the objections were arranged in pairs that did not correspond directly. For example, data about the existence of high-energy intermediate compounds (C#1) differed from data about the existence of membrane gradients (C-O#1), though both related to the intermediate energy state at issue. There was no one crucial experiment that would allow one to select one alternative while simultaneously rejecting the other. Similarly, facts about how membrane conductivity was affected by uncouplers were weighed opposite facts about specific uncouplers (#6). Though both objections were about "uncoupling," one was based on membrane properties, the other on chemical properties—different categories both conceptually and experimentally. That is, the pairs were mere analogs, not functional substitutes for one another. The functionally incongruent categories inhibited commensurable comparison of the two hypotheses.

Slater's lists of objections also exhibited some of the conceptual and domain asymmetries found in Racker's scoreboard. Many objections were phrased in terms of "no evidence for" or "no experimental support for" (C#1; C-O#1,3,4,7), "no explanation is given for" (C#3,6; C-O#6,8), or "takes insufficiently into account" (C#5; C-O#5). One does not find the phrases "evidence contradicts" or "the explanation does not match available data." Each hypothesis was challenged by absence of data, not outright error. Unconfirmed, theoretically predicted results were criticized indirectly through terms such as "unlikely" (C#2) or "unprecedented" (C-O#2). As in Racker's review, evidence for or against a hypothesis was assessed in terms of whether certain domain items were effectively mapped by the concepts or demonstrated experimentally (see also Allchin 1992a). While the domains of the two hypotheses overlapped in terms of fundamental claims about energy transfer, they diverged in views of the range of relevant phenomena. Slater's assessment thus resonates strongly with

Racker's. But his table differs markedly from the scoreboard in suppressing, rather than highlighting, these distinctions so that they fit into an either-or format for winner-take-all theory choice. Slater's paired figures and his lists of objections, then, were not parallel conceptually, despite the graphic organization and commentary. The striking juxtapositions in this case—paradoxically, perhaps—simply underscore a fundamental incommensurability between chemical and chemiosmotic hypotheses.

TABLE 2. *Objections to the C and chemiosmotic hypotheses*

C hypothesis	Chemiosmotic hypothesis
1. There is no evidence for the existence of the hypothetical A ~ C compounds in state-$_4$ mitochondria	1. There is no evidence for the existence of a membrane potential of sufficient magnitude in state-$_4$ mitochondria
2. A high-energy compound with a $\Delta G'_0$ value of hydrolysis of 17 kcal/mole is unlikely	2. A membrane potential of 370 mV is unprecedented in either artificial or natural membranes
3. No explanation is given for the multiplicity of electron carriers in the respiratory chain	3. There is no experimental support for alternate hydrogen and electron transfer in the respiratory chain
4. An *ad hoc* hypothesis (the proton pump) is necessary to explain energy-linked cation uptake	4. There is no experimental evidence for the translocation of H$^+$ in the absence of cation
5. This hypothesis takes insufficiently into account the fact that the energy-transducing reactions take place in membranes	5. This hypothesis takes insufficiently into account recent advances in our knowledge of the chemical properties of haemoproteins
6. No explanation is given for the fact that uncouplers increase the electrical conductivity of artificial membranes	6. No explanation is given for the fact that some uncouplers are not proton conductors
7. An oligomycin- and uncoupler-sensitive ATP-P$_1$ exchange reaction is found in pro-mitochondria lacking a respiratory chain	7. There is no experimental support for the postulated diffusible X- and IO-
8. There is no site specificity for reaction with ADP, or for the action of uncouplers or inhibitors of oxidative phosphorylation	8. No explanation is given for kinetics of ADD-induced oxidation of ubiquinone

Figure 3. Slater's (1971) table of objections (p. 52).

Slater's analysis is remarkable because at every turn it tends to betray the presence of incommensurability while virtually refusing to acknowledge it. Slater, too, adhered to a polar, either-or, winner-take-all orientation to theory choice or theory competition. Indeed, Slater's sensitivity to the competitive framework is evident in his scrupulous fairness. The Super Bowl reminds us that the outcome of competition is legitimized in part by the "objectivity" embodied in rules of fair play.
Fairness—comparing similar cases by similar standards—is part of justifying the winning team's victory. In science, similarly, each hypothesis must be given a "fair hearing" or a "fair chance" to prove itself, if a comparative assessment is to be justified. Note the role of "referees" as ajudicators in both contexts (Sindermann 1982, 3). The "'rules' governing the '*game of science*'" (Lakatos 1978, 140-43) must ensure that all evidence will be evaluated and each hypothesis weighed according to a uniform method. Slater's conceptual shoehorning can thus be viewed as an effort to fit noncorresponding concepts into parallel categories for "fair" comparison. Slater imposed symmetry where the domains were, in fact, asymmetric. His review is a tribute to fair

competition between hypotheses in science framed in the Super-Bowl model. But in this case, "fairness" emerged by suppressing or avoiding the problems posed by incommensurable hypotheses.

Racker's scoreboard, by contrast, implicitly acknowledged the incommensurability by treating the hypotheses as incompatible, integrated wholes (*sensu* Kuhn 1962; see Hoyningen-Heune 1993, 220-21). The scoreboard posed possibilities about the differentiation of domains that the format of Slater's diagrams and table did not—and perhaps could not—allow. Slater's analysis, in its failure to represent the nature of the debate effectively, shows more clearly why the shift in competitive frameworks was necessary. The incommensurability—incompatible hypotheses with diverging but still overlapping domains—was itself the signal that winner-take-all assumptions were open to reassessment.

One may pause to consider this perhaps counterintuitive conclusion: *the very incommensurability of two hypotheses leads us to challenge whether they are mutually exclusive.* In conventional philosophical interpretations, of course, incommensurability is the hallmark of mutually exclusive hypotheses or paradigms. Slater's and Racker's figures in tandem, however (together with the outcome of the controversy), show us that hypotheses that are conceptually incompatible may nevertheless be "compatible." They may justifiably coexist. One must focus on domains. One must consider how empirical contexts may be differentiated.

Using Kuhn's and Hanson's gestalt metaphor, perhaps, we have become accustomed to an image of either-or competition between complex (incommensurable) wholes as winner-take-all. In a frequent gestalt example, we see either a duck or a rabbit: both cannot exist simultaneously. We sometimes assume in a competitive framework, therefore, that only one can "win." The winner-take-all principle implies that if we accept the duck image, then we accept the duck image exclusively—and no rabbit image is permissible. (Conversely, we may choose the rabbit image exclusively.) However, we may easily imagine scenarios where we may differentiate the contexts in which duck and rabbit interpretations are appropriate, even for the "same" image (far example, in a collection of rabbit drawings versus duck drawings). Either-or principles still apply, here: the final solution is not a hybrid image, half-rabbit, half-duck. The two interpretations remain distinct and incompatible (incommensurable) in the sense that their meanings are non-recombinable and resistant to hybridization. Winner-take-all assumptions break down, however, when we can specify contexts that justify each (holistic) interpretation. The whole rabbit and the whole duck may possibly both be accepted, each in clearly differentiated domains. Indeed, we often do flip happily back and forth between the two gestalts in an extended time frame (differentiating, for example—however weakly—in time). The gestalt example illustrates again that incompatible or 'incommensurable' theories need not be mutually exclusive. The either-or framework, so prominent in the Super Bowl and in Slater's review, leads us to view theories as polar opposites—as rivals—and to view choice as eliminative: only one will win. The outcome of the ox-phos debate, however, foreshadowed in Racker's scoreboard, is a stunning example of a "violation" of assuming a winner-take-all principle. In some cases—here, where domains are notably asymmetrical—two incompatible (incommensurable) hypotheses, such as the chemical and chemismotic hypotheses, may each still be justified.

To summarize, then, where alternative concepts are commensurable—that is, functionally intersubstitutable—or where whole hypotheses are incompatible but coincide in their domains, the either-or, winner-take-all framework applies well. Where hypotheses are based on incompatible concepts and, at the same time, their domains diverge, then a framework of differentiating domains is more appropriate. Incommensurability (exemplified in Slater's review) and asymmetric domains (revealed in some of Slater's phrases and captured more fully in Racker's scoreboard) are thus two contextual clues for abandoning the "Super-Bowl" mode of competition—and for focusing instead on sorting domain items. While incommensurability may tend to lead us to polarize two hypotheses in either-or terms and to regard them as mutually exclusive, it may instead be the very signal that competition is no longer winner-take-all.

4. Differentiation and Strategies for Resolution

Debate on ox-phos continued to unfold, and resolution to the controversy became clearer with additional findings. Racker was thus able to present a "revised scoreboard" one year later (after Racker and Horstman 1972, 15):

REVISED SCOREBOARD

	Hypothesis	
	Chemiosmotic	Chemical
Role of membrane	+	-
Model systems	-	-
Uncouplers and ionophores	+	-
Proton translocation	+	-
Topography of oxidation chain	+	-
K+ and Ca++ transport	-	+
The Painter and Hunter experiments	-	+
Exchange reactions	-	+

Racker still assumed an either-or, winner-take-all orientation, though now he felt that "the balance is shifting in favor of the chemiosmotic hypothesis." If one disregards the sum-total approach, though, one notices, more fundamentally, that new categories had been added to the scoreboard. The domains (or data domains) or evidence had become, literally, more finely resolved. "Ion transport" had been divided by type of ion into "proton translocation" and "K^+ and Ca^{++} transport"; this was critical because formerly ambiguous, confusing evidence now sorted itself more neatly, in Racker's view, between the two hypotheses. New categories of phenomena had also been added: topography of the oxidation chain and the Painter-Hunter experiments—again, contributing positively to each hypothesis. In Racker's revised scoreboard, one sees the growing articulation or differentiation of the evidence: the chemiosmotic hypothesis now explained the role of the membrane, ionophores and topography, for example—while the chemical hypothesis explained exchange reactions and transport of potassium and calcium ions. While the structure of the sports-like scoreboard still asked us to compare right versus left columns, the significant distinction was, in fact, between top rows and bottom rows. The revisions to the scoreboard all clarified the distributed of the data—and indicated more clearly how to reconcile the two hypotheses and to differentiate their once overlapping claims. Epistemically, then, we need to focus on how the differentiation occurs, how we move from Racker's first scoreboard to its revision, or perhaps even how we frame the first scoreboard.

Some recent philosophical approaches, focusing on experiment, have noted the significance of teasing fact from meaningless data, or fact from backgrond "noise": distinguishing fact from artifact (Galison 1986; Latour and Woolgar 1979). In the ox-phos debate, however, the significant experimental task was teasing fact from fact.

One central role of experiments in resolving the disagreement was to sort or partition the evidence. Given a Bayesian framework, for instance, one focuses less on comparing rival hypotheses and more on comparing different ("rival"?) categories of evidence. In differentiating domains, one construes the critical variable in the expression P(H/E) as the evidence (E), not the hypothesis (H). "Super-Bowl" principles compare $P(H_1/E)$ v. $P(H_2/E)$. Instead, the more important comparison may be expressed as $P(H/E_1)$ v. $P(H/E_2)$. The task is to characterize the boundaries of E properly (E = E_1, E_2, ... or E_n?). Experience from the ox-phos case, at least, suggests that "crucial tests"—those that embodied an either-or strategy and aimed to decide between two hypotheses—were relatively unsuccessful. The complexity and incommensurabilty evident in Slater's review may suggest why. Rather, researchers laid "claim" to certain domain items through demonstrations—that is, through concrete examples that the hypothesis "mapped" causal relationships in the appropriate domain (Allchin 1992a). Demonstrations were thus needed over a wide domain to sort the "territory." This case illustrates how the experimental resolution of domains (differentiation) is ultimately coupled to the resolution of disagreement. The controversy was resolved, in both senses of the word.

Diagnosing such occasions is crucial. The diagnostic tool as a strategy lies on the cusp between the context of discovery and context of justification. The intent is to justify, but the occasion is one where knowledge is incomplete and where discovering further information may be helpful before justification is complete or stabilized. Characterizing the proper competitive framework allows one to organize available information and ask whether differentiation may be possible—and to identify exactly where further information would be valuable. In short, a carefully framed debate articulates, and in a sense justifies, an ensuing research agenda. Strategic thinking about competition can guide us from Racker's first scoreboard to his revised scoreboard—and from there, to the resolution of controversy.

When the Ox-Phos Controversy finally ended, there was no single winner as there is in the Super Bowl. Both hypotheses had "survived" the competition, but their boundaries or scope had been dramatically reshaped. There was no "one-best" theory. There was no exclusive "winner." Rather, there were several interrelated, though still quite distinct, theories or models, each with its own scope, domain or "niche." What had once been viewed as competing, mutually exclusive, irreconcilable hypotheses became, through the sorting of evidence, complementary theories. The distribution of evidence that guides such sorting can thus be as important as the overall "score" of the evidence itself. The process of differentiating domains—vividly captured in Racker's "scoreboards" of competing hypotheses—must thus play a significant role in any complete model of theory choice in science.

Note

[1] Of course, one may generalize the pattern of differentiation beyond the ox-phos episode. Consider, for example, the conflict between phlogistonists and "anti-phlogistonists" in the late 18th century, typically cast as one of the most dramatic examples of either-or, winner-take-all competition in history (e.g., Kuhn 1962, Chap. 7). We have often assumed the concepts of oxygen and phlogiston were mutually exclusive, but late phlogistonists often accepted the discovery of oxygen. Further, their concerns about the generation of heat and light in burning, ignition, phosphorescence, electricity, and the relationship of animal heat and coal to plants and the sun as a source of light—all aspects of what we would call energy—were all warranted

(Allchin 1992b). Lavoisier, of course, focused on naming elements and studying reactions with a balance. Thus, there was a crude differentiation of the domains of matter and energy (even for combustion itself)—though the problems in the second category were largely not tractable at the time. Though the concepts of oxygen and phlogiston may have fallen in separate paradigms (see Kuhn), they could nonetheless be accommodated through differentiation.

References

Ackerman, R.J. (1985), *Data, Instruments and Theory*. Princeton: Princeton University Press.

Allchin, D. (1990), "Paradigms, Populations and Problem Fields: Approaches to Disagreement", in *PSA 1990*, Volume 1, A. Fine, M. Forbes and L. Wessels (eds.). East Lansing: Philosophy of Science Association. pp.53-66.

_ _ _ _ _ _. (1992a), "How Do You Falsify a Question?: Crucial Tests versus Crucial Demonstrations", in *PSA 1992*, Volume 2, D. Hull, M. Forbes and K. Okruhlik (eds.). 1: East Lansing: Philosophy of Science Association. pp.274-88.

_ _ _ _ _ _. (1992b), "Phlogiston After Oxygen", *Ambix* 39(3): 110-16.

Franklin, A. (1986), *The Neglect of Experiment*. Cambridge: Cambridge University Press.

Galison, P. (1986), *How Experiments End*. Chicago: University of Chicago Press.

Goodman, N. (1965), *Fact, Fiction and Forecast*, 2nd ed. Indianapolis: Bobbs-Merrill.

_ _ _ _ _ _ _. (1976), *Languages of Art*. Indianapolis: Hackett.

_ _ _ _ _ _ _. (1978), *Ways of Worldmaking*. Indianapolis: Hackett.

Howson, C. and P. Urbach. (1989), *Scientific Reasoning: The Bayesian Approach*. La Salle: Open Court.

Hoyningen-Huene, P. (1993), *Reconstructing Scientific Revolutions: Thomas S. Kuhn's Philosophy of Science*. Chicago: University of Chicago Press.

Hull, D. (1988), *Science as a Process*. Chicago: University of Chicago Press.

Kuhn, T.S. (1962), *The Structure of Scientific Revolutions*. Chicago: University of Chicago Press.

_ _ _ _ _ _. (1992), "The Trouble with the Historical Philosophy of Science", Cambridge, Mass.: Harvard University, Dept. of the History of Science.

Kukla, A. (1992), "Ten Types of Scientific Progress", in *PSA 1990*, Volume 1, A. Fine, M. Forbes and L. Wessels (eds.). East Lansing: Philosophy of Science Association. pp.457-66.

Lakatos, I. (1970), "Falsification and the Methodology of Scientific Research Programmes", in *Criticism and the Growth of Knowledge*, I. Lakatos and A. Musgrave (eds.). Cambridge: Cambridge University Press. pp.91-195.

_____. (1978), "Popper on Demarcation and Induction", in *Philosophical Papers*, Volume 1, J. Worrall and G. Currie (eds.). Cambridge: Cambridge University Press. pp.139-67.

Latour, B. and S. Woolgar. (1979), *Laboratory Life*. Princeton: Princeton University Press.

Laudan, L. (1977), *Progress and Its Problems*. Berkeley: Univeristy of California Press.

_____. (1990), *Science and Relativism*. Chicago: University of Chicago Press.

Laudan, R., L. Laudan and A. Donovan. (1988), "Testing Theories of Scientific Change", in *Scrutinizing Science*, A, Donovan, L. Laudan and R. Laudan (eds.). Baltimore: Johns Hopkins University Press. pp.3-44.

Niiniluoto, I. (1990), "Measuring the Success of Science", in *PSA 1990*, Volume 1, A. Fine, M. Forbes and L. Wessels (eds.). East Lansing: Philosophy of Science Association. pp.435-46.

Popper, K. (1975), "Rationality of Scientific Revolutions", in *Problems of Scientific Revolutions*, R. Harre (ed.). Oxford: Oxford University Press, pp.72-101.

Racker, E. (1970), "Function and Structure of the Inner Membrane of Mitochondria and Chloroplasts", in *Membranes of Mitochondria and Chloroplasts*, E. Racker (ed.). New York: Van Nostrand Reinhold. pp.127-71.

Racker, E. and Horstman, L.L. (1972), "Mechanism and Control of Oxidative Phosphorylation', in *Energy Metabolism and the Regulation of Metabolic Processes in Mitochondria*, Myron A. Mehlman and Richard W. Hanson (eds.). New York: Academic Press, pp. 1-25.

Rowen, L. (1986), *Normative Epistemology and Scientific Research: Reflections on the "Ox-Phos' Controversy, A Case History in Biochemistry*. Ph.D. dissertation, Nashville: Vanderbilt University.

Shapere, D. (1984), "Scientific Theories and Their Domains", in *Reason and the Search for Knowledge*. Dordrecht: D. Reidel, pp. 273-324.

Sindermann, C. (1982), *Winning the Games Scientists Play*. New York: Plenum.

Slater, E.C. (1971), "The Coupling Between Energy-Yielding and Energy-Utilizing Reactions in Mitochondria", *Quarterly Review of Biophysics* 4: 35-71.

Weber, B. (1991), "Glynn and the Conceptual Development of the Chemiosmotic Theory: A Retrospective and Prospective View", *Bioscience Reports* 11(6).

Ecological Explanation and the Population-Growth Thesis[1]

Kristin Shrader-Frechette

University of South Florida

1. Introduction

Many ecologists have questioned the status of proposed ecological "laws," charging that they are devoid of empirical content and predictive power or that they are tautological, trivial, circular, or nontestable (Peters 1976, 1978, 1991; Simberloff 1983, 1984; Simberloff and Boecklen 1981; Strong and Simberloff 1981; Strong and Levin 1979; see also Williams 1970; Ferguson 1976; Rosenberg 1978; Stebbins 1978; Caplan 1978; and Shrader-Frechette and McCoy1990). As a result, a number of ecologists have urged their colleagues instead to "study ... real organisms," rather than worry about untestable theories and laws (Van Valen and Pitelka 1974, 925; McIntosh 1982, 23).

In this essay we investigate the epistemological status of a prominent "ecological law" in order to gain insights about the characteristics of ecological explanation. Admittedly, laws may not be the chief criterion for progress in ecological science, and "many important questions about theorizing are exposed when we shift our focus away from specific hypotheses [and laws] that can be readily tested" (Taylor 1989, 122). Nevertheless, prominent scientists such as Murray (1979, 1986) and Loehle (1988) have claimed that there are at least three "ecological laws"—concerning competitive exclusion, population growth, and energy flow through trophic levels. As formulated by Murray (1986, 150; see Loehle 1988, 101), the thesis of competitive exclusion is that "competing populations cannot coexist indefinitely." The population-growth thesis is that "populations in finite environments cannot grow indefinitely" (Murray 1986, 156; see Loehle 1988, 101). The energy-flow thesis is that "the energy flow through a trophic level is less than that through the preceding level and greater than that through the subsequent trophic level" (Murray 1986, 157; see Murray 1979; Loehle 1988, 101). Because the epistemological status of competitive exclusion has already been discussed extensively (see, for example, Cody 1968; Connell 1983; Diamond 1975; Diamond and Gilpin 1982; Gilpin and Diamond 1982, 1984; Hutchinson 1959; Peters 1976, 1991; Pyke 1982; Schoener 1974, 1983, 1984; Shrader-Frechette 1990; Simberloff 1984; Strong 1982a; Strong and Levin 1979; Strong and Simberloff 1981; Strong et al. 1984), in this essay, we shall address the population-growth thesis.

2. The Population-Growth Thesis

The thesis has a long history. At the end of the 18th century, Thomas Malthus suggested that human populations are limited by natural checks, an idea that gave Darwin "a theory by which to work" (Barlow 1958), a theory that, according to Haeckel, formed the "groundwork of ecology" (McIntosh 1985, 8). "The earliest law commonly used in ecology" (McIntosh 1985, 268), the thesis is a population-level variant of "Liebig's Law" or "the law of the minimum." Liebig (1803-1873), an agricultural chemist, held that the growth of a plant or the rate of a process is limited by the slowest factor. McIntosh (1985, 268) claims that this early "law" went through a number of changes, gained new eponyms, and is still encountered in some textbooks.

In 1838, Verhulst formulated the famous logistic equation of population growth, heralded by Pearl and Reed (1920; Pearl 1925, 1927; Kingsland 1981) as the "law" of population growth and as comparable to Boyle's and Kepler's laws (McIntosh 1985, 148-152, 172). The logistic equation describes the growth of a population increasing at a rate determined by the intrinsic rate of increase (r) and the population size (N) to an environmentally set maximum (K): $dN/dt = Nr(K-N)/K$. It describes a sigmoid curve asymptotically approaching K (see Verhulst 1838, 1845; Kingsland 1982; Peters 1991, 54ff.). Formulating more specific, physiological variants of the population-growth thesis, C. Hart Merriam (1894) spoke of the "fundamental law" that animal distribution was controlled by temperature. Hopkins (1920) proposed the "bioclimatic law" or "Hopkins's Law," asserting that a biotic event lagged 4 days per degree of latitude northward, 5 degrees of longitude eastward, and 400 feet of altitude in spring and early summer. Livingston and Shreve (1921), for example, argued that it was a "law of plant geography that the existence, limits, and movements of plant communities are controlled by physical conditions" (McIntosh 1985, 269).

A central problem with many variants of the population-growth thesis, especially the physiological variants such as Hopkins's Law, is their reliance on problematic assumptions. Two of the most common assumptions are (1) that population growth sometimes can be described as a function of only one variable, such as temperature; and (2) that the effects of environmental variables are adequate to explain the growth or distribution of species and populations in their normal habitats (see Orians 1962). Although such simplistic assumptions suggest that specific physiological variants of the population-growth thesis are false, their falsity nevertheless indicates that some variants of the thesis are testable. It is possible to test, for example, whether a particular instance of population growth is a function of only one variable such as temperature. Hence, there is at least one sense—although perhaps a derived one—in which the general thesis of population growth (from which the physiological variants came) is testable. Another testable variant of the thesis is Humphreys' (1979) claim that one can predict annual population production on the basis of population respiration rates. Such observations about testable variants of the population-growth thesis, however, tell us little about the epistemological status of the general thesis formulated by Murray and Loehle.

3. Is the Population-Growth Thesis A Priori or Tautologous?

One problem with the general thesis is that it appears untestable, as Andrewartha and Birch (1954; see also Sinclair 1989, 202) assert. Perhaps it is a tautology in the sense that it excludes no logical possibilities and merely repeats a concept. Consider, for instance: "populations in finite environments cannot grow indefinitely" (Murray 1986, 156; see Loehle 1988, 101). It seems logically impossible for a "population in a finite environment" to be an infinite population because populations, by definition, are located in spatio-temporal environments and, by definition, the environment described in the thesis is "finite." Thus a "population in a finite environment" is a finite population. If so, then the thesis appears to be formulable as: "finite populations cannot grow indefinitely." But what cannot grow indefinitely is finite. Therefore, the thesis seems again reformulable as a tautology: "finite populations are finite" (Peters 1991, 39).

As Quine (1953, 1960) pointed out, however, there is no unproblematic test for whether a proposition is a priori or tautologous. And, as Elliott Sober (1984, 61-85) wisely noted in his discussion of fitness, even if there are a priori truths, it is difficult to recognize them because propositions don't wear their epistemological status on their sleeves. The population-growth thesis, for example, is not obviously a tautology, for the same reasons that "all bachelors are unmarried" is not a tautology. The logical terms the thesis contains do not alone suffice to show that it is true (Sober 1984, 64). Rather, once one understands the logical, *as well as the nonlogical*, terms the thesis contains, one can see that it must be true. But is the proposition—"populations in finite environments cannot grow indefinitely"—true a priori? Quine rejected the apriority of any proposition on the grounds that every belief is open to empirical revisability. If he is right—and I think he is, because reasonable persons might be convinced that their concepts were wrong (see Quine 1953, 1960; Sober 1984; Kripke 1972; Putnam 1970)—then the problem with the apparent apriority of the population-growth thesis may not be unrevisability, but rather that no further sense experience, beyond mastering the constituent concepts, is required to determine its truth.

Mastering the constituent concepts of any scientific proposition, however, is difficult (Sober 1984), especially in ecology (Haila 1986, 379ff.). Thus, although the population-growth thesis asserts that environmental factors limit population growth, it is extremely difficult to master concepts such as "environment," "limiting factor," or even "population growth." Brandon (1990, 47), for example, notes that although the environment is the sum total of biotic and abiotic (physical) factors external to an organism, the sense of "environment" relevant to ecologists is only the external factors that actually affect the organisms in question in a particular way. Mastering the concept of "limiting factors" is equally problematic. Because such factors may limit a population, but only rarely, knowing them could be difficult. *Second*, because distributions are limited more often by conditions that are regularly suboptimal (rather than lethal), the small deltas associated with such conditions may be difficult to spot, especially for a short period or for systems already exhibiting much noise. *Third*, sub-optimal environmental conditions may not manifest themselves as explicitly suboptimal, except that they may alter the outcome of a biological interaction between the relevant species and other species. *Fourth*, suboptimal conditions often interact, so that it is difficult to isolate a single environmental condition that limits a population. *Fifth*, the negative consequences of suboptimal environmental conditions frequently influence other factors, so that it is often impossible to isolate them. Finally, *sixth*, because a species found at the edge of its range occupies patches in which conditions are closest to those found in the center of its range, the judgment (about whether a population has decreased in an area) is in part a function of the spatio-temporal scale used. If factors such as these six make it difficult to master the concepts central to the population-growth thesis, then one cannot unequivocally claim that no further sense experience, beyond mastering the constituent concepts, is required to determine the truth of the thesis (see Begon et al. 1986, 60-61). And if not, then it is not obvious that the population-growth thesis is a priori.

Someone might claim that the thesis is a priori, however, on the grounds that there is no obvious procedure by which it appears testable. But if the population-growth thesis does not wear its epistemological status on its sleeve, then we ought not put much stock in the claim that there is no "obvious" procedure for testability. Rather, we ought to try to classify or imagine the various ways that a thesis might be tested (Lloyd 1987, 277ff.). As Sober (1984, 69) points out, physicists before the time of non-Euclidean geometry might have claimed that there was no obvious procedure by which the thesis—that the angles of a triangle were exactly 180 degrees—was testable. Hence, although imaginability sets the limits of neither testability nor possibility, it might suggest ways of achieving testability of the population-growth thesis. One could imagine, for example, a small population in a spatially large, finite environment with great recycling capacity. If the rate of growth of the population were slow, and its resource needs were very limited and nonspecific, then for so long as community ecologists could measure population growth, it might appear to be indefinite. Such a situation might occur if the small population were growing but continually cut back by a series of random disturbances.

The apriority of the population-growth thesis is also problematic because it cannot be determined by appeal to the contents of a single interpretation of the thesis. In fact, imagining alternative interpretations is exactly what is at issue (see Sober 1984, 71). When we test the thesis that populations in a finite environment cannot grow indefinitely, we really must test our thesis by means of consequence laws that describe its effects, just as we must supplement analytic models with various theories before evaluating them (Haila 1986, 378; Sober 1984, 71). If our earlier consideration—of imagining a population that might be able to grow indefinitely—is plausible, then the consequences set by the population-growth thesis need not be defined a priori. Also, if Quine (1953, 1960) is right that one tests hypotheses and consequences together, then when a population in a finite environment appears to grow indefinitely, one could either redefine the concept of "finite environment" or modify the consequence laws in terms of which the effects of population growth in finite environments are known. Modifying the consequence laws might take the form of appending conditions to the consequences under which growth is said not to be indefinite. For example, the conditions might specify the size of the population and its habitat, the rate of growth of the population over time or space, whether the population appears to exhibit a random walk through time, or whether sub-populations in a large, patchy environment become extinct and then recolonize, particularly at the edge of a species range.

One of the main lessons to be learned—from revising the concepts and consequence laws, in the process of testing the population-growth thesis—is that we cannot evaluate apriority unless the contents of the thesis are accompanied by various physical assumptions. One cannot determine apriority, for example, without assessing what is local and contingent about alleged a priori theses, without assessing their complexity and variety (see Taylor 1986, Lloyd 1987), without investigating the theoretical contexts in which they are imbedded (see Sober 1984, 73; Quine 1953, 1960; Peters 1978, 761). For example, in the case of natural selection, one could specify the causes of differential adaptness for a group of organisms in a given environment and therefore obtain instances of the schematic "law" of natural selection which, on its own, has no empirical content (Brandon 1990, 139). Similarly, one could specify the causes of definite and indefinite rates of growth for a specific population, for a given time and environment, and therefore obtain instances of the schematic population-growth "law" which, on its own, has no empirical content. Thus, although this thesis may be used in some contexts that are a priori, it is not *merely* a priori.

4. Heuristic Power of the Thesis

Because the population-growth thesis is not merely a priori, it is not devoid of explanatory power. Rather it helps us describe "the limits of possibility" (Lewontin 1968; see Wangersky 1978) and shows us the logical consequences of our assumptions (Levin 1981b; see also Hutchinson 1978; May 1977). It also provides a model that may be made empirically relevant by specification of constants and unknowns (see Peters 1991, 60). Even if "populations in finite environments" = "populations that cannot grow indefinitely" ($P = A$), and even if P cannot explain *why* A, P may explain *what* the nature of A is. For example, P may explain a relationship defined in terms of stochastic, rather than deterministic, properties (see Sober 1984, 75).

If population growth is a probabilistic disposition, then greater relative population growth can regularly occur in less limited environments, but chance can intervene and break that regular connection. Such an intervention seems plausible, both because of the "imagined interpretations" suggested earlier and because adaptedness is likely a probabilistic disposition whose connection with actualized fitness can be broken by chance (see Brandon 1990, 45-46). In other words, finite environments may not be totally vacuous as explanations of different rates of population growth, because the improbable—e.g., extinction and recolonization of sub-populations—sometimes occurs. In fact, if Andrewartha and Birch (1954; see Sinclair 1989, 203) are correct, then density-independent factors might change sufficiently frequently that at least some sub-populations remain extant and thus able to recolonize other areas, especially at the edge of a species' range, where sub-populations have become extinct. If they are right that populations can persist without regulation and show a random walk through time (see Reddingius 1971; Reddingius and Den Boer 1970; Ehrlich et al. 1972), then it sometimes might be false to say that populations in finite environments cannot grow indefinitely. Perhaps some of them can grow indefinitely, if the growth is slow enough and measured over a long enough period. Perhaps some of them can, if there is stochasticity in both the density-dependent and the density-independent variables (see Strong 1984, 1986). As Sinclair (1989, 212) emphasizes, density-dependent effects may not lead invariably to dampening oscillations because some factors may cause stable limit cycles or increasing fluctuations, as occur in some predator-prey models for insect-parasitoid interactions.

Admittedly, the fact that populations in finite environments may have a very low probability of experiencing indefinite growth does not explain much. Nevertheless, examining alternative interpretations of the population-growth thesis can increase our understanding by providing strategies for new domains (Cooper 1990, 175; Kitcher 1989). Just as Brandon (1990, 140, 158) argues that natural selection is a "schematic law," "an organizing principle ... that structures natural-selectionist explanations," that "serves the role of systematic unification of the theory of natural selection," so also one could argue that the population-growth thesis provides a similar general schema. Moreover, without a "schematic law" such as the population-growth thesis, ecologists would have only unconnected low-level theories about growth in particular populations in particular environments. A schematic law, however, might help us pinpoint specific areas in which the thesis is more or less probable or has more or less utility and explanatory power. The possibility of developing criteria for differential growth rates of populations suggests that certain instantiations of the population-growth thesis have significant heuristic power.

5. Some Objections

Practicing ecologists may object that meta-level distinctions about heuristic power and apriority are purely semantic (see Murray 1989; Loehle 1990) and have little utility. They also may admit that by manipulating certain physical assumptions (e.g., expanding the spatial scale, reducing the temporal scale) one could falsify (or at least avoid confirming) the population-growth thesis in a particular situation. They might argue, however, that the falsifiable thesis-plus-assumptions is merely a "neighborhood" population-growth thesis, whereas the "global" population-growth thesis (holding over the long term) is a tautology. Ecologists might not care whether neighborhood versions of the thesis avoid tautology and might claim that progress in ecology depends on whether the thesis is reliable in a "global" sense.

On the contrary, determining the conditions under which proposed laws hold or not is one way of refining ecological theses and making them more particular and more capable of generating risky predictions (Popper 1968, 1959; Krebs 1980). Moreover, if the empirical core of a theory like that of population growth is in existential statements, rather than in falsifiable universal generalizations, then these conclusions about apriority will have consequences for the way that we do ecology. They suggest that the often-criticized, black-box, mathematical models in ecology may not be so bad as alleged (see, for example, Simberloff 1982, 83; Strong 1982b, 256; Gray 1987; Levin 1981a; see also Kingsland 1985), and that we need not emphasize only modest inductive generalizing, field work, and natural history (see, for example, Dayton 1979; Simberloff 1982; Strong 1984; Van Valen and Pitelka 1974; Wiens 1983). More generally, our conclusions about the population-growth thesis suggest that a top-down approach to ecological explanation (for example, Kitcher 1989; Cooper 1990), as well as the much-touted bottom-up account (Salmon 1989), may be useful.

Another objection to the population-growth thesis might be that it is trivial. Populations are dependent upon resources, and no resources are limitless; therefore, as Darwin recognized, all populations are limited. The interesting and non-trivial questions, according to some ecologists, are (1) to what extent (if any) does a *particular* environmental factor limit a population? (2) To what extent (if any) does such a factor affect equilibrium positions? (3) To what extent (if any) do environmental factors restrict populations by means of "limitation," "regulation," or "control"? (4) To what extent (if any) is a particular community regulated by density-dependent factors such as intraspecific competition, as opposed to predators or parasites? (5) To what extent (if any) is a particular population regulated by density-independent factors? (6) To what extent (if any) is a particular population regulated by density-dependent factors that act together, as opposed to those that compensate for each other? (7) To what extent (if any) is a particular population regulated by extrinsic or intrinsic factors? (8) To what extent (if any) are there multiple stable states in populations (see Sinclair 1989, 199-203, 232)?

Admittedly such questions suggest that the population-growth thesis, *alone*, is trivial and uninteresting because, over the long term, it is obviously true. However, if our earlier remarks (see Haila 1986, 382ff.; Lloyd 1987, 277-278, 291; Sober 1984, 73; Duhem 1914; Quine 1953, 1960; Hempel 1966, 23)—about not examining theses in isolation—are correct, then their claims about triviality may be technically correct but beside the point. It is not so easy to tell whether the population-growth thesis, as imbedded in a particular theoretical context, is true or false. It is not so easy to tell, for example, whether the evolutionary ecologists are correct in rejecting climate as a significant regulating factor for particular populations, or whether the functional ecologists are correct in accepting it (Orians 1962, 260). And if not, then the population-growth thesis, as imbedded in a particular theoretical context, is not trivially true. Likewise, if the thesis—together with a number of specific physical assumptions—sometimes is false, as we suggested earlier (see, for example, Andrewartha and Birch 1954; Sinclair 1989, 203; Reddingius 1971; Reddingius and Den Boer 1970; Ehrlich et al. 1972), then it is not trivially true.

6. Conclusions: What Epistemology Tells Us about Ecology

The difficulties in distinguishing between populations showing a random walk through time and those exhibiting extremely complex forms of density-dependent or density-independent regulation suggest that, although the population-growth thesis, taken alone, is general and nonempirical, it is neither trivial nor tautological. And if not, then perhaps the traditional positivistic account of theory structure is wrong (Brandon 1990, 150; see Kitcher 1989; Cooper 1990). Perhaps community ecology includes universal generalizations, such as the population-growth thesis, that themselves are not empirical but which, when appended to physical assumptions, give us theories that, as a whole, are empirical and testable. In other words, although the population-growth thesis may be a "schematic law" without predictive power, when it is suitably instantiated, as "law"-cum-physical-assumptions, it may have predictive power.

In part because it has been so difficult to obtain empirical regularities in ecology (see, for example, Roughgarden 1983), and because so many ecological situations have been described as "unique" (see, for example, Strong 1982b, 255; Peters 1991, 172-175, 211ff.), we may need new methods to deal with ecological phenomena, schematic laws, and with the primacy of specific cases and theoretical contexts (see Shrader-Frechette and McCoy 1994, 1993).

Notes

[1] The author is grateful to biologists E. D. McCoy and Peter Taylor and philosopher Bruce Silver for constructive criticisms of an earlier version of this essay. Thanks also to the National Science Foundation for Grant DIR-91-12445, "Laws and Explanation in Community Ecology," that supported work on this project. Remaining errors are the responsibility of the author.

References

Andrewartha, H. and Birch, L. (1954), *The Distribution and Abundance of Animals*. Chicago: University of Chicago Press.

Barlow, N. (ed.) (1958), *The Autobiography of Charles Darwin*. New York: Norton.

Begon, M., Harper, J., and Townsend, C. (1986), *Ecology: Individuals, Populations, and Communities*. Sunderland: Sinauer.

Brandon, R. (1990), *Adaptation and Environment*. Princeton: Princeton University Press.

Caplan, A. (1978), "Tautology, Circularity, and Biological Theory", *The American Naturalist* 111: 390-393.

Cody, M.L. (1968), "On the Methods of Resource Division in Grassland Bird Communities", *The American Naturalist* 102: 107-148.

Connell, J.H. (1983), "On the Prevalence and Relative Importance of Interspecific Competition: Evidence from Field Experiments", *The American Naturalist* 122: 661-696.

Cooper, G. (1990), "The Explanatory Tools of Theoretical Population Biology", in A. Fine, M. Forbes, and L. Wessels (eds.), *PSA 1990*, vol. 1. East Lansing: Philosophy of Science Association, pp. 165-178.

Dayton, P. (1979), "Ecology", in R. Livingston (ed.), *Ecological Processes in Coastal and Marine Systems*. New York: Plenum, pp. 3-18.

Diamond, J.M. (1975), "Assembly of Species Communities", in M. Cody and J. Diamond (eds.), *Ecology and the Evolution of Communities*. Cambridge: Harvard University Press, pp. 332-344.

Diamond, J.M. and Gilpin, M. (1982), "Examination of the "Null" Model of Connor and Simberloff for Species Co-Occurrences on Islands", *Oecologia* 52: 64-74.

Duhem, P. (1914), *The Aim and Structure of Physical Theory*. Princeton: Princeton University Press.

Ehrlich, P., Breedlove, D., Brussard, P., and Sharp, M. (1972), "Weather and Regulation of Subalpine Populations", *Ecology* 53: 243-247.

Ferguson, A. (1976), "Can Evolutionary Theory Predict?" *The American Naturalist* 110: 1101-1104.

Gilpin, M. and Diamond, J. (1984), "Are Species Co-Occurrences on Islands Non-Random, and are Null Hypotheses Useful in Community Ecology?" in D. Strong, D. Simberloff, L. Abele, and A. Thistle (eds.), *Ecological Communities: Conceptual Issues and the Evidence*. Princeton: Princeton University Press, pp. 297-315.

_____. (1982), "Factors Contributing to Non-Randomness in Species Co-Occurrences on Islands". *Oecologia* 52: 75-84.

Gray, R. (1987), "Faith and Foraging", in A. Kamil, J. Krebs, and H. Pulliam (eds.), *Foraging Behavior*, New York: Plenum, pp. 69-140.

Haila, Y. (1986), "On the Semiotic Dimension of Ecological Theory", *Biology and Philosophy* 1:337-387.

Hempel, C. (1966), *The Philosophy of Natural Science*. Englewood Cliffs: Prentice-Hall.

Hopkins, A. (1920), "The Bioclimatic Law", *Journal of the Washington Academy of Science* 10: 34-40.

Humphreys, W. (1979), "Production and Respiration in Animal Populations", *Journal of Animal of Ecology* 48: 427-453.

Hutchinson, G. (1978), *An Introduction to Population Ecology*. New Haven: Yale University Press.

_____. (1959), "Homage to Santa Rosalia", *The American Naturalist* 93: 145-159.

Kingsland, S. (1985), *Modeling Nature*. Chicago: University of Chicago Press.

_____. (1982), "The Refractory Model: The Logistic Curve and the History of Population Ecology", *The Quarterly Review of Biology* 57: 29-52.

_____. (1981), *Modelling Nature*. Ph.D. Thesis, Toronto: University of Toronto.

Kitcher, P. (1989), "Explanatory Unification and the Causal Structure of the World", in P. Kitcher and W. Salmon (eds.), *Scientific Explanation*, Minnesota Studies in the Philosophy of Science. Minneapolis: University of Minnesota Press, pp. 410-506.

Krebs, J. (1980), "Ornithologists as Unconscious Theorists", *Auk* 97: 409-412.

Kripke, S. (1972), "Naming and Necessity", in D. Davidson and G. Harman (eds.), *Semantics of Natural Language*. Dordrecht: Reidel.

Levin, S. (1981a), "Mathematics, Ecology, Ornithology", *Auk* 97: 422-25.

_____. (1981b), "The Role of Theoretical Ecology in the Description and Understanding of Populations in Heterogeneous Environments", *American Zoologist* 21: 865-875.

Lewontin, R. (1968), "Introduction", in R. Lewontin (ed.), *Population Biology and Evolution*. Syracuse: Syracuse University Press, pp. 1-4.

_____. (1966), "Is Nature Probable or Capricious?" *BioScience* 16: 25-27.

Livingston, B. and Shreve, F. (1921), *The Distribution of Vegetation in the United States as Related to Climatic Conditions*, No. 284. Washington, D.C.: Carnegie Institution of Washington.

Lloyd, E.A. (1987), "Confirmation of Ecological and Evolutionary Models", *Biology and Philosophy* 2: 277-293.

Loehle, C. (1990), "Philosophical Tools: Reply to Shrader-Frechette and McCoy", *Oikos* 58: 115-119.

_____. (1988), "Philosophical Tools: Potential Contributions to Ecology", *Oikos* 51: 97-104.

May, R. (1977), "Mathematical Models and Ecology", in C. Goulden (ed.), *The Changing Scenes in the Natural Sciences*. Philadelphia: Academy of Natural Sciences, pp. 189-201.

McIntosh, R. (1985), *The Background of Ecology*. Cambridge, MA: Cambridge University Press.

_____. (1982), "Some Problems of Theoretical Ecology", in E. Saarinen (ed.), *Conceptual Issues in Ecology*. Dordrecht: Reidel, pp. 1-62.

Merriam, C. (1894), "Laws of Temperature Control of the Geographic Distribution of Terrestrial Plants and Animals", *National Geographic Magazine* 6: 229-238.

Murray, B.G. (1989), "Review of Shrader-Frechette and McCoy", Personal Communication, p. 7.

_____. (1986), "The Structure of Theory and the Role of Competition in Community Dynamics", *Oikos* 46: 145-158.

_____. (1979), *Population Dynamics: Alternative Models*. New York: Academic Press.

Orians, G. (1962), "Natural Selection and Ecological Theory", *The American Naturalist* 96: 257-264.

Pearl, R. (1927), "The Growth of Populations", *The Quarterly Review of Biology* 2: 532-548.

_____. (1925), *The Biology of Population Growth*. New York: Knopf.

Pearl, R. and Reed, L. (1920), "On the Rate of Growth of the Population of the United States Since (1970) and Its Mathematical Representation", *Proceedings of the National Academy of Sciences* 6: 275-288.

Peters, R.H. (1991), *A Critique for Ecology*. Cambridge: Cambridge University Press.

_____. (1978), "Predictable Problems with Tautology in Evolution and Ecology", *The American Naturalist* 112: 759-762.

_____. (1976), "Tautology in Evolution and Ecology", *The American Naturalist* 110: 1-12.

Pyke, G. (1982), "Local Geographic Distributions of Bumblebees Near Crested Butte, Colorado: Competition and Community Structure", *Ecology* 63: 555-573.

Popper, K. (1968), *Conjectures and Refutations*. New York: Harper and Row.

_____. (1959), *The Logic of Scientific Discovery*. New York: Harper and Row.

Putnam, H. (1970), "Is Semantics Possible?", in H. Putnam (ed.), *Mind, Language, and Reality*. Cambridge: Cambridge University Press, pp. 139-152.

Quine, W.V.O. (1960), *Word and Object*. Cambridge, MA: MIT Press.

_____. (1953), *From a Logical Point of View*. Cambridge: Harvard University Press.

Reddingius, J. (1971), "Gambling for Existence", *Acta Biotheoretica, Suppl. 1* 20: 1-208.

Reddingius, J. and Den Boer, P. (1970), "Simulation Experiments Illustrating Stabilization of Animal Numbers by Spreading of Risk", *Oecologia* 5: 240-284.

Rosenberg, A. (1985), *The Structure of Biological Science*. Cambridge: Cambridge University Press.

_____. (1978). "The Supervenience of Biological Concepts", *Philosophy of Science* 45: 368-386.

Roughgarden, J. (1983), "Competition and Theory in Community Ecology", *The American Naturalist* 122: 583-601.

Salmon, W. (1989), "Four Decades of Scientific Explanation", in P. Kitcher and W. Salmon (eds.), *Scientific Explanation*, Minnesota Studies in the Philosophy of Science. Minneapolis: University of Minnesota Press, pp. 3-219.

Schoener, T. (1984), "Size Differences Among Sympatric, Bird-Eating Hawks: A Worldwide Survey", in D. Strong, D. Simberloff, L. Abele, and A. Thistle (ed.), *Ecological Communities: Conceptual Issues and the Evidence*. Princeton: Princeton University Press, pp. 254-281.

_____. (1983), "Field Experiments on Interspecific Competition", *The The American Naturalist* 122: 240-285.

_____. (1974), "Resource Partitioning in Ecological Communities", *Science* 185: 27-39.

Shrader-Frechette, K. (1990), "Interspecific Competition, Evolutionary Epistemology, and Ecology", in N. Rescher (ed.), *Evolution, Cognition, and Realism*. New York: University of Pittsburgh Center for Philosophy of Science Series in Philosophy of Science, University Press of America, pp. 47-63.

Shrader-Frechette, K. and McCoy, E. (1994), "Applied Ecology and the Logic of Case Studies", *Philosophy of Science* 61, June Issue, in press.

_____. (1993), *Method in Ecology: Strategies for Conservation*. Cambridge: Cambridge University Press.

_____. (1990), "Theory Reduction and Explanation in Ecology", *Oikos* 58: 109-114.

Simberloff, D. (1984), "Properties of Coexisting Bird Species in Two Archipelagoes", in D. Strong, D. Simberloff, L. Abele, and A. Thistle (eds.), *Ecological Communities: Conceptual Issues and the Evidence*. Princeton: Princeton University Press, pp. 234-258.

_____. (1983), "Competition Theory, Hypothesis Testing, and Other Community Ecological Buzzwords", *The American Naturalist* 122: 626-635.

_____. (1982), "A Succession of Paradigms in Ecology", in E. Saarinen (ed.), *Conceptual Issues in Ecology*. Dordrecht: Reidel, pp. 139-153.

Simberloff, D. and Boecklen, W. (1981), "Santa Rosalia Reconsidered", *Evolution* 35: 1206-1228.

Sinclair, A. R. (1989), "Animal Population Regulation", in J.M. Cherrett (ed.), *Ecological Concepts*. Oxford: Blackwell, pp. 197-241.

Sober, E. (1984), *The Nature of Selection*. Cambridge: MIT Press.

Stebbins, G. (1978), "In Defense of Evolution", *The American Naturalist* 111: 386-390.

Strong, D. (1986), "Density-Vague Population Change", *Trends in Ecology and Evolution* 1: 39-42.

_____. (1984), "Density-Vague Ecology and Liberal Population Regulation in Insects", in P. Price, C. Slobodchikoff, and W. Gaud (eds.), *A New Ecology*. New York: Wiley, pp. 313-327.

_____. (1982a), "Harmonious Coexistence of Hispine Beetles on *Heliconia* in Experimental and Natural Communities", *Ecology* 63: 1039-1049.

_____. (1982b), "Null Hypotheses in Ecology", in E. Saarinen (ed.), *Conceptual Issues in Ecology*. Dordrecht: Reidel, pp. 245-360.

Strong, D., and Levin, D. (1979), "Species Richness of Plant Parasites and Growth Form of Their Hosts", *The American Naturalist* 114: 1-22.

Strong, D. and Simberloff, D. (1981), "Straining at Gnats and Swallowing Ratios: Character Displacement", *Evolution* 35: 810-812.

Strong, D., Simberloff, D., Abele, L., and Thistle, A. (eds.) (1984), *Ecological Communities: Conceptual Issues and the Evidence*. Princeton: Princeton University Press.

Taylor, P. (1986), "Dialectical Biology as Political Practice", in L. Levidow (ed.), *Science as Politics*. London: Free Association Books, pp. 81-109.

_____. (1989), "Revising Models and Generating Theory", *Oikos* 54: 122-126.

Van Valen, L. and Pitelka, F. (1974), "Commentary: Intellectual Censorship in Ecology", *Ecology* 55: 925-926.

Verhulst, P. (1845), "Recherches mathematiques sur la loi d"accroissement de la population", *Memoirs de l" Academie Royale de la Belge* 18: 1-38.

_____. (1838), "Notice sur la loi que la population suit dans son accroissement", *Correspondence Mathematique et Physique* 10: 113-121.

Wangersky, P. (1978), "Lotka-Volterra Population Models", *Annual Review of Ecology and Systematics* 9: 189-218.

Wiens, J. (1983), "Avian Community Ecology", in A. Brush, and G. Clark (eds.), *Perspectives in Ornithology*. Cambridge: Cambridge University Press, pp. 355-403.

Williams, M. B. (1970), "Deducing the Consequences of Evolution", *Journal of Theoretical Biology* 29: 343-385.

Defending Robustness:
the Bacterial Mesosome as a Test Case[1]

Sylvia Culp

Western Michigan University

1. Introduction

In "Facts, Artifacts, and Mesosomes: Practicing Epistemology with the Electron Microscope" (1993), Nicolas Rasmussen presents an account of the mesosome, a bag-shaped membranous structure first observed by electron microscopists in gram-positive bacteria,[2] as a test case for robustness. Rasmussen argues that robustness as a criterion for the reality or artifactuality of experimental results fails this test; for, after 15 years of interpreting electron micrographs of specimens prepared by a variety of techniques as supporting evidence for the existence of mesosomes, electron microscopists ultimately concluded that mesosomes were an artifact produced by these preparation techniques. According to Rasmussen, electron microscopists did not use robustness to "decide reliably between conflicting observations" (ibid., 231). Moreover, Rasmussen also claims that research on mesosomes "resembles that of high flux gravity waves as described by Harry Collins, especially in the obviousness of 'experimenter's regress' or logical circularity entailed in the evaluation of experiments according to their results..."(ibid., 256).[3]

I argued in a previous paper, "Objectivity in Experimental Inquiry: Breaking Data-Technique Circles", that experimentalists use and are warranted in using robust bodies of data to break the 'experimenters' regress'. Dependence on both idiosyncratic and shared theoretical presuppositions can be eliminated from interpretations of raw data (for example, electron micrographs) to the extent that the resulting data is comparable with data produced by other independently theory-dependent techniques. I worked out a theory of robustness that was inspired by and is offered as a supporting theory for Wimsatt's concept of robustness.[4] I used this theory to argue that the degree of robustness for a body of data is determined by 1) the size of the set of techniques producing the comparable data; 2) the degree to which theories used for raw data interpretations by one of the techniques in the set are independent from the theories used by all the other techniques in the set; and 3) the degree to which the raw data interpretations for each technique in the set are theory-dependent.

Rasmussen's critique is aimed at the claims of Hacking (1983), Franklin (1986, 1990) and Jardine (1986, 1991) that experimentalists use and are warranted in using

multiple techniques to determine the reliability of observations made with scientific instruments. My theory of robustness is susceptible to the same criticisms. In this paper I will argue that Rasmussen's test case for robustness is based on an incomplete, and, therefore, misinterpreted reading of the research literature on the mesosome.
Moreover, I will argue that a more complete reading of the literature shows that the mesosome "ended up an artifact after some fifteen years as a fact" *because* the body of data indicating that bacterial cells do not contain mesosomes was more robust than the body of data indicating that they do. Mesosomes were not consistently observed when electron microscopists attempted to observe mesosomes both by varying conditions with already established sample preparation techniques and by using newly developed sample preparation techniques. Moreover, after 15 years of research on the biochemistry of the mesosome, biochemists produced a robust body of data showing that there were no biochemical differences between mesosomal and cytoplasmic membranes.

Table 1

Four Specimen Preparation Techniques for Gram-positive Bacterial Cells

	Stage I	Stage II	Stage III	Electron Microscope	Raw Data
R-K	chemically prefix	chemically fix	1)dehydrate 2)embed 3)section 4)stain	bombard with electrons under vacuum	electron micrographs
F-F	none	cryofix	1)fracture 2)replicate	bombard replicate with electrons under vacuum	same
F-S	none	1)ultrarapid cryofix 2)cryosubstitute	1)dehydrate 2)embed 3)section 4)stain	bombard with electrons under vacuum	same
CU	none	ultrarapid cryofix	section	bombard with electrons under vacuum at extremely low temp.	same

2. Ultrastructural and biochemical identification of bacterial mesosomes

Mesosomes were first described by Chapman and Hillier (1953) as inclusions, which they named peripheral bodies, in dividing cells. Throughout the 1950s electron microscopists worked to develop preparation techniques for bacterial cells. By the end of this decade, the Ryter-Kellenberger (R-K) technique, had become the standard technique for preparing bacterial specimens with "well preserved" cell envelopes and nucleoplasms (Ryter and Kellenberger 1958). When using this technique, electron microscopists prepared samples by prefixing growing cells with the addition of osmium tetroxide at a final concentration of 0.1% to the culture medium ten minutes prior to harvesting by centrifugation. Cells were fixed by resuspending them in a buffer containing 1% osmium tetroxide and then embedded in polyester resin after a series of dehydration steps.[5] To assist in the analysis of this and other specimen preparation techniques I will refer to three stages: Stage I, a prefixing stage where cells in culture are treated to prepare then for fixation; Stage II, a fixing stage where cells are either chemically or physically fixed; and Stage III, a sectioning stage where specimens thin enough to be subjected to electron microscopy are prepared (see Table 1).

Fitz-James (1960) gave the name 'mesosome' to the membranous organelle he observed in thin sections prepared by the R-K technique and proposed that they were involved in spore- and cell-wall synthesis. At various times during the 1960s and early 1970s, bacteriologists proposed other possible functions for the mesosome: that they were involved in nucleosome division or in the release of extracellular enzymes, or were sites of oxidation and reduction. Considerable effort was expended in trying to identify a biochemical difference between the mesosomal and cytoplasmic membranes (for a review of this research see Greenawalt and Whiteside 1975).

Table 2
Observation of Large Centralized Mesosomes:
Higgins's and Daneo-Moore's Dispute with Nanninga

Mesosomes	Technique	Stage I Prefixing	Stage II Fixing	Stage III Sectioning
Nanninga (1968, 1969, 1971)				
YES	R-K	GA (4°C, 60min)	OsO4	thin section
	F-F	NONE	cryofix with or w/o glycerol	cryofracture
		grow in glycerol or stand o.n. at 4°C		
NO		NONE		
Higgins & Daneo-Moore (1974)				
YES	R-K	OsO4 or GA (26°C, 60min)	OsO4	thin section
			cryofix with or w/o glycerol	cryofracture
	F-F	centrifuge cells	cryofix with glycerol	
NO			cryofix w/o glycerol	
Higgins, Tsien & Daneo-Moore (1976)				
YES	F-F	GA (37°C, 3min) GA (3°C, 120min) centrifuge cells	cryofix	cryofracture
NO		GA (3 C, 10min) filter cells		
Parks, Dicker, Conger, Daneo-Moore, Higgins (1981)				
YES	F-F	GA (37°C, 60min)	cryofix	cryofracture
NO		X-irradiate before GA (37°C, 60min)		

3. Nanninga's dispute with Higgins and Daneo-Moore

In 1968 Nanninga published a paper reporting that mesosomes were observed in cells that had been prepared for electron microscopy by the newly developed freeze-fracturing (F-F) technique. When using this technique, electron microscopists prepared samples by freezing the cells in a small block that they then cracked apart by fracturing along paths of local weaknesses including weaknesses determined by cellular structures such as the lipid bilayer of cellular membranes. As originally developed, the F-F technique differed from the R-K technique at all three stages (Table 1).

In 1971, however, Nanninga reported that he did not observe mesosomes when he used some variations of the F-F technique. He continued to observe mesosomes when he used other variations of the F-F technique or when he used the R-K technique (Table 2). Although he did not explicitly state that mesosomes were an artifact, his 1971 paper ends with the claim that "it can be said that some care is needed in drawing conclusions concerning the structure of mesosomes in chemically fixed material" (ibid., 223).

Three years later Higgins and Daneo-Moore (1974) disputed Nanninga's results. They reported that few mesosomes were found when unfixed cells were prepared for electron microscopy by the F-F technique, but that numerous large mesosomes were found when the cells were fixed in osmium tetroxide or glutaraldehyde (GA) or treated with glycerol before freezing. (Table 2). Higgins and Daneo-Moore argued that, when cells are fractured after being cryofixed without a cryoprotectant (a substance that protects cells against ice crystal damage) such as glycerol, their mesosomes are rendered invisible because they cross-fracture, i.e., the plane of the fracture does not follow a surface layer of the mesosomes. Moreover, they argued, that when the cells are chemically fixed or are cryofixed with a cryoprotectant, (modifications at Stages I and II) their mesosomes could become visible either 1) because mesosome surface fracturing is increased, or 2) because the membrane is reorganized "so that mesosomes can be observed in cross fracture". In other words, at Stage III mesosomes are seen because either there are more surface fractures instead of cross fractures or there are just as many cross fractures but the modifications at Stages I and II allow them to be seen.

Two years later, however, Higgins, Tsien and Daneo-Moore (1976) reported that they found only a few small mesosomes when they repeated their experiments with modifications at Stage I of specimen preparation: chilling the cells before prefixation in glutaraldehyde or concentrating the cells by filtration rather than centrifugation. On the other hand, as before, they found numerous large mesosomes when these cells were not chilled before being prefixed or were prefixed for a long time at the low temperature (Table 2). They concluded that "the structure seen in fixed cells...is apparently an artifact produced by fixation" (ibid., 1522) and they proposed two possible models for explaining the production of these artifactual structures. In their first model, small mesosomal precursors in the cytoplasm were aggregated when cells were prefixed in osmium tetroxide or glutaraldehyde or when large amounts of glycerol were added. In their second model, mesosomes were formed by the invagination of the cytoplasmic membrane when the cell wall contracted during the prefixation step (see below for a test of this second model).

Rasmussen claims that research interest in mesosomes waned after the 1976 paper by Higgins, Tsien and Daneo-Moore; "some time around 1975 (the date of the last review article on mesosomes), the majority opinion shifted to artifactuality" (ibid., 255). Rasmussen attributes diminished interest to "the loss of support from its biochemical constituency". I agree that research interest in mesosomes did wane in the mid-70's. Moreover, this diminished interest probably was due, at least in part, to the failure of biochemists to identify a biochemical difference between mesosomal and cytoplasmic membranes. Two of these biochemists, Salton and Owen, concluded in a 1976 review that "the possibility that the mesosomes are artifactual structures must now be seriously considered, especially as extensive searching for the past two decades has failed to reveal a unique function for the structures" (ibid., 476). Rasmussen claims that

> one major factor making the mesosome vulnerable to demolition was that biochemists were dissatisfied with the disunity of their findings...[so that] given two negotiable bodies of evidence and the apparatus for evaluating them on opposite pans of the balance, and the one side no longer secured by a set of biochemists trying to purify the mesosome and to validate their prior findings, the scale could have tipped either way (ibid., 255-56).

And, he evokes the 'experimenter's regress to argue that the biochemists' failure did not tip the scale toward rejecting the mesosome as an artifact because their failure could be explained away. i.e., that unlike electron microscopists, biochemists had not developed the techniques necessary for producing "the expected" data on the mesosome. Rasmussen concludes, that for electron microscopists the scale tipped toward rejecting the mesosome not because the biochemists' failure gave it a lower degree of robustness but simply because the biochemists stopped working on it.

In trying to identify a biochemical marker that could distinguish mesosomal membranes from the cytoplasmic membrane biochemists used a variety of techniques. Thus, by the mid-70s there was a robust body of data showing that the mesosomal and cytoplasmic membranes were biochemically indistinguishable. In the next section I will examine research on the mesosome that Rasmussen did not, for the most part, include in his account. I will argue that by using new specimen preparation techniques and variations on the established techniques, electron microscopists produced a body of data on the mesosome that was comparable with the biochemists' data. Moreover, I will argue that this body of data was more robust in indicating that mesosomal and cytoplasmic membranes were *not* distinguishable than in indicating that they were.

4. Electron microscopists use robustness as a criterion for the reality or artifactuality of mesosomes

Beginning in the early 1970s and ending in the mid 1980s seven groups of electron microscopists published papers about the effect of various specimen preparation techniques on the appearance of mesosomes in electron micrographs. The four main specimen preparation techniques are outlined in Table 1. Nanninga's and Higgins's and Daneo-Moore's experiments have already been discussed (Table 2). In 1971 Silva reported that, when the prefixing stage of the R-K technique was modified, mesosomes were not observed in three of four strains of gram-positive bacteria (Table 3). Silva concluded that from "the data available it is difficult to conclude which, if any, of these different patterns of membrane ultrastructure is real" (ibid., 227) so that "other preparative techniques...might be important controls of chemical fixation" (ibid., 231). Five years later Silva et. al., (1976) reported that, when the R-K technique was used to prepare samples, large complex and centralized mesosomes were only observed in cells that had been prefixed in osmium tetroxide. In addition, more mesosomes were observed the longer the cells were prefixed. More importantly, they showed that the membranes of these cells were damaged during this prefixation step (by measuring the quick and extensive leakage of intracellular potassium) and that large centralized mesosomes were found when these cells were subjected to three other membrane-damaging pretreatments, moist heat, phenethyl alcohol (PA), or nitroblue tetrazolium (NT) (Table 3).

In 1974 by Fooke-Achterrath, et., al. also reported that, when bacterial cells were chilled before being prefixed in osmium tetroxide during preparation by the R-K technique, only one or two small mesosomes were observed in each cell. When the cells were not chilled before being prefixed or when cells were chilled but then warmed before being prefixed in osmium tetroxide they observed numerous large centralized mesosomes. Moreover, when they prepared specimens by the F-F technique they observed only a few small peripheral mesosomes unless they fixed the cells in osmium tetroxide before freezing (Table 3). They concluded that in "a still unknown manner, large, artifactual, membranous bodies—'technikosomes'—arise in the chromosomal area after addition of fixative... 'Technikosomes' have erroneously been considered to be mesosomes" (ibid., 270).

Table 3
Observation of Large Centralized Mesosomes:
the Ryder-Kellenberger and Freeze-Fracture Techniques

Mesosomes	Technique	Stage I Prefixing	Stage II Fixing	Stage III Sectioning
Silva (1971)				
YES	R-K	OsO4 (35°C,10min)	OsO4	thin section
NO		NONE		
Foote-Achterrath, Lickfeld, Reusch, Aebi, Tschope, Menge (1974)				
YES	R-K	OsO4 (26°C,10min)	OsO4	thin section
	F-F		cryofix with glycerol	cryofracture
NO	R-K	OsO4 (4°C,10min)	OsO4	thin section
	F-F		cryofix with glycerol	cryofracture
		NONE	cryofix w/o glycerol	
Silva, Sousa, Polonia, Macedo, Parente (1976)				
YES	R-K	OsO4 (35°C,10min), moist heat,PA or NT	OsO4 or GA	thinsection
NO		NONE		

In 1981, Daneo-Moore's and Higgins's groups reported experiments aimed at testing the second model they had proposed in their 1976 paper. Since the appearance of large central mesosomes was correlated with the contraction of the nucleoid, they had proposed that nucleoid contraction during glutaraldehyde fixation at 37°C might pull portions of the cytoplasmic membrane into the cytoplasm. To test this model they irradiated bacterial cells with X-rays before fixing them in glutaraldehyde (Table 2). By breaking the chromosomal DNA at multiple sites they hoped to release any nucleoid attachments to the cytoplasmic membrane. They found that as they increased the dosage of X-rays that the number of cells with large central mesosomes decreased as the molecular weight of chromosomal DNA and the amount of DNA attached to the cytoplasmic membrane decreased and as the nucleoid became more diffusely organized. They concluded that large central mesosomes were an artifact caused by *invagination* of the part of the cytoplasmic membrane attached to the nucleoid during glutaraldehyde fixation.

Ebersold, Cordier and Luthy reported in 1981 that no mesosomes were observed in cells that had been prepared with a new technique, Freeze-Substitution (F-S) (Tables 1 and 4). Specimens prepared with this technique are rapidly frozen in a propane jet (to prevent the formation of ice crystals that might damage the specimen) and then incubated in an organic solvent containing the same fixatives used in the R-K technique. Ebersold, Cordier and Luthy reported that there were "striking morphological differences" between chemically fixed and cryofixed cells. In cryofixed fixed cells the cytoplasmic membrane and the cell wall "remained intimately associated" while in chemically fixed cells there were spaces between the cytoplasmic membrane and the cell wall. Four years later, Hobot, et. al. (1985) also reported that mesosomes could not be found in cells prepared by the F-S technique (Table 4). Ryter and Kellenberger (developers of the R-K technique) are two of the authors of this paper; interestingly, they conclude that

> there is now better experimental evidence for the nonexistence of mesosomes than the earlier results, obtained by cryofracturing, which had led to a similar result....After decades of research, no functions have been appointed to the mesosomes. It is therefore not so sad to contemplate that whether it is pleasant or not, mesosomes (as we are used to seeing them in beautiful sections of gram-positive cells) are artificially produced (ibid., 969-70).

Finally, in 1983 Lickfeld et. al. used yet another newly developed technique, Cryoultramicrotomy (CU) (Table 1) to look for bacterial mesosomes. Specimens pre-

Table 4

**Observation of Small Peripheral Mesosomes::
the Freeze Substitution and Cryoultramicrotomy Techniques**

Mesosomes	Technique	Stage I Prefixing	Stage II Fixing	Stage III Sectioning
Ebersold, Cordier, Luthy (1981)				
YES	R-K	NONE	OsO4 & GA	thin section
NO	F-F		ultrapid cryofix OsO4/GA substitution	
			ultrapid cryofix	cryofracture
Dubochet, McDowall, Menge, Schmid, Lickfeld (1983)				
YES	R-K	OsO4	OsO4	thin section
	F-H		ultrapic cryofix	
NO		NONE		cryosection
Hobot, Villiger, Escalg, Maeder, Ryter, Kellenberger (1985)				
YES	R-K	OsO4	OsO4 or GA	thin section
NO	C/F-S	NONE	ultrapid cryofix OxO4 substitution	

pared with this technique are grown in glucose (for its ability to act as a cryoprotectant), placed in a microtome specimen holder, and immediately projected into liquid ethane. The specimen is kept in liquid nitrogen until sectioned with a cryoultramicrotome. Under these conditions ice crystals cannot form inside the cells but the hydration of the specimen is maintained. Lickfeld et. al. reported that they did not observe mesosomes in these amorphous, unstained, frozen-hydrated sections. They did report, however, that they could observe mesosomes in sections prepared by the R-K technique (Table 4).

In summary, to observe cells with mesosomes these electron microscopists had to prepare specimens by the R-K technique (notice that in Tables 3 and 4 each group used the R-K technique to show that they were competent producers of cells with mesosomes), or by subjecting them to membrane damaging treatments (including osmium tetroxide or glutaraldehyde at room temperature or above, centrifugation, overnight in the cold, moist heat, phenethyl alcohol, or nitroblue tetrazolium) before fixation. On the other hand, to observe cells without mesosomes they could use the F-S or CU techniques or variations of the R-K or F-F techniques (ones that do not include prefixing or include prefixing with osmium tetroxide or glutaraldehyde at 4°C).

5. Robustness defended

By 1985 electron microscopists had used variations on four different specimen preparation techniques to observe that gram positive bacterial cells did not contain large centralized mesosomes and in some cases small peripheral mesosomes. In interpreting the raw data (splodges on electron micrographs) produced when each technique is used, electron microscopists depended on a set of theories, $\{Ti\}$, to support the claim that gram positive bacterial cells either do or the claim that these cells do not contain mesosomes (Table 5). Thus, for each of these techniques, the interpretation of raw data was heavily theory-dependent.

Let's use the criteria presented at the beginning of this paper to work out the degree of robustness for the bodies of data on mesosomes. First, with the development of variations on established techniques and newer techniques the set producing comparable data (including data from the biochemists working on mesosomes) that could be used to support mesosomal membranes as distinct from the cytoplasmic membrane stopped growing while the set that could be used to reject a distinction got larger.

Second, for the set of techniques that could be used to reject the mesosome, there is a higher degree of independence among the theories used to interpret electron micrographs than for the set of techniques that could be used to support the mesosome, i.e., members of this set all depend on theories about the effects of chemical fixation or cryoprotectants (see Tables 2-4).

All four of the techniques that could be used to reject the mesosome do share a dependence on one theory, the theory-of-gram positive bacteria cell ultrastructure (Tx). Even though these techniques share a dependence on Tx, however, they are somewhat independent because they do not share a dependence on the same parts of Tx. If Tx is regarded as a collection of propositions, then it is possible to identify a sub-collection of these propositions, Tx2, that includes just those propositions in Tx that are not independent of $\{Ti\}$ (Kosso 1989; Culp). For the four techniques at least some of the propositions in Tx2 are different (Table 5). Thus, the four techniques are not dependent in exactly the same way on Tx.

Third, for each technique in each set, the sub-collection of propositions, Tx2, does not intersect with another sub-collection of propositions from Tx called Tx1. This sub-collection contains those propositions that can be confirmed by either observing that gram-positive bacteria cells either do or observing that they do not contain mesosomes. Thus, for each technique, support for an electron micrograph interpretation is theory-dependent but this support is not dependent only on the theory that these cells have mesosomes or the theory that these cells do not have mesosomes.

Contrary to Rasmussen's interpretation of the research literature on the mesosome, the mesosome was rejected as an artifact because experimentalists used robustness as a criteria for evaluating their data. Moreover, they were warranted in concluding that the mesosome was an artifact *because* when multiple independently theory-dependent techniques were used the more robust body of data did not support the mesosome. Thus, robustness as a criterion for the reality or artifactuality of experimental results passes this test case.

Table 5

Theory-dependence of Techniques Used to Prepare Gram-positive Bacteria for Electron Microscopy

Ryder-Kellenberger	Freeze-Fracture	Freeze-Substitution	Cryoultramicrotomy
Tx1: The sub-collection of propositions belonging to the theory-of-gram-positive bacteria cell ultrastructure, Tx, that could have been confirmed by the data produced after each of these techniques is used to prepare specimens for electron microscopy • Gram positive bacteria cells have large central mesosomes • Gram positive bacteria cells have small peripheral mesosomes			
{Ti}: Set of theories used in interpreting raw data (electron micrographs) to support either the claim that gram positive bacterial cells do or the claim that they do not have mesosomes.			
• chemical fixation • dehydration • embedding • sectioning • staining • electron irradiation	• [cryofixation -OR- chemical fixation and freezing] • fracturing • replication • electron irradiation	• ultrarapid cryofixation • cryosubstitution • dehydration • embedding • sectioning • staining • electron irradiation	• ultrarapid cryofixation • sectioning with a cryoultramicrotome • electron irradiation
Tx2: The sub-collection of propositions belonging to the theory-of-gram-positive bacteria cell ultrastructure, Tx, that not independent of {Ti}. These propositions are about the *preservation of ultrastructure* after:			
chemical fixation, dehydration, embedding, sectioning, staining, and then electron irradiation.	[cryofixation -OR- chemical fixation and freezing], fracturing, replication, and then electron irradiation	ultrarapid cryofixation, cryosubstitution, dehydration, embedding sectioning, staining, and then electron irradiation.	ultrarapid cryofixation, sectioning with a cryoultramicrotome, and then electron irradiation.

Notes

[1] I would like to thank Arthur Falk and William Wimsatt for valuable comments and Donald McQuitty for help with the bacteriology literature. An early draft of this paper was presented to the Heraclitean Society at Western Michigan University. This work was supported by a New Faculty Research Support Program Grant and by funds from the Faculty Research and Creative Activities Support Fund, Western Michigan University.

[2] The Gram stain is used in light microscopy to divide bacteria into four groups: gram-positive, gram-negative, gram-nonreactive, and gram-variable. A bacterium is placed in one of these groups depending on how its cell wall reacts with the Gram stain. The cell wall of Gram-positive bacteria is a thick layer of peptidoglycan, a large polymer of two alternating sugar units cross-linked by short peptide chains, that is closely attached to the outer surface of the cytoplasmic membrane. The cell wall maintains the characteristic shape of a bacterial cell and prevents the cell from bursting when fluids flow into the cell; however, it is porous and plays only a minor role in regulating the entry or exit of materials. The cytoplasmic membrane regulates the movement of materials into and out of the cell. The cytoplasmic membrane is composed of phospholipids that form a bilayer; the outer layer is in contact with the cell wall and the inner layer is in contact with the cytoplasm. Some of the protein molecules interspersed among the lipid molecules extend through the bilayer to form pores or to act as carriers for materials being transported across the membrane. Other protein molecules are embedded in one or the other layer; these proteins perform a variety of cellular functions (Black 1993).

[3] Collins (1985) and Collins and Pinch (1993) have named the circle between experimental technique and the data produced by it the "experimenters' regress". According to Collins experimentalists cannot determine if a technique is "working" other than by the production of the "expected data". Collins argues that for high flux gravity waves "what the correct outcome is depends upon whether there are gravity waves hitting the Earth in detectable fluxes. To find this out we must build a good gravity wave detector and have a look. But we won't know if we have built a good detector until we have tried it and obtained the correct outcome!" (1985, 83-4)

[4] See Wimsatt (1981) for his fullest treatment of robustness. Wimsatt argues that robustness (as "triangulation" using a variety of independent ways of getting a result) is a criterion for determining the reality, objectivity or non-artifactuality of knowledge claims.

[5] Cells are fixed so as to preserve their structure, prevent the loss of cellular constituents, and protect them during subsequent treatments. Osmium tetroxide acts as a fixative by cross-linking membrane and cytoplasmic proteins with each other and with lipids. Glutaraldehyde, another chemical fixative that will be mentioned below, cross-links proteins and is often used in conjunction with osmium tetroxide because it penetrates the cell more quickly (Hayat 1989).

References

Black, J.G. (1993), *Microbiology: Principles and Applications*. 2d ed. Englewood Cliffs: Prentice Hall.

Chapman, G.B. and Hillier, J. (1953), "Electron Microscopy of Ultra-thin Sections of Bacteria", *J. Bacteriol.* 66: 362-373.

Collins, H.M. (1985), *Changing Order*. London: SAGE Publications.

_ _ _ _ _ _ _ . and Pinch, T. (1993), *The Golem*. Cambridge: Cambridge University Press.

Culp, S. (forthcoming) "Objectivity in Experimental Inquiry: Breaking the Data-Technique Circle"

Dubochet, J., McDowall, A.W., Menge, B., Schmid, E.N. and Lickfeld, K.G. (1983),, "Electron Microscopy of Frozen-Hydrated Bacteria", *J. Bacteriol.* 155: 381-390.

Fitz-James, P.C. (1960), "Participation of the Cytoplasmic Membrane in the Growth and Spore Formation of Bacilli", *J. Biophysic. Biochem. Cytol.* 8: 507-529.

Fooke-Achterrath, M., Lickfeld, K.G., Reusch, Jr., V.M., Aebi, U., Tschope, U., and Menge, B. (1974), "Close-to-life Preservation of *Staphylococcus aureus* Mesosomes for Transmission Electron Microscopy", *J. Ultrastruct. Res.* 49: 270-285.

Franklin, A. (1986), *The Neglect of Experiment*. Cambridge: Cambridge University Press.

_ _ _ _ _ _ . (1990), *Experiment Right or Wrong*. Cambridge: Cambridge University Press.

Greenawalt, J.W. and Whiteside, T.L. (1975), "Mesosomes: Membranous Bacterial Organelles", *Bacteriol. Rev.* 39: 405-463.

Hacking, I. (1983), *Representing and Intervening*. Cambridge: Cambridge University Press.

Hayat, M.A. (1989), *Principles and Techniques of Electron Microscopy: Biological Applications*. 3d ed. Boca Raton: CRC Press, Inc.

Higgins, M.L. and Daneo-Moore, L. (1974), "Factors Influencing the Frequency of Mesosomes Observed in Fixed and Unfixed Cells of *Streptococcus faecalis*", *J. Cell Biol.* 61: 288-300.

_ _ _ _ _ _ _ ., Tsien, H.C. and Daneo-Moore, L. (1976), "Organization of Mesosomes in Fixed and Unfixed Cells", *J. Bacteriol.* 127: 1519-1523.

Hobot, J.A., Villiger, W., Escaig, J., Maeder, M., Ryter, A. and Kellenberger, D. (1985), "Shape and Fine Structure of Nucleoids Observed on Sections of Ultrarapidly Frozen and Cryosubstituted Bacteria", *J. Bacteriol.* 162: 960-971.

Jardine, N. (1986), *The Fortunes of Inquiry*. Oxford: Clarendon Press.

_ _ _ _ _ _. (1991), *The Scenes of Inquiry*. Oxford: Clarendon Press.

Kosso, P. (1989), *Observability and Observation in Physical Science*. Dordrecht: Kluwer Academic Publishers.

Nanninga, N. (1968), "Structural Features of Mesosomes (Chondroids) of *Bacillus subtilis* After Freeze-Etching", *J. Cell Biol.* 39: 251-263.

_ _ _ _ _ _ _. (1969), "Preservation of the Ultrastructure of *Bacillus subtilis* by Chemical Fixation as Verified by Freeze-Etching", *J. Cell Biol.* 42: 733-743.

_ _ _ _ _ _ _. (1971), "The Mesosome of *Bacillus subtilis* as Affected by Chemical and Physical Fixation", *J. Cell Biol.* 48: 219-224.

Parks, L.C., Dicker, D.T., Conger, A.D., Daneo-Moore, L. and Higgins, M.L. (1981), "Effect of Chromosomal Breaks Induced by X-Irradiation on the Number of Mesosomes and the Cytoplasmic Organization of *Streptococcus faecalis*", *J. Mol. Biol.* 146: 413-431.

Rasmussen, N. (1993), "Facts, Artifacts, and Mesosomes: Practicing Epistemology with the Electron Microscope", *Stud. Hist. Phil. Sci.* 24: 227-265.

Ryter, A. and Kellenberger, E. (1958), "Etude au microscope electronique de plasmas contenant d l'acide desoxyribonucleique. I-Les nucleoides des bacteries en croissance active", *Z. Naturforsch.* 13B: 597-605.

Salton, M.R.J. and Owen, P. (1976), "Bacterial Membrane Structure", *Ann. Rev. Microbiol.* 30: 471-482.

Silva, M.T. (1971), "Changes Induced in the Ultrastructure of the Cytoplasmic and Intracytoplasmic Membranes of Several Gram-positive Bacteria by Variations in OsO4 Fixation", *J. Microsc.* 93: 227-232.

_ _ _ _ _ _., Sousa, J.C.F., Polonia, J.J., Macedo, M.A.E. and Parente, A.M. (1976), "Bacterial Mesosomes Real Structures or Artifacts?", *Biochimica et Biophysica Acta* 443: 92-105.

Wimsatt, W.C. (1981), "Robustness, Reliability, and Overdetermination", in M.B. Brewer and B.E. Collins (eds), *Scientific Inquiry and the Social Sciences*. San Francisco: Jossey-Bass Publishers, pp. 124-163.

Part II

EXPLANATION, INDUCTION, AND LINGUISTIC REPRESENTATION

Explaining Brute Facts

Eric Barnes

Southern Methodist University

1. Introduction

I hope to convince you that one way of testing the mettle of a theory of scientific explanation is to inquire what that theory entails about the status of brute facts. In what follows I briefly consider the nature of brute facts, and then survey several contemporary accounts of explanation vis a vis this subject. These include the Friedman-Kitcher theory of explanatory unification (Friedman (1974), Kitcher (1981) and (1989)), Peter Lipton's (1991) account of the nature of explanatory loveliness, and the causal theory of event explanation developed by Paul Humphreys in his (1989). It is my view that each of these accounts of explanation entails (or at least lends itself to) a false view about the nature of brute facts: according to each account, a brute fact is unexplainable and thus represents a scientific 'mystery', where by this expression I mean some manner of a lack of scientific understanding. It's my view that brute facts need not be thought to represent any gap whatsoever in scientific understanding, and that indeed brute facts are explainable; hence each of these accounts of explanation is in one way or another deficient according to me. This point, however, is a mere symptom of a deeper underlying problem—our thinking about the very ideas of explanation and understanding is heavily influenced by our predilection to use as examples certain cases of explananda regarded by practitioners in the field as typical. I propose to ponder brute facts as a way of helping us to recognize our dependence on the idiosyncracies of such examples, recognition of one's dependence being as always the first step toward recovery.

2. What is a brute fact?

If Sally were to ask Bobby 'Why is the sky blue?', Bobby might reply 'It just is!' Bobby's response is not as question begging as it appears, for Bobby means to imply that the blueness of the sky is merely a brute fact—a fact about the world for which there is no further reason or explanation. But to say of a fact F that it is brute is to make an ambiguous claim: one might mean that F is a fact whose causal history is hidden—it is a fact which, as of the present moment, must simply be accepted without explanation until its explanation is one day revealed. Thus the Boyle-Charles law of gases might have been regarded as a brute fact in this sense until the law was reduced to the kinetic theory of gases, at which time it was (in this sense) no longer brute. I shall call such facts *epistemically brute facts*, where a fact merits this label simply because its explanation (which we assume exists) remains unknown. I mean to distinguish epistemically brute facts from facts which are *ontologically brute*: facts with no explanatory basis beyond themselves. Ontologically brute facts would presumably include the ultimate laws of physics, the fact that the universe contains a particular amount of matter or energy rather than some other amount, or the fact that the universe exists at all. These facts are ontologically brute just in case there is no underlying reason for their existence—they are simply the fundamental facts about reality from which all other facts in some sense derive. It suffices to simply stretch our imagination to conceive of a possible world consisting almost entirely of brute facts—imagine a universe that consisted of nothing but brightly colored explosions which each occurred for no reason whatsoever. Let's now consider how various accounts of explanation handle brute facts of both types.

3. Explanatory Unification

Although the theory of explanatory unification has received an extensive development in the writings of Philip Kitcher ((1981), (1989)), I believe the fundamental motivations of this theory are best described by Michael Friedman in his 1974 article "Explanation and Scientific Understanding". The fundamental idea underlying the theory of explanatory unification is that those theories are genuinely explanatory which serve to reduce the number of unexplained facts that scientists are forced to accept in their total picture of nature. A scientist who asks why the Boyle-Charles law is true might adopt as an explanation thereof the kinetic theory of gases—but if we don't understand why the kinetic theory itself is true, Friedman asks, why does the adoption of the kinetic theory amount to any kind of gain in scientific understanding? All that has been done, he argues, is that scientists have replaced one brute fact with another. The reason the kinetic theory did increase scientific understanding is that it did not entail only the Boyle-Charles law, but other empirical gas laws as well, including Graham's law of diffusion and facts about specific heat capacities. Thus Friedman argues that the kinetic theory, like all true scientific explanations, works to increase our understanding "by reducing the total number of independent phenomena that we have to accept as ultimate or given. A world with fewer independent phenomena is, other things being equal, more comprehensible than one with more." (1974, p.15)

In (Barnes 1992a) I attribute to Friedman (and criticize on grounds independent of issues I discuss here) what I deem 'the Unificationist Thesis of Understanding', which claims that scientific understanding of the world is inversely proportional to the degree of unification in our world picture. What needs to be emphasized for our current purpose is that science possesses a higher degree of unification the fewer 'independent phenomena' scientists are forced to accept as given— but by 'independent phenomena' Friedman apparently simply means 'epistemically brute facts'—such facts are independent and unexplained simply because there are no deeper, more unifying theories available from which such facts can somehow be derived. What furthermore deserves careful attention is his interesting claim that some possible worlds are more understandable than others: a world with fewer phenomena that are ultimately irreducible to any other phenomena, i.e. a world with fewer ontologically brute facts, is, "other things being equal", more understandable on Friedman's view than a world with a greater number of ontologically brute facts. My point is that Friedman (and I gather Kitcher as well) thus clearly views ontologically brute facts as unexplainable scientific mysteries—the fewer ultimate mysteries a world happens to contain, the more understandable that world is—the absence of mystery being construed as equivalent to the presence of understanding. This view follows immediately, it would appear, from the Unificationist Thesis of Understanding, a thesis that continues to strike me as supplying the fundamental motivation for the theory of explanatory unification (see Barnes (1992b) for an independent criticism of Kitcher's version of explanatory unification).

4. Lipton on Explanatory Loveliness

The literature falling under the rubric of 'inference to the best explanation' continues to grow. Peter Lipton's 1991 book *Inference to the Best Explanation* is, I believe, the most thorough attempt to articulate a model of inference based on the assumption that we infer those hypotheses that strike us as scoring highest in terms of explanatory virtue vis a vis some data. The primary task facing proponents of such a model is to articulate the nature of explanatory virtue—what makes one putative explanation 'better' than a competitor? It is on this subject that Lipton has a number of interesting things to say.

Lipton's starting point is the claim that the aim of explanation is understanding. A good explanation is thus one that would, if it were true, offer a substantial measure of understanding of whatever data it purports to explain. Lipton refers to such an explanation as one which offers a high degree of potential understanding and (equivalently) as one that demonstrates the virtue of 'explanatory loveliness'. Lipton proposes to measure the explanatory loveliness of a putative explanation by the application of various criteria. A complete survey of Lipton's loveliness criteria cannot be provided here; for present purposes we will note just one of them. This is the mechanism criterion of explanatory loveliness: among those competing explanations of some effect compatible with the total evidence, Lipton argues, we prefer those "that would mark causes we can link to the effect by some articulated mechanism..." (Lipton 1991, 118) Theories entailing some data D that fail to describe the mechanism by which D is produced fall short in their provision of potential understanding of D, and are thus explanatorily unlovely—and hence, for Lipton, unlikely. One might claim that attempting to explain why Joshua is a bold leader by noting the fact that he is a Leo suffers from the failure of this putative explanation to meet the mechanism requirement—the theory offers no mechanism which links his August birthdate to his personality, and hence would not (even if true) substantially increase our understanding of Joshua's leaderly qualities. Hence (in part) our disposition, Lipton would argue, to regard the astrological explanation as unlikely and thus not infer it. The kinetic theory of gases qua explanation of the Boyle-Charles law does seem to meet the mechanism criterion, for the theory that gases are collections of infinitely elastic particles does provide a mechanistic account of why the pressure, volume, temperature, and mass of gases are related in the way described by the Boyle-Charles law—hence the explanatory loveliness of the kinetic theory vis a vis the mechanism criterion, and hence (in part) our disposition to regard it as likely true.

What does Lipton's account of explanatory loveliness entail about the status of brute facts? Confining our attention to just the mechanism criterion, it seems clear that ontological brute facts end up as unexplainable and thus mysterious once again. For of course there is no mechanism underlying the truth of such a brute fact—were there such a mechanism the fact would not be ontologically brute. But Lipton's position is that our knowledge of the mechanism which underlies fact F is (partially) constitutive of the very notion of what it means to understand F, so it follows that Lipton must regard ontologically brute facts as forever and ultimately not understood, viz., mysterious (as he regards non-mechanistic explanations in general as offering less potential understanding than competing mechanistic ones.) So Liptonian loveliness leads us to the same conclusion about the unexplainable status of brute facts as did unificationism. Let's consider one more contemporary account of explanation.

5. Humphreys on Causal Explanation

Paul Humphreys' 1989 book *The Chances of Explanation* argues that the canonical form of a request for an explanation of a specific event is 'What is the explanation of Y in S at t?' where 'Y', 'S', and 't' are terms referring to, respectively, a property or change in property, a system, and a trial. An appropriate explanation will be "Y occurred in S at t because of Φ despite ψ"; 'Φ' is a (nonempty) list of terms referring to contributing causes of Y; and 'ψ' is a (possibly empty) list of terms referring to counteracting causes of Y. Humphreys emphasizes that the explanation itself consists of just the causes to which the elements of Φ jointly refer. ψ is not part of the explanation proper. He implies that where Φ is empty we simply have no explanation of Y's occurrence.

What does Humphreys' model of causal explanation imply about the status of brute facts? We note first of all that his model is tailored to apply to just specific events. His model is not designed to accommodate the explanation of scientific laws—insofar as we think of brute facts as including laws (e.g. the ultimate laws of physics) this model may seem to simply not apply. Let's note in response simply that there is no reason in principle that specific events cannot be brute if they occur for no reason whatsoever (recall the example noted in Section 1 of a world containing just uncaused explosions). (Of course, the fundamental idea of explaining P by citing P's contributing causes might well be extended to the explanation of laws, if one is willing to count the derivation of such laws from more fundamental theories as constitutive of a description of the laws' 'causes'.) What would Humphreys' model imply about the status of ontologically brute specific events? Appropriately enough, they are not susceptible to causal explanation a la Humphreys, for there is *ex hypothesi* no set of contributing causes to constitute the explanation Φ. In the absence of any alternative mode of explanation, such facts end up as entirely unexplainable. Humphreys' (1989) is rather silent on the subject of understanding, so I don't know whether Humphreys is prepared to count any unexplainable fact as representing a lack of scientific understanding—and thus count ontologically brute facts as mysterious—but such a consequence is at least compatible with Humphreys' analysis of explanation here. Significant here is the absence of any basis in his account on which to distinguish an epistemically brute fact—one whose contributing causes are simply unknown—and an ontologically brute fact—one with no contributing causes whatsoever. For both types of brute facts, Φ is simply the empty set. In the absence of some further analysis, it would seem that if epistemically brute facts are 'mysterious' in virtue of the corresponding Φ being empty, it may be rather difficult to avoid assigning the same status to ontologically brute facts, given that their corresponding Φs will also be empty.

6. Do Brute Facts Represent a Lack of Scientific Understanding?

The answer is: epistemically brute facts, yes (by definition), but ontologically brute facts, not necessarily. Let O be some ontologically brute fact. Now the fact that O is a fact may represent a lack of scientific understanding if scientists don't know that O is ontologically brute; they may suspect it to be the product of some explanatory basis or other but take themselves to be ignorant of this basis. But if O is known by scientists to be ontologically brute, then it seems to me that O represents no mystery whatsoever—O is simply partly constitutive of way the world is. I would conjecture that the tendency to take ontologically brute facts as mysterious, a tendency that is implicit or explicit in the accounts of explanation discussed above, stems from a failure to distinguish epistemically brute facts and ontologically brute facts. Both types of brute facts have no apparent explanatory basis, hence their apparent similarity, but in the one case, the explanatory basis is hidden, in the other (one might say) non-existent. Our tendency to conflate them might also arise from a tendency to think that ontologically brute facts must have some explanatory basis or other—a theist might insist that which facts are ontologically brute depends on God's will, but a fact which has an explanation in God's decree is of course not ontologically brute after all. Now our understanding suffers a gap just in case there is some hidden explanatory basis for a fact we know to hold true—where there is no hidden explanatory basis, and we know this, there is nothing lacking in our understanding—for there is no explanation that we fail to have. All possible worlds, I thus want to maintain, are equally understandable in principle, for any world is thoroughly understood just insofar as it contains no explanatory bases which are hidden to scientists. (I don't want to deny that some worlds are easier to understand than others—the more ontologically brute facts are out there to discover, the greater the task of reaching complete understanding.) Thus, e.g., a world containing nothing but ontologically brute facts is perfectly understood once it is known that all facts therein are indeed ontologically brute. Such a world may be hopelessly messy, but not cloaked in mystery.

I see no reason not to go further and claim that a correct theory of explanation ought to entail that brute facts are perfectly explainable: their explanation consists of the stipulation that such facts are ontologically brute. In one sense, this view is compatible with a broadly causal account of explanation, for in describing the fact as ontologically brute one provides a correct account of its relevant causal history—it is a fact with no causal history beyond its status as ontologically brute.

7. Our Theories of Explanation Reconsidered

Let us now return to the accounts of explanation surveyed above and attempt to make sense of just how they go wrong. In that the unificationist theory of explanation is committed to the view that the more 'independent phenomena' exist, the less we understand the world, the unificationist is clearly committed to the view that a world with many ontologically brute facts is less understandable than one with more, and this is a thesis I have argued is clearly false, depending for its plausibility (I suggest) on its failure to distinguish ontologically and epistemically brute facts. I confess that understanding this point diminishes my enthusiasm for the Unificationist Thesis of Understanding on which the theory of explanatory unification is based. But an interesting question remains—why does explanatory unification strike so many intelligent persons as plausible? Here I want to suggest that its plausibility might derive in part from the fact that many famous examples of scientific explanation have turned out to fit the unificationist pattern perfectly well—Kitcher's (1989 437-448) cites several classic examples of argument patterns that are at once highly unifying and explanatory, including Mendelian genetics, Darwinian Evolutionary Theory, Dalton's theory of the chemical bond, and others. Insofar as these are the kinds of examples of explanation we consider as exemplars, the theory of explanatory unification will gain in plausibility, for it will indeed seem right to say that explanation proceeds by the reduction of independent phenomena (or, on Kitcher's program, reducing the number of argument patterns) in the scientific image of the world. But this program runs the risk of producing a picture of explanation that is tied more tightly to these examples than its proponents are prone to suggest, and may misleadingly imply that they have isolated hereby the very nature of explanation. It also runs the risk of seeing explanatory virtue in theories where there are really other virtues at work—such as the virtue of providing a systematized world picture that is easier for human subjects to master, or the straightforwardly quantitative virtue of just explaining a lot of phenomena (regardless of how unified the explanatory basis is).

I want to suggest something similar about Lipton's account of explanatory loveliness and potential understanding, and Humphreys' causal model of event explanation.

Lipton means to construe the very idea of understanding some fact F as constituted (in part) by describing a mechanism underlying F—thus my argument that ontologically brute facts must be regarded by Lipton as not understood. But the plausibility of this move depends also, I suggest, on a tendency to run together ontologically and epistemically brute facts—an epistemically brute fact will be mysterious insofar as it's the product of some unknown mechanism, so it does represent a gap in understanding—but this just doesn't hold for ontologically brute facts. The plausibility of Lipton's mechanism criterion of explanatory loveliness, like the theory of explanatory unification, holds so long as we are looking at what we take to be a set of typical examples of explananda—those that result from mechanistic causes. Considering ontologically brute facts takes us outside this set to reveal the dependence of the theory's plausibility on the idiosyncracies of these examples, however important and typical they may otherwise be. As to Humphreys' model, I've suggested that it lends itself to the view that specific events which are ontologically brute are not understood, for it seems to hold that the only way to explain an event is to describe its various contributing causes, so that if we hold that the only way to understand the event is to possess a true explanation of it, ontologically brute facts are incapable of explanation and thus mysterious. And here I repeat the same moral: a causal model of specific event explanation seems right so long as we construe our explananda to be events with causal histories (be these histories deterministic or stochastic). Where explananda are not the product of such histories, it need not follow that they cannot be explained—such is the case for ontologically brute facts.

8. Conclusion

The moral I am urging is not new, although it's clear by now that I regard the moral as one that is still worth preaching, given the current literature. In his book *Scientific Explanation and the Causal Structure of the World* Wesley Salmon concluded with the following statement:

> It is my view that attempting to give a logical characterization of scientific explanation is a futile venture, and that little of significance can be said about scientific explanation in purely syntactical or semantical terms. I believe, rather, that what constitutes adequate explanation depends crucially upon the mechanisms that operate in our world. In all of this there is—obviously—no logical necessity whatsoever. (1984, 240)

Salmon is making the point that we must beware the temptation to think that there is some one thing that scientific explanation—or scientific understanding—is—a temptation that philosophers are particularly susceptible to. What good explanations look like will depend heavily on what structure the world happens to have—e.g., if it is a unified structure, then unifying explanations may increase our understanding, but a highly fragmented theory will best enable our understanding if the world if highly fragmented.[1] The same goes for the hypothesis that events are the products of mechanisms, or causal histories.

For my own part, I retain the view that there is a single thing that explanation is, a 'thing' that has been described quite well by Salmon himself under the rubric of what he deems the ontic conception of scientific explanation—in the following passage Salmon defines the ontic conception while explaining his own preference for a causal theory of explanation:

> Scientific explanation, according to the ontic conception, consists in exhibiting the phenomena-to-be-explained as occupying their places in the patterns and

regularities which structure the world. Causal relations lie at the foundations of these patterns and regularities; consequently, the ontic conception has been elaborated as a *causal conception* of scientific explanation. (1984, 239)

What I want to note about the ontic conception is that it's clearly compatible with the view I adopt above that brute facts are explainable. For one exhibits the place of an ontologically brute fact in the patterns and regularities which structure the world by demonstrating that it is an ontologically brute fact—and thus has no further causal history. Thus exhibited, the ontologically brute fact is explained in the ontic sense. The passage just quoted mirrors the views of Humphreys: in recognizing causal relations to be crucial in the world's structure, Salmon and Humphreys equate empirical explanation with stipulation of contributing causes. But the ontic conception does not entail that causal history is essential for explanation—this is part of Salmon's point.

I have offered these remarks about explanatory unification, Lipton's account of explanatory loveliness, and Humphreys' causal theory of explanation in the spirit of criticism, but of course what I am urging entails that there are surely substantial virtues attributable to each of these accounts as well. They do well insofar as the model of explanation they propose respects the world's actual structure. They err to the extent that they claim to isolate the very ideas of explanation or understanding— this is the conclusion to which our consideration of brute facts has led.

Note

[1] In a section of his (1989) entitled 'But What if the World Isn't Unified?' (494-497) Kitcher apparently addresses this issue. He concludes that whatever picture of the world's 'basic mechanisms' (read: ontologically brute facts) is eventually accepted as correct, that picture must be the most highly unifying of the competing world pictures, given his Kantian view that our picture of the world's causal structure derives from considerations about which world picture is most highly unifying. But one might admit this and hold to the possibility that as scientific knowledge increases, the degree of unification in our world picture decreases if in investigating the world's causal structure we encounter a proliferation of basic mechanisms, where the corresponding gain in the number of events susceptible to explanation is not sufficiently high to offset the diminishing effect on unification attributable to the growing number of mechanisms. This decreasing unification would coincide with an increase in understanding, and this does violate the spirit of explanatory unification.

References

Barnes, E. (1992a), "Explanatory Unification and Scientific Understanding", in Hull, D., Forbes, M., and Okruhlik, K. (eds.) *PSA 1992*, Vol. I, 3-12.

_____. (1992b), "Explanatory Unification and the Problem of Asymmetry", *Philosophy of Science 59*: 558-571.

Friedman, M. (1974), "Explanation and Scientific Understanding", *Journal of Philosophy 71*: 5-19.

Humphreys, P. (1989), *The Chances of Explanation: Causal Explanations in the Social, Medical, and Physical Sciences*. Princeton: Princeton University Press.

Kitcher, P. (1981), "Explanatory Unification", *Philosophy of Science 48*: 507-531.

_____ . (1989), "Explanatory Unification and the Causal Structure of the World", in P. Kitcher and W.C. Salmon, (eds.) *Minnesota Studies in the Philosophy of Science*, Vol. 13, *Scientific Explanation*, Minneapolis: University of Minnesota Press, pp. 410-505.

Lipton, P. (1991) *Inference to the Best Explanation*. London and New York: Routledge.

Salmon, W. (1984), *Scientific Explanation and the Causal Structure of the World*. Princeton: Princeton University Press.

Scientific Explanation:
From Covering Law to Covering Theory

Fritz Rohrlich

Syracuse University

1. Introduction

The premise of the present paper is the primacy of understanding over explanation. The question "is the act of explaining a goal in itself, or is it a means to an end?" is here answered by "explaining is a pragmatic concept and serves the purpose of responding credibly to a request for understanding." Explaining is secondary, understanding is primary.

Given this view of scientific explanation, one is faced with the received view and the recent literature based on it. Models of explanation cast into the covering law format do not necessarily ensure understanding, nor do they necessarily provide sufficient information for credibility.

After explicating this dissatisfaction (Section 2) I shall first clarify what I mean by scientific theory, law, and model (Section 3) since these concepts are crucial for the covering *theory* model proposed in Section 4. How this new model can satisfy the requirements of understanding and provide grounds for credibility is left for the subsequent Sections 5 and 6. In the concluding remarks (Section 7) I can only touch on the big issue of the implications of a theory's cognitive level and its relation to scientific truth.

The received view has been syntactic in nature focussing on the logic of the explanatory process (following logical positivism). But other features are at least as important for understanding. The new model of explanation is in the spirit of a semantic view of scientific theories (Giere 1988, Suppe 1989) but not anti-realist as in van Fraassen's constructive empiricism (van Fraassen 1980).

2. Objections to the covering law model of scientific explanation

Since the middle of this century the philosophical literature on scientific explanation has been dominated by ideas first expressed by Hempel and Oppenheim (1948). These authors proposed a model that is a formalized description of a scientific explanation. It claims that scientific explanation is essentially an argument (a logical inference) based on one or more given scientific laws and antecedent conditions. This "covering law" model was originally deductive (deductive-nomological model) and was later extended to inductive situations involving probabilistic considerations (statistical-relevance and other models). But the basic idea that explanation is provided by the logic of an argument based on a given law remained. I shall argue that this idea of an explanation is insufficient to ensure understanding, and that it does not permit a judgment whether to simply accept that explanation or to actually believe it.

The enormous literature that developed subsequently (for a review see Salmon 1989) also includes various criticism. But the dissatisfaction with the covering law model that I wish to express is largely of a different nature.

That the emphasis on the logical structure of explanation is not satisfactory has been felt before. Thus, Salmon (1984, 91) says "We have, perhaps, gone too far in our efforts to treat all philosophical problems in the 'formal mode' rather than the 'material mode'." My contention goes in the same direction.

I claim that the key deficiency of the covering law model is that it forces explanation into the straight-jacket of an argument based on a law (or laws) and a set of antecedent conditions. Positing such laws and conditions *without further justification* does not necessarily provide for either understanding or credibility. First, the questioner is implicitly told not to question the authority of science: the covering law explanation provides no justification for the posited law even though it may not at all appear credible. Secondly, the covering law model does not involve explication of the underlying ontology: the explanation may involve an entirely different ontology from that of the question. And finally, - and that is perhaps most important - the emphasis of the explanation is not on the "causal mechanism" or, more generally, on the way the world *functions*. Yet, it is generally accepted that understanding is largely based on "the causal structure of the world" (Salmon 1984).

In fact, the posited covering law may not even be relevant for the explanation. While the inference is correct, no understanding can result. Humphreys 1989b gives an example: when the barometer shows a drop in pressure a storm is approaching; given this "causal law" (empirical regularity) the inference from a given drop in pressure does not explain (i.e. does not provide understanding) why a storm is approaching. The following examples provide further insight into these deficiencies.

A problem of credibility: a bottle of water when put into the freezer will crack. The covering law explanation posits the law that water *expands when cooled* to freezing temperature. That is incredible since everyone knows that gases *expand when heated* and so do metals (copper, etc.); and mercury is well known to expand in every thermometer when heated rather than cooled. Some questioners may even have the more sophisticated (though somewhat oversimplified) notion that increased heat means increased molecular motion which in turn leads to larger average spacing between molecules.

Why, then, should water expand when cooled rather than when heated? Even if the above explanation is understood logically, its premise appears incredible, and it will therefore at best be only accepted and will not be believed.

A problem of understanding: keeping within the same field of science, consider the question why copper expands when heated. The logic of the covering law explanation is impeccable: all metals expand when heated; copper is a metal; thus, copper expands. But that argument does not explain because it only reduces the explanandum to a more general law without providing understanding. What is lacking here is the cause for the expansion. Causal arguments do provide understanding; but arguments that reduce the explanandum to a more general claim do not. Thus, covering law explanations do not necessarily provide understanding and therefore are not necessarily explanations.

Given the above objections, my dissatisfaction with the covering law model for a *scientific* explanation can be further characterized by saying that one requires higher standards. A scientific explanation should also answer such implied questions as: how credible is the law on which it is based? Under what assumptions (on ontology, idealizations, etc.) is the given answer valid? How well established is the theory on which the answer is based? Are there alternative theories that cannot be dismissed?

3. Laws, theories, and models

In what follows it is essential to make a clear destinction between laws, theories, and models based on a given theory. I must therefore specify what I mean by these terms.

A scientific law is a proposition that can be cast into the form $\forall x\, (Fx \supset Gx)$. A scientific theory on the other hand, is a complex scheme involving various components. These include an ontology, a set of central terms, a set of principles, a semantics, often a mathematical structure, and a domain of validity. Some authors even include in their concept of "scientific theory" the accumulated body of empirical evidence that supports the theory.

Within a given theory, laws can be deduced which then hold within that theory's domain of validity. But the questioner puts more weight on how well established that theory is than on its domain size.

A model in a scientific theory is an idealization of the actual state of affairs that can be treated within the framework of that theory. These models generally involve counterfactual descriptions. But idealizations can be weakened to lead to "more realistic" models.

Laws are frequently derived from a given model within a theory. These laws then hold only under the idealized conditions underlying that model. These conditions are the "ceteris paribus" conditions that are often cited (see for example Cartwright 1983). They must be included in a scientific explanation if it is not to be misleading. The reasonableness of these conditions affects the credibility of the explanation.

4. The covering theory model

A scientific explanation based on the covering theory model involves three steps: the choice of the theory, the specification of the model, and the inference process. Only in the third step does one find common ground of both the covering law and the covering theory model.

Consider a well-known matter as an example. Take the why-question: why does the earth go around the sun in an ellipse with the sun in a focal point rather than in the center? The covering *law* model simply posits Kepler's first law of planetary motion, adds the subsumption "the earth is a planet", and concludes the explanandum. No understanding results from this. It is strictly a requirement for blind acceptance of the authority of science. Where do these laws come from? Why should I believe them? How accurately do they describe the state of affairs? The covering law model provides answers to none of these questions. Yet, the answers to them are essential for two reasons, first for understanding and second for the important decision on whether to believe the explanation or only to just accept it. Let us now look closer at the above three steps of the covering *theory* model.

The first step: *theory choice*. In the above example an appropriate gravitation theory must be identified. One chooses Newtonian rather than Einsteinian gravitation theory because, while the results of the former are only an approximation to those of the latter (valid for low energy), it suffices for the purpose at hand. This theory choice provides the central terms, principles and laws that may be needed for the explanation, and also establishes its *cognitive level*: Newton's and Einstein's theories are on different cognitive levels implying entirely different ontologies (Rohrlich 1988). No such issue enters in the covering law model. Yet, the philosophical understanding of a scientific explanation requires it.

The second step: *model choice*. Consider just the sun and one planet thus neglecting the existence of all the other planets (first counterfactual feature); assume the mass of the planet to be negligible compared to the mass of the sun (second counterfactual feature); assume the size of both the sun and planet to be negligible compared to the distance betwen them such that they can both be treated as point masses (third counterfactual feature). This model specification establishes the particular *idealized ontology* within the cognitive level of Newtonian theory. Such idealizations are not necessarily made explicit in the covering law model. But they are needed for understanding and credibility.

The simplifying assumptions of a model identify the improvements that can be made on it in order to make its predictions more accurate. But, in our example, when any one idealization is weakened, the improved model will no longer permit the derivation of Kepler's first law. Thus, the specification of the model also provides the best approximating assumptions under which the explanandum holds. In fact, Kepler may not have obtained his laws of planetary motion if he had had available more accurate observations. Such knowledge on the limited validity of the questioner's tacit assumption contributes greatly to the success of the explanation.

The third step: *inference*. One uses the available technical machinery of the theory, the fundamental equations as they apply to the chosen model and the mathematical derivations that lead to the desired result (in our example, the earth's orbit and the sun's location within it.) This is a formal matter: mathematics and logical inference. If that together with the posited law(s) and condition(s) were all that is needed for a satisfactory explanation, the covering law model may still be adequate.

But, these formal steps hide a qualitative picture of how the world functions (in this case, within the cognitive level of Newtonian dynamics); one deduces a *causal mechanism* in which gravitational forces and conservation laws lead to the characteristics of the earth's orbit and its location relative to the sun. Depending on the questioner's technical knowledge, that qualitative part may be the only part that is important for his (her) understanding. It is certainly necessary.

The philosophically important points in the three steps of the covering theory model of explanation are the following. In the first step, the theory choice, one specifies the cognitive level on which the explanation takes place. Cognitive levels of different "depth" are often characterized by the names of corresponding scientific disciplines: macroscopic physics, atomic and molecular physics, elementary particle physics, or, zoology, physiology, molecular biology. On each cognitive level scientific theories are developed which are competent on that particular level only.

The specification of a cognitive level must be done first; one must first indicate whether an explanation can be given on the cognitive level on which the question is asked or whether a different cognitive level is necessary. The question why a glass of water cracks when put into the freezer cannot be explained on the level of everyday experience on which the question is asked but rather requires the deeper level of atomic and molecular forces. This is why the covering law explanation for this question fails.

In the second step, the model choice, one specifies the ontology that is required in order to deduce the facts implied in the question; this step contains the counterfactual features of the model, its idealizations and correspondingly an idealized ontology. Within this framework, the covering theory explanation thus specifies the conditions under which the state of affairs implied in the explanandum are actually true. A satisfactory explanation of the above question on the location of the sun within the earth's orbit must state that Kepler's laws are only true within the specified ontology. As soon as the idealized conditions are relaxed, they become false.

Finally, in the third step, the "inference", one learns *how the world is claimed to function* according to the chosen model. In many cases, this functioning ("the way nature works") can be expressed as a causal mechanism (see Section 6 below). It is exactly here where understanding is produced. In the covering law model on the other hand, causal mechanism are often only implied and are not always explicated. Thus, the logic of the argument is by itself simply insufficient.

A final example: Hempel and Oppenheim (Hempel 1965, 246) give as their first example a mercury thermometer. The question is to explain why, when immersed into hot water, the mercury column drops slightly before it rises. The laws posited are that both mercury and glass expand when heated, that the former has a larger expansion coefficient, and that heat is transferred from the heat bath to the interior of the thermometer by conduction.

The structure of the thermometer is an antecedent condition. Indeed, the mechanism that causes the phenomenon can be inferred from this. But it is not part of the logical structure of the argument. Why not explicate that crucial mechanism? Why not state how nature functions in this particular instance? That mercury expands when heated is generally known; but that the glass expands too is likely to be ignored (this important revelation to the questioner leads to the "aha! effect"); and the additional fact that it expands (a) at a much smaller rate and (b) earlier than the mercury column is absolutely essential. Thus, the mechanism involves a *time sequence of events* which accounts for the explanandum. It is that sequence that yields understanding; the laws used are necessary as auxiliaries to the argument but the time sequence of events is necessary for understanding: to wit, the heat bath at first makes only the glass tube expand; as a result, the mercury column can form a cylinder of slightly larger radius; this makes the mercury height drop; after that, the heat reaches the mercury and it expands much more rapidly than the glass cylinder; thus leading to a rapid increase in the mercury height despite its (very small) increase in radius.

5. Credibility

In the covering law explanation it may be the case that no underlying theory exists from which the law can be derived. In that case, the law may be based directly on empirical evidence with no theory structure involved (empirical conjecture). The justification of such a law clearly makes the explanation based on it less credible than if that explanation were based on a law derivable from a theory.

Let us now assume that one deals only with explanations whose underlying laws come from a scientific theory, i.e. these laws are either fundamental principles of that theory ("light in a vacuum always moves with the same universal speed irrespective of the reference frame") or are derivable from them. Then, if the scientific law posited in a covering law explanation is not by itself credible, its credibility must be reduced to the credibility of the underlying theory and of the assumed model.

The issue of the credibility requirements of a scientific theory is too big a subject matter to be treated here. Nor does this matter have to be raised in any detail in a typical scientific explanation. What does have to be provided, however, is some indication of the credibility of the theory used: is it barely more than a recent hypothesis that is competing with alternatives? Is it a new theory that has been confirmed only in recent years? Is it a mature theory shaped by many hands and confirmed by a large body of empirical evidence over a considerable period of time? Or is it perhaps an established theory? (See Rohrlich and Hardin 1983). A scientific explanation without such information provides little ground for credibility.

Approval of a theory as credible does not necessarily imply approval of any particular model within that theory because the credibility of the counterfactual assumptions of that model is in question. As discussed earlier, that credibility depends on the idealizations in the particular model. Therefore, at least the most important of these idealizations must also be made known to the questioner.

6. Understanding

The notion of understanding is often deferred to psychology. But if the goal of explanation is understanding, one cannot ignore the philosophical aspects of "understanding" in an attempt to explicate explanation. In this spirit Hempel and Oppenheim wrote:

A class of phenomena has been scientifically understood to the extent that they can be fitted into a testable, and adequately confirmed, theory or a system of laws. (Hempel 1965, 329)

This is a sophisticated version of the feeling that an explanation must "make sense" or must be "sensible" in that it "can be fitted into" a world view. *Sensibility* can thus provide for understanding.

Another requirement for understanding that has been discussed widely since Friedman (1974) is *unification*. His suggestion has been modified and expanded by Kitcher (1989 and earlier references quoted there). No doubt, there are many cases in which understanding is provided by unification. But, surely, there also exist explanations that yield understanding and that do not involve unification. The above explanation of Kepler's laws from Newtonian gravitation theory by itself involves no unification. The fact that Newtonian gravitation theory unifies the explanation of many different phenomena (planetary motion, the tides, the motion of comets, etc.) greatly contributes to the credibility of that theory but does not add to the understanding of these phenomena. It follows that unification is not a necessary requirement for understanding.

It is widely believed that *causal relations* are a necessary requirement for understanding (see *e.g.* Salmon 1984). Causal relations are here understood to include probabilistic causation and, as recently proposed by Humphreys (1989a), also includes causation in aleatory explanations. The latter include probabilities both favoring as well as disfavoring the occurrence.

However, also this requirement is not sufficiently general. It excludes quantum mechanical phenomena. One must therefore extend it to explanations of phenomena which are based on potentialities (propensities).

In classical natural sciences, explanations requested are typically for the purpose of understanding the *causal* functioning of such things as machines or biological systems. One wants to know how they "work" be they predictable (exactly or probabilistically) or unpredictable. A scientific explanation is to provide the explication of that functioning. Scientists do not usually ask for a cause but for the way things function. In fact, it can become rather awkward or even impossible to phrase a question about the functioning of a process in terms of a why-question (asking for the cause) rather than a how-question (asking for the sequence of events).

It is now only a relatively small step to extend the inquiry of the classical sciences to that of quantum mechanics. One asks for the sequence of events - not necessarily causally or probabilistically related but possibly related by potentialities. The term "functioning" of nature encompasses all of these alternatives.

One concludes that the third step in the covering theory perception of explanation, the inference from the chosen model, provides *understanding by explication of the functioning* of the world. On the classical level this can often be expressed in terms of causal mechanisms; but it is much more general than that since quantum phenomena cannot be ignored.

7. Concluding remarks

As the example of the cracked glass of water in the freezer shows, a question on one cognitive level may require an explanation on a different level. This is consistent with the view of the existence of mutually complementing ontologies for a complete description of the world (Rohrlich 1988). But it conflicts with the still widely held view of reductionism. According to the latter, the ontology of a "less fundamental" theory is fully reducible to the ontology of a "more fundamental" one. Therefore, explanations across levels should never be necessary.

This view of a hierarchy of theories according to fundamentality is closely connected to the claim of a convergent realism. But theories are hierarchical in scale at best, and even this is not strictly true: quantum mechanics, for example, which is a microscopic theory is necessary to explain the macroscopic phenomena of superconductivity. So the failure of convergent realism raises the question of the truth of a scientific theory and of the explanation based on it. Which ontology tells "how the world really is"?

There exists a naive view of science in which all scientific laws and descriptions are judged as strictly true or false. This view leads to the disastrous consequence that all present scientific knowledge may be false since the ultimate theory (the only true one) has yet to be found. But scientific statements can be judged true or false only within the validity limits of the corresponding theory. These validity limits are known quantitatively at least for established theories (Rohrlich and Hardin 1983) and are often qualitatively known for not yet established ones. They always belong to a particucur cognitive level and appropriate ontology. The truth of a scientific theory can therefore be justified only by the empirical evidence gathered on the particular cognitive level of that theory. Evidence from a different such level cannot be held against it. *Nor can one level of ontology be fully reduced to another* (Rohrlich 1989).

The ontology of an established theory (say, Newtonian gravitation theory) that was "superseded" (by Einstein's gravitation theory) continues to be valid for the perception of the "superseded" level. The old theory, its cognitive level and its ontology is in fact essential and *supplements* the new one (mathematical reduction may exist for *some* theories, but there are always different central terms and a different semantics.) Cognitive levels and ontologies therefore do not supersede one another: they complement one another representing different faces of the same part of the world.

In its barest outlines, science seeks to identify the furniture of the world, its properties, and the way it functions. Correspondingly, scientific explanation must first identify this furniture of the world, i.e. it must first specify the perception within which the explanation is to take place: its cognitive level and its ontology. (Some philosophers speak of context dependence.) It requires the choice of a scientific theory and a suitable model within that theory. Only then can the explanation proceed to deduce from first principles the way things are and the way things function on that level. And since every scientific theory has its validity limits and every model its idealizations, these validity limits and idealizations must be indicated in order for the questioner to decide on its credibility.

References

Cartwright, N. (1983), *How the Laws of Physics Lie*. Oxford: Clarendon Press.

Friedman, M. (1974), "Explanation and Scientific Understanding", *Journal of Philosophy,* 71: 5-19.

Giere, R.N. (1988), *Explaining Science*. Chicago: University of Chicago Press.

Hempel, C.G. (1965), *Aspects of Scientific Explanation*. New York: The Free Press.

Hempel, C.G. and P. Oppenheim (1948), "Studies in the Logic of Explanation", in *Philosophy of Science* 15: 135-75. Reprinted in Hempel (1965).

Humphreys, P.W. (1989a), *The Chances of Explanation*. Princeton: Princeton University Press.

_____. (1989b), "The Causes, Some of the Causes, and Nothing but the Causes", in P. Kitcher and W.C. Salmon (1989), pp.283-306.

Kitcher, P. (1989), "Explanatory Unification and the Causal Structure of the World", in P. Kitcher and W. Salmon (1989), (eds.), *Scientific Explanation*. Minneapolis: University of Minnesota Press, pp. 410-505.

Rohrlich, F. (1988), "Pluralistic Ontology and Theory Reduction in the Physical Sciences", *British Journal on the Philosophy of Science,* 39, 295-312.

_____. (1989), "The Logic of Reduction: The Case of Gravitation", *Foundations of Physics*, 19, 1151-1170.

Rohrlich, F. and Hardin, L. (1983), "Established Theories", *Philosophy of Science*, 50, 603-17.

Salmon, W.C. (1984), *Scientific Explanation and the Causal Structure of the World.* Princeton: Princeton University Press.

———— . (1989), *Four Decades of Scientific Explanation.* Minneapolis: University of Minnesota Press.

Suppe, F. (1989), *The Semantic Conception of Theories and Scientific Realism.* Cambridge: Cambridge University Press.

van Fraassen, B.C. (1980), *The Scientific Image.* Oxford: The Clarendon Press.

Why There Can't be a Logic of Induction[1]

Stuart S. Glennan

Butler University

Carnap's attempt to develop an inductive logic has been criticized on a variety of grounds, and while there may be some philosophers who believe that difficulties with Carnap's approach can be overcome by further elaborations and modifications of his system, I think it is fair to say that the consensus is that the approach as a whole cannot succeed. In writing a paper on problems with inductive logic (and with Carnap's approach in particular), I might therefore be accused of beating a dead horse. However, there are still some (e.g., Spirtes, Glymour and Scheines 1993) who seem to believe that purely formal methods for scientific inference can be developed. It may still then be useful to perform an autopsy on a dead horse when establishing the cause of death can shed light on issues of current concern.

My intention in this paper is to point out a problem in Carnap's inductive logic which has not been clearly articulated, and which applies generally to any inductive logic. My conclusion will be that scientific inference is inevitably and ineliminably guided by background beliefs and that different background beliefs lead to the application of different inductive rules and different standards of evidentiary relevance. At the end of this paper I will discuss the relationship between this conclusion and the problem of justifying induction.

1. The Task of an Inductive Logic

An inductive logic is a calculus which allows one to determine the relevance of some set of evidence to establishing the truth of an empirical hypothesis. Carnap conceives the problem of defining such a calculus to be one of defining a conditional probability measure, $c(h,e)$. The parameters h and e are sentences in a scientific language representing the hypothesis and the evidence respectively; $c(h,e)$, is meant to measure the probability of the hypothesis given the evidence. Once one has chosen a particular c, the evidentiary relationship becomes completely formal. Since there are infinitely many possible c's, the problem is to choose one which captures our intuitions about evidentiary relations.

Carnap's strategy for choosing a c function has two steps. First Carnap proposed a set of adequacy conditions that any c function must satisfy. These adequacy conditions are meant to explicate our intuitions about inductive inference. They include the requirement that a c function define a fair betting system, and the requirement that the function should allow one to "learn from experience." The adequacy conditions, while greatly narrowing one's choice of c function, can be satisfied by infinitely many different functions. Carnap's second step therefore was to propose a particular probability measure, c^*, that satisfied the adequacy conditions.[2]

Carnap's analysis of inductive inference has been criticized on a variety of grounds. Perhaps the most important involve the justification of Carnap's choice of a particular c function. First, given that the adequacy conditions can be satisfied by a wide range of functions, there seems to be no reason to prefer c^* to other functions satisfying those conditions. Second, Carnap can offer no justification of his adequacy conditions other than that they capture our intuitions about inductive inference. Like Hume's principle of the uniformity of nature, Carnap's adequacy conditions seem to be principles which can be established neither by logic nor experience.

The presumption behind these objections is that Carnap's c^* has succeeded in capturing our intuitions about inductive inference, and that the philosophical problem is that these intuitions cannot be justified. Even those who argue that c^* is an incorrect reconstruction of our intuitions about induction leave open the possibility that an alternative to c^* could be offered, and that then only the problem of justification would remain. If these were the only sorts of objections that could be posed, then further work on inductive logic would be warranted. The problem of the justification of induction could be dismissed as a pseudo-problem, and problems with the particular choice of c function could be met by introducing c functions which meet more refined adequacy conditions.

My objection is more fundamental, for it suggests not merely that a particular c function is inadequate, or that our choice of c functions cannot be justified on empirical or logical grounds, but rather that the very same evidence can lead us to make different inductive inferences, depending upon the context in which we make the inference. If I am right, then there can be no logic of induction, because our inductive inferences inevitably depend upon background beliefs which we have concerning the causes of the observations we use as evidence, and which cannot be completely captured in the calculus.

I will argue for my conclusion by analyzing a particular thought experiment which Carnap himself discussed. Carnap used this thought experiment to argue for the superiority of his c^* over another possible c function, $c\dagger$.[3] The crucial difference between these two functions is that c^* allows one to "learn from experience," while $c\dagger$ does not. I will argue, contrary to Carnap, that the reason why c^* seems more plausible than $c\dagger$ as a way to make predictions about the outcome of his thought experiment is not that c^* is *universally* rationally preferable to $c\dagger$, but rather that Carnap (and the reader) has made certain implicit assumptions about the nature of the mechanism producing the experimental results. Different assumptions about the experimental mechanism make $c\dagger$ more plausible than c^*.

2. Two Thought Experiments

In a non-technical article on inductive logic (Carnap 1955), Carnap attempted to make c^* plausible by illustrating its application to a simple experiment involving drawing balls from an urn. We are to imagine an experimental setup in which there are four balls in an urn, each of which has one of two mutually exclusive properties, say being white or black. Suppose that we draw balls one by one from the urn. For each draw from the urn, our inductive method will assign a probability to the hypothesis that the ball will be a certain color. For instance, our inductive method will give the probability that the first ball will be white, or the probability that the third ball will be white, given that the first two were black.

Suppose we make four draws from the urn. There are sixteen possible outcomes, which we can denote by a sequence of Bs and Ws. For example, BWWW represents the outcome of drawing a white ball followed by three black balls. We can completely specify a c function for this experiment by assigning a prior probability to each of these possible outcomes. Carnap calls these possible outcomes *individual distributions*. So long as the sixteen values sum to one, the c function will satisfy the minimum condition of being a probability measure. If there really is a logic of induction, then there should be some assignments that are better than others, and ideally one assignment which we can offer reasons to prefer.

Carnap of course thinks that some assignments are better than others. To make his point he considers two possible assignments which I shall call method † and method *.[4] Both of them apply the classical principle of indifference to the set of possible outcomes, but they do so in different ways. In the first case, Carnap assumes that, in the absence of further evidence, each of the sixteen individual distributions are equiprobable, assigning them probability 1/16. This is the probability assignment that would be made by the measure $c\dagger$. In the second case, we apply the principle of indifference twice. First we aggregate the individual distributions according to the number of black balls they contain. Carnap calls these aggregates *statistical distributions*. There are five such distributions, each of which we assign the probability 1/5. Next, we apply the principle of indifference to the individual distributions comprising each statistical distribution, assuming that the probability of 1/5 is divided evenly among each of the constituent individual distributions. This method yields probability assignments identical to those that would be made by $c*$. The assignments of prior probabilities to individual distributions are summarized in the table below:[5]

Once we have specified the prior probabilities, it is possible to measure the probability of any hypothesis given any evidence. To calculate the prior probability of a hypothesis, one merely sums the probabilities for all of the individual distributions in

Statistical Distributions		Individual Distributions	Method †	Method *	
Black	White		Probability of Individual Distributions	Probability of Statistical Distributions	Probability of Individual Distributions
4	0	BBBB	1/16	1/5	1/5
3	1	BBBW	1/16	1/5	1/20
		BBWB	1/16		1/20
		BWBB	1/16		1/20
		WBBB	1/16		1/20
2	2	BBWW	1/16	1/5	1/30
2	2	BWBW	1/16		1/30
		BWWB	1/16		1/30
		WBBW	1/16		1/30
		WBWB	1/16		1/30
		WWBB	1/16		1/30
1	3	BWWW	1/16	1/5	1/20
		WBWW	1/16		1/20
		WWBW	1/16		1/20
		WWWB	1/16		1/20
4	4	WWWW	1/16	1/5	1/5

which the hypothesis holds. To calculate the probability of a hypothesis given some evidence, one finds the sum of probabilities for individual distributions in which both the evidence and hypothesis hold, and divides it by the sum of probabilities for individual distributions in which the evidence holds.

Let H1 be the hypothesis that the first ball picked is white and H2 the hypothesis that the first two balls picked are white. Using the above table we can calculate for each method the probability of H1, H2 and H2 given H1:

Hypothesis	Method †	Method *
H1	1/2	1/2
H2	1/4	1/3
H2 given H1	1/2	2/3

Both of these methods yield coherent sets of expectations. The problem is to decide which to apply. Carnap argued for method * over method † because method * allows one to learn from experience. We can illustrate this point by reference to the table above. Method † says that the probability that the next ball picked will be white is 1/2, regardless of what picks have gone before. On the other hand, according to method *, the probability of a second white ball given a first white ball is higher (2/3) than the probability of the first ball being white (1/2). It is important for the subsequent argument to notice that these methods are *formal* in the sense that the values calculated do not depend upon of our interpretation of the B's and W's.

The requirement that an inductive method allow us to learn from experience is one of Carnap's adequacy conditions. Carnap states the principle as follows:

Inductive thinking is a way of judging hypotheses concerning unknown events. In order to be reasonable, this judging must be guided by our knowledge of observed events. More specifically, other things being equal, a future event is to be regarded as the more probable, the greater the relative frequency of similar events observed so far under similar circumstances. This *principle of learning from experience* guides, or rather ought to guide, all inductive thinking in everyday affairs and in science (Carnap 1955, 286).[6]

Elsewhere (Carnap 1945, §16) Carnap argues for this claim using Reichenbach's argument that a method which allows one to learn from experience is rational, because only such a method will in the long run allow one to improve one's ability to predict things in the world, supposing that the world is predictable at all. Much of the appeal of Carnap's c^* however derives from its intuitive plausibility as applied to concrete cases like the urn experiment. If the first three balls that we draw out of the urn are white, it would seem foolish (at least according to Carnap's intuitions) to insist that the probability of the next one being white was only 1/2.

I believe we have been cheated here. We have been asked to endorse the principle of learning from experience largely on the basis of its plausibility as applied to particular thought experiments like this one. It is possible however to construct a formally identical thought experiment which yields quite different intuitions. Suppose that rather than drawing four balls from an urn, our experiment consists of flipping a coin four times. Let "W" stand for our getting a heads on a particular toss, and "B" stand for our getting tails. Then, using the same nomenclature as in the first experiment, we can describe a particular sequence of coin tosses by a sequence of letters such as "WWWB". Using table one, we can apply the method discussed above to calculate inductive probabilities. Let H1' be the hypothesis that the first coin toss turns up heads, and H2' be the hypothesis that the first two coin tosses turn up heads. Then we get the same probabilities as we calculated for H1 and H2. I think that our intuitions strongly suggest that we should ignore evidence from the first coin toss in predicting the outcome of the second coin toss; that is, we should choose method † over method * as a method for predicting the outcome of this experiment.

We are left in the following intolerable situation: We have two experiments which are described by structurally identical languages, and whose spaces of possible outcomes are isomorphic. We nevertheless have strong intuitions on the basis of information not formally described in our language that make us choose method * for the first thought experiment, and method † for the second. These experiments show that there is no inductive method that applies to all situations. We choose an inductive method appropriate to a particular situation on the basis of background knowledge or beliefs about that situation.

I will discuss several responses to these objections in the next section. For the moment though, I would like to consider the character of the background beliefs which make method * intuitively plausible in one case, but not in the other. In the first case, the experiment involves drawing balls from an urn. What we do in the experiment is to sample from an underlying distribution. We do not know what this distribution is, but we assume that it is fixed. If we believe that our sampling method is unbiased, we expect our sample to approximate (within some margin of error) the underlying distribution. Because of our understanding of the causal mechanism producing the evidence, we believe that past experience (our sample) should be a guide to future expectations.[7]

In the second case, the situation is quite different. Rather than thinking that there is an underlying distribution from which we sample, we think of the evidence as being generated by a sequence of independent coin tosses. What happens on the second trial has nothing to do with what happens on the first, beyond the mere fact that we used the same coin. What matters is only our judgment that we are dealing with a fair coin; and even if we believe that our coin is not fair, we should, in virtue of the independence of the trials, assign the same probability of heads to each trial.

3. Carnapian Rejoinders

I would like to consider two rejoinders which Carnap might offer to the claim that these thought experiments show that there is no logic of induction, in the sense of no uniquely determined c function. The first is to maintain heroically that method * is the correct method even for the second experiment. The second is to argue that the apparent inadequacy of method * results from our failure to take into account all relevant evidence.

Before turning to these rejoinders, I would like to indicate more precisely what challenge has been posed by my argument that there are situations in which method † is the right method. Carnap admits that there is a continuum of inductive methods from which we can not single out one *a priori*. One might think that in indicating that we cannot prefer method * for the coin tossing experiment, I am holding Carnap to a standard which he himself did not hold. This is not, however, the case. First, I am not arguing that we cannot justify the choice of method * over method † on *a priori* grounds; rather, I am suggesting that there are sound reasons related to our background knowledge about experimental setups which dictate the choice of method † over method *. Furthermore, according to Carnap my preference for method † over method * cannot be licensed either by experience or subjective whim because it violates *a priori* axioms for *all* acceptable c functions.[8]

The heroic defense of method * – Carnap's first line of defense is to argue that, contrary to appearances, method * is the correct method to apply to the coin toss case. One could construe the coin toss experiment as an experiment to determine whether the coin is fair. After all, how can we know that a coin is fair except by experience? The problem with this rejoinder is that it should take many more than four trials to shake our confidence in the fairness of the coin. In the first experiment, we identify the probability of a black ball with the underlying distribution of black balls, but in the second, we do not identify the probability of a heads with the distribution of heads in a sequence of experiments.

I am not denying that in the long run experience might lead us to doubt the fairness of the coin. If one gets 100 heads in a row, then it would be plausible to doubt the fairness of the coin. The difficulty with Carnap's reconstruction of the situation is that it does not show the way in which such evidence is brought to bear. There is no single method of induction which we apply uniformly to our experience. Rather, given antecedent beliefs about the nature of the mechanisms producing the states of affairs we take as evidence, we choose a particular inductive method. If our expectations are repeatedly not borne out by experience, then at some ill-defined point, we begin to doubt our beliefs about the mechanisms which bring about those states of affairs. We eventually consider using different inductive methods.

The requirement of total evidence – A second and related way in which Carnap could seek to undermine my counterexample would be to invoke what he calls the principle of total evidence:

> For an application of inductive logic by an observer X at a certain time t the following holds: … If X wishes to apply a principle or theorem of inductive logic to his knowledge situation then he must use as evidence his total observational knowledge $K(t)$ (Carnap 1963, 972).

Presumably there must be some observational evidence for our belief that the mechanism producing the sequence of experimental outcomes is of a certain kind. It may well be that the probability of H2' given H1' *alone* is 2/3, but when we consider as evidence both H1' and our evidence for the fairness of the coin, we could very well get a probability of H2' around 1/2. The failure of method * to give a plausible value comes not from a defect in or limit in the applicability of that method, but rather from a failure to consider all relevant observational evidence.

There is something correct in this rejoinder. If we have any reason to believe that the experimental outcomes are being produced by a fair coin, it is because we have at some time had evidence to that effect. It would not be fair dealing to expect an inductive method to give us the intuitively "right" answer without taking this evidence into account. The requirement of total evidence is, however, implausibly strict, and is not acceptable even as a rational reconstruction of good science. Leaving aside the objection that it is unclear what would count as total evidence, it is simply not the case that we ever could (or should want to) take into account all observational evidence in evaluating each hypothesis. Furthermore, when we take all evidence into account, we will in general invoke more than observations. We will also invoke background theories which refer to unobserved (and unobservable) entities.

The fundamental mistake in Carnap's view is that Carnap conceives of inductive inference as a process of formal calculation. He believes that at least ideally we should assess the probability of a hypothesis h by calculating the value of our preferred c function with h and our total observational evidence $K(t)$ as arguments. My claim is that the connection between h and $K(t)$ must be mediated by background theory. In order to bridge the logical gap between h and $K(t)$ we must invoke some set of background hypotheses T. We believe T in light of $K(t)$ and further background hypotheses T', and so on. We see why we must invoke background theory by considering cases like the experiments described above where implicit assumptions about the mechanisms producing the evidence determine what method for calculating the probability of a hypothesis we should use.[9]

The view I am suggesting is reminiscent of Glymour's (1980) account of bootstrapping. Glymour argues that we must use auxiliary hypotheses to make the logical connection between evidence and hypothesis. The difference between his view and mine is that Glymour thinks that all auxiliary hypotheses can be made to explicitly enter into one's calculations of evidentiary relevance, while I am arguing that they may implicitly guide one's choice of an inductive method as well. This difference is significant, because if Glymour is right, it should be possible to define a three-place relation $c(h, e, T)$ between hypothesis and evidence with respect to some body of background theory. This would still be a logic of induction, albeit a more complex one than Carnap envisioned.

There are reasons why this proposal cannot succeed. First, it would be impossible to actually write down the rules for a plausible three place c function. Given that different background theories determine different relations between hypothesis and evidence, writing down the three place function would in essence require one to list the two place c functions determined by each possible background theory. However, there are infinitely many such theories, so we could never write them all down. Such a function, if it exists, would not be computable (at least in any practical sense). Furthermore, we will run into the same problem as we did for two place c functions of determining the relation between evidence and hypothesis in the absence of background theory (i.e., for null T). If all our theoretical claims are ultimately based only upon our inductive method and our total observational knowledge, then there must be some theories which do not require background theory for confirmation. A three place c function must consequently determine probabilities of h given e and null T. But what values should it choose — those given by c^* or $c\dagger$, or some other set of values? My analysis suggests that in the absence of any background theory there is no answer to this question. Prior probabilities cannot be assigned on *a priori* grounds, but only on the basis of hypotheses concerning the mechanisms producing the evidence in question.

4. The Justification of Induction

Carnap belongs to a tradition beginning with Hume that regards the justification of induction with considerable indifference. While members of this tradition admit that principles of inductive reasoning do not admit of either empirical or deductive justification, they point out that we use such principles with confidence and practical success. The point of skepticism about induction is merely to exorcise our rationalistic pretensions. Beyond that point, they believe we can and must ignore it.

Carnap says surprisingly little about the justification of induction, but what he does say supports the theory that he took it to be a pseudo-problem. In Carnap 1945 he remarks that "the situation [regarding induction] has sometimes been characterized by saying that a theoretical justification of induction is not possible, and hence that there is no problem of induction. However, it would be better to say merely that a justification in the old sense is not possible" (Carnap 1945, §16). The new kind of justification involves showing (1) that the logic we have given accords with our inductive practices; and (2) that these practices are guaranteed a certain measure of success (provided that the universe is at all predictable). We might rest comfortably at this point so long as we believe that there is more or less a single set of intuitive judgments about inductive inference upon which we can and do successfully rely.

The point of my thought experiments is to challenge this assumption. They provide an example of structurally identical experiments which produce conflicting intuitions concerning what can be inferred from their outcomes. My challenge is analogous to the challenge that Goodman posed with his new riddle of induction. The new riddle of induction replaced the question "What justifies our principle of induction?" with "Why is it that we choose some inductive principles instead of others?" Like Goodman's new riddle, my argument shows how formal properties of the language describing hypothesis and evidence are insufficient to answer this question.

A true logic of induction would be a calculus which we ascertain *a priori* and which we can use to calculate the weight which our experience lends to any hypothesis. There is no such *a priori* method. Our empiricism must be more thoroughgoing, allowing that even our methods of inference are determined by empirical considerations. Up to a certain point Carnap is sympathetic to this idea. In discussing the reasons for choosing one among the continuum of inductive methods (by choosing a particular value for a parameter λ), Carnap remarks:

> An inductive method is ... an instrument for the task of constructing a picture of the world on the basis of observational data and especially of forming expectations of future events as a guidance for practical conduct. X may change this instrument just as he changes a saw or an automobile, and for similar reasons. ...[A]fter working with a particular method for a time, he may not be satisfied and thereby look around for another method. ... Here, as anywhere else, life is a process of never ending adjustment; there are no absolutes, neither absolutely certain knowledge about the world, nor absolutely perfect methods of working in the world (Carnap 1952, 55).

This remark is characteristic of what Howard Stein has called Carnap's dialectical attitude. My suggestion is that we must extend this attitude further. Empirical considerations guide more than our choice of a single parameter; they infect all of our assumptions about the significance of the evidence we collect. Our inductive methods cannot claim the title of inductive logic because we choose our inductive methods on the basis of our understanding of the world which we investigate.

Notes

[1] I would like to thank Erich Reck and Mike Price for comments on earlier drafts of this paper, and Howard Stein for discussions on Carnap's views on induction.

[2] Carnap discusses adequacy conditions in a number of places, e.g., Carnap 1950. Kemeny (1963) has provided a succinct list which Carnap endorsed. For a definition of c^*, see Carnap 1945 or the appendix to Carnap 1950. Carnap ultimately gave up c^* (Carnap 1963, 974), but the revised function he proposed suffers from the same difficulties that I shall discuss here.

[3] $c\dagger$, which Carnap attributes to Wittgenstein, is defined in Carnap 1952, §13.

[4] These methods correspond to $c\dagger$ and c^*, but are defined only for the experiment under discussion. They should not be confused with c^* and $c\dagger$ which are functions defined for arbitrary sentences in a first order language.

[5] This table is based upon a figure from Carnap 1955.

[6] Page references to Carnap 1955 refer to the version reprinted in Brody and Grandy 1989.

[7] For a discussion of the nature of causal mechanisms and their significance for confirmation see Glennan 1992.

[8] Specifically, it violates what Carnap calls the axiom of instantial relevance and the axiom of convergence (Carnap 1963, 976). These axioms in turn are justified by the condition that an inductive method must allow us to learn from experience.

[9] In chapter 5 of Glennan 1992, I argue that the background knowledge needed to legitimate inference concerns the mechanisms which produce the states of affairs we take as evidence. See also Helen Longino's argument for the ineliminability of background theory in evidentiary reasoning (Longino 1990, chapter 3). Longino offers examples of situations in which differing background assumptions make the same states of affairs take on different evidential significance.

References

Brody, B.A., and Grandy, R.E. (ed.) (1989), Readings in the Philosophy of Science. 2nd ed. Englewood Cliffs, NJ: Prentice Hall.

Carnap, R. (1945), "On Inductive Logic", *Philosophy of Science*. 12: 72-97.

_____. (1950), *Logical Foundations of Probability*. Chicago: University of Chicago Press.

_____. (1952), *The Continuum of Inductive Methods*. Chicago: University of Chicago Press.

_____. (1955), *Statistical and Inductive Probability*. Brooklyn, N.Y.: Galois Institute of Mathematics and Art. Reprinted in Brody and Grandy 1989.

_____. (1963), "Replies and Systematic Expositions", in P. Schilpp (ed.), *The Philosophy of Rudolph Carnap. The Library of Living Philosophers*, Vol. XI. LaSalle, IL: Open Court, 859-1016.

Glennan, S. S. (1992), "Mechanisms, Models and Causation". Ph.D. Dissertation, the University of Chicago.

Glymour, C. (1980), *Theory and Evidence*. Princeton: Princeton University Press.

Kemeny, J. (1963), "Carnap's Theory of Probability and Induction", in P. Schilpp (ed.), *The Philosophy of Rudolph Carnap. The Library of Living Philosophers*, Vol. XI. LaSalle, IL: Open Court, 711-738.

Longino, H E. (1990), *Science as Social Knowledge: Values and Objectivity in Scientific Inquiry*. Princeton: Princeton University Press.

Spirtes, P., Glymour, G., and Scheines R. (1993), *Causation, Prediction, and Search. Lecture Notes in Statistics*, Vol. 81. New York: Springer-Verlag.

Earman on the Projectibility of Grue

Marc Lange

University of California, Los Angeles

In *Bayes or Bust?*, John Earman (1992, 104-113) attempts to express in Bayesian terms a sense of "projectibility" in which it is logically impossible for "All emeralds are green" and "All emeralds are grue" simultaneously to be projectible. I argue that Earman overlooks an important sense in which these two hypotheses cannot both be projectible. This sense is important because it connects projectibility to lawlikeness,[1] as Goodman intended. Whether this suggests a solution to Goodman's riddle remains unsettled, awaiting an account of lawlikeness. I explore one avenue that might prove illuminating.

1. Earman's Senses of Projectibility

Suppose we discover that object a is G, where our background beliefs K include that Ea. Earman notes that by Bayes's theorem, hypothesis H ("All E's are G") is confirmed—i.e., $pr(H|Ga,K) > pr(H|K)$—if $0 < pr(H|K)$, $pr(Ga|K) < 1$. Hence, instances confirm "All emeralds are green" *and* "All emeralds are grue," since "it is much too draconian to suppose that ... unprojectible hypotheses are to be initially and forever condemned to limbo by receiving zero priors" (Earman 1992, 105). So, Earman asks, in what sense do the green and grue hypotheses differ in "projectibility"? Let the G_i be the claims that various objects (i=1,2,...), already believed E, are G. Earman defines two senses of "projectible"[2]:

"H is weakly projectible in the future-moving sense, relative to K" iff $pr(G_{n+1}|G_1,...,G_n,K)$ approaches 1 as n approaches infinity.

"H is strongly projectible in the future-moving sense, relative to K" iff $pr(G_{n+1},...,G_{n+m}|G_1,...,G_n,K)$ approaches 1 as n and m approach infinity.

But, Earman notes, $pr(H|K) > 0$ guarantees projectibility in these senses; green *and* grue are so projectible. Now let the G_i extend infinitely in both directions. Earman defines

"H is weakly projectible in the past-reaching sense, relative to K" iff for any n, $pr(G_{n+1}|G_n,G_{n-1},...,G_{n-j},K)$ approaches 1 as j approaches infinity,

(and an analogous sense of "strong projectibility"). In taking this limit, we fix the individual being confirmed to be G, unlike in the future-moving senses. Clearly, if K includes that this (n+1)th individual is the first emerald examined after the year 2000, a contradiction results if "All emeralds are green" *and* "All emeralds are grue" are projectible in the past-reaching sense. Earman concludes that here, finally, is a sense in which green and grue differ in projectibility.

Projectibility concerns the capacity of instances to confirm H's accuracy to unexamined case(s). This was Goodman's concern; Earman quotes Goodman's (1983, 69) remark that genuine confirmation occurs "only when the instance imparts to the hypothesis some credibility that is conveyed to other instances." But to *which* other instances? Goodman denies that the grue hypothesis undergoes genuine confirmation, yet accepts that by discovering a given emerald before 2000 to be grue (i.e., green), we confirm an unexamined emerald *before* 2000 to be grue (i.e., green). Apparently, the grue hypothesis is unprojectible because by discovering a given emerald before 2000 to be grue, we don't confirm an unexamined emerald *after* 2000 to be grue (i.e., blue). For Goodman, "All Es are G" undergoes genuine confirmation only when an instance imparts to it some credibility that is conveyed to another E *no matter what else* (within some suitably generous range) is believed about that other E. Earman's senses of projectibility obscure this feature. Earman characterizes various senses of "projectibility *relative to K*"; Earman's concern is whether instances of "All Es are G" make arbitrarily likely to be G object(s) believed, under some *fixed* K, to be E. Goodman's concern is whether an instance would, under *every* K (in some suitably generous range), confirm to be G each other object believed under that K to be E.

That is, Goodman doesn't relativize "projectible" to background beliefs. Suppose we believe that as it happens, all men in the lecture audience have the same number of older brothers. Then an instance establishes "All men in the lecture audience have two older brothers." But despite the capacity of instances to confirm the accuracy of this hypothesis in unexamined cases, Goodman deems this hypothesis unprojectible. Apparently, a "projectible" hypothesis can be confirmed to hold of unexamined cases even without the benefit of background beliefs about physical accidents.

Let's be more careful. I'll deem an object to be "grue" at a time before 2000 iff it is green then; at other times, an object is "grue" iff it is blue then. The grue hypothesis is "All emeralds at all times are grue," i.e., "All temporal slices of emeralds are grue." Projectibility concerns confirmation by instances. To discover an "instance" of the grue hypothesis is to discover that some object, already believed an emerald slice, is grue. According to Goodman (1983, 89ff.), discovery of the "instance" doesn't include discovery of *when* this slice exists; the time doesn't belong to the instance. So what makes the grue hypothesis unprojectible isn't that by discovering an emerald slice *before* 2000 to be grue, we don't confirm an unexamined emerald slice after 2000 to be grue—because the instance doesn't include that the examined slice precedes 2000.

When an instance confirms "All E's are G," it may increase our willingness to predict, of some unexamined E, that it is G. Whether the instance does so may depend upon what else we already believe about the unexamined E. To characterize this dependence, we must consider the various properties P where we might already believe of some unexamined E that it is E and P. We must identify those P's such that (roughly) if Ea&Eb&Pb (and its logical consequences) is all we believe, we'd confirm Gb by discovering Ga. Whether we actually believe there is an E possessing P is irrelevant to whether, *if* we did, the evidence would confirm that individual to be G. (For instance, though we haven't identified any emerald slices after 2000, the discovery that some emerald slice preceding 2000 is grue doesn't increase my confidence that a *posited* post-2000 emerald slice—say, the first one examined—is grue.) The grue hypothesis is unprojectible because the discovery that a given emerald slice (no mention of the time) is grue doesn't confirm to be grue an object believed only to be an unexamined emerald slice possessing the following property: It exists during or after 2000 iff the examined slice precedes 2000.

Can't H's projectibility be Ga's capacity to confirm that an object is G, for any object where we already believe exactly the logical consequences of its being E, its belonging to a history (consistent with our beliefs about the laws) that includes Ea, and—for some property P belonging to a certain generous set—its being P? Earman briefly entertains roughly this possibility, but says (1992, 242): "I am unconvinced that there is any workable and useful notion of homogeneous confirmation to be had. And I am convinced that the tools already developed are adequate to capture the valid lessons to be drawn from Goodman's examples." I'll argue otherwise by presenting a sense of "homogeneous confirmation" and showing that it captures one of Goodman's concerns (that Earman doesn't discuss): projectibility's relation to lawlikeness. I'll suggest that this relation may help to resolve Goodman's riddle.

2. Projectibility as the Capacity to undergo Homogeneous Confirmation

I'll define the "projectibility" of H ("All emerald slices are G") as H's capacity to be confirmed "homogeneously" by the discovery that Ga, where Ea ("a is an emerald slice") is already believed. Roughly, by discovering an emerald slice to be G, we confirm H "homogeneously" iff for any property P in a certain range, we confirm an object to be G if we already believe only that it is an emerald slice and P, belongs to a history (consistent with our beliefs about the laws) in which the object being tested for G-ness is an emerald slice, and their logical consequences.

For any property P in what range? Let L be the claims "...is a law" we believe, and let R be L, the claims "...is not a law" we believe, the claims "...is lawlike" and "...is non-lawlike" we believe concerning claims we don't believe false, and the claims we believe about which claims (that we don't believe false) may state laws if certain others do. I presume we don't already believe H true (or, perforce, a law-statement) and don't already believe H false; otherwise, we wouldn't be entertaining H as a hypothesis. I presume R&H&Ea&Eb is consistent; otherwise, we wouldn't be trying to check E's for compliance with H.[3] Property P must be "suitable," meaning that it must satisfy:

(S1) None of these is a logical truth:
(Eb&Pb&Ea&R) ⊃¬Ga,
(Eb&Pb&Ea&R) ⊃ Ga,
(Eb&Pb&Ea&R) ⊃ Gb;

(S2) If we believe H may be physically necessary (i.e., if we believe neither "H is physically necessary" nor "H isn't physically necessary"), then Eb&Pb&Ea&R&(H is physically necessary) is consistent;

(S3) If we believe H isn't physically necessary, then Eb&Pb&Ea&L&H is consistent.[4]

By S1, the property of being green isn't a suitable P when H is the green hypothesis, for if Pb is Gb, Pb⊃Gb is a logical truth. So to confirm H homogeneously by discovering that object a (already believed to be an emerald) is green, we aren't required to do the impossible: to confirm to be green an emerald we already believe green. If we believe H lawlike, then by S2, we aren't required to confirm to be green an emerald we already believe non-green or an emerald b such that (b is green) ⊃ (H is not a law). If we believe H non-lawlike, then by S3, we aren't required to confirm to be green an emerald we already believe non-green or alongside a non-green emerald.

Roughly, when the discovery of an instance confirms "All emerald slices are G," it does so *homogeneously* exactly when it confirms to be G any object where prior to this discovery, we believed exactly that this object is an emerald slice, is P (for some suitable P), belongs to a history in which all our beliefs about the laws obtain and the object under examination is an emerald slice, and their logical consequences.

I haven't said what it means to "confirm" such an object to be G. Suppose we believe that no actual emerald slice possesses P. Then, seemingly, Ga cannot confirm Gb, given Eb and Pb and Ea; pr(Gb|Ga,Eb,Pb,Ea,R) and pr(Gb|Eb,Pb,Ea,R) are undefined, since pr(Eb,Pb) = 0. To avoid this result, I invoke a Bayesian view: that our current degrees of belief derive by conditionalization from an original probability distribution; even if there was no moment, when the agent's doxastic life began, at which her degrees of belief were the unconditional probabilities in this original distribution, talk of such a distribution is perhaps needed to rationally reconstruct the present. In defining "homogeneous" confirmation, I wish to speak of Ga "confirming" Gb, where *b* is posited to be E and P, and I want no trouble from our belief that as a matter of fact—a fact that, for all we know, may be physically accidental—no actual E is P. Accordingly, I'll identify this "confirmation" with pr(Gb|Ga,Eb,Pb,Ea,R) > pr(Gb|Eb,Pb,Ea,R), *where these probabilities belong to the original distribution*. R includes the claims we currently believe to be law-statements, but (since for all we know, it is a physical accident that no E is P) not all the evidence that led us to conclude that no E is P. So the probabilities in this inequality can be well-defined, since if P is suitable, Ga&Eb&Pb&Ea&R is consistent.

Suppose evidence V confirms that hypothesis H ("All E's are G") is true. Suppose V ⊃ Ga is a logical truth, where before discovering V, we believed Ea and didn't believe Ga. By discovering V, we confirm H *homogeneously* if and only if, for any suitable property P,

pr(Gb|Ga,Eb,Pb,Ea,R) > pr(Gb|Eb,Pb,Ea,R) (1)

holds in the original probability distribution.

3. Green and Grue Cannot Both Be Confirmed Homogeneously

Paralleling Earman, I'll now show it impossible for the green and grue hypotheses simultaneously to be confirmed homogeneously. Let evidence V be that object *a*, already believed an emerald slice (Ea), is green (Ga) and exists before 2000 (Ta). To confirm the green hypothesis homogeneously,

pr(Gb|Ga,Eb,Pb,Ea,R) > pr(Gb|Eb,Pb,Ea,R) (2), from (1)

must obtain for any suitable P, whereas to confirm the grue hypothesis (i.e., (x)(Ex ⊃ (Tx⊃Gx & ¬Tx⊃Bx)), using "G" for green and "B" for blue) homogeneously,

pr(Tb⊃Gb,¬Tb⊃Bb|Ta⊃Ga,¬Ta⊃Ba,Eb,Pb,Ea,R) >
pr(Tb⊃Gb,¬Tb⊃Bb|Eb,Pb,Ea,R) (3), from (1)

must obtain for any suitable P. By selecting P properly, we can render these two inequalities incompatible. Bearing in mind what makes grue unprojectible, I'll let Pb be (Ta & ¬Tb & ¬Bb⊃Gb). I presume—as in any discussion of grue—that R is such that P is suitable for both hypotheses,[5] and Bb⊃¬Gb. To confirm the green hypothesis homogeneously,

pr(Gb|Ga,Eb,Ta,¬Tb,¬Bb⊃Gb,Ea,R) >
pr(Gb|Eb,Ta,¬Tb,¬Bb⊃Gb,Ea,R) (4), from (2)

must obtain, and to confirm the grue hypothesis homogeneously,

$$pr(Tb{\supset}Gb, \neg Tb{\supset}Bb | Ta{\supset}Ga, \neg Ta{\supset}Ba, Eb, Ta, \neg Tb, \neg Bb{\supset}Gb, Ea, R) >$$
$$pr(Tb{\supset}Gb, \neg Tb{\supset}Bb | Eb, Ta, \neg Tb, \neg Bb{\supset}Gb, Ea, R) \qquad \text{from (3)}$$

must obtain. This last is equivalent to

$$pr(\neg Gb | Ga, Eb, Ta, \neg Tb, \neg Bb{\supset}Gb, Ea, R) >$$
$$pr(\neg Gb | Eb, Ta, \neg Tb, \neg Bb{\supset}Gb, Ea, R),$$

which is incompatible with (4), and (4) is a necessary condition for the green hypothesis to be confirmed homogeneously.[6]

4. Homogeneous Confirmation and Lawlikeness

We've identified a sense in which the green and grue hypotheses cannot both be "projectible." I'll now argue that this sense allows us to retain Goodman's connection between the projectibility of a hypothesis and its lawlikeness. I'll argue that we cannot confirm a hypothesis homogeneously if we believe it non-lawlike. Suppose we believe H ("All E's are G") non-lawlike; this belief is part of R. Take "Pb" to mean that if Gb, then H states a law. Because R includes "H is non-lawlike," it is a logical truth that (R&Pb) ⊃ ¬Gb; thus, pr(Gb|Pb,R) = 0. Hence, (1) cannot be satisfied. It follows that if P is suitable, H cannot be confirmed homogeneously.

And P is suitable. Consider S1. Eb&Ea&R is consistent, so Eb&Pb&Ea&R is consistent unless Eb&¬Gb is incompatible with R. This incompatibility would require that R logically entail "All Es are G," which would require that we already believe H states a law. But we believe H doesn't state a law, so Eb&Pb&Ea&R is consistent. Since Eb&Ea&R doesn't necessitate Ga and doesn't necessitate ¬Ga, neither does Eb&Pb&Ea&R. Since Eb&Pb&Ea&R is consistent and logically entails ¬Gb, it doesn't logically entail Gb. So P satisfies S1. P satisfies S3 iff Eb&Pb&Ea&H is compatible with the lawhood of the claims we believe state laws. Eb&Ea&H is consistent with their lawhood, otherwise we wouldn't be trying to check cases of H in order to confirm it. So Eb&Pb&Ea&H is consistent with their lawhood if "H is a law" is consistent with their lawhood. As a hypothesis we are entertaining, H must be consistent with the truth of the claims we believe state laws. I believe that "H is a law" must then be consistent with the lawhood of these claims, since I believe it logically possible for any deductively closed subset of the regularities to be the subset containing all and only the physically necessary regularities. That is, I contend that among the logically possible worlds involving all of the same particular facts, there is—for each deductively closed set of regularities—a world in which that set contains exactly the physically necessary regularities.

That physical necessity can be distributed in any way among the regularities (so long as the logical consequences of physically necessary regularities are physically necessary) has been suggested by Armstrong (1983, 71) and is encouraged by Jackson's (1977, 4) talk of "the Hume world" in which all particular facts agree with those in the actual world but every regularity is accidental. (Admittedly, this intuition runs counter to certain regularity accounts of law, such as David Lewis's. Roughly, he believes that the law-statements are the generalizations in the deductive system of truths having the best combination of strength and simplicity.) I accept this intuition, and see it manifested in the following fact. It is standardly held (following Goodman, though there are dissenters, notably Lewis) that laws are intimately related to counterfactuals. Roughly, the counterfactual "Had p obtained, the laws wouldn't have been any different" is true for any non-nomic p consistent with the laws.[7] For example, had no human being in 1993 believed $E=mc^2$, or had the universe contained only a proton in uniform motion forever, the laws wouldn't have been different. (The distinction between laws and initial conditions supports this view; one can consistently imagine possible worlds with the same laws as the actual world and any initial conditions consistent with the laws.) But if the non-nomic facts in a logically possible world rendered it logically impossible for a given regularity in that world to be physically necessary there, then we'd expect some non-nomic p, consistent with the actual laws, to be such that in one of the closest p-worlds, the laws differ from those in the actual world. For example, Lewis's account of law apparently has the counterintuitive consequence that had the universe contained merely a proton in uniform motion forever, then the laws would have been different, e.g., the laws governing particle collisions and "All robins' eggs are blue" would not have been laws.

Since P is suitable, we can confirm H homogeneously only if we believe H may be lawlike.

5. Grue, Natural Laws, and Physical Necessities

I've elaborated a sense in which we cannot "project" the grue hypothesis if we believe it non-lawlike. But wouldn't our prior belief in its non-*lawlikeness* be as "draconian" as our according its *truth* a zero prior? The answer depends on what lawlikeness *is*. I cannot here offer an account of lawlikeness. But I'll suggest that it is not unpromising to approach Goodman's riddle by trying to find a conception of lawlikeness that deems the grue hypothesis a non-starter as a law-statement.

The most striking difference between the grue and green hypotheses is that, once we've observed many instances of both and no exceptions to either, no one lacking an ulterior philosophical motive even thinks of the grue hypothesis. The green hypothesis comes to nearly everyone's mind very quickly and strikes nearly everyone as plausible. This is simply a fact about how we react in this situation, about which correlations jump out at us when we ascertain instances. IQ tests ("Give the next member of the sequence 2,4,6,8") and ostensive definitions ("This, that, and that over there are 'green'") presuppose that in such situations, all persons agree very nearly in their reactions—as Wittgenstein has emphasized. I wish to entertain the possibility of a connection between lawlikeness and our reactions.

Suppose any logically possible E either possesses or lacks property Q. Imagine our evidence consists exclusively of E's known to possess Q and found to be G. It might be that no matter how many E's this evidence includes, we would never react by taking "All E's are G" seriously; without evidence of E's known not to possess Q, this hypothesis would jump out at no one in the manner of the correct answer to "Give the next member of the sequence 2,4,6,8." Now imagine our evidence consists entirely of E's known to lack Q and found to be G. It might again be that no matter how many E's this evidence includes, "All E's are G" would never jump out at us. In that event, I propose, "All E's are G" cannot state a law, even if it is true. Clearly, these conditions are satisfied by the grue hypothesis, where Q is the property of preceding 2000. On this proposal, then, we could justly condemn to zero-prior limbo the hypothesis that "All emeralds are grue" states a law.

I cannot properly defend this rough proposal here; I mention it principally to add some substance to my contention that by ignoring homogeneous confirmation, and thus projectibility's connection to lawlikeness, Earman fails to capture a feature of Goodman's riddle that may be important to its resolution.[8] But I acknowledge two concerns one might immediately raise regarding this proposed connection between lawlikeness and our reactions. First, one might object that this proposal makes the natural laws too subjective. Obviously this depends upon how much subjectivity the laws can countenance. The proposed connection between lawlikeness and our reactions doesn't entail that there are no undiscovered laws or that we make certain generalizations lawlike by believing them so to be (as Goodman may maintain). Nor does it obviously entail that had our reactions been sufficiently different, the laws would have been different. This last is a counterfactual, and laws have a special persistence under counterfactuals.

A second objection to this connection between lawlikeness and our reactions is that it is *ad hoc*: Apart from the fact that it might account for the non-lawlikeness of such weird hypotheses as "All emeralds are grue," it has no motivation in scientific practice. I disagree; here's one phenomenon saved by this connection. Arbitrary conjunctions of laws—e.g., "Anything that is a raven or a dove is black if a raven, white if a dove"—are intuitively not laws. As Hempel and Oppenheim (1965, 273 n. 33) noted, the conjunction of Kepler's and Boyle's laws lacks the explanatory power we expect of a law. A connection between lawlikeness and our reactions could account for this. The hypothesis about ravens and doves would not jump out at us, so long as we believed we've checked only ravens or only doves, no matter how many. Rather, we must believe we've checked ravens and doves, just as the grue hypothesis would not strike us forcefully until we'd discovered instances both before and after 2000.

Although the raven-dove generalization doesn't state a law, it is physically necessary.[9] Some claims (e.g., "All diamonds in Pittsburgh are carbon") that we believe non-lawlike, but perhaps physically necessary, can be homogeneously confirmed (for when "All diamonds are carbon" is homogeneously confirmed, each suitable diamond in Pittsburgh is confirmed to be carbon). So our earlier result, that we cannot confirm a generalization homogeneously if we believe it non-lawlike, isn't quite right. In arguing for this result, I contended that the conjunction of Eb with the claims we believe state laws, which doesn't entail ⌐H, also doesn't entail "H doesn't state a law." I argued that it is logically possible for any deductively closed set of regularities to be the one containing exactly the physically necessary regularities, and by *supposing* that all physically necessary generalizations state laws, I concluded that it is logically possible for H and the generalizations we believe state laws all to state laws. But I've now rejected this supposition.

Can the earlier argument be repaired? Yes—to show that we cannot confirm H homogeneously if we believe H is not *physically necessary* (i.e., H is an accidental generalization) if H is true. Take "Pb" to mean "If Gb, then H is physically necessary"; to show this P suitable, use the fact that it is logically possible for any deductively closed set of regularities to contain exactly the physically necessary regularities.

That we cannot confirm a generalization homogeneously, if we believe it is not physically necessary, seems initially to help little with the grue problem. Though my earlier remarks perhaps suggested that "All emeralds are grue" (like the hypothesis covering ravens and doves) is non-lawlike, they didn't suggest that the grue hypothesis would, if true, be an accidental generalization. Like the hypothesis concerning ravens and doves, the grue hypothesis might be made physically necessary by two distinct sets of laws (perhaps "All emerald slices before 2000 are green," "All emerald slices not preceding 2000 are blue"), one consisting of those essential for necessitating that emerald slices preceding 2000 are grue, the other consisting of those essential for necessitating that the other emerald slices are grue.

Maybe there's a way out. Apparently, if the grue hypothesis is physically necessary, then by checking only emerald slices before 2000 (or ones after 2000), we will not make jump out at us the hypotheses stating the unknown laws responsible for the physical necessity of "All emeralds are grue." The connection between lawfulness and our reactions suggests that if the grue hypothesis is physically necessary, then not only doesn't it state a law, but the (presumed unknown) laws helping to make "If *c* is an emerald slice before

2000, then *c* is grue" physically necessary are distinct from those responsible for the physical necessity of "If *c* is an emerald slice after 2000, then *c* is grue." This suggests that we cannot confirm the grue hypothesis homogeneously, because an instance doesn't confirm to be grue another emerald slice whose grueness would result from unknown laws distinct from those responsible for the grueness of the examined case.

That is, the connection between lawlikeness and our reactions would explain why, if the grue hypothesis is true, none of the unknown laws responsible for the colors of emerald-slices before 2000 is responsible for the colors of emerald-slices after 2000. By examining only emerald-slices before (after) 2000, we'd never be led to the unknown laws governing the colors of emerald-slices after (before) 2000. Rather, we'd be misled; false hypotheses would jump out at us, as would occur if we tried to predict the colors of doves entirely from the colors of ravens. So even if we believe that the grue hypothesis may be physically necessary, we believe that we'd be misled if we treated the colors of emerald-slices before 2000 as bearing upon the colors of emerald-slices after 2000. Therefore, we don't treat an instance *a* of the grue hypothesis H as confirming the grue-ness of an unexamined case *b* posited as possessing property P, where "Pb" means "Ta & ¬Tb & (H is true ⊃ H is physically necessary)." We therefore don't confirm the grue hypothesis homogeneously.

Goodman once (1966) held a similar view: that we consider "All emeroses are gred" unprojectible because we believe it non-lawlike, though we also believe it may be physically necessary. We believe that if this hypothesis is true, the laws responsible for its holding of emeralds are distinct from those responsible for its holding of roses.

To properly defend my conjecture that lawlikeness is connected to our reactions, I must analyze lawfulness. This task I gladly leave for another occasion.

Notes

[1] "H is lawlike" means (H is true) ⊃ (H states a natural law). "H is non-lawlike" means (H is true) ⊃ (H does not state a law).

[2] Earman says "projec*ta*ble"; I retain Goodman's spelling.

[3] I'm presuming R consistent. Otherwise, we must exploit some manoeuvre from the logic of inconsistency if we are to identify confirmation with pr(H|evidence, background) > pr(H|background).

[4] I haven't space to motivate S1-3 beyond showing them able to serve my purposes here. In an unpublished paper, I show that they aren't *ad hoc* because they account for the distinctive relation of laws to counterfactuals. Roughly: It's often held (see (Lange 1993b)) that if we believe H states a law, we preserve H under any counterfactual antecedent Eb&Pb in a certain generous range (e.g., compatible with H and the other law-statements), whereas if we believe H an accidental generalization, there are some counterfactual antecedents in an analogous range under which we don't preserve H. The relevant ranges coincide with the Eb&Pb's with suitable P. Consider Popper's accidental generalization H: "All moas die before age fifty." Since we believe H accidental, we don't preserve H under the antecedent "Had there been a moa, aged 49 years 364 days, in a virus-free environment, without pneumonia, not standing before a moving vehicle, etc." If even this moa dies before age fifty, H must be a law; we mean this antecedent to be "Had there been a moa possessing property P," where "Pb" means "If *b* dies before age fifty, H states

a law." I want the suitable P's to be those in antecedents "Had there been a moa possessing P" where the reason we (don't) believe H preserved under every such antecedent is that we (don't) believe H a law-statement. So the above P must be suitable when R includes "H is non-lawlike." This precludes S1's including "(Eb&Pb&Ea&R) ⊃ ¬Gb" as a fourth line, or S3's requiring that Eb&Pb&Ea&*R*&H be consistent.

[5] I'll show that we cannot confirm a hypothesis homogeneously if we believe it non-lawlike. So P is suitable iff P satisfies S1 and S2. P does so for the green [grue] hypothesis iff our beliefs about the laws, that *a* is an emerald slice preceding 2000, *b* is an emerald slice after 2000, and *b* is green if not blue, doesn't logically entail that *a* is green [grue], that *a* is not green [grue], that *b* is green [grue], or that the green [grue] hypothesis doesn't state a law.

[6] The incompatibility here doesn't result from any peculiarity of grue's; the hypotheses (x)(Ax⊃Bx) and (x)(Cx⊃Dx) can't simultaneously be confirmed homogeneously if P is suitable for the former and Q for the latter, where both "Pb" and "Qd" mean that Aa, Ab, Cc, Cd, (Ba iff Dc) and (Bb iff¬ Dd). Thanks to Byeong Yi for discussion here.

[7] I discuss putative counterexamples to this view in (1993b).

[8] Salmon (1963) suggests that the grue hypothesis is unjustified because "grue" can't be defined ostensively. This suggestion has merit; I, too, exploit the fact that instances preceding 2000 don't suffice for grue to jump out at us. But while I admit that "grue" is learned only after "green" is learned ostensively, I don't see why this fact renders the green hypothesis better supported by the evidence. On my proposal, our reactions help to *make it the case* that certain claims state laws. I cannot properly defend this proposal here.

[9] See my (1993a) for other examples of non-lawlike physical necessities.

References

Armstrong, D. (1983), *What Is A Law of Nature?*. Cambridge: Cambridge University Press.

Earman, J. (1992), *Bayes or Bust?*. Cambridge: Bradford.

Goodman, N. (1966), "Comments", *Journal of Philosophy* 63: 328-331.

_ _ _ _ _ _ _ . (1983), *Fact, Fiction, and Forecast*. Cambridge: Harvard.

Hempel, C.G. (1965), *Aspects of Scientific Explanation*. New York: Free Press.

Jackson, F. (1977), "A Causal Theory of Counterfactuals", *Australasian Journal of Philosophy* 55: 3-21.

Lange, M. (1993a), "Lawlikeness", *Noûs* 27: 1-21.

_ _ _ _ _ . (1993b), "When Would Natural Laws Have Been Broken?", *Analysis* 53: 262-269.

Salmon, W. (1963). "On Vindicating Induction", *Philosophy of Science* 30: 252-261.

A Representational Reconstruction of Carnap's Quasianalysis

Thomas Mormann

Universität München

1. Introduction

According to general wisdom, quasianalysis belongs to the large family of Carnap's ingenious, but finally failed contributions to epistemology and philosophy of science. In this paper I want to show that this is not the case. Rather, Carnapian quasianalysis is to be considered as a promising theory of a representational constitution of scientific objects. That is to say, I intend to embed Carnap's approach of quasianalytical constitution in the framework of a general theory of meaningful representation (cf. Mundy (1986)).

The outline of this paper is as follows: In section 2 I recall the basics of the quasianalytical approach, taking into consideration not only the well-known account in "Der Logische Aufbau der Welt" (*Aufbau*) but also a rather unknown first version of quasianalysis ("Quasizerlegung") which Carnap developed in an unpublished manuscript written in 1923. This paper deserves attention not only for philosophico-historical reasons, rather it contains quite a lot of interesting features of the quasianalytical approach which do not appear in the *Aufbau* account. In section 3 I reformulate Carnap's account of quasianalysis in the framework of a representational theory of similarity measurement. This allows us to consider the theory of quasianalysis as a special case of a general theory of structural representation. In section 4 it is shown how Goodman's objections against the feasibility of any quasianalytical account may be defused in the new framework. As an application of representational quasianalysis, in section 5 I sketch how Quine's thesis of empirical underdetermination of theories may be elucidated in the framework of a representational quasianalysis.

2. Carnap's Quasianalysis of 1923

Carnap distinguishes that there are two essentially different ways of describing a set of elements. The first way is to say what are the properties or parts every element has. This method he calls the method of *individual* description. The second way is to tell what are the relations between the elements. It might be called the method of *relational* description. The relational description has the advantage of being an *internal* description, it does not go beyond the set it intends to describe: the elements of the set in question are *not* decomposed in parts that (usually) do not belong to that set. Rather, the relational description characterizes them by appropriate subsets of Cartesian products of the basic set itself. On the other hand, the method of relational description has the drawback of being rather clumsy. Thus it would be desirable to have a method which transforms a relational description in a handier individual one but keeping the virtue of being an immanent description thereby joining the advantages of both. This is the method of quasianalysis. The term "quasianalysis" ("Quasizerlegung") appears for the first time in an unpublished manuscript of 1923 which has the programmatic title "Quasi-Analysis - A Method to order non-homogenous sets by means of the theory of relations". It is a purely formal theory that might be considered as a generalization of the well-known Russell-Whitehead theory of equivalence classes. Carnap describes the task of quasianalysis as follows:

> Suppose there is given a set of elements, and for each element the specification to which it is similar. We aim at a description of the set which only uses this information but ascribes to these elements quasicomponents or quasiproperties in such a way that it is possible to deal with each element separately using only the quasiproperties, without reference to other elements. Carnap (1923, 4)

The ascription of quasiproperties is not arbitrary, of course, but should obey four basic conditions (cf. Carnap (1923, 4-5):

(C1) If two elements are similar they coincide in at least one quasiproperty.
(C2) If two elements are not similar they do not coincide in any quasiproperty.
(C3) If two elements a and b are similar to exactly the same elements, i.e., if they have the same similarity neigborhood, they have the same quasiproperties.
(C4) There is no quasiproperty which can be removed such that the conditions (C1) - (C3) are still satisfied.

As Carnap observes, these axioms are consistent and independent of each other. In the *Aufbau* only the conditions (C1) and (C2) appear. It is useful to have formal definitions of the concepts used in the following. A similarity structure, denoted by $(S,\sim)$, is to be a set S endowed with a similarity relation. A similarity relation is a reflexive and symmetric, but not necessarily transitive, relation $\sim S \times S$. The similarity neighborhood of x $\in$ S is denoted by $co(x) := \{y \mid x \sim y\}$. Most often our examples are *finite* similarity structures, i.e., the underlying set S is a finite set. Then it can be conveniently represented by a finite numbered graph such that two elements define an edge if and only if they are similar as displayed in the following example (cf. Carnap (1923, 5):

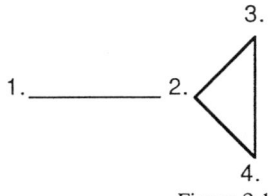

Figure 2.1

This graph is to be interpreted as a similarity structure with underlying set S = $\{1,2,3,4\}$ where 1 and 2 are similar, 2,3,4 are similar to each other, and no other pairs of (different) elements are similar to each other. A quasianalysis of a finite similarity structure $(S,\sim)$ can be succinctly described by a list (cf. Goodman 1954, Ch. VI): as above denote the elements of S by natural numbers 1,2, ..., n. The quasiproperties are denoted by a,b,c,... Then a quasianalysis of (2.1) can be given as a list of the following kind: {1.a, 2.ab, 3.bc, 4.bc} to be read in the obvious way, to wit, 1 has the quasiproperty a, 2 has the quasiproperties a and b etc. Sometimes it's convenient to combine lists and graphs in the following way:

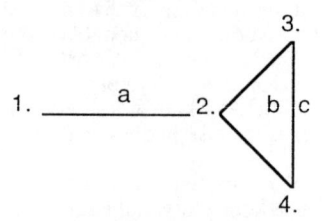

Figure 2.2

For later purposes let us mention one famous example, which may be called "Goodman's triangle":

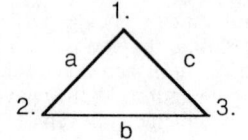

Figure 2.3

It is a quasianalysis in the sense of the simplified definition of the *Aufbau*, i.e., it satisfies (C1) and (C2). However, it is not a quasianalysis according to the original definition since it does not satisfy (C3).

3. Quasianalysis (QA) in the framework of a representational theory of similarity measurement

Now I embark on the task of reformulating QA in the framework of a representational theory of similarity measurement. This will enable us to exploit some interesting analogies of QA with the representational theory of measurement. The starting point is the following representational reformulation of quasianalysis:

(3.1) Definition.

(i) A *weak* quasianalysis of $(S,\sim)$ is a map $f: S \rightarrow 2^Q$ which satisfies the following properties:

(1) $s \sim s' \Rightarrow f(s) \cap f(s') \neq \emptyset$

(2) $f(s) \cap f(s') \neq \emptyset \Rightarrow s \sim s'$

(ii) A *strong* quasianalysis is a weak quasianalysis which satisfies the following two further conditions:

(3) co (x) = co (y) ⇒ f (x) = f (y)

(4) No elements of Q can be removed, unless the resulting f does not satisfy at least one of the conditions (1) - (3).

In (3.1) I apparently have introduced the set Q of quasiproperties as independent of the set S. As the reader will remember, according to Carnap one of the main virtues of quasianalysis is that it does allow us to consider the quasiproperties as derived, i.e., set theoretically constructed entities. Quasianalysis in the sense of (3.1) can be considered as an *immanent* description in the following way:

(3.2) Lemma

Let f: S → 2^Q be a (weak or strong) quasianalysis. Denote the power set of S by Po(S). Define f*: S → $2^{Po(S)}$ by f*(s) := {q*| q ∈ f (s)}. Then f* is a (weak or strong) quasianalysis.

(3.2 gives rise to the equivalence relation of *extensional equivalence*: two quasianalysis f: S → 2^Q and f′:S → $2^{Q'}$ are *extensionally equivalent* iff f* = f′*. In the following I'll consider quasianalysis only "up to extensional equivalence". This means, I essentially work with "immanent" quasianalysis in Carnap's sense. In the *Aufbau* Carnap introduced the distinction between quasianalysis of the first and the second kind which can be rendered precise as follows:

(3.3) Definition

A (weak, strong) quasianalysis of the *first kind* is a (weak, strong) quasianalysis f: S → 2^Q for which f*(q) for each q ∈ Q satisfies the following two requirements:

(i) (x) (y) (x,y ∈ f * (q) ⇒ x ~ y)
(ii) (x) (x ∉ f * (q) ⇒ ∃ y (y ∈ f * (q) and x ≁ y)

If these conditions are not satisfied f is said to be of the *second kind*. A subset of S satisfying (i) and (ii) is a called a *similarity circle*. The set of similarity circles is denoted by SC(S).

Stated informally the condition of (3.3) requires that an element x which is similar to all y having the quasiproperty q also has the quasiproperty q, or, in Carnap's own terms it requires that the extension of each quasiproperty is a *similarity circle* (see Carnap (1928, § 70 f). As can easily be verified, Goodman's triangle is a weak quasianalysis of the second kind.

We may consider the set f(s) of quasiproperties of s as a *model* of s, i.e., a quasianalysis is a kind of *theoretical representation*: the elements s of the similarity structure are represented by their models f(s), and the similarity relation "~" is represented by the set theoretical relation of intersection. Of course, this representation is not arbitrary but has to satisfy certain conditions of adequacy, to wit, the conditions (3.1) (1) - (4).

Considering the quasianalysis of a similarity structure as a representation immediately leads us to ask the following questions: Does a *Representation Theorem* hold, i.e., given a similarity structure (S,~), is there a quasianalytical representation f: S → 2^Q? This question was already positively answered by Carnap in 1923.

More interesting is whether an *Uniqueness Theorem* holds: having established the existence of a quasianalytical representation for all similarity structures, the natural question arises whether it is "essentially" unique? Carnap knew very well that weak quasianalysis usually are *not* unique. All the authors who criticized the quasianalytical approach only treated weak quasianalysis, and usually they considered its non-uniqueness as a fatal blow. As far as I know nobody has ever treated the uniqueness question for quasianalysis which satisfy something like (C1) - (C4) except Brockhaus (1963). To show that in general a strong quasianalysis of the first kind is not unique it suffices to give a counter-example. The smallest I've been able to find is the following one:

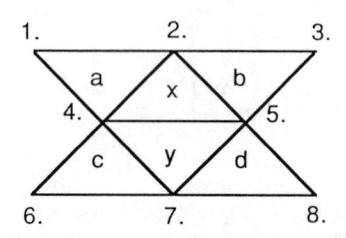

Figure 3.4

As one easily verifies this similarity structure has *two* essentially different quasianalytical representations satisfying (C1) - (C4): one has the quasiproperties a,b,c,d, and x, while the other has a,b,c,d, and y. A quasianalysis with fewer properties does not exist.

Thus, in general, even for strong quasianalysis of the first kind, not to mention weak quasianalysis, a *Uniqueness Theorem* does *not* hold. Does this show that the quasianalytical approach is doomed to fail, as many authors maintained? I don't think so. One way out is to switch to a quasianalysis of the second kind thereby eventually reaching uniqueness. This path is beset with certain difficulties which I cannot discuss in this paper. Another more promising rout is trying to find a special class C of similarity structures such that all members of C have a *unique* quasianalysis. In this paper I deal only with the second path starting with a theorem which we owe to Brockhaus (1963).

(3.5) Theorem

A similarity structure (S,~) has a unique quasianalysis f: S → 2^Q of the first kind iff (S,~) has the following property: there is a set SC(S,2) ⊆ SC(S) satisfying the following requirements:

(i) ∪ SC(S,2) = S

(ii) for T_i∈ SC(S,2) there are two (not necessarily different) elements x_i, y_i∈ T_i such that co(x_i) ∩ co(y_i) = T_i.

The proof is lengthy but elementary, i.e., it does not use any new concepts or methods that would not have been available to Carnap in 1923. Counter-examples show that the condition that f is of the first kind cannot be removed.

The content of this theorem can be formulated as the statement that a strong quasianalysis of the first kind is unique iff each of its quasiproperties is *generated extensionally* by at most two elements. To get a feeling for this condition let us make the following remarks:

(1) If (S,~) is a transitive similarity structure then, obviously its quasiproperties, i.e., its equivalence classes, all are generated by *one* x_q. However, the reverse is not true: there are a lot of similarity structures (S,~) which satisfy this condition but are not equivalence structures. The smallest example is given by the following similarity structure:

$$1._____2._____3.$$

(2) All examples of similarity structures encountered in the literature satisfy the condition of (3.5). Hence they have a unique strong quasianalysis of the first kind.

(3) In the counter-example (3.4) either the quasiproperty x or y belongs to each strong quasianalysis, and x and y have exactly *three* generators.

(3.5) gives the motivation to characterize similarity structures according to how many generators are needed for their strong quasianalysis:

(3.6) Definition.

A similarity structure (S,~) is *of the nth-order* iff there is a set $SC(S,n) \subseteq SC(S)$ satisfying the following requirements:

(i) $\cup \, SC(S,n) = S$

(ii) for $\in T_i \, SC(S,n)$ there are $x_{i1}, \ldots x_{in} \in T_i$ such that

$$co(x_{i1}) \cap \ldots \cap co(x_{in}) = T_i$$

After this preparatory definition we are able to succinctly express the main result of this section as follows:

(3.7.) Theorem.

A similarity structure (S,~) has a unique strong quasianalysis of the first kind if and only if it is of the first or second order.

Summarizing we may say that due to (3.7) similarity structures of the second kind indeed provide a "natural" realm where the quasianalytical approach works even if we rely on the most severe requirement of uniqueness. This realm is strictly larger than the class of transitive similarity structures which can be considered as the genuine field of the Russell-Whitehead method of equivalence classes. Thus, Carnapian quasianalysis actually is a working generalization of the latter.

4. Criticism of Criticisms of Quasianalysis

Almost all the authors who have dealt with the formal aspects of Carnap's quasianalysis have followed Goodman's criticism launched against this approach in Goodman (1954). The only exceptions known to me are Brockhaus (1963), Moulines (1991) and Proust (1984). In our representational reconstruction of quasianalysis the basic line of Goodman's criticism can be reconstructed as follows: given a similarity structure (S,~), one singles out a certain distribution of properties $f_G: S \to 2^P$ as the "real" one. The only conditions imposed on f_G are (C1) and (C2). f_G might be considered as God's distribution chosen by Him for some reason we mortals don't know.

Then, according to Goodman, the task of quasianalysis is to reconstruct f_G from relational information only, i.e., only from the information contained in the extensional list of the similarity relation "~". Of course, this is in general not possible because for weak quasianalysis a *Uniqueness Theorem* does not hold. The simplest and most famous example is Goodman's triangle mentioned above.

We would be better off if we'd required God's property distribution to satisfy not only (C1) and (C2) but to be a strong quasianalysis of the first kind. This plot would allow us to get rid of Goodman's triangle (and a lot of other "counter-examples") since it does not satisfy (C3) and is not of the first kind. However, Goodman and his followers might try harder confronting us with similarity structures of the third or higher order which definitively do not possess a unique strong quasianalysis (even of the first kind). In this case, it would be mere luck if our property distribution f coincided with f_G. A first objection to this strategy of disavowing the quasianalytical approach could claim that similarity structures of higher kind ($n \geq 3$) are too complicated so that they might be found in nature. This contention is supported by the fact that till now in the literature no similarity structure of the third or of higher order has been discussed.

But I think we can do better: let us grant that there might be "natural" similarity structures of higher order. Even then the thesis that Carnap's quasianalytical approach is doomed to fail is drawn much too hastily. It is only justified as long as we accept that the main goal of quasianalysis is to reconstruct a pre-given property representation f_G. When we challenge this premiss the perspective of the quasianalytical approach doesn't look that bleak anymore. This objection has been put forward by Proust (cf. Proust (1984). According to her, Goodman realistically misunderstands the very intentions of the quasianalytical approach: "Goodman's objections ... reestablish in spite of him the fiction of an omniscient God capable of controlling through originary intuition, that is, without construction, what the constitution derives from its extensional data." (Proust (1984, 299) In a less picturesque language this just amounts to challenge the legitimacy of a "real" property distribution $f_G: S \to 2^P$ as the one and only guiding star. Instead one should take seriously the quasianalytical perspective: if we have no other means of constituting properties than through the quasianalysis of our elementary experiences, it might very well happen that these experimental data are *not* sufficient to single out a uniquely determined "objective" property distribution. This amounts to admitting the possibility of the empirical underdetermination of a quasianalytical representation of a similarity structure as a *theory* of that structure, or so I want to argue in the next section.

5. Application: The Thesis of Underdetermination in the Framework of Quasianalysis

The thesis of the empirical underdetermination of theories maintains that there are incompatible theories which are empirically equivalent. Usually the underdetermination thesis has been studied in the standard approach which considers theories as sets of sentences. Without arguing for it, I propose a structural approach which considers quasianalysis of similarity structures as prototypes of empirical theories. This claim is in line with Carnap's contention, put forward in the *Aufbau*, that we might conceive the world as a huge similarity structure (cf. *Aufbau* § 27). More precisely, this can be spelt out as follows: the theory's domain of data is a similarity structure $(S,\sim)$. A map $f: S \to 2^Q$ (not necessarily a quasianalysis) is a kind of *theory* of this structure in the following sense: The "theory-map" f represents the data s by conceptual models f(s) which are bundles of quasiproperties, relations of data are represented by relations of their models. For example, suppose that for s, s'$\in$ S we have $f(s) \cap f(s') \neq \emptyset$. This is to say that the theory f claims the following *observation categorical* to be true: "Whenever x = s and y = s' then x and y are similar to each other". Whether all universal sentences of this type are true depends on whether f is a structure preserving map, i.e., whether it satisfies $(s \sim s' \Leftrightarrow f(s) \cap f(s') \neq \emptyset)$, i.e., (C1) and (C2). Considering a theory f as adequate iff all the observation categoricals implied by it are true, we get that a quasianalysis of the world's similarity structure can be considered as an empirically adequate theory of that world. Now the underdetermination thesis claims that a theory "is bound to have empirically equivalent alternatives which, if we were to discover them, we would see no way of reconciling by reconstrual of predicates". (Quine 1975. 327). Empirically equivalent alternatives should be "equally good", i.e., their theoretical virtues such as simplicity, economy etc. should be roughly the same (cf. Bergström (1993, 335). In the quasianalytical approach this is captured by the requirements (C3) and (C4). Quine's "reconstrual of predicates" can be reconstructed as follows: Assume we have two quasianalysis f: $S \to 2^Q$ and f': $S \to 2^{Q'}$ of the same similarity structure $(S,\sim)$. Then the quasianalytical counterpart of a reconstrual of predicates is an appropriate map g: 2^Q $2^{Q'}$ which makes the following diagram commutative:

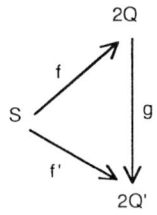

$$\begin{array}{ccc} & & 2^Q \\ & \nearrow^f & \\ S & & \downarrow g \\ & \searrow_{f'} & \\ & & 2^{Q'} \end{array}$$

Figure 5.1

Looking at the example (3.4) it is easy to see that there is no map from the set Q = (a,b,c,d,x) to the set Q' = {a,b,c,d,y} of rival quasiproperties which renders (5.1) commutative. This means the quasianalytical systems based on Q and Q', respectively, are incompatible. Now, depending on the contingent structure of the world we are ready to prove or to disprove the underdetermination thesis:

If the world $(S,\sim)$ happens to be a similarity structure of the first or second order the underdetermination thesis is wrong: according to (3.5) there is one and only one empirically adequate quasianalytical theory of the world.

On the other hand, if the world happens to be a similarity structure of higher order (at least of order three) the underdetermination thesis is true. At least this is the case as long as we don't find other theoretical virtues which allow us to establish a ranking between different strong quasianalysis of the first kind.

References

Bergström, L. (1993), "Quine, Underdetermination, and Skepticism", *The Journal of Philosophy* 90: 331-358.

Brockhaus, K. (1963), *Untersuchungen zu Carnaps Logischem Aufbau der Welt*. Unpublished PhD dissertation, University of Muenster, Germany.

Carnap, R. (1923), *Die Quasizerlegung, Ein Verfahren zur Ordnung nichthomogener Mengen mit den Mitteln der Beziehungslehre*. Unpublished Manuscript RC-081-04-01, University of Pittsburgh.

_____., (1928), *Der Logische Aufbau der Welt. (Aufbau)*. Hamburg: Meiner.

Goodman, N. (1954), *The Structure of Appearance*. Indianapolis: Bobbs-Merrill.

Moulines, C.U. (1991), "Making Sense of Carnap's *Aufbau*" *Erkenntnis* 35: 263-286.

Mundy, B. (1986), "On the General Theory of Meaningful Representation", *Synthese* 67: 391-437.

Proust, J. (1984), "Quasi-analyse et reconstruction logique du monde", *Fundamenta Scientiae* 3: 285-303.

Quine, W.V.O. (1975), "On Empirically Equivalent Systems of the World", *Erkenntnis* 9: 313-328.

Part III

SPACETIME AND RELATED MATTERS

Locality/Separability:
Is This Necessarily a Useful distinction?[1]

James T. Cushing

University of Notre Dame

1. Introduction

In the philosophy of science, we are to assess critically and on their intrinsic merits various proposals for a consistent interpretation of quantum mechanics, including resolutions of the measurement problem and accounts of the long-range Bell correlations. In this paper I suggest that the terms of debate may have been so severely and unduly constrained by the reigning orthodoxy that we labor unproductively with an unhelpful vocabulary and set of definitions and distinctions. I first review this situation and how we arrived there. Then I present an alternative conceptual framework, free of many of the standard conundrums, and ask why we seem unwilling to pursue it as a serious possibility.

2. The Standard Conceptual Background

I begin by sampling how some of the central tenets of the standard, or "Copenhagen", theory of quantum mechanics came to be formulated. As is typical for his writing on broader philosophical issues, Niels Bohr's pronouncements on the interpretation of quantum mechanics are often difficult to understand and at times just plain opaque. In a 1925 article, written after Heisenberg's matrix-mechanics formulation had been published, Bohr gave his own view on the new mechanics. There (Bohr 1934, 48-51) he tells us that it was formulated in terms of directly observable quantities only, required a renunciation of mechanical models in space and time and indicated an inherent limitation on our means of visualization. The difficulties to which Bohr referred here were the failures to give a classical-mechanical account of the interactions between atoms and light. As a consequence, Bohr suggested that these fundamental revisions in our way of representing the world *may* be necessary, but he did not yet foreclose alternatives. In the version of his Como lecture published in *Nature* in 1927, Bohr took what can easily be read as a positivist (perhaps an operationalist) stance when he wrote that the quantum postulate "implies a renunciation as regards the causal space-time co-ordination of atomic processes" (Bohr 1934, 53). For him causality meant the conservation laws and one cannot simultaneously determine experimentally the energy-momentum and the space-time coordinates of an object, so that the simultaneous reality of these properties can be called into question. A spacetime representation and causality, which were characteristic of classical theories, became for him complementary but exclusive means of description (Bohr 1934, 54-56). By this time (1927) Bohr cited the Heisenberg uncertainty relations as further evidence of the *impossibility* of simultaneously having a causal (*i.e.*, conservation laws or sharp values of energy and momentum) and a space-time (*i.e.*, sharp locations in space-time) description of atomic processes. These are his own (perhaps somewhat odd) definitions of 'causal' and 'space-time'. One could certainly put another gloss on those terms and still hope for a causal description in a space-time background. In a 1929 lecture recounting the development of quantum mechanics, Bohr (1934, 108) told his audience that "only by a conscious resignation of our usual demands for visualization and causality" is it possible to advance. Not only was this what *did* happen, but it had now become the only *possible* way to progress.

In his *Physics and Philosophy* (1958, 44), Werner Heisenberg laid out quite specifically the finality (or completeness) of *the* interpretation of quantum mechanics. He claimed that the concepts and language of classical physics, while limited by the uncertainty relations, were essential to describe experimental results and could not be improved upon. This was a prohibition against any change in the structure of the interpretation. How did he argue for it? He claimed that probabilities play an ineliminable role due to the uncertainty relations, but that these probabilities were of a fundamentally different type from those encountered in classical physics. Quantum-mechanical probabilities represented not the course of events in time, but rather the *tendency* for events. A connection with reality could be made only when "a new measurement is made to determine a certain property of the system" (Heisenberg 1958, 46). Here Heisenberg introduced the notion of *potentia* for one actual event or another, but there were to be *no* actual events until measurement. Using the familiar *gedanken* gamma-ray microscope, he then described the result produced in attempting to observe an electron in its atomic orbit and, of course, the electron was disturbed in the process. He concluded that "there is no orbit in the ordinary sense" and that use of such a concept "in quantum theory · · · would be a misuse of the language which · · · cannot be justified" (Heisenberg 1958, 48). As he pointed out immediately thereafter, the central question, which he left open for the moment, was whether this prohibition against orbits was about epistemology or ontology (to use his own terms). In explaining the time evolution of the probability function, Heisenberg (1958, 49) used the change of this function upon observation (*i.e.*, the collapse of the wave packet) to conclude *quite generally* that "the space-time description of the atomic events is complementary to their deterministic description." Such a claim has force only if one already grants that a probabilistic description is, in principle, the best one can have. In attempting to block any possible *description* of a system between measurements or observations – "why one would get into hopeless difficulties if one tries to describe what happens between two consecutive observations" – Heisenberg (1958, 50-51) rehearsed the double-slit experiment and showed that one must necessarily be frustrated in any attempt to determine through which slit a single light quantum went. From this he concluded (1958, 52 and 54) that the concept of the quantum-mechanical "probability function does not allow a description of what happens between two observations" and that "therefore, the transition from the 'possible' to the 'actual' takes place during the act of observation." Nowhere is it *proven* that alternatives are not possible (*i.e.*, that the indeterminism of quantum mechanics is actually ontological, rather than simply epistemological). A *consistent* interpretation has been presented, but not necessarily the *only possible* one compatible with observations. The extent to which these arguments could constitute a "proof" would depend upon the acceptance of a positivistic view. That Heisenberg did accept just such a view he made explicit when he discussed (1958, 129-130) the "impossibility" of interpretations truly alternative to the Copenhagen one.

Finally, I turn to Max Born's characterization of the Copenhagen interpretation. In his introductory popular book *The Restless Universe* (the first edition of which appeared in 1936), Born summarized the lessons of wave mechanics in terms of his own statistical interpretation. There he stated (1951, 155) that "in the quantum theory it is the *principle of causality*, or more accurately that of *determinism*, which must be dropped and replaced by something else" (emphases in original). The only way he saw (1951, 157) out of "the dilemma [the dual entity of wave and particle] · · · [is] the *statistical interpretation* of wave mechanics [according to which] *the waves are waves of probability*" (emphases in original). For Born, these waves, apart from their objective reality, had to have something to do with the subjective act of observation. With this new form of causality, only the *probability* of subsequent events was to be governed by exact laws (Born 1951, 163-164). Born took the only physically meaningful questions to be those that could be answered operationally. Hence, for him, the trajectory of an atomic particle during interaction or scattering was not a proper topic of discussion for quantum mechanics. In the 1936 edition of his *Atomic Physics*, Born explicitly said (1936, 84-85) that we are not justified in speaking of the existence of a "particle" unless we can determine both its position and its momentum simultaneously and exactly. Here he laid down a criterion for what is and is not justified, but there was no independent argument that one *must* (logically, or on pain of contradiction) accept this stricture. Bohr's principle of complementarity was of central importance for Born.

This brief overview certainly is not an exhaustive catalogue of various Copenhagen interpretations. However, one can see here a certain set of common commitments: complementarity, completeness of the description (in terms of the state vector or probability amplitude), a prohibition against any possible alternative causal description in a space-time background and a positivistic attitude. Henceforth, I use the expression 'Copenhagen interpretation' in this sense.

Applications of the formalism of quantum mechanics to (idealized) position-momentum measurements, double-slit arrangements and the like have often been taken to lead to a picture, or interpretation, in which definite space-time trajectories cannot be maintained, specific possessed values of observables (such as position) are not possible at *all* times, event-by-event causality must be abandoned, the process of measurement assumes a central and highly problematic role in nature (*i.e.*, the projection postulate or collapse of the wave function) and the passage to a classical limit (in terms of an underlying physical ontology) defies any coherent description. An examination of the formalism in specific Einstein-Podolsky-Rosen-Bohm (EPRB) type experiments shows the nonseparable nature of the theory and this gives rise to correlations that may imply the existence of nonlocal influences between spatially separated regions (really, at space-like separations). On the Copenhagen interpretation of quantum mechanics, physical processes are arguably, at the most fundamental level, both inherently indeterministic and nonlocal. On this reading, the ontology of classical physics is taken to be dead. This becomes especially clear in the measurement process as described by standard quantum theory.

3. An Alternative Framework[2]

However, I ask that we now consider a quantum theory that is conceptually radically different from, yet that has all of the observational consequences of, the standard formulation outlined in the previous section. David Bohm (1952) began with the nonrelativistic Schrödinger equation and, by means of a mathematical transformation, rewrote the basic dynamics of quantum mechanics in a "Newtonian" form. More specifically, he first expressed the wave function $\psi(\mathbf{x}, t)$ in the polar form $\psi = R \exp(iS/\hbar)$ where

R(x, t) and S(x, t) are *real* functions. (This can *always* be done for any complex function ψ.) In terms of the phase S(x, t) of ψ, a velocity field v(x, t) is defined by the "guidance" condition $v = (1/m)\nabla S$. The dynamical equation for the motion of this particle of mass m (and momentum $p = mv$) then becomes

$$dp/dt = -\nabla(V + U) \tag{1}$$

Here V is the usual classical potential energy and U is the so-called "quantum potential" that is given in terms of the wave function ψ as $U = -(\hbar^2/2m)(\nabla^2 R/R)$. Classical mechanics can be seen to emerge when $U \ll V$. Furthermore, there is no measurement problem since one essentially "discovers" where a particle is upon observation and no "collapse" of the wave function ever occurs.[3] Position is, ultimately, the *only* observable in Bohm's theory.[4]

Notice that ψ plays two conceptually very different roles here: first, as determining the influence of the environment on the microsystem and, second, as determining the probability density P. It is not a *logical* or an *a priori* necessity that the same function need play both of these roles. In fact, the *primary* conceptual role for ψ in Bohm's theory is the first of these. That is, the Schrödinger equation and the guidance condition *alone* would constitute a perfectly complete and coherent system of mechanics and, in a sense, *are* the essence of Bohm's theory (Dürr, Goldstein and Zanghi 1992). This would be a thoroughly deterministic scheme and no concept of probability would be needed, although there is certainly room for such a concept in this framework. Bohm's quantum-potential formulation and standard quantum mechanics are empirically equivalent only as long as

$$P = |\psi|^2 \tag{2}$$

In 1953 Bohm examined a few specific cases to show that, even if initially $P \neq |\psi|^2$, then random interactions with other systems could drive P to its quantum equilibrium value, $P \to |\psi|^2$. Once Eq.(2) is satisfied, then the continuity equation for P and Schrödinger's equation for ψ guarantee that Eq.(2) will continue to be satisfied. So, it is reasonable to ask under what conditions a system can be driven to this equilibrium distribution. It is important to appreciate that, in standard quantum mechanics, P is *by definition* equal to $|\psi|^2$ since the very meaning of ψ is a *probability* amplitude. This reflects the choice made (*cf.* Section 2 above) of taking probability to be an irreducible and ineliminable central tenet of standard quantum mechanics.

Recently, Antony Valentini (1991a, 1991b) has given an insightful discussion of the relations among the quantum equilibrium distribution for P, the uncertainty relations and the possibility of superluminal signaling, all within the framework of Bohm's theory. He shows that random subquantum interactions can drive the system to quantum equilibrium. This is conceptually very similar to the way we typically envision the molecules in a large sample of gas being driven toward the (Maxwell-Boltzmann) equilibrium distribution through random interactions among the gas molecules themselves. Even if all of the gas molecules in a room were initially squeezed into a small volume in one corner of the room, they would quickly (once released) diffuse throughout the room and reach an equilibrium distribution.

In the present case, the equilibrium distribution is characterized by Eq.(2) for individual particles. Valentini shows (1991b) that the impossibility of instantaneous signaling ("signal locality") and the (Heisenberg) uncertainty principle are related and both turn on the equilibrium-distribution condition of Eq.(2). On this view, the world would be fundamentally nonlocal in its structure, yet possess signal locality as a *contingent* fact once equilibrium has been reached. Valentini's arguments build on Bohm's (1952, 1953) earlier ones and on Bohm's original insight that the uncertainty relations obtain *only* when Eq.(2) is satisfied. He also proves that once the wave function for the "universe" (which contains N particles) satisfies the condition of quantum equilibrium of Eq.(2), then the wave function for any individual subsystem drawn from the larger system will also satisfy this equilibrium condition *for measured values*. (In general a subsystem does not have a wave function, but it does under the conditions that must obtain for a measurement or an observation.) Once the universe is itself in quantum equilibrium, then, for the ensemble of subsystems that eventually have a wave function ψ, the probability P of outcomes of this ensemble of observations must be given by Eq.(2).

Valentini (1991b) also demonstrates with specific calculations that, if we have two interacting systems, whose overall state is an *entangled* one, then a change on one subsystem (accomplished, say, by modifying the Hamiltonian for that subsystem) can produce an instantaneous change in the probability distribution P for the other system *if and only if* Eq.(2) is *not* satisfied for the interacting system. He establishes a no-signaling theorem when Eq.(2) obtains, but illustrates how to signal *instantaneously* for an entangled state when Eq.(2) is violated. He is also able to relate Eq.(2) to the Heisenberg uncertainty principle. Bohm (1952) previously showed that Eq.(2) leads to that principle, while Valentini (1991b) exhibits a counterexample when Eq.(2) fails. Here we have a realistic ontology of actually-existing particles and events in which some remarkable features of our world, such as the uncertainty relations and no instantaneous signaling, are given a unified and essentially understandable explanation.

4. Casting Off the "Copenhagen" Legacy

The majority view in the foundations community today is that a key element in folding together the principles of special relativity and of standard quantum mechanics is a careful distinction between locality and separability. I want to suggest, on the contrary, that there may be no meaningful physical difference in our actual world between the terms 'nonlocal' and 'nonseparable'. *Logical* distinctions do exist between them and I begin by laying out some basic terminology common in the literature on Bell's theorem and its implications. The context for these comments is a standard EPRB type correlation experiment. At spatially separated stations, experimenters can make choices of instrument settings (labeled by i and j, for each station respectively) and the possible outcomes of these measurements are denoted by x and y. In this notation, $p(x, y \mid i, j)$ represents the joint probability of obtaining a result (or outcome) x at one station when the setting (or parameter) i has been chosen by the experimenter there, while the result y obtains at the other station when setting j has been chosen there.

In John Bell's celebrated theorem, a set of parameters λ (the "hidden variables") is assumed to give a complete state specification for the system. At the fine-grained level, the appropriate joint probabilities are denoted as $p_\lambda(x, y \mid i, j)$. The λ are distributed according to a normalized function $\rho(\lambda)$ so that the observed joint probabilities $p(x, y \mid i, j)$ are given as $p = \int \rho(\lambda) p_\lambda d\lambda$. Bell assumed that there could be no instantaneous influences (of *any* kind) from one region on another that is spatially separated from it. On the basis of such locality, Bell assumed that the joint p_λ could be factored as[5]

$$p_\lambda(x, y \mid i, j) = p_\lambda(x \mid i) \, p_\lambda(y \mid j) \qquad (3)$$

where the $p_\lambda(x \mid i)$ and $p_\lambda(y \mid j)$ are the probabilities for outcomes at one station irrespective of what happens or is done at the other station. Equation (3) is sometimes termed *factorizability*. A Bell inequality then follows directly from this. To give away the end of my story early on, let me say that when I use the term 'nonlocality' I am referring to a violation of Bell's version of locality.

Having defined my 'nonlocality', I should now explain why I do not make use of several distinctions that are possible. Jon Jarrett (1984) has shown that factorizability is implied by the conjunction of two other properties: Jarrett locality and Jarrett completeness. *Jarrett locality*, also termed 'parameter independence' by Abner Shimony (1986) and simply 'locality' by Don Howard (1985), means that the single distribution (or marginal) of outcome x at one station is independent of the choice j made at the other station

$$p_\lambda(x \mid i, j) = p_\lambda(x \mid i) \qquad (4)$$

A similar condition holds for $p_\lambda(y \mid i, j)$. If this is violated (and it is *not* in quantum mechanics), then there *could* be a conflict with special relativity, in the following sense. Jarrett argued that, *provided* we could control the parameters λ, then superluminal communication would be possible with the marginals (hence, Shimony's *controllable* nonlocality). Quite simply, if Eq.(4) were violated, the statistics gathered at one station for outcome x would depend not only upon the choice i made there, but also upon the choice j made at the distant station. By prior arrangement (of what choice one experimenter would make if a given occurrence did or did not take place at his station), the two spatially separated experimenters could communicate via this correlation and do so instantaneously. The caveat about being able to control the parameters λ is crucial, as Jarrett himself (1984, 573 and 587) admitted.

However, for Bohm the "hidden" variables are the initial positions $\{x_0\}$ of the particles and they are *not* under our control. Let me recall that, *in Bohm's theory*, "instantaneous signaling [via the quantum-mechanical probabilities] is possible *if and only if* $P_0 \neq |\psi_0|^2$" (Valentini 1991b, 4). Such signaling does require an *entangled* state. Since for Bohm *all* measurements are ultimately position measurements, the only probabilities relevant to this discussion are the $P(x_1, x_2, \cdots, x_n)$ (*i.e.*, probabilities of finding various particles at certain *positions*). Our inability to "beat" the Heisenberg uncertainty relations prevents us from controlling the $\{x_0\}$ well enough to signal. I emphasize again "that the uncertainty principle holds if $P = |\psi|^2$, but is generally violated otherwise" (Valentini 1991b, 5). Quantum equilibrium (*i.e.*, $P = |\psi|^2$) is the key to placing (Bohmian) nonlocality beyond our control. While there are such superluminal influences, these cannot, in Bohm's theory, be used to send a signal (*i.e.*, to communicate). Does this mean the special theory of relativity is violated or not? Well, if the first *signal* principle means no superluminal *communication* (or signaling), then no, but if it means no superluminal *influences*, then yes. I return to this question below. Bohm's theory is nonlocal in the sense of violating Eq.(4) at the level of the hidden variables, but that need not generate any *empirical* conflict with special relativity.[6]

Jarrett completeness (Shimony's outcome independence and Howard's separability) means that the conditional probability for outcome x at one station is independent of the measurement outcome y at the other station

$$p_\lambda(x \mid i, j, y) = p_\lambda(x \mid i, j) \qquad (5)$$

That is, this conditional probability becomes a marginal. However, a violation of Eq.(5) (which *is* the case for quantum mechanics) is usually taken not to allow superluminal signaling.[7] Shimony has labeled a violation of Eq.(5) *uncontrollable nonlocality*. It should be clear that Bohm's theory does satisfy Eq.(5) at the level of the hidden variables (*i.e.*, it is a completely deterministic theory). If specification of (λ, i, j) completely determines (in principle) the outcomes (x, y), then additional conditioning on y is superfluous and cannot affect the probability (which can be only 0 or 1 in a deterministic theory). The point of Jarrett's analysis is that if Jarrett locality [Eq.(4)] and Jarrett completeness [Eq.(5)] hold, then Eq.(3) follows at once. So, (Howard's) locality and separability together imply factorizability.

In Bohm's nonlocal, deterministic theory, Eq.(4) is violated and Eq.(5) respected *at the level of the hidden variables*. However, these conditions need not hold for the actually accessible (or integrated) probabilities. For Bohm, the "experimental" probabilities are just those of standard quantum mechanics. A theory *local* on the fine-grained (*i.e.*, λ) level *must* be *local* at the observable (*i.e.*, integrated) level because the relation between the joints and the marginals is a linear one and this linearity is preserved under the integration over λ. However, a theory *separable* at the λ-level *need* not be *separable* at the observable level just because the conditional p(x | i, j, y) is *not* simply the integrated p_λ(x | i, j, y). That is, the conditional is not linearly related to the joint and the marginal but rather as

$$p(x | i, j, y) \equiv p(x, y | i, j)/p(y | i, j) = \int p_\lambda(x, y | i, j) \rho(\lambda) d\lambda / \int p_\lambda(y | i, j) \rho(\lambda) d\lambda$$
$$\neq \int [p_\lambda(x, y | i, j)/p_\lambda(y | i, j)] \rho(\lambda) d\lambda = \int p_\lambda(x | i, j, y) \rho(\lambda) d\lambda \quad (6)$$

So, it is not clear that distinct locality and separability concepts are particularly useful *in this case*.[8] If one were to *insist* upon such terminology here, then the p_λs of Bohm's theory would be nonlocal and separable, while the (integrated) ps would become local and nonseparable.[9] This simply reflects the fact that separability at the fine-grained or λ level does *not* (necessarily) imply separability at the observable level, even though locality at the λ level *does* imply locality at the observable level. Similarly, nonlocality (in Jarrett's sense) at the fine-grained level *need not* imply nonlocality at the observable level. From the perspective of Bohmian mechanics, being in quantum equilibrium (P = |ψ|²), rather than a locality/separability distinction, is the relevant condition for no signaling. While there can be nonlocality for a *single* particle (via the quantum potential), there is further entanglement when the state vector is not separable. What should this be termed: nonlocality or nonseparability? Why not just nonlocality? That is basically what I suggest.

On this question of the proper concept of and terminology for nonlocality, it is useful to return to Bell's 1964 paper. There he worked with expectation values, not with probabilities. The basic quantities in his proof were the actual outcomes A_λ(i, j) and B_λ(i, j). Notice that, as one might expect in a *deterministic* theory, A_λ does not depend upon B_λ or vice versa. On the basis of (Bell) locality, he then required that A_λ(i, j) = A_λ(i) and B_λ(i, j) = B_λ(j). This then led to his famous inequality (Bell 1964). What would separability even *mean* for the A_λ(i, j) and B_λ(i, j)? Here we have a locality condition in the sense of Eq.(4). But separability, according to Eq.(5), is to have something to do with independence of the *outcome* at the *other* station. However, the A_λs do not depend upon the B_λs (and vice versa). In terms of the physical quantities A_λ and B_λ, locality seems to be *the* relevant concept.[10] In the case of Bohm's *deterministic* theory, it is not helpful to frame the question of (Bell) locality in terms of a conjunction of Jarrett's or Howard's locality and separability. Hence, I use simply the term 'locality' (*i.e.*, loosely speaking, Bell locality) and 'nonlocality' (perhaps "Bell nonlocality?").

Once what I claim to be a fairly benign type of nonlocality has been accepted in Bohm's theory, several long-standing difficulties, such as the measurement problem and the existence of a classical limit, simply evaporate. Many of the more nearly standard attempts at resolving the measurement problem (*e.g.*, the modal interpretations, spontaneous reduction theories and the decoherence program) go through contorted manipulations and arguments to obtain (at least effective) "collapse" and yet, for EPR type correlations, they must still face up to (some type of) nonlocality. One can reasonably ask what the motivation is for these projects when nonlocality *alone* is sufficient to resolve so many of the standard conundrums of quantum mechanics.

Notes

[1] Partial support for this work was provided by the National Science Foundation under Grant Nos. DIR89-08497 and SBE91 21476.

[2] An extensive discussion of Bohm's program, its philosophical implications and the status of nonlocality can be found in Cushing (1994).

[3] See Bohm (1952, 176-178) and Cushing (1994, Chapter 4) for details on measurement in Bohm's theory.

[4] *All* other observables (such as spin, etc.) become constructs that we employ to "tell a story" about observed events.

[5] Bell's actual argument was framed in terms of observables and their expectation values, rather than in terms of probabilities. I return to this difference shortly. However, general philosophical discussions typically work with the probabilities and I follow this line of presentation in order to get directly to the distinctions I am interested in.

[6] Bohm's program also has an empirically adequate extension to the relativistic and quantum-field-theory domains, as discussed in Cushing (1994, Chapter 10).

[7] Jarrett's (1984, 580-581) original argument against the possibility of signaling with a violation of Eq.(5) amounts essentially to the observation that a deterministic hidden-variables theory would necessarily be complete so that a violation of completeness would require a nondeterministic theory. However, so the argument goes, in such a theory the outcomes are not under the control of the experimenter and therefore cannot be used to send a signal. In fact, Shimony (1986, 192) discusses a possibility of signaling via a violation of Eq.(5), but not superluminally. Jones and Clifton (1993) go further and argue that a violation of Eq.(5) could be used for *superluminal* signaling, but in their model a violation of Eq.(5) actually implies a violation of Eq.(4). In that case, one might view this as just another case of the possibility of signaling with nonlocality.

[8] I do not say this to diminish in any way the importance of these distinctions in *general* philosophical analyses of hidden-variables issues. I do feel, however, that it remains an open question whether or not separability is a *physically* meaningful concept (apart from locality).

[9] One can actually see how this comes about in the EPRB thought experiment by explicitly calculating the relevant correlations in terms of the initial positions of the "electrons" and then integrating over these coordinates to obtain the quantum-mechanical correlations. Some of the necessary details can be found in Dewdney, Holland and Kyprianidis (1987). In Bohm's theory, of course, $\rho = \rho(\lambda; i, j)$.

[10]That Bell chose to work directly with the As and Bs may be related to the fact that he had been familiar with Bohm's theory and much impressed with it as a counterexample to Copenhagen dogma (Bell 1964, 1966). Since Bohm's theory is deterministic, an explicit separability condition in terms of the A_λs and B_λs would seem superfluous. Of course, one could say that Bell had already assumed separability in writing $A_\lambda(i, j)$ rather than $A_\lambda(i, j, B_\lambda)$. I claim that, for Bell, 'locality' included both (Jarrett or Howard) locality *and* separability. Hence, I suggest that we speak simply of (Bell) nonlocality. This is discussed at length in Cushing (1994, Chapter 4).

References

Bell, J.S. (1964), "On the Einstein Podolsky Rosen Paradox", *Physics* 1: 195-200.

– – – – –.(1966), "On the Problem of Hidden Variables in Quantum Mechanics", *Reviews of Modern Physics* 38: 447-452.

Bohm, D. (1952), "A Suggested Interpretation of the Quantum Theory in Terms of 'Hidden' Variables, I and II," *Physical Review* 85: 166-179, 180-193.

– – – – –.(1953), "Proof That Probability Density Approaches $|\psi|^2$ in Causal Interpretation of the Quantum Theory", *Physical Review* 89: 458-466.

Bohr, N. (1934), *Atomic Theory and the Description of Nature*. Cambridge: Cambridge University Press.

Born, M. (1936), *Atomic Physics*. New York: G. E. Stechert & Co.

– – – – –.(1951), *The Restless Universe*. New York: Dover Publications.

Cushing, J.T. (1994), *Quantum Mechanics: Historical Contingency and the Copenhagen Hegemony*. Chicago: The University of Chicago Press.

Dewdney, C., Holland, P.R. and Kyprianidis, A. (1987), "A Causal Account of Non-Local Einstein-Podolsky-Rosen Correlations," *Journal of Physics* 20: 4717-4732.

Dürr, D., Goldstein, S. and Zanghi, N. (1992), "Quantum Mechanics, Randomness, and Deterministic Reality", *Physics Letters A* 172: 6-12.

Heisenberg, W. (1958), *Physics and Philosophy*. New York: Harper & Row.

Howard, D. (1985), "Einstein on Locality and Separability", *Studies in History and Philosophy of Science* 16: 171-201.

Jarrett, J.P. (1984), "On the Physical Significance of the Locality Conditions in the Bell Arguments", *Nous* 18: 569-589.

Jones, M.R. and Clifton, R.K. (1993), "Against Experimental Metaphysics", in *Midwest Studies in Philosophy* 18. Notre Dame, IN: University of Notre Dame Press, pp. 295-316.

Penrose, R. and Isham, C.J. (1986), *Quantum Concepts in Space and Time*. Oxford: Oxford University Press.

Shimony, A. (1986), "Events and Processes in the Quantum World", in Penrose and Isham, pp. 182-203.

Valentini, A. (1991a), "Signal-Locality, Uncertainty, and the Subquantum H-Theorem. I", *Physics Letters A* 156: 5-11.

--------.(1991b), "Signal-Locality, Uncertainty, and the Subquantum H-Theorem. II", *Physics Letters A* 158: 1-8.

Spacetime and Holes[1]

Carolyn Brighouse

Occidental College and University of Southern California

1. Introduction

Here I describe and defend a version of space-time substantivalism. My account is meant partly as a reply to an argument against a particular substantivalist position that can be found both in John Earman and John Norton's "What price spacetime substantivalism? The hole story", and Earman's *World Enough and Spacetime*. Their argument, the Hole Argument, purports to show that substantivalism leads to a radical form of indeterminism within a class of theories that includes our best theory of spacetime, namely, General Relativity (GTR). The hole argument contends that the substantivalist must view diffeomorphic models of a spacetime theory as representing genuinely distinct physical situations. It is this contention I argue against.

The version of substantivalism I present, although not a traditional account, incorporates the important features of traditional substantivalist views. The hole argument is directed against a form of substantivalism that Earman and Norton call "manifold substantivalism"; the manifold substantivalist takes the manifold in the models of a spacetime theory as the basic object of predication. They would contend that this at least is a necessary commitment of any traditional substantivalist view. The view I present is manifold substantivalist in just this sense. I take substantivalism to be realism about spacetime points or regions. Spacetime points or regions exist, not as logical constructions out of matter, but in their own right. Substantivalism is motivated by the fact that modern spacetime theories take the notion of a field seriously, where this is most plausibly characterized as an assignment of properties to each point of spacetime (Field 1980, 35 and Field 1989, 181). My claim is that there is a manifold substantivalist position that is consistent with a deterministic reading of GTR.

A word about the situation confronting the substantivalist in the face of the hole argument is necessary to set the context of this discussion; I will assume familiarity with the details of that argument. The central claim is that the substantivalist has to view diffeomorphically related models as representing distinct situations. I shall call such models Leibniz equivalent. This leads to a threat to determinism if the diffeomorphism relating such models is the identity on any given region. Earman and Norton show how to construct such "hole diffeomorphs" in the context of GTR.

There are at least two ways for the substantivalist to reply to the argument. The argument implicitly uses a particular formulation of determinism; one could argue that this formulation is not the most natural one, suggesting an alternative formulation that is not violated by the "hole diffeomorphs" in question. The second response would be to deny that even substantivalists need hold that Leibniz equivalent models represent distinct situations. This could be done in one of two ways; the first would be to maintain that Leibniz equivalent models represent the same situation; the second would be to claim that, of a given class of Leibniz equivalent models, at most one represents a possible physical situation. This second attitude is motivated by the generally accepted view that a substantivalist cannot accept that Leibniz equivalent models represent the same situation. I will examine these two attitudes in more detail in section 3 below. One might think these responses independent; this is a mistake. If one denies that diffeomorphic models represent different situations, one will hold that the formulation of determinism presupposed in the hole argument is not the most natural formulation.

2. Determinism

If a response that claims the formulation of determinism in the hole argument is not the most natural is to be adequate, any alternative formulation will at least have to capture the intuitive idea of determinism that Norton and Earman consider important for local spacetime theories. Clearly some brand of laplacian determinism is what they have in mind, whereby agreement (of some kind) on a given region is sufficient to ensure agreement (of some kind) throughout spacetime. Consider the following definition of determinism:

D A spacetime theory, T, is deterministic iff for any given models $< M, O_i >$ and $<M, O'_i>$ of T, if those models are Physically Equivalent before time t (or, at t) then they are Physically Equivalent for all times.[2]

Everyone can agree that **D** adequately expresses the intuitive idea of determinism. The issue as to whether a theory satisfies **D** will depend on which relation between models defines when two models are physically equivalent.

There is room for dispute here. We are defining determinism in terms of models, but want this to reflect agreement or difference in the physical situations these models represent. We want a "coarse grained" enough definition of physical equivalence of models so that we are not committed to saying that the physical situations these models represent are distinct, when they differ in some representationally insignificant, but perhaps mathematically significant, way; and a "fine grained" enough definition so that different physically significant features that these models represent are not glossed over.

The relationist will hold that Leibniz equivalence is sufficient for physical equivalence: Two models will be physically equivalent if there is a diffeomorphism d:M -> M such that for all O_i in one model, and O'_i in the other $d*O_i = O'_i$. Norton and Earman would like to claim that a substantivalist is committed to there being a more fine grained relation between models than Leibniz equivalence that defines physical equivalence. I take their claim to be that a manifold substantivalist is committed to models being physically equivalent just in case the diffeomorphism, d, is the identity map over the entire manifold. Thus hole diffeomorphs meet the substantivalist requirement for physical equivalence outside the hole; within the hole, however, physical equivalence fails. Hence the failure of determinism.

So the issue of proposing an alternative definition of determinism reduces to whether any interesting substantivalist can accept that Leibniz equivalent models represent the same physical situation. A substantivalist who accepts that, will hold that two models are physically equivalent just in case they are diffeomorphically related, and as a result will not consider hole diffeomorphs as violating determinism.

If one were to accept that Leibniz equivalence is sufficient for physical equivalence, then one could consider D as a special case of a broader definition of determinism. D only considers models whose manifolds share the same base set of points. It is clear, however, that given the completeness condition satisfied by local spacetime theories there will be models $<M,O_i>$ and $<M',O'_i>$, whose manifolds do not have identical base sets. Someone who holds that Leibniz equivalence is sufficient for physical equivalence may want to count diffeomorphisms between such models as sufficient for physical equivalence. Such a substantivalist is committed to the view that the particular objects (points, quadruples of reals, etc.) that constitute the base set of a manifold are not representationally significant features of a model. This is an attractive position to hold. Modern presentations of GTR leave unspecified the members of the base set of a differentiable manifold, the important features of the manifold being its differential structure.

3. Attitudes towards Leibniz Equivalent models

It is clear that the hole argument loses its force if a substantivalist can hold that Leibniz equivalent models represent the same situation. Earman and Norton think that no manifold substantivalist can hold such a view. Two substantivalist positions have been adopted in light of the hole argument, the proponents of each share Earman and Norton's conviction that a substantivalist should not claim that Leibniz equivalent models represent the same situation. As a result both positions hold that the best attitude a substantivalist can have towards Leibniz equivalent models is that at most one represents a possible physical situation.

The first of these two positions can be found in a paper by Tim Maudlin "The Essence of Space-Time". Maudlin takes the hole argument as an argument for essentialism. We are to view the metric properties of actual spacetime points as essential properties (Maudlin 1988, 86). His claim is that if one accepts this then one will hold that at most one of a class of diffeomorphic models of a spacetime theory assigns the right metric properties to spacetime points, and so at most one model will represent a possible physical situation. Part of the motivation for his metric essentialism comes from a desire to be able to give the traditional substantivalist assessment of the Leibniz shift argument. He thinks that any adequate substantivalist must side with a Newtonian in holding that Leibniz shifted universes represent different possible worlds (Maudlin 1988, 84). It is for this reason that he holds that a substantivalist should not claim that Leibniz equivalent models represent the same situation. The metric essentialist can endorse such a response, since the metric in Newtonian spacetime is flat.

There are various reasons for thinking that Maudlin's metric essentialism is not the best solution to the hole problem. It renders certain counterfactuals as false: if the sun had had extra mass then the curvature of spacetime around the sun would have been different. Such a counterfactual would be false if the metric properties of spacetime points are taken to be essential properties. Of course a counterpart theoretic evaluation of such counterfactuals could be given, even by the metric essentialist. To do so would involve picking some constraints on what counts towards counterparthood, other than metric properties of points and denying transworld identity for the points concerned. The counterfactual will then be evaluated as true in virtue of the counterpart of the actual region of spacetime around the sun (in some class of worlds most similar to this one) having different metric properties. In fact Maudlin appears to want to endorse a counterpart analysis in order to account for such counterfactuals (1988,

90). It seems unattractive, at least for an essentialist, to have to evaluate such counterfactuals as true in virtue of counterparts that do not even have essential properties in common with the objects they are counterparts of. Being forced to resort to a blend of essentialism and counterpart theory seems to weaken the case for metric essentialism; it begins to look like rather an ad hoc solution to the hole argument. Counterpart theory, on its own, may be a more attractive alternative.

The other such response to the hole argument can be found in a paper by Jeremy Butterfield, "The Hole Truth". His position is motivated in part by the sorts of problem that an essentialist encounters when trying to deal with counterfactuals. Butterfield is some version of a counterpart theorist. He claims that to deny transworld identity, i.e. to claim that a given point (or of course object) is an inhabitant of just one world "will clearly secure (that at most one of a Leibniz equivalent class of models represents a world)"(Butterfield 1989, 23). He holds that the properties encoded by the fields, O_i, in the models determine the counterpart relation for points or regions of spacetime. He is thus in a position to give a counterpart theoretic evaluation of the counterfactuals that caused problems for the metric essentialist, without the further commitment to metric properties being essential to spacetime points. A further claimed advantage of the account is that when faced with the Leibniz shift argument, such a substantivalist is able to say of the shift translation that it does not produce another possible world. That such a shifted world is not a possible world is supposedly secured by the denial of transworld identity for spacetime points.

I take the claim that a denial of transworld identity secures that at most one of a class of Leibniz equivalent models represents a possible physical situation to be a central feature of his position. However, it is not immediately obvious how a denial of transworld identity secures this. Issues concerning transworld identity are issues about possible worlds, rather than about models that purport to represent those worlds. Butterfield does try to motivate the claim that some models of GTR fail to represent a world. He points out that some models will have manifolds whose base sets consist not of points, but of objects that have essential properties that could not be had by points. Such models, on Butterfield's view, could not represent a world. Thus a necessary condition for a model of GTR to represent a possible world for Butterfield, is that the base set of the manifold consist of actual points of spacetime. This, however, is still not sufficient for his purposes. If a denial of transworld identity is to secure that at most one of a Leibniz equivalent class of models represents a world, Butterfield must endorse the stronger claim that a model can represent a world only if its manifold consists entirely of the spacetime points *of that world*. This, coupled with the claim that no point inhabits more than one world, and the claim that no two (nonidentical) models can represent the same world, will secure that at most one of a Leibniz Equivalent class of models represents a world.

This is an extremely strong restriction to impose on when a model represents a world, so strong as to almost require that the model be identical to what it represents. One would be inclined to say that the objects in the base set of the manifold can *represent* spacetime points even if they are not, or do not have the essential properties of, spacetime points. In fact, as mentioned earlier, I would endorse the view that even models whose base sets are disjoint may still represent the same world. When standardly defined, the base set of a manifold is left unspecified. Accepting such a restricted account of when a model represents a world is, however, sufficient to blunt the force of the hole argument.

Norton suggests a further consideration that might lead one to think that the Maudlin and Butterfield stance towards Leibniz equivalent models is counterintuitive. (Norton 1988, 63) Someone who thinks that at most one of a diffeomorphic class of models represents a possible situation is going to face the problem of distinguishing real models from impostors. The problem being that there are no physically significant features of the models that would lead one to conclude that one represents, but the other fails to represent a world. Thus, even though in principle ones spacetime theory is deterministic, when confronted with two Leibniz equivalent models that are identical up to some time, nothing about the theory will tell you which model really represents the future. I take the fact that there are no physically significant features of the models that would lead one to pick one over another as representing a world to lend support to the claim that each of the models represent the same world. Each model could serve equally well to represent a given world, W (where W is a particular member of the class of qualitatively identical worlds that it is claimed a substantivalist believes are represented by the class of models in question). In fact no considerations could lead one to hold that any particular model failed to represent W.

My intention was not to claim that one couldn't use the Butterfield or Maudlin replies to blunt the force of the hole argument, but merely to illustrate that such responses appear unattractive.

4. Leibniz Equivalence, Possible Worlds and Models

It is clear that the hole argument is an argument against substantivalism only if the substantivalist is committed to viewing Leibniz equivalent models as representing distinct situations. There is certainly inductive evidence to believe that a substantivalist is committed to this; it has been the assumption of every substantivalist since Newton. In the Leibniz Clarke correspondence, when confronted with the shift argument, Clarke maintained, in the face of a violation of the principle of sufficient reason and the principle of identity of indiscernibles that the substantivalist view this world and its shifted counterpart as genuinely distinct. But the substantivalist has more at stake when faced with the Hole argument so it is important to assess exactly why it has been thought that a substantivalist cannot accept that Leibniz equivalent models represent the same world. As should be clear, the issue hinges on whether or not a substantivalist is committed to a particular way of identifying spacetime points across possible worlds (Talk of possible worlds is sometimes a useful aid in understanding modal statements. About whether possible worlds exist I wish to remain neutral).

In both the hole and the shift arguments it is assumed that a substantivalist identifies spacetime points across the possible worlds in question independently of any of their qualitative properties. It is assumed that independently of the properties of points coded by the fields O_i the issue of identifying points across possible worlds has been settled. The substantivalist who then wants to claim that Leibniz equivalent models represent the same world has to deal with a given spacetime point having different properties in worlds that she wants to say are the same. But realism about spacetime points, just as realism about any other entities, does not commit you to a transworld identification thesis that is independent of the qualitative properties of an object.

That realism does not commit you to such a thesis is just as well, since as illustrated by Paul Horwich (1978), arguments similar to those we've seen against the existence of spacetime points, if successful, can be used against the existence of any entities you like: Consider electron A and electron B in this world, where A and B are qualitatively distinct, for example A is part of a helium atom and B is part of a hydrogen atom. Imagine a world, distinct, but qualitatively identical to this world, in which Electron A and Electron B have been switched throughout their entire history. If the Leibnizian is to insist that these worlds are the same world, in virtue of the worlds taken as a whole being qualitatively identical, then this must be seen as at least as powerful an argument against electrons as the shift or hole arguments are against spacetime points. Of course there is a Leibnizian response to this argument against electrons. We should consider the qualitative properties of the electrons in the two descriptions of the world. Electron A in the one shares all the properties, including relational properties to other qualitatively described objects, of electron B in the other. Applying the identity of indiscernibles to the worlds, assuming that there is no complete symmetry of the world with respect to qualitative properties that maps A onto B, leads us to conclude that electron A according to the first description is just electron B according to the second description. They are just two different ways of talking about the same world.

The Leibnizian response in the electron case shows that a realist about electrons can accept that, what are in effect, Leibniz equivalent worlds are really just different ways of talking about the same world. But if a realist about electrons can say this, then there appears to be no reason why a realist about spacetime regions cannot give the analogous response in the Hole and shift arguments. So it would seem that a substantivalist can accept that Leibniz equivalent models represent the same world.

Hartry Field (Field 1989, 40) has some discussion of the Leibniz shift argument where he suggests that the most natural thing to say of the shifted universes is that they represent the same world. His claim is that we tend, as a matter of convention, to use the identity of indiscernibles as applied to possible worlds. I think that this is basically correct. What a substantivalist should say about the way we individuate spacetime points or regions across possible worlds is that we individuate according to qualitative similarity. If we do so then all hands can agree that the hole diffeomorphs and shifted universes just represent the same world, independent of whether they think spacetime points exist.

If such a substantivalist is a realist about possible worlds then there is a well known framework within which such a reply, or at least a similar, but equally adequate reply, is obviously the right one: namely, David Lewis's view of possible worlds. The Lewisian substantivalist will be a counterpart theorist, and is going to view the counterpart relation as guided by qualitative constraints. However, unlike Butterfield, the Lewisian will not hold that at most one of a class of Leibniz equivalent models represents a possible world. Rather, he will view the worlds represented by Leibniz equivalent models as physically equivalent. Moreover, he has good reason for doing so since the counterpart of any given point in any of the qualitatively indiscernible worlds will have all the same qualitative properties as that point has. Thus the hole argument does not give rise to an indeterminism problem for the Lewisian substantivalist.

There seem to be prima facie objections to this position. Firstly, is it really legitimate for a substantivalist to give such a "Leibnizian" response to the hole and shift arguments. Given that it has been commonly thought, since Leibniz, that substantivalism was incompatible with viewing Leibniz equivalent models as representing the same world, one might think that the position urged here is really some kind of disguised relationism. For example one might claim that a consequence of using a qualitative criterion to individuate across possible worlds, is that spacetime no longer has an existence independent of its contents. But, so the argument would go, substantivalism just is the view that spacetime points exist independently of the matter that inhabits them.

In the introduction substantivalism was characterized as realism about spacetime points or regions. One can characterize substantivalism as the view that spacetime has an existence independent of its contents provided that it is clear what the independence claim amounts to. Usually what is meant by independence is some kind of counterfactual independence. We are inclined to say that I have an existence independent of my hair colour just in case we are prepared to accept that I could have existed had I had different coloured hair. On the view of substantivalism I have been developing, spacetime has an existence independent of its contents in precisely this sense. The desk in front of me could have been situated five metres to the left; in the counterfactual situation where it is so situated the region of spacetime that it actually occupies has different properties. The reason we are prepared to accept such counterfactuals as true is that in the counterfactual situation in which my desk is five metres to the left, the counterpart of the actual region of spacetime in front of me has many of the same properties in the two situations, for example its relations to other objects. It is precisely because of these shared properties that we consider it, rather than some other region of spacetime, to be the counterpart of the actual region. Thus provided what is meant by independence here is counterfactual independence this version of substantivalism can be characterized by saying that spacetime has an existence independent of its contents.

A second objection to this position might be that it renders certain modal statements false. Doesn't it make sense to say that this region of spacetime could have had all the qualitative properties (including relational ones to other qualitatively described objects) that that region has, or that this electron and that one could have been switched throughout their entire history? We are as strongly inclined, at least prephilosophically, to think that such claims are true, as we are to think that any others are true. However, adopting the identity of indiscernibles at the level of worlds, appears to commit one to the view that no two worlds differ in the way required for these claims to come out true.

I don't think that the objection has much force, however I think that how one replies to it will depend on whether one is a realist about possible worlds. Both the realist and the non realist can dig their heels in and maintain that such modal claims are just false. This seems unattractively dogmatic. For the realist, I take it that an account of the truth conditions of such modal claims will have to be given in terms of something other than possible worlds. David Lewis (1986) outlines how such an account might go, in terms of "this worldly" counterparts.

The non realist, i.e. someone who considers possible worlds fictions, but nonetheless of heuristic value, can claim that we can talk about possible worlds in such a way that these modal claims do come out true. If one does so, however one will have to drop the convention that we use the identity of indiscernibles as applied to worlds. On this view one would claim that we can make sense of some alternative counterpart relation that allows us to talk of possible worlds in the way required by such modal claims, however, we cannot thereby infer that electrons or spacetime points don't exist, by using a qualitative criterion for identifying worlds. You can, if you want, talk about possible worlds in either way, to explicate different senses of possible, but you should not conflate these two senses of possible.

If one is to take this line, and claim that we adopt different conventions when talking about possible worlds, something has to be said about why we should adopt the identity of indiscernibles as applied to worlds when discussing determinism. As Norton and Earman have pointed out, however, in modern texts on GTR it is accepted that Leibniz equivalent models represent the same situation when considering determinism (Earman and Norton 1987, 11 and Earman 1989, 186). Earman and Norton take this to be evidence for the claim that substantivalism is at odds with the way physicists talk about determinism, however, as we have seen, a substantivalist is not committed to the claim that Leibniz equivalent models represent different possible worlds. Thus it seems natural for the substantivalist to adopt this convention when discussing the hole argument.

So the substantivalist is not committed to denying that Leibniz equivalent models are physically equivalent. One can consistently hold that spacetime points or regions exist while also maintaining that diffeomorphic models represent the same physical situation, or that shifted worlds are just different ways of talking about the same world. I urge that the substantivalist accept that Leibniz equivalent models represent the same situation. Qualitatively identical, or alternatively, isomorphic possible worlds, should be viewed as the same world.

I turn now to an objection, raised by Earman, to the view that identity across possible worlds respects isomorphism. Earman's claim is that the view that identity follows isomorphism is "in general incoherent if "identity" means strict identity and if isomorphism is not unique." (1989, 198). Worlds may exhibit symmetries such that multiple isomorphisms exist. If $I_1:W \rightarrow W'$ and $I_2:W \rightarrow W'$ are two such isomorphisms this view forces us to conclude that $I_1(i) = I_2(i)$ for i in W, which may not be the case, if the two isomorphisms are distinct.

It is not clear, however, whether there being no reason to pick one isomorphism over another, in some cases, when individuating across possible worlds, is sufficiently strong to force us to conclude that we should individuate across possible worlds by using something other than isomorphism. When discussing similar problems with qualitative counterpart relations Lewis suggests that context determines which of the competing counterpart relations we should use, to make sense of particular modal statements. Different contexts, different counterpart relations; furthermore he sees this as an advantage of his account. I am inclined to agree with much of his discussion, however, in cases of symmetries of worlds it is not clear how appeal to context will help. Despite this I don't see the above objection as forcing individuation across possible worlds to be explicated in terms of some further, nonqualitative criteria. Confronted with a world with multiple symmetries, and hence multiple isomorphisms as candidates for individuating across possible worlds, it is a mistake to think that there must be some further feature of the worlds in virtue of which one isomorphism is the right one. It is not a deficiency of the qualitative view of possible worlds, it rather accords with our intuitions that there is no fact of the matter as to which one of two qualitatively similar regions we are to consider as the counterpart of a given region. Which region to pick may be indeterminate, but further criteria for picking one over the other are not available.

So the hole argument and it's ancestor the shift argument are not sufficiently strong to force a rejection of substantivalism. The version of substantivalism I have been endorsing is a form of manifold substantivalism. A manifold substantivalist can accept that Leibniz equivalent models represent the same situation, and thereby escape the Hole argument, without any threat of being inconsistent.

Notes

[1] I would like to thank Hartry Field, Frank Arntzenius, John Norton, Seth Crook, Gordon Belot and Tom Cuda for helpful comments on earlier drafts.

[2] Following Earman (1989, 179) I shall assume that we are dealing with models that possess a cauchy surface, so that there is a global time function $t:M \rightarrow R$ such that t increases as one moves in the future direction along any timelike curve, and the level surfaces of t are all cauchy surfaces.

References

Butterfield, J. (1989), "The Hole Truth." *British Journal for the Philosophy of Science 40*:1-28

Earman, J. (1986), *A Primer on Determinism*. Dordrecht: D.Reidel

_____. (1989), *World Enough and Spacetime*. Cambridge, Massachusetts: MIT Press

Earman, J. and Norton, J. (1987), "What Price Spacetime substantivalism? The hole story." *British Journal for the Philosophy of Science 38*:515-525

Field, H. (1989), *Realism, Mathematics and Modality*. Oxford:B.Blackwell

_____. (1980), *Science Without Numbers* Oxford:B.Blackwell

Horwich, P. (1978), "On the Existence of Times, Space and Space-Times"*Nous 12*:396-410

Lewis, D. (1986), *On the Plurality of Worlds* Oxford:B.Blackwell

Maudlin, T (1988), "The Essence of Spacetime." *Proceedings of the Philosophy of Science Association Vol.2*:82-91

Norton, J. (1988), "The Hole Argument." *Proceedings of the Philosophy of Science Association. Vol.2*:56-64

Non-Turing Computers and Non-Turing Computability[1]

Mark Hogarth

University of Cambridge

1. Introduction

Building on an idea by Pitowsky (1990), David Malament (private communications), Hogarth (1992) and Earman and Norton (1993) have shown how it is possible to perform computational supertasks—that is, an infinite number of computational steps in a finite span of time—in a kind of relativistic spacetime that Earman and Norton (1993) have dubbed a *Malament-Hogarth* spacetime[2].

Definition 1 A spacetime (M,g) is *Malament-Hogarth* just when there is a future endless curve $\lambda \subset M$ with past endpoint and a point $q \in M$ such that
$\int_\lambda ds^2 = \infty$ and $\lambda \subset J^-(q)$.

(Hereafter, the symbols "q" and "λ" are assumed to have the properties they have in Definition 1. I shall also speak of a "λ-curve".)

Various examples of Malament-Hogarth (hereafter, *M-H*) spacetimes are given in Hogarth (1992), but the following artificial example from Earman and Norton (1993) is perhaps the simplest. Start with Minkowski spacetime (R^4, η) and choose a scalar field Ω on M such that $\Omega=1$ outside a compact set $C \subset M$ and Ω tends rapidly to infinity as a point $r \in C$ is approached. The spacetime (R^4-r, $\Omega^2\eta$), depicted in Figure 1, is then M-H. (Although the region inside C *appears* quite small, it is in fact as large as the complement of C.)

Hogarth (1992) and Earman and Norton (1993) show how in a M-H spacetime, e.g., (R^4-r, $\Omega^2\eta$), one might solve, e.g., the Goldbach conjecture. From a point $p \in \lambda$ launch a Turing machine along λ that is primed to first check if 2 is the sum of two primes, then likewise to check 4, then 6, and so on, *ad infinitum*. The Turing machine is also primed to signal to q if and only if it finds a counter-example to the conjecture, and then to halt operations. Since an observer, e.g. O in Figure 1, can travel from p to q in a finite span of proper time, she can discover the truth of the conjecture before her day's out. Fermat's last theorem is cracked in a similar fashion.

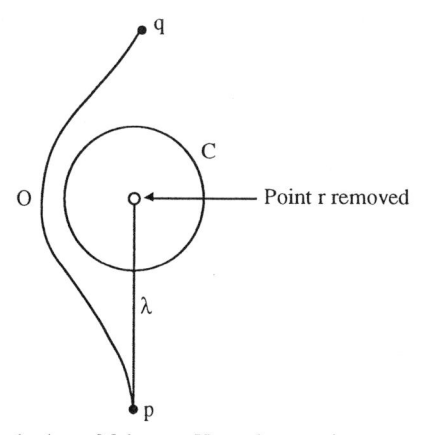

Figure 1. A toy Malament-Hogarth spacetime.

2. Solving the Turing Unsolvable

I move now to the objectives stated in the abstract. To simplify matters in this section, no account will be taken of the physical plausibility of the spacetimes under consideration or of the behaviour of the matter fields they support. Moreover, it will be assumed that every spacetime permits Turing machines of any size to operate unproblematically and that communication to future events is always possible. Note, however, that Hogarth (1992) has shown that M-H spacetimes cannot be globally hyperbolic (so they violate strong cosmic censorship), and that Earman and Norton (1993) have shown that in many M-H spacetimes photons travelling between some events suffer infinite blue shifts (which may indicate horizon instability).

A problem is said to be *Turing solvable* if there is a Turing machine (operating according to a finite instruction set, but having access to an infinite memory store) that can solve the problem after a finite number of steps. In this paper I shall say, somewhat informally, that a problem is *solvable in a spacetime (M,g)* if there is an observer O⊂M who can initiate a procedure which is comprised of only Turing machines and ordinary communication devices and which will deliver the problem's solution to O after a finite span of O's proper time.

Thus the Goldbach conjecture and Fermat's last theorem are both solvable in the M-H spacetime in Figure 1, and indeed in any M-H spacetime. These two problems are Turing solvable (because in both cases the solution can be written into the finite program of a TM), so they do not prove a difference between this more general solvability and Turing solvability. I will now give a proof, by showing how a problem that is known to be Turing unsolvable is demonstrably solvable in any M-H spacetime. In fact I will give two examples: (see Boolos and Jeffery 1989.)

(1) *The halting problem.* This is the problem of deciding if an arbitrarily given Turing machine, TM, will or will not eventually halt. Working in a M-H spacetime (M,g), adopt the following procedure. Launch TM along λ⊂M, having first primed TM to signal to q∈ M if and only if TM halts. The question will then be settled at q.

(2) *The decision problem for first-order logic.* This is the problem of deciding the validity or invalidity of an arbitrary sentence of first-order logic. First recall that

there is Turing machine, TM, that will halt after a finite number of steps if and only if a given sentence S of first-order logic is valid (ibid., p. 142). So, working in a M-H spacetime (M,g), launch TM along $\lambda \subset M$, having first primed TM to signal to $q \in M$ if and only if TM halts. Upshot: a signal at q means the sentence is valid, no signal at q means the sentence is invalid.

In general, the *decision problem* for a property P is Turing solvable if there is both a Turing machine TM that will halt after a finite number of steps if and only P holds and a Turing machine TM´ that will halt after a finite number of steps if and only if P does not hold. If only one (or both) of the pair TM, TM´ exists, then the decision problem for P is said to be *partially Turing solvable*. Problems (1) and (2) above are clearly of this kind. It is now evident that:

Result 1. Any decision problem that is partially Turing solvable is solvable in a M-H spacetime.

Attention is now turned to the *decision problem for arithmetic in the standard model*. In what follows, the word "sentence" is used as shorthand for "sentence in the language of arithmetic". The results below are standard (ibid.).

(i) Deciding (i.e. establishing whether true or false) arbitrary sentences in arithmetic is not partially Turing solvable. (This is a version of Gödel's first incompleteness theorem.)

(ii) Deciding arbitrary quantifier-free sentences is Turing solvable.

(iii) There is a Turing machine that will, in a finite number of steps, translate an arbitrary sentence S into a coextensive (i.e. a sentence with the same truth-value) sentence, S´, in prenex form (all quantifiers occurring at the extreme left).

(iv) There is a Turing machine that will, in a finite number of steps, translate an arbitrary sentence S (written in prenex form) containing two juxtaposed quantifiers of the same type into a coextensive sentence, S´, with one quantifier of that type in place of the previous two.

(v) If $\forall n S(n)$ is a sentence or $\exists n S(n)$ is a sentence, then the set of sentences $\{S(1), S(2), S(3),...\}$ is recursively enumerable (that is, there is a Turing machine that will generate S(n) for any given n).

Because of (iii), it may be assumed that all sentences are in prenex form.

Because of (ii) and (v), it is clear that deciding arbitrary sentences of either the form $\exists n S(n)$ or $\forall n S(n)$, where S is quantifier-free, is partially Turing solvable. So by Result 1, they are both decidable in any M-H spacetime. This fact, together with (iv) above, implies that arbitrary purely existential or purely universal sentences in arithmetic are decidable in any M-H spacetime. (Incidentally, the Goldbach conjecture and Fermat's last theorem can both be stated as purely universal sentences.)

But because of (i), Result 1 cannot be used to show that arithmetic is decidable in any M-H spacetime. Indeed, although it cannot be "proved" that arithmetic is undecidable in, e.g., the M-H spacetime in Figure 1, there is no obvious way to construct a procedure in this particular spacetime that will decide even an arbitrary $\forall \exists$ type sentence. A more subtle kind of M-H spacetime is required, as I will now show. The following analysis contains a new piece of terminology. If a spacetime (M,g) contains non-intersecting open regions O_i, i=1,2,... such that (1) for all i $O_i \subset I^-(O_{i+1})$ and (2) there is point $q \in M$ such that for all i $O_i \subset I^-(q)$, then the O_is are said to form a *past temporal string*, or just *string* for short. See Figure 2 (i).

(i) A temporal string.

(ii) A SAD_2 spacetime.

(iii) A SAD_3 spacetime.

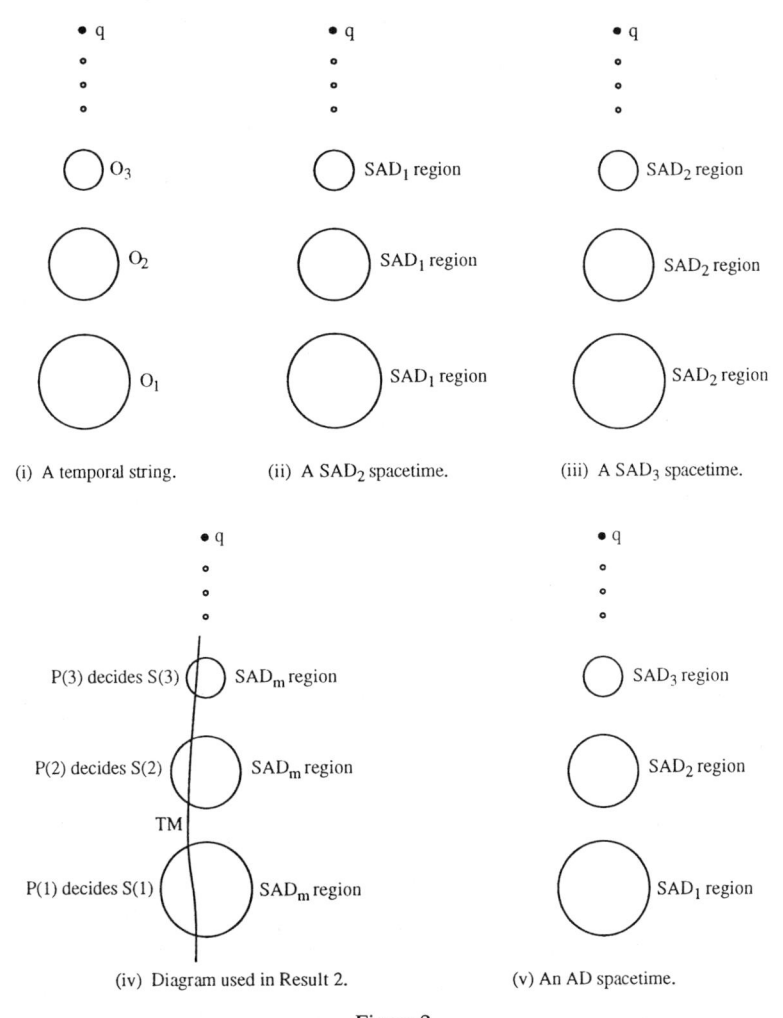

(iv) Diagram used in Result 2.

(v) An AD spacetime.

Figure 2

Definition 2. A spacetime (M,g) is an *nth-order arithmetical sentence deciding* (denoted SAD_n) spacetime if the n conditions contained in the following scheme are satisfied.

If n=1, (M,g) is a M-H spacetime.
If n>1, (M,g) admits a string of SAD_{n-1} spacetimes.

According to Definition 2, a SAD_1 spacetime is a M-H spacetime, a SAD_2 spacetime is a spacetime that contains a string of SAD_1 spacetimes (Figure 2 (ii)), a SAD_3 spacetime is a spacetime that contains a string of SAD_2 spacetimes (Figure 2 (iii)), and so on.

The efficacy of SAD spacetimes to decide sentences in arithmetic derives from the following result.

Result 2. Let (M,g) be a SAD_{m+1} spacetime and let $\exists nS(n)$ and $\forall nS(n)$ be two sentences in arithmetic. Suppose that for each $n \geq 1$, $S(n)$ is decidable in any SAD_m spacetime.

Then $\exists nS(n)$ and $\forall nS(n)$ are both decidable in (M,g).

Proof. This consists of showing how appropriately chosen and appropriately located hardware can be used to decide $\exists nS(n)$ and $\forall nS(n)$. Part of this involves the Turing machine, TM, of (v) above travelling along a λ-curve that picks up one hour of proper time in each SAD_m component (this is possible because every such component admits a λ-curve), as depicted in Figure 2 (iv). TM is primed to generate $S(1)$ in the lead up to the first component and to signal that sentence to the first component. TM is also primed to generate, for $n>1$, $S(n)$ in the (n-1)th component and to signal that sentence to the nth component.

Now let P(n) denote the procedure that decides $S(n)$. The procedure for deciding $\exists nS(n)$ consists of adding to each P(n) transmitting devices and receivers which operate as follows.

P(1) signals to q and P(2) if and only if $S(1)$ is true.
For $n>1$, P(n) signals to P(n+1) if and only if P(n-1) signals to P(n).
For $n>1$, P(n) signals to q and P(n+1) if and only if $S(n)$ is true and P(n) has not received a signal from P(n-1).

This procedure ensures that a *single* signal is sent to q if and only if there is an n such that $S(n)$ is true. (The signal is actually sent by P(m), where m is the smallest integer for which $S(m)$ is true.) Upshot: a signal at q means $\exists nS(n)$ is true, no signal at q means $\exists nS(n)$ is false.

The procedure for deciding $\forall nS(n)$, given below, is similar except this time a single signal at q means $\forall nS(n)$ is false, no signal at q means $\forall nS(n)$ is true.

P(1) signals to q and P(2) if and only if $\neg S(1)$ holds.
For $n>1$, P(n) signals to P(n+1) if and only if P(n-1) signals to P(n).
For $n>1$, P(n) signals to q and P(n+1) if and only if $\neg S(n)$ holds and P(n) has not received a signal from P(n-1).

Thus $\exists nS(n)$ and $\forall nS(n)$ are seen to be decidable in (M,g). QED

We have seen already how single quantifier sentences can be decided in SAD_1 (=M-H) spacetimes. By Result 2, double quantifier sentences can be decided in SAD_2 spacetimes. Applying Result 2 again shows that triple quantifier sentences can be decided in SAD_3 spacetimes. And continuing in this way we see that n-tuple quantifier sentences can be decided in SAD_n spacetimes.

In fact, these different order SAD spacetimes can be fitted into a single spacetime.

Definition 3. A spacetime (M,g) is an *arithmetic deciding* (AD) spacetime just when (M,g) admits a string of open regions $O_1, O_2, O_3,...$ such that for each $n \geq 1$, (O_n, g) is a SAD_n spacetime.

In Figure 2(v), an observer can decide an arbitrary sentence S by communicating it to the SAD_n region that decides sentences of that order. Arithmetic is therefore decidable in an AD spacetime.

It is natural to wonder how the computing hardware necessary to decide all of arithmetic gets installed in the AD spacetime. What follows is a *prima facie* reasonable method of performing that task.

The various paths of the hardware through spacetime are represented by worldlines, and the idea of the method is to begin with one worldline at the initial event p in Figure 3 and to have a process of worldline branching that results in each SAD_1 component being populated by a λ-curve and every other component of every string being populated by at least one worldline. (Recall that the SAD_1 spacetimes accommodate Turing machines travelling on λ–curves, while all the other components accommodate communicating devices à la Result 2.)

(1) A worldline that meets a SAD_n component, n>1, must bifurcate, with one worldline branch extending to the next component of that string and the other branch entering the component and extending to the first component of the string on the "next level down".

(2) Two kinds of worldline must follow the first available λ-curve: the one that enters the first SAD_1 component and any one that is constrained by (1) to enter a SAD_1 component.

Figure 3 illustrates the process at work on the SAD_1 and SAD_2 stages of the AD spacetime. The tree-like property is an obvious attraction, but this system also guards against one potential disaster, namely, that an infinite amount of hardware mass might be forced to reside in a compact set. I omit the formal proof of why this cannot happen. It relies on the fact that if a region R contains worldlines whose lengths sum to infinity, then it must contain at least one future endless curve. But according to Proposition 6.4.7 in Hawking and Ellis (1973), this can only occur if R is non-compact or violates strong causality. Part of the reason why I chose from the outset to consider only strongly causal spacetimes was to ensure that in this case R must be non-compact.

Admittedly, there remains the worry that an unbounded mass might reside in a region which is non-compact but of finite volume. I am not sure whether or not this can happen. (Are there any SAD_1 spacetimes with finite volume?) In any case, the example in the next section suffers no such pathology.

We have also shown that the hardware for the AD spacetime can be built up using a finite set of instructions. Roughly: start at an event p, then manoeuvre and bifurcate according to (1) & (2), while creating communication devices and the Turing machines of (iv), and having them operate according to Result 2. One other instruction states that the instructions of the previous sentence must be issued to each new piece of hardware.

A word about terminology. In what follows, a *SAD spacetime* will be used as a generic term to cover SAD_n spacetimes of all orders and AD spacetimes. (Of course in this case SAD=SAD_1, but the term "SAD_1" sounds rather specific.) Also, a *simple* SAD_n spacetime is a SAD_n spacetime that is not SAD_{n+1}. Finally, a SAD_n (respectively, SAD, AD, etc.) *computer* will refer to a computing device whose underlying spacetime is SAD_n (respectively, SAD, AD, etc.).

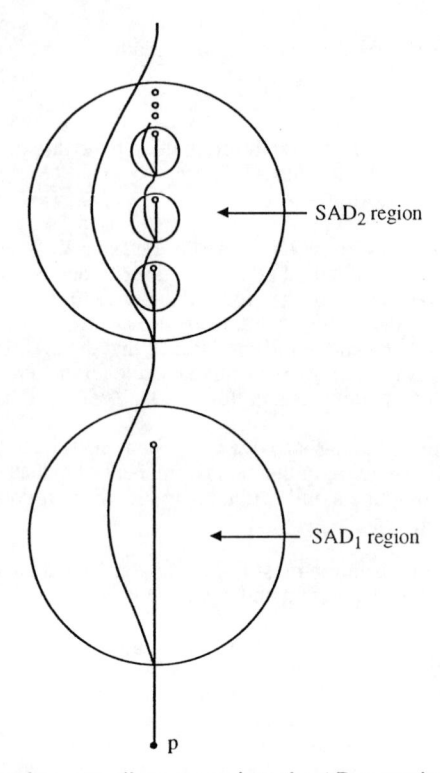

Figure 3. The "hardware tree" grows up into the AD spacetime.

3. An Example of an AD Spacetime[3]

Start with Minkowski spacetime (R^4, η) and choose a compact set $C \subset R^4$. Now draw a closed inertial line segment $v \subset C$. About this v, define regions $O_1, O_2, O_3,...$ with inclusion relations appropriate for all the strings of an AD spacetime, in such a way that v intersects every component, as depicted in Figure 4. Then choose a scalar field Ω on M such that $\Omega=1$ outside C and Ω tends rapidly to infinity as the line v is approached. Remove v. Then (R^4-v, $\Omega^2 \eta$) is an AD spacetime because O_1 is a SAD_1 spacetime, O_2 is a SAD_2 spacetime, and so on. Moreover, it not difficult to show that (despite appearances!) every component of every string has infinite volume.

Although the corresponding AD computer consists of an infinite number of infinitely large regions, each with its own communication devices and Turing machines, it can still be contained in a box, e.g. the one depicted in Figure 4, with *finite* spatio-temporal surface area. So in this regard this AD computer is no different to an ordinary desktop computer.

I do not yet have a proof, but I think that anti-de Sitter spacetime (see Hawking and Ellis 1973, p.132) is probably an AD spacetime. This spacetime is usually not viewed as a reasonable solution of the Einstein equations, but the fact that its structure is so simple hints, perhaps, that reasonable AD solutions may indeed exist.

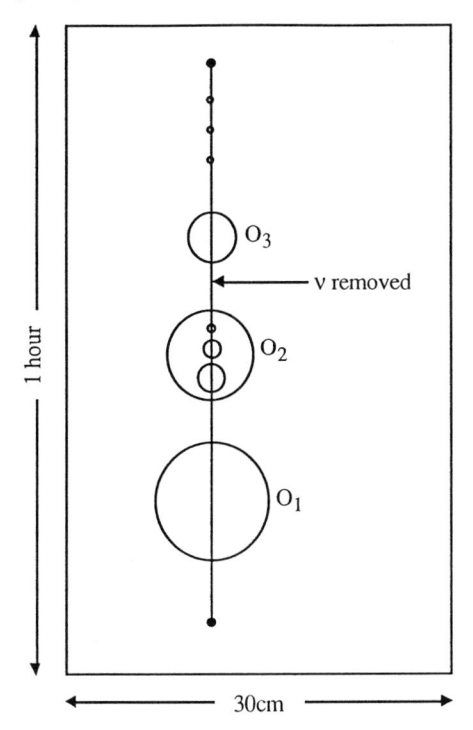

Figure 4. An AD spacetime.

4. The Impact on Church's Thesis

A typical way of stating Church's thesis (CT) is the following: In an ideal world the limit of computation is exactly captured by Turing computability. As it stands, this statement of CT is vague: it has no truth-value since we are not told which set of worlds, W, the "ideal" worlds are chosen from. Different choices of W give different truth-values, but I want to choose a W which gives CT a real fighting chance of success. (In that way, the doubt I manage to cast on CT will not be seen as a hollow victory.) This means W should not be the set of logically possible worlds; for the "Zeus machine" in Boolos and Jeffery (1989, p. 14) is perfectly consistent and is able to perform Turing unsolvable tasks. Nor, on the other hand, should W be a set of worlds all of which are very similar to our universe in such matters as possessing only a finite amount of material (as our universe might) or being temporally finite (as our universe might be). For that would again make CT fail—this time because the arbitrarily massive Turing machines entailed by Turing computability would be impossible, leaving the ideal computing limit somewhere short of Turing computability.

Thus the set W should be neither very large (all logically possible worlds) nor very small (worlds very similar to our own). A middle way is needed. One's initial reaction might be to take W as the set of worlds that are "beefed up" versions of our world, i.e.

worlds based on our world but with enough added space, time and material to allow the realisation of a Turing machine. But of course this is naive, for spacetimes like the Robertson-Walker k=+1 (big bang, big crunch) model, which is a good candidate for the cosmological structure of our universe, cannot be spatio-temporally extended in any natural way and cannot be plied with an unbounded amount of material. Another and more promising proposal is that W should be the set of physically possible worlds, i.e. worlds that share our universe's fundamental laws of physics, but not necessarily our universe's boundary conditions. Of course given our ignorance of what these laws are, this proposal provides at best a working characterisation of W. Nevertheless, this somewhat ill-defined W results in a version of CT that is physically relevant, is not refuted by the hypothetical Zeus machine, and does not fail on account of material limitation—assuming of course that material limitation is not a requirement of physical law.

Moreover, this CT stands firm against two putative counter-examples that have appeared in the literature (Grünbaum 1967, Earman and Norton 1993). Both are machines that can perform an infinite number of computations in a finite time. In one version the machine's components are accelerated to infinite speed in a finite time; in the other, the components shrink continuously to nothing in a finite time. But neither machine seems to cut any ice with CT, for whatever the laws of physics in fact are, the requirements of speed and shrinkage presumably violate them.

But is this CT true? Well, it will be if there is at least one physically possible Turing machine *and* if there are no physically possible non-Turing computers (=SAD computers in this context). The first part of the condition is almost certainly true because there are some thoroughly reasonable solutions of the Einstein equations (e.g. the Robertson-Walker k=0 big bang model) which have both spacetime enough and matter enough to realise a Turing machine. What about the second part of the condition, that all SAD computers are physically impossible? This seems *prima facie* false because there are known simple SAD_1 solutions of the Einstein equations with spacetime enough for the SAD_1 hardware. The Reissner-Nordström black hole solution (see Hogarth 1992) is a case in point. Admittedly this is a vacuum solution, but its existence suggests that simple SAD_1 is a not uncommon property among reasonable models, including ones with infinite material. Against this however is the fact that all SAD spacetimes violate various versions of cosmic censorship. That is to say, physicists have conceived of laws that would effectively outlaw SAD spacetimes and *a fortiori* SAD computers. But of course it is not yet known whether *any* kind of censorship laws really exist, still less whether one exists that will outlaw SAD computers and save CT. Far be it from me to predict the outcome of the debate on the cosmic censorship hypothesis, but I will note that, in concluding his survey of the subject, Earman (1994) remarks that "it is much too soon to pronounce the cosmic censorship hypothesis dead, but the prognosis is not particularly cheerful". Perhaps then the same could be said of CT.

At this juncture, advocates of CT might try to advance their case from a quite different angle, by appealing to the fact that several apparently independent explications of the concept of computability—by Markov, Church, Post, Kleene etc.—have all been found to be exactly equivalent to Turing computability. Of course, I concede this equivalence. But as Horowitz (1992) has rightly stressed, all these explications rest on a shared assumption: that it is impossible to perform a computational super-task. Thus the very assumption that threatens Turing computability, also threatens these other explications. Any attempt, therefore, to base a case for CT on this equivalence is futile: if one explication falls, they all fall.

To summarise, the physically possible computing limit does not "hold sway above the flux", like the concepts of pure mathematics, but is firmly tied to some contingent and as yet unknown facts about the world. I have suggested that this elusive limit will extend at least as far as the Turing machine, but that it may extend yet further, to the simple SAD$_1$ computer and perhaps even to the AD computer. Indeed the limit point could even lie somewhere between these last two computers, e.g. at some simple SAD$_n$ computer, where n>1. In any case, this particular issue can only be settled with a deeper understanding of singularities and the status of the cosmic censorship hypothesis.

5. Towards a Theory of Non-Turing Computability

In his famous paper of 1937, Alan Turing set out the details of what he believed to be the most general computer. The result was surprisingly simple. Just a machine comprised of an ordinary mechanical device[4] and a infinite supply of paper tape for memory storage. This Turing machine is intuitive, for sure, but it is also fantastic: it possesses an unbounded quantity of material and is capable of operating for all eternity. This might lead us to worry that such a machine is physically impossible (how, otherwise, could it be set apart from the incredible shrinking computer of the previous section?), but for Turing and most of his followers there was no such concern. They simply grasped the idea of a finite computing device, closed their eyes, and extrapolated like mad.

Now Turing's intuition was, I presume, based on Newtonian spacetime. Had his intuition been deepened by exposure to SAD$_1$ spacetimes then he might have arrived at the SAD$_1$ computer in Figure 1. Of course that is only a guess. But it does show that if the intuitive approach of Turing and his followers—roughly, the global mechanics of the machine are unproblematic—is applied not to (the now defunct) Newtonian spacetime but to a spectrum of relativistic spacetimes, the result is not only Turing machines but also various kinds of non-Turing computers. This provides the initial impetus for elaborating a theory of non-Turing computability. The theory's *raison d'être* is further discussed in the final section. But now I turn to some results and conjectures of the theory itself.

We first recall that Turing computability theory ordinarily begins with an argument aimed at showing that any number of Turing machines in any configuration can always be mimicked by a single appropriately programmed Turing machine. The theory then proceeds in terms of this one abstract and easily characterised machine, and thereby manages to transcend irrelevant hardware details. I now use this approach with the simple SAD$_1$ computer and the AD computer (i.e. the least and the most powerful of the SAD computers, roughly speaking). Thus, an *ideal* simple SAD$_1$ computer is a simple SAD$_1$ computer that can mimic any other simple SAD$_1$ computer; and an *ideal* AD computer is defined analogously. Now consider the following five claims.

(a) The SAD1 computer underpinned by the spacetime in Figure 1 and fitted with a single Turing machine that follows the l-curve is an ideal simple SAD1 computer. I shall refer to this particular computer as the *naked Turing machine*.

(b) All partially Turing solvable problems are solvable by the naked Turing machine. They included: all Turing solvable problems, the Halting problem, first-order predicate logic (i.e. Hilbert's Entscheidungsproblem), Diophantine problem (i.e. Hilbert's tenth problem) and the word problem for semi-groups.

(c) Arithmetic is not decidable by the naked Turing machine.

(d) A decision problem that is not partially Turing solvable is not solvable by the naked Turing machine.

Statements (b), (c) and (d) are true ((b) is Result 1, (c) and (d) are easily proved), but statement (a) is a conjecture. My evidence for (a) derives only from the fact that all my attempts to construct a simple SAD_1 computer that cannot be mimicked by this putative ideal computer have failed.

Now to AD computers.

(e) The AD computer underpinned by the spacetime in Figure 4 and fitted with the AD solving hardware described in Section 2 is an ideal AD computer. I shall refer to this particular computer as the *multi-string computer*.

(f) Arithmetic is decidable by the multi-string computer.

(g) The multi-string computer can mimic any computer underwritten by a relativistic spacetime and containing only Turing machines and simple communications devices. In other words, the multi-string computer is the "ideal relativistic computer".

Statement (f) is true, but statements (e) and (g) are conjectures. Because (g) is stronger than (e), I will only try to justify the former. Two arguments support its case. The first is that *if* SAD_1 computers are the basic building blocks of relativistic computers (as they seem to be), and *if* forming strings is the best method of connecting computers together (as it seems to be), then the multi-string computer is king because it possesses strings of every order. The other reason is that all my various attempts to construct a machine that cannot obviously be mimicked by this computer have failed. For example, one can show that a single multi-string computer can mimic a string of multi-string computers; it can also mimic a countably infinite number of multi-string computers working "in parallel". I know of no problems that are not solvable by the multi-string computer. But some are sure to exist.

6. Concluding Dialogue

Frank, who works on the theory of computability by means of Turing machines, reckons these new computers are not worth the candle. Isabel disagrees.

Frank I like the idea of these non-Turing computers, but frankly I can't see them catching on. They're just too, well, fantastic.

Isabel Surely a Turing machine is fantastic. At least, an infinitely massive device capable of computing to eternity sounds pretty fantastic to me.

Frank Well that's one way of putting it. I prefer to think of a Turing machine as just the natural extension of an everyday computer.

Isabel In a way it is. And that's why the hardware of these new computers is chosen to be essentially nothing but Turing machines. A simple SAD_1 computer, for example, is just a Turing machine to the past of a point. Or in more picturesque terms, it's a naked Turing machine.

Frank Yes, O.K., when I said the non-Turing computers are fantastic I didn't mean the hardware so much as the spacetime supporting the hardware. Of course I believe in Turing machines; that's my job! No, my unease stems from the wild spacetimes you employ.

Isabel They are not all wild. For example, there is a spacetime that represents a charged black hole which is SAD_1.

Frank Yes, but surely the spacetime underlying our universe is not like that. These solutions are just idealisations.

Isabel That's beside the point. You don't want to rubbish a hypothetical computer—Turing or non-Turing—simply because it can't fit into our universe. If you do, you'll leave your precious Turing machine to the mercy of the cosmologists, because according to one of their theories, the universe and all it contains, will crunch to nothing in a few billion years. Your Turing machine would be cut-off in mid-calculation!

Frank O.K. I grant you that the particular spacetime structure of our universe has little bearing here. But isn't it true that while lots of really nice spacetimes can house Turing machines, all the spacetimes that might house non-Turing computers, including that black hole spacetime you just mentioned, are in some way grossly unphysical? I've heard you say yourself that these spacetimes are prone to infinite photon blue shifts, horizon instabilities, and the red pencil of the cosmic censor. They don't seem to have a hope.

Isabel On the contrary, they have lots of hope. For one thing, the problems you mention are probably just aspects of the putative cosmic censor; they're not additional problems. And the verdict on whether there is a cosmic censor could go either way: the jury is still out. I take it that you're prepared to accept that. In that case, am I to understand that you want to argue against non-Turing machines solely because the spacetimes that underwrite them may fall prey to a censor?

Frank Pretty much, yes.

Isabel Then let me put this to you. Just suppose that tomorrow we read in *Nature* that a new law of physics has been discovered that forbids spacetime to extend to temporal infinity. It would be a kind of "temporal infinity censor", if you like. Now that would censor your Turing machines, and so my question is: would the fact of this censor destroy the Turing machine's *raison d'être*? I mean: be honest, would you quit your line of research?

Frank Well probably not.

Isabel And I say "fair enough". But just as the prospect of temporal censorship need not affect your attitude towards Turing computers, so also the prospect of cosmic censorship need not put us off non-Turing computers.

Notes

[1] I would like to thank Gordon Belot, George Boolos, Rob Clifton, John Norton, Adrian Stanley and particularly Jeremy Butterfield for their helpful suggestions.

[2] I shall follow the standard notational conventions of Hawking and Ellis (1973). All spacetimes are assumed to be causally well-behaved in the sense that they satisfy strong causality.

[3]There are other known AD spacetimes, but this beautifully simple example is due to John Norton.

[4]In Turing (1937), the "mechanical device" is actually a man faithfully executing a finite set of instructions.

References

Boolos, G.S. and Jeffrey, R.C. (1989). *Computability and Logic*. Cambridge: Cambridge University Press.

Earman, J. and Norton, J. (1993). "Forever is a Day: Supertasks in Pitowsky and Malament-Hogarth Spacetimes" *Philosophy of Science* 5: 22-42.

Earman, J. (1994). *Bangs, Crunches, Wimpers, and Shrieks: Singularities and Acausality in Relativistic Spacetimes.* Forthcoming book.

Grünbaum, A. (1967). *Modern Science and Zeno's Paradoxes*. London: George Allen and Unwin.

Hawking, S.W. and Ellis, G.F.R. (1973). *The Large Scale Structure of Space-Time*. Cambridge: Cambridge University Press.

Hogarth, M.L. (1992) "Does General Relativity Allow an Observer to View an Eternity in a Finite Time?" *Foundations of Physics Letters* 5:173-181.

Horowitz, T. (1992) "Computability as a Physical Modality" Unpublished manuscript.

Pitowsky, I. (1990) "The Physical Church Thesis and Physical Computational Complexity", *Iyyun* 39: 81-99.

Turing, A.M. (1937) "On Computable Numbers, with an Application to the Entscheidungsproblem", *Proc. Lond. Math. Soc.* 42: 230-65; a correction 43, 544-6

Part IV

PHILOSOPHY OF CHEMISTRY

Spectrometers as Analogues of Nature

Daniel Rothbart

George Mason University

Traditional empiricist epistemologies have raised major philosophical obstacles to a clear understanding of instruments. Empiricist philosophers have given scant attention to instruments as a separate topic of inquiry because the reliability of instruments is presumably reducible to the epistemology of common sense experience. Instruments function presumably to magnify our physiologically limited sensory capacities by linking the specimen's sensory properties to accessible empirical data; such data are then validated by the same empiricist standards used to access ordinary (middle-sized) phenomena. Thus, no epistemic insight is revealed by studying instrumentation *per se*, according to empiricists.

Critics of empiricism have championed the cause that all sensory data are theory-laden. Yet even this dictum has the effect of minimizing the philosophical significance of instruments. This is because many critics of empiricism are working within the empiricist's distinction between the subjectivity of theory and the apparent objectivity of data. Such a distinction assumes a scientifically naive understanding of the design of scientific instruments. Once we overcome an empiricist conception of instrument design, we may realize that the theory-laden character of data does not imply the inherent failure (subjectivity, circularity, or rationalization) of instruments to expose nature's secrets. Rather than a warrant for its subjectivity, this theory-laden character of data is inseparable from the instrument's proficiency.

I argue that the success of many modern instruments is based on their design as analogical replicas of those natural systems which are familiar to scientists at a given time. Progress in designing proficient instruments is generated by analogical projections of theoretical insights from known terrain to an incomplete system. Instrumentation enables scientists to expand theoretical understanding to otherwise hidden domains. (Some of these themes are treated in considerably more detail in Rothbart and Slayden (1994).

After exploring this analogical function of instruments (Section 1), the nature of instrumental data is discussed (Section 2), followed by an explicit rejection of both skepticism and naive realism (Section 3). In the end I argue for an experimental realism which avoids any theory-neutral access to the fundamental analogies of nature.

1. Instruments as Replicas of Nature

In his *Data, Instruments, and Theory*, Robert Ackermann (1985) provides a rather sophisticated reformulation of empiricist doctrines, within the philosophical framework of evolutionary epistemology. Facts are organized into socially sanctioned data domains. Theories evolve in ways that best adapt to the environmental niches of "facts". For Ackermann the primary function of scientific experiments in which instruments are used is to break the line of influence from theory to fact, that is, to constrain the subjectivity of interpretation by the intersubjectivity of fact (Ackermann 1985, 128). Instruments are needed as epistemic intermediaries between theories and data domains, because instruments depersonalize experimentation. No single scientist can legitimately alter the experimental results.

Without exploring these general epistemic themes, the proposed rationale for instruments is considerably oversimplified. Contrary to the empiricists' claims, extraordinary phenomena do not become ordinary by sensory magnification but do become ordinary by projection of powerful theoretical insights. The reliability of most modern scientific instruments rests on the analogical underpinnings of instrumental design.

Typically, a modern scientific instrument is designed as a complex system of many action/reaction events; the specimen is "activated" by physically responding to humanly generated prodding, pushing, and poking within the experimental setting. Rather than observing the specimen in its passive state, the experimental scientist detects the physical reactions of the specimen to complex experimental stimuli. Such a reaction is called the experimental phenomenon of interest. The experimental phenomenon represents a local event of intersection between the specimen's structure and humanly manufactured experimental conditions. (Even the term "between" retains the Cartesian metaphor of sharply separating the internal from external realms. This metaphor underlies most ordinary language.) The experimental phenomenon is not generated exclusively by external physical structure and not generated exclusively by internal conceptualization. All experimental properties instrumentally detected are tendencies, or conditional manifestations of the specimen, to react to certain experimental stimuli. The specimen has tendencies which are manifested, only if certain humanly designed experimental conditions are realized and obstructing conditions do not materialize (Harré 1986, Chapter 15).

The physical sequence of events from specimen structure to data readout is understood theoretically as a technological analogue to multiple natural systems, based on background physical processes. Within instrumental engineering, such a sequence of real physical processes is intended to parallel select patterns found in natural systems. Thus, the instrument's success at exposing unknown properties of nature is tied directly to the capacity of scientists to extend analogue models of natural events to certain artificial experimental contexts.

Consider the basic design principles for absorption spectrometers, commonly used for identification, structure elucidation, and quantification of chemical substances. Modern absorption spectrometers were designed from the analogical projection of physical processes concerning the photoelectric effects of light.

Within an absorption spectrometer a beam of electromagnetic radiation emitted in the spectral region of interest passes through a monochromator, which is a series of optical components such as lenses and mirrors. This radiation then impinges on a sample. The monochromator isolates the radiation from a broad band of wavelengths to a continuous selection of narrow band wavelengths. These wavelengths can be held constant, or they can be scanned automatically or manually.

Depending on the sample, various wavelengths of radiation are absorbed, reflected, or transmitted. That part of the radiation which passes through the sample is detected and converted to an electrical signal, usually by a photomultiplier tube. The electric output is electronically manipulated and sent to the readout device, such as a meter, a computer, a controlled video display, or a printer/plotter. From this perspective the absorption spectrometer functions as a complex system of detecting, transforming, and processing information from an input event, typically an instrument/specimen interface, to some output event, typically a readout of information.

The interaction of electromagnetic radiation and a specific chemical sample is unique. The "fingerprint" of this interaction is revealed by the absorption spectrum over the entire electromagnetic energy continuum, and thus the interaction provides vital information about a specimen's molecular structure. Some of the most convincing evidence about atomic and molecular structure has been obtained by spectral analysis.

The success of spectral analysis is based on the following physical principle about atomic structure: if a specimen absorbs a certain wavelength of light (the wavelength corresponding to a particular energy), then that absorbed energy must be exactly the same as the energy required for some specific internal change in the molecule or atom. Any of the remaining energies in the light spectrum which do not match a specific change are "ignored" by the substance, and these energies are then reflected or transmitted. (The absorbed light energy causes such changes as atomic and molecular vibrations, rotations, and electron excitation.) As a result of the absorption, a specially designed instrument may detect an energy change, which we may "sense" in some cases as heat, fluorescence, phosphorescence, or color. Thus, the signal that is detected can expose the molecular structure of the specimen in terms of the specific patterns of absorbed and reflected/transmitted light energies.

Which instrument type and radiation source should be chosen for studying a particular specimen? This problem requires extensive knowledge of the range of chemical structures, the different types of spectrometers, the electronic processes, and the measurement of the resultant spectra. The designer's articulation of channel conditions, as well as the experimenter's operation, include complex modeling from electromagnetism, optics, atomic theory, chemistry, and geometry. Consider how the various stages of energy transformation throughout the instrument are represented iconically by various power flow models. A thorough understanding of the spectrometer requires a major segment of the physical sciences in general, a point which Hooker illustrates within the design of the Wilson cloud chamber for testing particle reactions (1987, 116).

The informational output of the absorption spectrometer centers on the electromagnetically understood energy absorbed by the specimen. Because such spectrometers are designed by analogy to the photoelectric effects, the conception of energy detected within the spectrometer is analogically derived in part from light beams consisting of discrete photons. When a flash of light is observed with a photomultiplier and displayed on an oscilloscope, the observed signal is a set of impulses (Bair 1962, 15). Such photoelectric signals from a flash of light function as the data-constituting analogue to the conception of energy detected in absorption spectrometers.

Thus, the empiricist's dictum that scientific instruments extend the limited sensory capacity distorts the inherent rationale for spectrometers: spectrometers function to expose the specimen's underlying physical structure by technological analogy to natural symmetries. Access to unknown properties of the specimen's structure occurs by theoretical extension of familiar physical systems. The technology exposes the specimen's unknown attributes by generating a moment of theoretical intersection between the actual and the possible, that is, between the familiar theories functioning externally to the experiment and the hypothetical models which presumably replicate the specimen's structure.

The instrument can expose previously hidden physical properties by cross-fertilization from known physical symmetries to the incomplete systems under investigation. This cross-fertilization motivates scientists to project parameters from the known models of natural phenomena to the inadequate models of physical processes underlying instrument design. As further documentation, C.T.R. Wilson designed the cloud chamber not as a particle detector but as a meteorological reproduction of real atmospheric condensation. As Peter Galison and Alexis Assmus document, meteorology in the 1890s was experiencing a "mimetic" transformation, in which the morphological scientists began to use the laboratory to reproduce natural occurrences (1989, 225-274). The mimeticists, as Galison and Assmus call these scientists, produced miniature versions of cyclones, glaciers, and windstorms. John Aitken's dust chamber, recreating the effects of fogs threatening England's industrial cities, directed Wilson's design of the cloud chamber (Galison and Assmus 1989, 265). Wilson transported the basic components of the dust chamber (the pump, reservoir, filter, values, and expansion mechanics) to his cloud chamber for the reproduction of thunderstorms, coronae, and atmospheric electricity.

J.J. Thompson and the researchers at the Cavendish laboratories gave the "same" instruments a new theoretical rationale. Rather than imitating nature, Thompson intended to take nature apart by exploring the fundamental character of matter. For their matter-theoretic purposes, scientists at Cavendish became indebted to Wilson's artificial clouds for revealing the fundamental electrical nature of matter. Galison and Assmus write:

> As the knotty clouds blended into the tracks of alpha particles and the 'thread-like' clouds became beta-particle trajectories, the old sense and meaning of the chamber changed. (1989, 268)

For 20th century physicists the formation of droplets were replaced by the energies of gamma rays, the scattering of alpha particles, and discovery of new particles. Wilson and the matter physicists proffered rival theoretical interpretations of the chamber's physical structure, interpretations derived from distinct physical analogues. Thompson and Wilson employed different instruments from one another.

2. Reading Instrumental Data

The reliability of instrumental data rests on a wide-ranging set of practices within the instrument's design, function, and operation. Ian Hacking's dictum that we see with a microscope and not through one (1983, Chapter 11) applies perfectly to spectral analysis. For many modern instruments the benchmark for the data's reliability is almost never direct visual perception, although sensory experiences are required for manipulating the apparatus. Within spectral analysis the prevalence of visual data, e.g., the yellow from a sodium flame, has been replaced in modern spectrometers by discursive readouts. Typically, the experimenter reads the graphic displays, the digital messages, or the coded charts directly from the instrumental readout. The computer controlled video display and the more common printer/plotter, for example, always convey information through a language that is accessible to the trained technician. For example, a photomultiplier readout device transforms the radiant energy of a signal into electrical energy, while simultaneously multiplying the generated current approximately one million times. It would even be possible for data to emanate from a computerized synthesizer producing an audio readout in English sentences.

Trivially, such digital readouts for modern instruments include a sensory component. The data from certain instruments, such as the bubble chamber, require the perceptual act of "seeing" images on photographic plates. But frequently the tracks take place much too quickly to be detected by an unaided visual system, and many trajectories of electrically neutral particles produce no visual record at all. Moreover, the data analysis frequently involves complex measurement techniques, with various statistical and curve fitting procedures. Visual images are far less relevant for data reliability than are lengthy and complex procedures of information extraction (Woodward 1989, fn. 35, 459). Such instruments constitute readable technologies because they communicate information directly to skilled "observers", as Patrick Heelan argues (1983).

The very determination of the number of spectral lines requires active manipulation of the apparatus, rather than the passive mental inspection of the content of sensory experience. The sharpness of the image is determined by the instrument's resolving power, which represents a limit of the separation of wavelengths into distinct images. How much drop in intensity must there be between adjacent spectral peaks for the lines to be considered resolved? Lord Rayleigh required of resolution an intensity drop of at least 19% between peaks. But any of several answers may be given, although most scientists require at least a 10% drop between peaks. In practice scientists operationalize the notion of resolution by actively adjusting the calibrations, manipulating the instrument's components, and changing the meter readings. No ideal value can be given for a single parameter in isolation from others. Instrument design and operation must be understood as combinations of conditions and combinations of circumstances ranging across distinct domains of inquiry.

No instrumental datum is secured to a non-patterned bedrock of empirical attributes. The information conveyed by data for most instruments is inherently structured into units of relatively low levels of abstraction. In general, "facts" are created by applying fine-grained conceptual systems, *ceteris paribus,* to the coarse-grained deliverance of our perceptual systems (Harré 1986, 176). Such information is conveyed by data structures, which are localized conceptual patterns representing series of detected signals. Of course data structures are theory-laden, but this declaration should not be used to resurrect the empiricist's issues surrounding the purity, or lack of purity, of sensory perception.

The conceptual grid defining each data structure is supplied by analogical connections to distinct structures. The construction of a particular data structure is conceptually driven by analogical projections from external conceptual grids, typically found in models of natural systems. The fine tuned conceptual grid of each data structure is partially generated by analogical projections from some distinct instrumental findings, as Harré has shown (1986, 175). Consequently, such analogue projections of such grids are data-constituting when the analogues define the limits and possibilities of an instrument's informational capacity.

Data-constituting analogies are quite prevalent throughout the history of optics. In his 1675 "Hypothesis of Light" Newton describes experiments on the "musical spectrum of light". As Penelope Gouk documents (1986), he discovers that the light spectrum can be divided into the same arithmetic ratios underlying a particular musical scale. Although initially categorizing light into five major colors, the musical scale of light motivates Newton to add two more colors. The spectrum's division into seven colors became obvious to Newton from the analogy to the seven tone musical scale, expressing the traditional names for a rising major scale from an eleventh century octave. Just as the musical sounds are generated by periodic vibrations of varying wavelengths in the air, so too the sensations of colors are produced by aethereal vibrations.

The intensity of vibrations generating red is causally analogous to a particular tone, "for the Analogy of Nature is to be observed" (Newton 1959).

3. Overcoming the Skeptic's Noise

How do we overcome skepticism concerning the capacity of instruments to reveal the specimen's physical properties? The skeptic is correct, although trivially, that no signal is completely unequivocal, that no mapping from data structure to specimen structure is one-to-one, and that the signal-to-noise ratio is never indefinitely high. Nevertheless, reliable channels of communication, based on background theories, can be in principle achieved. The signal is practically unequivocal, the mapping from data structure to specimen structure approaches one-to-one, and the signal-to-noise ratio can be maximized. The instrument's signals span extraordinary epistemic distances to a point within the experimenter's immediate access. The triple line sequence of spectral analysis is not an artifact of the scientists' conceptualizations. The signals from infrared detectors employed by astrophysicists to reveal newborn stars are not the complete fabrication of the experimenter's symbol system. The tracks of alpha particles within a bubble chamber are not fictitious concoctions by self-deluding scientists. A skeptic shows signs of neurosis if the channel conditions underlying the instrument are repeatedly checked beyond necessity (Drestke 1981, 115-116).

Furthermore, within instrument design the analogue system functions as an idealized prototype that is projectible onto the physical systems under scrutiny. Technological obstacles to instrumental progress frequently require radical conceptual breakthroughs. Such breakthroughs transform the "inconceivable" into the manifestly plausible by overthrowing antiquated scales of conceptual vision. Such analogue systems can acquire normative force for understanding the entire domain of inquiry. Yet the discovery of fresh analogies, and new prototypes, does not always require a monolithic overhaul of the entire scientific enterprise, as is suggested by a Kuhnian-type paradigm shift. A newly discovered analogy typically yields a localized transformation of some specific deficiency. No radical discontinuity of physical science is necessary from such prototypes.

Simultaneously, a prominent factor in judging a theory's success is its capacity to motivate instrumental progress. The theory's fertility is partially measured through the theory's capacity to generate instrument designs. Thus, a mutual dependence arises between instrumental design and theoretical progress: the instrument's design requires the complex combinations of various theoretical insights, and yet the theory's vitality is measured by its capacity to generate successful instruments. In this respect the internal/external distinction assumed above between the specimen's unknown parameters and the background theoretical models must be qualified.

The antirealist's dismissal of "unobservable" entities cannot be used to explain the success of absorption spectrometers. Such an instrument reveals the markings of reference through a maze of physical processes which define the signals' channel conditions. Existential continuity (Harré 1961, 54) extends from data structures to the specimen's unobservable physical properties. Existential commitment to the specimen's physical properties is signalled by the nomic nesting of physical processes within the data. A primary epistemic function of many modern instruments is to provide referential signals of physical attributes through such nomic nesting.

Nevertheless, the spectrometer's success inevitably reflects the human intervention of technological manipulation, conceptual segmentation, and implicitly theoretical ideas. Human intervention defines the immediate subject matter of instrumental inquiry in general. Again, scientific phenomena from absorption spectrometers are typically action/reaction events, which are artificially created. No pristine contact to the undifferentiated, and unmanipulated, realm of physical reality is attained from such instruments. The very domain of instrumental inquiry is typically a range of physical reactions which reflect the needs and capacities of human agency.

Finally, the success of spectrometers undermines the basis of convergent realism. The convergent realist promotes the scientific ideal of validating a unified ontological system, in order to capture the structure of an independent and undifferentiated physical reality. Such an ideal has been subject to the charge that no theory-neutral methodological measurement of rival theories has been discovered. As an additional charge against convergence, such a pure undifferentiated world is not even the immediate subject matter of many instruments. A goal of instrumentation is the systematic replication of the physical tendencies which are released under humanly generated enabling conditions. Such conditions are themselves dispositional properties which are exposed from some previous study. Human agency becomes pivotal to every level of instrumental inquiry. The interactionism within instruments is paradigmatic of scientific research generally. Throughout the history of science the conceptual foundations and premises of theories undergo continuous transformation, driven in part by powerful analogies of nature (Hesse 1980, 174; 1991).

References

Ackermann, R. (1985), *Data, Instruments, and Theory*. Princeton: Princeton University Press.

Bair, E. (1962), *Introduction to Chemical Instrumentation*. New York: Mcgraw-Hill.

Dretske, F. (1981), *Knowledge and the Flow of Information*. Cambridge, Massachusetts: MIT Press.

Galison, P. and Assmus, A. (1989), "Artificial Clouds and Real Particles," in Gooding, Pinch and Schaffer (eds.), *The Uses of Experiment*. Cambridge: Cambridge University Press, pp. 225-274.

Gouk, P. (1986), "Newton and Music: From the Microcosm to the Macrocosm," *International Studies in the Philosophy of Science: The Dubrovnik Paper I*, 1: 36-59.

Hacking, I. (1983), *Representing and Intervening*. Cambridge: Cambridge University Press.

Harré, R. (1961), *Theories and Things*. London: Sheed and Ward.

_____. (1986), *Varieties of Realism*. Oxford: Basil Blackwell.

Hesse, M. (1980), *Revolutions and Reconstructions in the Philosophy of Science*. Bloomington, Indiana: Indiana University Press.

_____. (1991), "Science, Beyond Realism and Relativism", in D. Raven, *et. al.*, (eds.), *Cognitive Relativism and Social Science*. London: Transaction Publishers, pp. 91-106.

Heelan, P. (1983), "Natural Science as a Hermeneutic of Instrumentation," *Philosophy of Science, 50:* 181-204.

Hooker, M. (1987), "On Global Theories", *A Realistic Theory of Science*. Albany, New York: SUNY Press.

Newton, I. (1959), "Newton to Oldenburg: 7 December 1675", in *The Correspondence of Isaac Newton,* Volume I: 1661-1675, H. W. Turnbull (ed.), Cambridge: Cambridge University Press, p. 376.

Rothbart, D. and S. Slayden (1994), "The Epistemology of a Spectrometer," *Philosophy of Science* 61: 25-38.

Woodward, J. (1989), "Data and Phenomena", *Synthese 79*: 393-472.

van Fraassen, B. (1980), *The Scientific Image*. Oxford: Clarendon Press.

Ideal Reaction Types and the Reactions of Real Alloys[1]

Jeffry L. Ramsey

Oregon State University

1. Introduction

Scientists must often solve analytically or computationally intractable problems. Philosophers of science have problematized the nature and role of laws, theories and evidence by examining how scientists use abstractions (Cartwright 1990), idealizations (Laymon 1985, Brezinski et. al. 1990), approximations (Cartwright 1983; Laymon 1989; Ramsey 1990, 1992), and visual images (Wimsatt 1991, Ruse 1991) in such situations. Here, I problematize the nature and role of concepts by examining ideal types as a response to intractability. I focus on the work of Carl Wagner, a former director of the Max Planck Institute for Physical Chemistry in Goettingen. Wagner used ideal types in his research on the oxidation of metal alloys during the middle third of this century.

I argue that: (1) natural scientists do use ideal typical concepts; (2) these concepts are distinct from 'standard' concepts; and (3) ideal typical concepts provide one means of connecting a tractable theoretical model with the experimental evidence. Weber (and others involved in the Methodenstreit) accepted (2) but denied (1) to separate the historical and natural sciences, whereas Hempel (1965) and Papineau (1976) accept (1) but deny (2) to unify them. In contrast, I argue for local identities of methodology due to similarities in the epistemology of the problem solving situation. The local identity I describe involves a natural scientist borrowing from a social scientist. (3) allows me to say something about how theoretical terms obtain meaning as a scientist 'articulates' (cf. Hacking 1983) an intractable theory with the evidence. In line with current work on concepts and categories (cf. Neisser 1987, Lakoff 1987), ideal typical concepts provide articulation *via* models and theoretical structures rather than through objective perceptual attributes alone. This fact allows me to draw out connections between conceptualization and experimentation in the natural sciences. In particular, ideal types allow scientists to discover new boundary conditions in phenomenologically complex situations. The process of articulation with ideal types also supports claims (Longino 1990, Giere 1988) that values inform the practice of the natural sciences.

I argue first that ideal types constructed to overcome intractability are relevantly different from 'standard' concepts. I use this interpretation to rebut Hempel's and

Papineau's claims. Following this, I demonstrate how the proper understanding provides a critical perspective on Wagner's practice and more generally on the use of ideal types to overcome intractability. Locating a robust example in the natural sciences will serve to rebut Weber's position. Throughout, I comment on the connection between conceptualization and experimentation and also on the role of values in the construction of ideal types in the natural sciences.

2. A Proper Understanding of Ideal Types

I appeal to a standard interpretation of ideal types (Burger 1987, Hekman 1983) in order to make relevant points about their structure, function, and axiology. Ideal types have a curious structure because they do not satisfy the applicative principle, i.e. "the simple principle that what is true of every is true of any" (Braithwaite 1953, 47). An ideal type is not a mirror of any reality. "It is even less fitted to serve as a schema under which a real situation or action is subsumed as one instance" (Weber 1949, 93). That is, ideal types do not stand to instances as general to specific. Like prototype concepts, it is not always possible to determine whether the ideal type applies to an instance. Unlike prototypes, however, ideal types are defined constructs and thus are characterized by a set of necessary and sufficient conditions.

How can ideal types be defined precisely yet fail to embody the applicative principle? In supposedly 'standard' concept formation (Weber 1949, Hempel 1952), the features of, e.g., trees are examined to see which properties all trees must possess in order to be a tree. The concept 'tree' is then defined by an ostensive definition listing the necessary and sufficient conditions; higher-order, or theoretical, concepts are defined using chains of definitions. Since the necessary and sufficient properties were discovered (or defined) on the basis of commonality, it is true of any one tree that it has the properties common to all trees. In contrast, "an ideal type is formed by the one-sided exaggeration (*Steigerung*) of one or several viewpoints and by the combination of a great many single phenomena (*Einzelerscheinungen*) existing diffusely and discretely, more or less present and occasionally absent..." (Weber 1949, 90). The exaggeration "is equivalent to the disregard of the causal influence of certain empirically operative factors" (Burger 1987, 126). That is, the necessary and sufficient conditions used to define ideal types are not common to all members of the class. Thus, the applicative principle can fail.

To make sense of this, consider first the exaggeration of one viewpoint: if the phenomena exhibit the properties or actions A_1, B_1, C_1, D_1, the ideal type X might be formed by disregarding C_1 and D_1. This viewpoint can then, if desired, be combined with other exaggerations. Suppose ideal type Y is formed by discarding C_2 and D_2 and keeping A_2 and B_2, and ideal type Z is formed similarly from the A_3, B_3, C_3 and D_3 properties. As an example, take the concepts of endo-, ecto- and mesomorphic body types. In this case, the applicative principle fails because almost every person is an interpolated mixture rather than a 'pure' body type. Most individuals can not be said to 'satisfy' the ideal type.

An ideal type can also be constructed from several viewpoints. Here, the properties A_1, B_2, C_3 and D_2 might be combined into a concept. The failure of the concept to embody any particular instance is then rather obvious: if the instance has none of those properties, the concept does not apply to the individual.

According to Weber, an ideal type functions retrodictively as a kind of pattern-matching. "The ideal-type is an attempt to analyze historically unique configurations or their individual components by means of...concepts.... It has the significance of a purely ideal limiting concept with which the real situation or action is compared and surveyed for the explication of certain of its significant components" (Weber 1949, 93). Ideal types allow us to make particular judgements about particular instances of behavior; in such cases, we are less interested in universal generalizations which stem from laws which are simultaneously general and empirical. (On a side note, Theobald (1976) has noted that explanation in chemistry has this character quite generally.) Ideal types can thus serve an important epistemological role through discovery of patterns in previously unanalyzable complexes of data.

The construction of ideal types involves values necessarily. "Order is brought into this chaos [of judgements about individual events] only on the condition that in every case only a part of concrete reality is interesting and significant to us, because only it is related to the cultural values with which we approach reality" (Weber 1949, 78). In terms of the simplified model above, values enter into ideal types because we choose, e.g., A_1 and B_1 for a reason. Ideal types do not reveal the 'essence' of phenomena; they are "neither historical reality nor even the 'true' reality" (ibid., 93). Ideal types are chosen for reasons not tied to discerning the 'natural' commonalities of phenomena. Different reasons will give rise to different ideal types.

I turn now to Hempel's and Papineau's arguments that ideal types have no distinct structure, function or value component. Both collapse all distinctions in order to maintain a unified methodology for the sciences.

Hempel (1965) correctly claims ideal types are used to order phenomena as 'more or less like this or that' and not to classify them 'as' this or that (cf. Burger 1987, 157). However, he incorrectly claims ideal types are non-semantical, explanatory concepts which have the status of hypotheses. According to Hempel, ideal types can serve their ordering purpose "only if they are introduced as interpreted theoretical systems" (Hempel 1965, 171). Interpretations link the theoretical structure with the data. The relations and not the idealized concepts themselves carry the empirical content. However, Weber noted that "ideal types are got from the actual...world..." (quoted in Burger 1987, 128). That is, the semantical relations between the elements of the concept are given by pre-theoretical experience and not by a separate interpretive step. For example, in setting up types of authority, Weber reasoned that "legitimation is important since simple observation shows that...he who is more favored feels the never-ceasing need to look upon his position as in some way 'legitimate'..." (Bendix 1971, 258-259).

Hempel misdiagnoses the semantical nature of ideal types because he thinks they "are intended to provide explanations [and] must be construed as theoretical systems embodying testable general hypotheses" (Hempel 1965, 166). However, as noted above, ideal types are largely retrodictive and pattern-discovery oriented and thus fail to meet the criteria for explanations laid down by Hempel. (Whether they meet the criteria of other models of explanation is an open question.) In addition, Weber says explicitly that an ideal type is "no hypothesis, but it offers guidance to the construction of hypotheses. It is not a description of reality but it aims to give unambiguous means of expression to such a description" (quoted in Burger 1987, 120-121). For Weber, concepts and the laws in which they appear are *not* to be "interpreted as hypotheses postulating certain invariant relationships between particular variables" (ibid.). Since Hempel thinks only 'standard' concept formation will produce laws which preserve the meaning of theoretical concepts (Hempel 1952) and ensure the applicative principle will hold (cf. Braithwaite 1953, 85), he has mistakenly interpreted ideal types as explanatory statements of invariant relationships.

Papineau notes that he "shall take it that a distinguishing characteristic of ideal type concepts is that they have no instances" (Papineau 1976, 137). However, Papineau has misdiagnosed how ideal types are constructed and thus overestimates the impossibility of their empirical instantiation. Like Hempel, Papineau appears to assume concepts must be formed by ostensively-based definitional chains in order to ensure the concept has meaning. Since ideal types have no instantiation, they can have no direct meaning. They get indirect meaning by functioning as Lakatosian negative heuristics.

Weber did claim the ideal type "cannot be found anywhere in reality" (Weber 1949, 90). However, "he did not want to convey the idea that they are free inventions of the imagination but merely that *they describe phenomena which in principle are empirically possible* but have failed to materialize as a result of factual circumstances" (Burger 1987, 124, emphasis added). If different factual circumstances obtain, it is conceivable that an actual empirical system will instantiate the meaning relations of the ideal type. This becomes more than an abstract possibility if we are able to construct an experimental situation which "materializes" only the causal factors assumed to operate in the pure ideal type. As we shall see when turning to Wagner's use of ideal types, this is perhaps a relevant difference between the cultural and physical sciences.

Papineau interprets ideal types as idealizations and is thus forced to think of them as models in which, if the assumptions could be relaxed, the ideal type would come to look more and more like the empirical phenomena it was constructed from. Despite the name however, ideal types are abstractions rather than idealizations. This is the import of Weber's claim noted earlier that ideal types are not descriptions—ideal or not—of reality; rather, they aim to give precise means to such a description. Very roughly, we can view an idealization as a deformation of existing properties (often by setting a property's value to zero) whereas an abstraction studies only a few of the properties in isolation (Cartwright 1990). Because ideal types are abstractions rather than idealizations, they are more like morphic body types than ideal gases. (I actually think ideal gases are not simply idealizations but will bypass that issue here.) For instance, we say that I am endomorphic with a tendency to mesomorphy. If we viewed ideal types as idealizations, we would view each type as a constraint to be satisfied. If I satisfied the ectomorphic type, we would then check to see if that I was also endomorphic, and so on. But of course we do not do work this way. The ideal types are not to be viewed as successive constraints to be satisfied, but as a mapping of the possible space of body types. In addition, we do not rule out *a priori* the possibility that someone will be an almost pure ectomorph, endomorph or mesomorph. Virtually "pure" instantiations are possible. Since ideal types are abstractions, it is—contra Papineau—possible to construct or discover instantiations of the ideal types.

If and when the type is instantiated, we can do experiments and gain knowledge of the theoretical model and of a new sector of empirical reality. In particular, we gain information how and under what conditions types behave. However, we gain only indirect knowledge of the domain which was used to construct the ideal type.

In sum, researchers use ideal types to comprehend phenomenologically complex situations. The ideal types have a prior semantical interpretation which derives from pretheoretical experience. An ideal-typical model is retrodictive and provides a means of assessing and ranking the various factors in any given individual case. Which factors are selected is a function of the cultural values of interest to the scientist.

3. The Method of Ideal Types in Metallurgy

During the 1930s and '40s, Carl Wagner resorted to an ideal-typical analysis because oxidation reactions of alloys were intractable with the analytical categories derived from the study of pure metals. Wagner once noted

Already my earliest work on the oxidation of alloys at high temperatures showed that each technical case involved a confusing abundance of individual problems. Starting from single observations I endeavored to pare away (*herauszuschaelen*) by means of thought simple boundary cases as so-called types and then to look for the experimental treatment of such real systems, which came close in some measure to the theoretical boundary cases. In case one knows the characteristic lawlike-behavior for a sufficient number of boundary cases, it is to be hoped that the behavior of technically interesting cases can be understood through the superposition (*die Ueberlagerung*) of the lawlike behaviors found for the boundary cases. In advantageous cases they could be described half-quantitatively. This way of working is something the same as a psychologist, who endeavors to understand and describe the wide diversity of human characters with the aid of character-types as theoretically exhibited (*erschauten*) boundary cases (Wagner 1958, 349).[2]

In this section, I argue that the understanding of ideal types developed in the previous section provides a much better critical perspective on Wagner's work than either Hempel's or Papineau's reinterpretations of the Weberian methodology. In particular, it allows us to understand how Wagner used ideal types to perform quantitative analysis and why he was able to discover empirical systems which instantiated the ideal types.

In the 1930s and 40s, oxidation reactions of alloys were analytically intractable. For a number of reasons, the analytical categories and mechanisms used in the study of pure metal oxidations did not transfer neatly to these complex alloy oxidations. First, in pure metal oxidations, only one metal is involved and it is found only in the oxide layer. That is, the oxidation produces a thin 'crust' of oxide. In alloys, in contrast, the oxide layer and the adjoining alloy layer can become enriched with the oxidizing metal. Also, elements from the alloy layer can diffuse out into the oxide layer. A number of factors including the mobilities of the cations in the oxide phases contribute to these processes.

Second, pure metals usually oxidize with the formation of a very simple boundary, i.e. the 'crust' layer contains an oxide of one chemical composition. In contrast, oxidizing alloys form a number of zones with two or more phases each. For instance, in a copper-nickel alloy, the oxide layer might have one zone of nickel oxide (NO) and copper (Cu) and another of nickel oxide and cupric oxide (Cu_2O). Other zones might contain different concentrations of the same oxides or oxides with the metals in different oxidation states. The result is a wide range of oxides with different compositions ($M_nN_yO_z$). A good (albeit imperfect) analogy is mixing various clays in various proportions before firing the clay to produce bricks. If you begin even with only three clays and allow the percentages of each to vary independently, the number of possible kinds of bricks is very large.

To further complicate matters, the variations are caused by numerous mechanisms. Wildly different mechanisms can give rise to qualitatively similar final compositions. Also, different mechanisms produce different experimentally measurable dependencies. For instance, even in pure metals, the movement of cations or anions between interstices in the lattice will affect the rate of reaction (with respect to, e.g., the partial pressure of oxygen) differently than will electron overconductance. Without knowing precisely how or with what effect, metallurgical chemists knew the the difficulties multiplied quickly when this mechanistic pluralism occurred in the large number of oxide layers in alloys.

Wagner claimed it did "not appear useful to develop a comprehensive system which would cover all" of the above difficulties (Wagner 1959, 773). Instead, he constructed two ideal types. These boundary cases were sufficiently "surveyable" (*uebersehbar*) and could be analyzed "half-quantitatively" (*halb-quantitativ*). The first ideal type is shown in Figure 1.

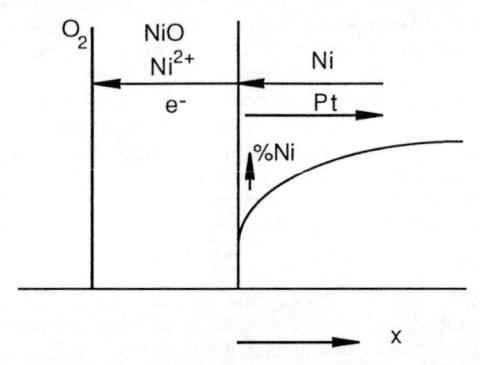

Figure 1. Oxidation of Nickel-Platinum Alloys (x = distance from the original surface). (From Wagner 1959)

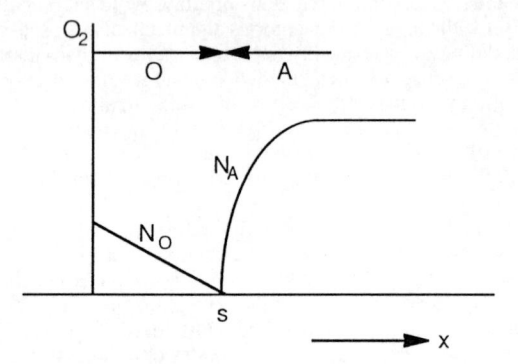

Figure 2. Diffusion processes during the inner oxidation of an alloy A-B with the precipitation of oxide AO_x at $x = s$

The non-noble metal (here, nickel) diffuses out to the oxygen layer resting on the surface of the alloy. There, the nickel reacts with the oxide to create a surface layer of oxide. In the second ideal type, shown in Figure 2, the oxygen is soluble in the alloy (here, a silver alloy) and so diffuses into the alloy before reacting with a non-noble metal such as copper, cadmium or aluminum. As a result, a metallic oxide rests within a non-oxidized metallic matrix.

Contra Hempel, Wagner constructed the ideal types semantically. He did not postulate relations among uninterpreted symbols and then give the symbols an intepretation. Rather, he used the empirical system at hand to generate possible categories of analysis. He appealed to the pre-theoretical observational material to make the following inductive conclusions: at small concentrations of non-noble metals and a high oxygen solubility, the second type would be favored; the first type would be favored at small oxygen solubilities and high concentrations of the non-noble alloy component. As it should with Weber's methodology, the meaning relations were constitutive of the ideal type and not supplied as an additional interpretive step. Even as a rational reconstruction, Hempel's analysis incorrectly identifies the direction of interpretation and the grounding for the process of interpretation.

For Wagner, these categories were abstractions rather than idealizations. The two types were "ideal" because he ignored many possible causes of deviation. At various points, Wagner listed no less than seven such disturbing causes. Some were chemical. For instance, he assumed the alloy formed a complete series of mixed oxides in which all valence states of the metal and oxidizing element were represented. (This minimizes reactions between oxides in the layers.) Some of the omitted causes were metallurgical phase problems such as 'fingers' rather than 'walls' of oxide growing into the alloy. Isolating the behavior of the alloys independently of all these disturbing factors allowed Wagner to exaggerate the behaviors of interest to him.

Wagner then made a methodologically significant move. In order to investigate how the simple pictures responded to different diffusion rates and alloy compositions, he needed to perform some mathematical analysis. Since calculations were still too difficult, he employed a series of simplifying idealizations. He simply assumed various conditions were satisfied. For instance, he assumed the increase in volume caused by uptake of oxygen was negligible and that the sample was thick enough to contain unreacted alloy after the reaction was complete. These simplified calculations allowed Wagner to discover possible boundary conditions for the analysis of the more complex systems.

Constraints on mathematical analysis in the sciences have typically focused on the existence of unsolvable equations or imprecise boundary conditions (Laymon 1989; Ramsey 1990, 1992; Wimsatt 1989). Both of those problems require knowledge of the existence of boundary conditions for their solution. However, not all problems arrive with boundary conditions (however imprecisely known). Ideal types provide one means of discovering such conditions and thus one way of exploring mathematically and analytically the extremes of physically possible processes. For instance, Wagner was able to derive the equation $D_A/D_O \ll N_O/N_A \ll 1$ to represent the formation of an outer oxide layer. D_A and D_O represent the diffusion constants of the metal and oxygen respectively, and N_O and N_A the mole fractions of oxygen and the metal respectively. In short, Wagner discovered a boundary condition which had not been known before. On the basis of previous experiments, chemists had been able to state only that "at small solubilities of oxygen and high concentrations of non-noble alloy components an outer oxide layer would be favored." The restatement not only clarified the relationship of the rough and ready experiential construct to the theoretical context of diffusion processes, it also allowed the researcher to design new experiments where the boundary conditions were either violated or approximately satisfied.

Importantly, the idealizations are epistemologically and methodologically independent of the abstractions used to construct the ideal types. For a quantitative analysis to be performed, the idealizations are conditions which even the messy phenomenological situations which Wagner confronted initially would have to meet. The abstractions apply only to a particular set of reactions constructed in the abstracting process itself. Their quantitative analysis is admittedly improved by ignoring causal disturbances, but that fact is separate from the improvement in mathematical treatment afforded by the idealizations.

Wagner then found some systems which he could study to understand the behavior of the ideal types. Contra Papineau, it is not impossible to instantiate ideal types. And unlike the historian, the laser physicist or the Hardy-Weinberg equilibrium theorist, Wagner and other metallurgists did not have to wrench nature into a new form to produce the systems. They were then able to do experiments on the naturally occurring systems and compare the results to the calculations performed on the ideal typical model. By learning more about the model, Wagner could then infer more precisely the behavior of the complex alloy systems.

It is always epistemologically and ontologically possible that an event or process will exhibit only the elements of the ideal type. Such systems serve a double epistemological function as instantiations of the ideal types and abstractions of the original complicated systems. Even though coincident, these functions are separate. Knowledge of the simple systems does not guarantee knowledge of the complicated systems. Wagner recognized this implicitly when he noted that the ideal typical constructs had to be superimposed (*ueberlagert*) onto other alloy systems. Emphatically, they were not to be treated as the underlying reality which persisted unabated in the complex systems.

Wagner repeatedly emphasized the technical value of the systems he studied. Corrosion by oxidation has long been an expensive problem confronting industry. Components of heat exchangers, petroleum refineries and ships have been made of cupro-nickel alloys for many years, and finding a way to prevent the corrosion of such was the subject of much research in the 1940s and 50s (cf. Leidheiser 1971, 85-87). Wagner recognized that most technically important alloys did not even begin to meet the conditions embodied in the ideal types. However, he was providing some handle (however weak) on technically important issues. I have no direct evidence that Wagner's inclusion of certain properties and behaviors in the ideal types was influenced by such pragmatic concerns; nonetheless, Wagner appealed to such problems in his work.

In sum, Wagner abstracted from pre-theoretical experience to construct ideal types. He then discovered some empirical systems which allowed him to understand more completely their behavior. With added knowledge of the model, he then returned to study the behavior of more complex alloys. By noting continually that many if not most of the technically interesting cases of the oxidation of alloys could not be analyzed directly with the ideal types, Wagner recognized that his ideal types provided only a means of interpolation and were not extrapolations of the underlying reality of the complex. Even so, Wagner's ideal typical analysis overcame some of the intractability which plagued the study of complex alloy oxidations.

4. Conclusion

I appealed to a standard interpretation of ideal typical concepts to pursue the claim that they have a distinct epistemological and functional status. I then argued that accounts of ideal types in the natural sciences which conflict with this interpretation should be rejected. Such accounts do not make sense of the practice of at least one natural scientist who used ideal types.

At the methodological level, a focus on ideal types as responsive to local problem situations implies similar problem situations should give rise to similar methodological perspectives. That is, a local similarity of strategy rather than a global unity of methodology is the result. If, e.g., ideal types appear more often in one science than another, that probably says more about the historical situation of the science and the kinds of problem its scientists are attacking than it does about the nature of scientific methodology *per se*. In and of itself, that seems to imply that the social sciences need not try to be more 'scientific.' Borrowing can occur in any direction. Methodology should be responsive to problem situations and not to the exigencies of philosophical ideals.

The considerations I have advanced also have epistemological and ontological consequences. In particular, a plurality of concept types seems warranted. Also, a specific amalgam of doctrines about meaning and concept formation should be rejected. Even though objective, perceptual cues are utilized in the construction of the ideal type, those cues do not exhaust the meaning or application of the concept. The ideal typical method involves abstraction and rearrangement of given properties into something quite novel. As a result, the concept will apply weakly or not at all to instances of the phenomenologically complex system. The novel complex of the ideal type may allow new features such as boundary conditions to be discovered. Nothing in the ideal-typical method precludes finding the abstracted constellation of properties in another empirical system. If such a system can be found, the pragmatically-based theoretical structure can be articulated more concretely with the world. In the process, the ideal typical concepts will be imbued with more meaning. Also, the model's adequacy will be enhanced.

However, the construction process does imply no essence will be embodied in the ideal type. Such concepts are constructed with particular purposes in mind. Given a different problem situation, a scientist can appeal to the same pre-theoretical experience to construct very different ideal types. Even if an isolated part of reality instantiates the ideal type, it does not follow that an underlying reality persists unabated in the complex systems. Ideal types are not isolated representations of reality; they are representations of an isolated, abstracted reality.

Notes

[1] This is a much changed version of a paper read at the 9th International Congress for Logic, Methodology and Philosophy of Science. Many thanks to the audience there for stimulating questions and comments. The original research was conducted as a Walter Rathenau fellow with the Verbund fuer Wissenschaftsgeschichte. Final revisions were completed while a post-doctoral fellow with the Center for the Philosophy of Science at the University of Minnesota. I wish to express my thanks to these institutions for their support and their hospitality.

[2] This and all subsequent quotations from Wagner's papers are my translations.

References

Braithwaite, R. (1953), *Scientific Explanation.* Cambridge: Cambridge University Press.

Brezinski, J., Coniglione, F., Kuipers, T.A.F., Nowak L. (eds.) (1990), *Idealization I: General Problems* and *Idealization II: Forms and Applications*. Rodopi: Amsterdam.

Burger, T. (1987), *Max Weber's Theory of Concept Formation*, New Expanded Edition. Durham, N.C.: Duke University Press.

Cartwright, N. (1990), *Nature's Capacities and Their Measurement.* Oxford: Oxford University Press.

_ _ _ _ _ _ _. (1983), *How the Laws of Physics Lie*. New York: Oxford University Press.

Giere, R. (1988), *Explaining Science*. Chicago: University of Chicago Press.

Hacking, I. (1983), *Representing and Intervening*. New York: Cambridge University Press.

Hekman, S. (1983), *Weber, the Ideal Type and Contemporary Social Theory*. Notre Dame, IN: Notre Dame University Press.

Hempel, C. (1965 [1952]), "Typological Methods in the Natural and the Social Sciences," in *Aspects of Scientific Explanation*. New York: The Free Press, pp. 155-172.

_ _ _ _ _. (1952), *Methods of Concept Formation in the Sciences*. International Encyclopedia of Unified Science, Vol. II, Number 7. Chicago: University of Chicago Press.

Lakoff, G. (1987), *Women, Fire and Dangerous Things*. Chicago: University of Chicago Press.

Laymon, R. (1989), "Cartwright and the Lying Laws of Physics," *Journal of Philosophy 86*: 353-372.

_ _ _ _ _ _. (1985), "Idealizations and the Testing of Theories by Experimentation," in P. Achinstein and O. Hannaway, (eds.), *Observation, Experiment and Hypothesis in Modern Physical Science*. Cambridge, MA: MIT Press, pp. 147-173.

Leidheiser, H. (1971), *The Corrosion of Copper, Tin and Their Alloys*. NY: John Wiley and Sons.

Longino, H. (1990), *Science as Social Knowledge*. Princeton: Princeton University Press.

Neisser, U. (ed.) (1987), *Concepts and Conceptual Development*. New York: Cambridge University Press.

Papineau, D. (1976), "Ideal Types and Empirical Theories," *British Journal for the Philosophy of Science 27*: 137-146.

Ramsey, J. (1992), "Towards an Expanded Epistemology for Approximations," in K. Okruhlik, A. Fine and M. Forbes, (eds.), *PSA 1992, vol. 1*. East Lansing, MI: Philosophy of Science Association, pp. 154-164.

_ _ _ _ _. (1990), "Beyond Numerical and Causal Accuracy," in A. Fine and M. Forbes, (eds.), *PSA 1990, vol. 1*. East Lansing, MI: Philosophy of Science Association, pp. 485-499.

Ruse, M. (1991), "Are Pictures Really Necessary? The Case of Sewall Wright's 'Adaptive Landscapes'," in A. Fine, M. Forbes and L. Wessels, (eds.), *PSA 1992, vol. 2*. East Lansing, MI: Philosophy of Science Association, pp. 63-78.

Theobald, D. (1976), "Some Considerations on the Philosophy of Chemistry," *Chemical Society Reviews 5*: 203-213.

Wagner, C. (1959), "Reaktionstypen bei der Oxydation von Legierungen," *Zeitschrift fuer Elektrochemie 63*: 773-782.

Wagner, C. (1958), "Die Aufgaben des Max-Planck-Instituts fuer physikalische Chemie," *Mitteilungen aus der Max-Planck-Gesellschaft zur Forderung der Wissenschaften*, pp. 347-350.

Weber, M. (1949), *The Methodology of the Social Sciences*, trans. E. A. Shils and H. A. Finch. Glencoe, IL: The Free Press.

Wimsatt, W. (1991), "Taming the Dimensions—Visualizations in Science," in A. Fine, M. Forbes and L. Wessels, (eds.), *PSA 1992, vol. 2*. East Lansing, MI: Philosophy of Science Association, pp. 111-135.

Has Chemistry Been at Least Approximately Reduced to Quantum Mechanics?

Eric R. Scerri

London School of Economics

1. Introduction

In order to discuss the question of the reduction of chemistry it will be necessary to begin with a brief review of what philosophers mean by reduction in science. I follow most authors on this subject by starting with the writings of Nagel (Nagel 1961). As is well known, Nagel stipulates that two formal conditions, namely connectability and derivability should be fulfilled in order to say that reduction of theory T_2 to theory T_1 has occurred. In addition he stipulates a non formal condition, that the primary or reducing science should be supported by experimental evidence.

Furthermore Nagel considers that reductions occur in two main varieties which he calls homogeneous and heterogeneous respectively. In homogeneous reduction the terms used by the reducing theory are also common to the theory to be reduced. For example Galileo's science of freely falling bodies was conceived of as separate from the mechanics of celestial bodies. Nagel claims that Newton's theory of mechanics absorbed or reduced both of these theories. The reduction in question is considered to be homogeneous since no new concepts are needed to describe motion in the Newtonian theory than were used in the older forms of mechanics. In heterogeneous reduction the distinctive traits of some subject matter are assimilated into those of a set of quite different traits. Nagel sees this type of reduction as problematical and worthy of further analysis unlike homogeneous reduction.

We may note in passing that the question of the reduction of chemistry would presumably fall into the heterogeneous category according to Nagel's scheme since, as many authors have pointed out, some typically chemical terms cannot be found in quantum mechanical language. For example Primas has written,

> Many calculations have been extremely sophisticated, designed by some of the foremost researchers in this field to extract a maximum of insight from quantum theory. For simple molecules, outstanding agreement between calculated and measured data has been obtained. Yet, the concept of a chemical bond could not be found anywhere in these calculations. We can calculate bonding energies without ever knowing what a bond is! (Primas, 1983, 5)

A number of criticisms of Nagel have appeared, which aim to sharpen the criteria for reduction. Some of these have been described as falling under the label of indirect reduction. For example Kemeny & Oppenheim claim that T_2 is not obtained in terms of T_1 as in Nagel's form of reduction. Instead one obtains identical observable evidence from T_2 and T_1 although T_1 can may predict more. (Kemeny & Oppenheim 1956)

Other authors deny that reduction of theories takes place at all. Popper would presumably argue that if a theory is refuted it can hardly be said to correspond to, or be derived from the reducing theory. In Popper's own words,

> Newton's theory unifies Galileo's and Kepler's. But far from being a mere conjunction of these two theories...it corrects them while explaining them. The original explanatory task was the deduction of the earlier results. It is solved not by deducing them but by deducing something better in their place. (Popper 1957, 33)

Then there are the radical critics of reduction. The early Kuhn holds that two competing theories cannot be compared because the terms used in each theory are incommensurable (Kuhn 1962). Also, for Kuhn all significant reductions are replacement reductions, where the reduced theory is replaced by the theory to which it reduces and not retained as a correct theory.

Meanwhile, in characteristic fashion, Feyerabend attacks both of Nagel's formal conditions for reduction (Feyerabend 1962). For example, Nagel considers that Galileo's laws of mechanics are reduced to Newton's laws in the sense that both conditions (connectability and derivability) are fulfilled.

With regards to connectability, Feyerabend claims that this feature often does not exist between successive theories. For example, in classical mechanics length is a relation that is independent of signal velocity, gravitational fields and motion of the observer. This meaning of length differs from the meaning of length in relativity theory in which it is dependent on all three of the above factors. Classical and relativistic length he says are incommensurable concepts. On the question of derivability, Feyerabend points out that in Galileo's laws of mechanics the acceleration of a freely falling body is constant whereas in Newton's mechanics, acceleration increases with decreasing distance from the earth. The two systems of mechanics are thus incompatible according to Feyerabend. His conclusion is that either (i) approximate reduction is possible but we must forego derivability as a condition for reduction, or (ii) reduction fails. Feyerabend eventually decides on the second option whereby reduction fails altogether.

The early Putnam has criticized Feyerabend on the question of derivability claiming that Nagel's requirement can be maintained provided that we accept approximations.

> It is perfectly clear what it means to say that a theory is approximately true, as it is clear what it means to say that an equation is approximately correct: it means that the relationships postulated by the theory hold not exactly, but with a certain specifiable degree of error. (Putnam 1965, 206-207)

I believe that Putnam's statement can be used to give a working definition of what constitutes approximate reduction and I will return to this statement in a later part of this article.

For the intervening sections however I wish to adopt an approach which I will call pragmatic reduction. I will attempt to examine the extent to which chemistry has been reduced in the terms used by physicists and chemists themselves. The most overt attempts at reduction in chemistry have been made through the use of Schrödinger's time independent equation. In the purest or ab initio approach the aim is to calculate the properties of atoms and molecules entirely from first principles, without recourse to any experimental input whatsoever.[1] Another part of current theoretical chemistry is concerned with semi-empirical approaches in which some experimental data is introduced into calculations. However the present survey will not consider such semi-empirical work.

The following two quotations may serve to give the essence of the *ab initio* approach. Firstly there is a little known remark made by Langmuir in the course of a popular lecture in 1921, before the advent of quantum mechanics.

These things mark the beginning, I believe, of a new chemistry, a deductive chemistry, one in which we can reason out chemical relationships without falling back on chemical intuition....I think that within a few years we will be able to deduce 90 percent of everything that is in every textbook on chemistry, deduce it as you need it, from simple ordinary principles, knowing definite facts in regard to the structure of the atoms. (Langmuir 1921)

Secondly, the much quoted passage from one of the founders of quantum mechanics, Dirac, who put the case for ab initio calculations somewhat optimistically,

The underlying laws necessary for the mathematical theory of a large part of physics and the whole of chemistry are thus completely known, and the difficulty is only that exact applications of these laws lead to equations which are too complicated to be soluble. (Dirac 1929)

2. Quantum Chemistry

The time-independent Schrödinger equation may be expressed in its most compact form as,

$$H\psi = E\psi$$

where H is the Hamiltonian operator concerning the kinetic and potential energy of the system, ψ is the wavefunction, which is a function of the coordinates of all the particles in the system and E is the observable energy of the system.

For a hydrogenic atom, that is a one-electron system such as the hydrogen atom, the He^{+1} ion or the Li^{+2} ion, the equation takes the form,

$$(-\hbar^2/2\mu \nabla^2 - Z e^2/r)\psi = E\psi$$

where $\hbar$ is Planck's constant divided by 2π and μ is the reduced mass of system. In the case of the hydrogen atom $\mu = m_n \cdot m_e / m_n + m_e$, ∇^2 or $(\partial^2/\partial x^2 + \partial^2/\partial y^2 + \partial^2/\partial z^2)$ is the operator for the kinetic energy, Z is the nuclear charge, e the electronic charge and r the distance between the nucleus and the electron.

The solution to this one-electron equation is exact and characterized by three integers n, ℓ and m, the quantum numbers (Pauling, Wilson 1935).

$$\psi_{n,\ell,m}(r,\theta,\varphi) = R_{n,\ell}(r) \cdot \Theta_{\ell,m}(\theta) \cdot \Phi_m(\varphi)$$

The ground state wavefunction, that is the solution corresponding to $n = 1$, $\ell = 0$ and $m = 0$, is found to be,

$$\psi_1 = 1/\pi^{1/2}(Z/a_0)^{3/2} e^{-\rho/2}, \text{ where } \rho = (2Z/na_0)r, \text{ and } a_0 = h^2/4\pi^2\mu e^2.$$

Another solution includes the case where $n = 2$, $\ell = 1$, $m = -1$,

$$\psi_{2p(-1)} = 1/4(2\pi)^{1/2} (Z/a_0)^{5/2} re^{-Zr/2a_0} \sin\theta \, e^{-i\varphi}$$

These so called 'orbitals' are related to electron clouds encountered in elementary chemistry courses. The energies permitted for a hydrogenic atom are found to be,

$$E = -\mu Z^2 e^4 / 2\hbar^2 n^2 = -Z^2/n^2 \cdot W_H$$

where W_H is the energy for the ground state of the hydrogen atom or 13.60 eV.

The time-independent Schrödinger equation for atomic, ionic or molecular systems containing two or more electrons does not yield exact solutions and approximation methods must be employed. To consider the simplest such atomic case, the helium atom, the time-independent Schrödinger equation for the system is,

$$(-\hbar^2/2m_e \nabla_1^2 - \hbar^2/2m_e \nabla_2^2 - Ze^2/r_1 - Ze^2/r_2 + e^2/r_{12}) \psi = E \psi$$

This differential equation cannot be solved by the method of separation of variables because of the presence of the term e^2/r_{12} which represents the inter-electronic distance. A good approximation method consists in the variation approach in which a trial function φ is chosen which contains variable parameters. It can be shown that the energy corresponding to this trial function E, is larger or equal to the exact ground state energy for the system E_0 (Pauling, Wilson, 1935).

$$E = \int \varphi^* H \varphi \, d\tau \geq E_0$$

where H represents the true Hamiltonian for the system.

For example, an approximate wavefunction for the helium atom can be taken to be,

$$\varphi = \varphi_1 \varphi_2 = (Z'^3/\pi) e^{-Z'r_1} e^{-Z'r_2}$$

that is the product of two one-electron functions, each containing a variable parameter Z'. Evaluation of these integrals and minimization with respect to Z' gives the energy for the ground state of the helium atom as -77.45 e.V. This result represents an error of approximately 2%.

Perhaps the most commonly used approximation in quantum chemistry is the Hartree-Fock method in which the wavefunction consists of an anti-symmetric product of one-electron functions, to take account of the permutation of electrons as dictated by the Pauli exclusion principle. It is assumed that each electron moves in the average field due to the nucleus and all the other electrons in the system. For the helium atom the required wavefunction takes the form of the determinant of a 2 x 2 matrix,

$$\psi = 1s(1)\alpha(1)1s(2)\beta(2) - 1s(2)\alpha(2)1s(1)\beta(1)$$

The so called Hartree-Fock equations represent a pseudo-eigenvalue problem which requires an iterative approach and for which the use of computers is ideally suited. The total energy of the helium atom calculated in this way shows an error of approximately 1.5% as compared with the experimental value.

The atomic energies calculated by the Hartree-Fock method typically show errors of approximately 1% when compared with experimental atomic energies. Such relative errors of 1% may not appear to be very significant, but since the energy of a typical atom in its ground state is about 1000 e.V., the absolute error represents about 10 e.V. This is of the same order of magnitude as a typical chemical bond. The Hartree-Fock method can therefore fail to predict chemical bonding and it becomes necessary to resort to more accurate methods of approximation in order to obtain chemically meaningful predictions. Nevertheless, the Hartree-Fock wavefunction serves as a point of departure in more elaborate approximation methods.

In the so called Configuration Interaction method more than a single determinant is used to represent the wavefunction of the system. The wavefunction for the helium atom, for example, is now represented by a linear combination of determinants,

$$\psi = c_1 D_1 + c_2 D_2 + \ldots\ldots$$

and the computational procedure consists in minimising the energy by variation of the mixing coefficients or c_i's. The additional determinants are formally excited states of the helium atom. Whereas in the helium atom the ground state configuration is $1s^2$, two excited configurations might be $1s^1 2s^1$ and $1s^1 2p^1$.

Other methods which go beyond the Hartree-Fock level of approximation include Cluster Methods and Many-Body Perturbation Theory (Wilson 1984). These approaches involve the introduction of repulsion effects due to simultaneous interactions between three, four, and even more electrons in the expansion of the wavefunction. One important drawback of cluster methods and many-body perturbation theory is that they are not variational. That is to say, the calculated energies no longer represent upper bounds and it is possible to obtain predictions in excess of 100% of the experimental values. Nevertheless, their use is capable of reducing the error in the calculation of the energy of the helium atom to something of the order of 10^{-3} %.

We might just pause at this point in order to take stock of the progress made in the light of the original question as to how successfully chemistry has been reduced to quantum mechanics. It has to be said that the calculation of the ground-state energy of an atom carried out completely from first principles and to an accuracy of 10^{-3} % does seem to provide an argument in favour of the reduction of atomic chemistry, at least with regards to the reduction of a physically measurable quantity. However, I believe that it is necessary to adopt a more critical attitude to such claims especially in view of the computational approaches which are used and which are examined further in the following section.

3. Convergence and Error Bounds

Provided that increasingly larger linear combinations of atomic orbitals are taken, the experimental energy of the atom may be approached ever more closely. This is not surprising due to completeness property of series expansions (Pauling, Wilson 1935). Authors in modern theoretical chemistry often make a virtue of being able to guess the correct procedure by a mixture of intuition and past experience. The usual means of proceeding with a calculation appear to be somewhat ad hoc in this sense.

The common approach used in all these approximations is one of expansion of the wavefunction for the system as an infinite series of one-electron functions. In using such a procedure it is essential that the series used in the expansion should converge to the function which it is meant to represent. It is often assumed that with a sufficiently flexible trial function the results will eventually converge to the exact solution. Otherwise convergence is checked, up to a point, by examining the results of successive approximations.

However, there is no guarantee that, although the experimental value is being approached, the next level of accuracy might not show a sudden divergence. Such occurences are not unknown in mathematical physics. Ideally, a general proof of convergence is required, which is independent of the data arising from any particular approximation in any particular case. The present lack of such convergence proofs in quantum chemistry must be recognized to mar any claim in favour of strict reduction.

From a pragmatic point of view however, it must be admitted that the convergence problem is no longer so pressing, since most modern computer packages such as the Gaussian series contain built-in convergence checking procedures.

The second problem which I wish to pint out is more serious. Although a variational calculation gives an upper bound to the exact solution it does not tell us how close we are to the true value. As is well known, error limits are demanded of experimental results as a matter of course. As Weinhold writes, perhaps one should also ask for a corresponding standard of reliability from the theoretical side (Weinhold,1972).

A general method of improving the situation would consist in finding a way to calculate both upper and lower bounds to the energy of any particular system under consideration. If this form of 'bracketing' were possible, it would endow quantum mechanics with a genuine power of prediction. The problem has been that whereas variation methods provide an upper bound to the energy as can readily be proved, sufficiently general and tractable method for determining lower bounds in cases applicable to atomic systems are not available. To sum up, what is needed in theoretical chemistry, is an independent non-empirical method of assessing the accuracy of the calculations[2].

4. The case of the CH_2 molecule

Although the early predictions made in quantum chemistry were generally unreliable, it has been forcefully argued that since 1970 quantum chemistry has "come of age" (Goddard, 1985). This arose from theoretical predictions on the geometry of the methylene molecule CH_2. This short lived and highly reactive molecule is unusual in having two unpaired electrons around the carbon atom as compared with the better known methane molecule in which all four of the outer electrons are said to be shared with electrons from four hydrogen atoms.

Various calculations carried out for methylene suggested that the molecule should be bent[3]. The molecule was first observed spectroscopically by Hertzberg who contrary to the theoretical predictions found it to have a linear shape. A more accurate treatment the following year by Bender and Schaefer put the angle in methylene at 135.1° (Bender, Schaefer, 1970). Three new experiments by independent groups finally confirmed (pace Popper) a bent geometry in methylene.

In 1971 Hertzberg re-examined his data and was forced to concede that he had previously been wrong and that his own experiment was also in keeping with a bent geometry. This change of mind on the part of Hertzberg has been frequently been exploited by Schaefer, who concludes that his own work represented a successful challenge against the findings of the world's leading spectroscopist.

I now turn to the claimed landmark paper of Bender and Schaefer. Firstly, to take up a general objection which was raised earlier, Bender and Schaefer do not produce any proof of convergence but merely examine convergence up to a certain point. In fact the authors applied computational methods developed earlier by Bender and Davidson but a key passage in this earlier paper betrays a rather serious drawback;

> The main difficulty in the selection of configurations...was of course the enormous number of possible configurations. In a typical calculation there are billions of configurations which can be formed with the correct symmetry. For an unfortunate choice...all of these might be equally important, but for a good initial guess only a few will really contribute to the wavefunction. (Bender, Davidson 1966, 2676).

Any procedure which relies on initial guesses must surely be judged to be essentially ad hoc and this raises doubts as to the extent to which chemical phenomena are being reduced in such supposedly ab initio calculations. Schaefer and co-workers also claim in a later paper that they set out to attempt to place error bars on the theoretical prediction of the bond angle. (McLaughlin, Bender, Schaefer, 1972). By employing an even larger basis set than the previous calculation, the authors estimate a bond angle of 134±2°. However, a detailed examination of the original source reveals a somewhat different picture for this claimed determination of error bars. To quote the authors:

> To aid our evaluation of the expected reliability of this 134° CH_2 angle, we point to the comparable first order calculations on the ground and excited states of NH_2 which yielded bond angles differing by 0.6 and 0.7 degrees from experiment values. In the light of these results and the H_2O results discussed above we estimate our theoretical bond angle of 134° is accurate to within 2°. (McLaughlin et al. 1972, 356-7)

Clearly, this approach represents an extrapolation from the application of the method from one molecule to that of another, and not a rigorous determination of error bars for the CH_2 molecule itself. In fact these calculations on CH_2 represent a perfect example of what Davidson has described as calibrated ab initio, as opposed to true ab initio work (Davidson 1984, 8-9)

5. The Si_2C story

In 1964 the molecule of Si_2C was first observed by infra-red spectroscopy (Veltner, McLeod, 1964). About twenty years later the fundamental Si-C symmetric stretching frequency was identified at 658 cm^{-1} (Kafafi et al. 1983). Shortly afterwards the molecule was studied theoretically and a value of 823 cm^{-1} was obtained for the symmetric stretching frequency (25% error). The authors did not however presume to challenge the assignment of the 658 cm^{-1} line (Grev, Schaefer 1985).

More recently another experimental group has identified the symmetric Si-C stretching mode with a new line at 840 cm^{-1} contrary to the earlier experiments (Presilla-Marquez, Graham 1991). In 1992 Schaefer and colleagues returned to the calculation to determine which line, the one at 658 cm^{-1} or at 840 cm^{-1}, is the true symmetric stretching mode. This provides an example of a state-of-the-art quantum chemistry calculation by one of the leading practitioners. The results obtained using various levels of approximation are tabulated below;

	$[\omega_1 - \nu_1 / \nu_1] \times 100$	$[\omega_3 - \nu_3 / \nu_3] \times 100$	Energy (eV)
TZ+ 2P SCF	1.4 %	17.9 %	- 615.64079
TZ + 2P CISD	2.4 %	9.2 %	
TZ + 2P CCSD	0.9 %	5.2 %	
TZ + 2P CCSD (T)	-2.6 %	1.2 %	
TZ + 2P + f SCF	-4.3 %	21.3 %	- 615.64462
TZ + 2P + f CISD	5.2 %	9.4 %	
EXT + 2P SCF	1.0 %	17.9 %	- 615.65611
EXT + 2P + f SCF	-2.9 %	20.4 %	- 615.65925

Table 1. Percentage errors in the calculation of the spectroscopic mode in question ω_1 as well as another fundamental mode ω_3.

TZ...	Triple Zeta.
2P...	Double Polarization.
SCF....	Self Consistent Field.
CISD...	Configuration Interaction. Single and Double Excitation.
CCSD...	Coupled Cluster. Single and Double Excitation.
(T)...	Triple Excitations included perturbatively.
+ f...	Also includes f functions on Si and C atoms.
EXT....	Extended Basis Set.

Several features of these results are significant.

(i) The addition of f orbitals on the silicon and carbon atoms, which usually improves agreement with experiment in these types of calculations, produces a worsening in the frequency error in three separate methods (SCF, CI and EXT SCF), although the energy shows improvement.

(ii) None of the above methods emerges as the clear winner in calculating fundamental modes from first principles. The outcome seems to depend on which particular mode is being considered.

(iii) Overall, the error in ω_1 strays considerably from one method to the next and even on going to more extended sets within the same method.

None of this suggests that we have a reliable method which can be applied systematically to a new molecule and finally, no error bars are computed in order to lend reliability to the calculated values. However, there is worse to come! A week or so after the Schaefer and Grev paper was published, Handy, another leading quantum chemist, presented some new results on the same molecule[4]. This author used an alternative approach called the density functional method which does not depend on solving the Schrödinger equation directly and which is becoming increasingly common in theoretical chemistry (Parr, Yang 1989). Handy announced the following results on the Si_2C molecule.

(i) At low levels of approximation the results are consistent with those of Schaefer and Grev above.

(ii) A more extended calculation causes the computed value of ω_1 to change to a lower value.

(iii) On addition of f functions to the Si and C basis sets, the value of ω_1 goes to 620 cm^{-1}, i.e., close to the discredited observation of 658 cm^{-1}!

This represents a flat contradiction of Schaefer and Grev results, all of whose methods attribute ω_1 to around 840 cm^{-1}. Once again this raises the question of which of the two observed lines should be assigned to the symmetric stretching mode. It should be recalled that this was precisely the question which had motivated the work of Schaefer and Grev. In more general terms, these findings on Si_2C do not say much for the reliability of current quantum chemistry, the claim that "quantum chemistry has come of age" or indeed the claim that chemistry has been reduced to quantum mechanics.

6. Conclusion

To return to the introduction, it will be recalled that the hope of any strict or exact reduction in the special sciences seems to have been abandoned and that all that remains is the possibility of approximate reduction. However, criteria for approximate reduction have not been put forward and the notion remains vague. The proposal here is that we should make use of an early Putnamian characterization of approximation in the context of theories. That is to say, an approximation is such that the relationships postulated by the theory hold not exactly, but with a certain specifiable degree of error.

As I have argued, errors are seldom computed by independent ab initio criteria in any of the calculations in theoretical chemistry which I discuss. Only the Self-Consistent Field calculations provide an upper bound whereas Many-Body Perturbation Theory and Coupled Cluster methods do not. More importantly perhaps, none of these methods computes a lower bound. As was remarked earlier the calculation of the ground state energies of atoms has been achieved to a remarkable degree of accuracy and similarly calculations on small or even medium sized molecules have given encouraging results. However, whether one can draw the conclusion that chemistry has been reduced rather depends on one's criteria of reduction. If we are to define approximate reduction as has been suggested in this paper then it must be concluded that *chemistry is not even approximately reduced to quantum mechanics*. The point I wish to emphasize is that we should not be misled by the apparent quantitative successes achieved and should appreciate the full nature of the approximation procedures employed.

Notes

[1] It should be mentioned that in the *ab initio* approach the values of experimentally determined fundamental constants such as Planck's constant, the velocity of light and the electronic charge are introduced. However, no experimental information on the particular system under investigation is permitted.

[2] Ramsey has made a similar plea in discussing approximations in general (Ramsey, PSA, 1990). I believe that the calculation of upper and lower bounds would provide the criteria which Ramsey seeks.

[3] For a detailed account of all the calculations on the methylene molecule as well as experimental results see Scerri, E.R., 1993, in 'Correspondence, Heuristics and Invariance, Essays in honour of Heinz Post', pp 45-61, eds, S. French, H. Kamminga, Boston Studies in Philosophy of Science, 148, Kluwer, Dordrecht.

[4]Lecture delivered by N.C. Handy, 'New Applications of Quantum Chemistry', Royal Society of Chemistry Symposium, Cambridge, 3rd December, 1992.

References

Bender, C.F., Davidson, E.R. (1966), *Journal of Physical Chemistry* 70: 2675.

_ _ _ _ _ _ _ _ _ _ _ _ _ _ _ . (1970), *Journal of the American Chemical Society*, 92: 4984.

_ _ _ _ _ _ _ _ _ _ _ _ _ _ _ . (1971), *Journal of Chemical Physics* 55: 4798.

Bolton, E.E., DeLeeuw, B.J., Fowler, J.E., Grev,R.S., Schaefer, H.F. (1992), *Journal of Chemical Physics* 97: 5586.

Davidson, E.R. (1984), *Faraday Symposia* 19: 7.

Dirac, P.A.M. (1929), *Proceedings of the Royal Society of London*, A123, 714 .

Feyerabend, P. (1962), "Explanation, Reduction and Empiricism", in H.M. Feigl, G. Maxwell eds., *Minnesota Studies in Philosophy of Science*, Vol III, Minneapolis: Univ. of Minnesota Press.

Goddard, W.A. (1985), *Science* 227: 917.

Grev, R.S., Schaefer, H.F. (1985), *Journal of Chemistry* 82, 4126.

Kafafi, Z.H. Hague, R.H., Fredin, L., Margrave, J.L. (1983), *Journal of Chemical Physics* 87: 797.

Kemeney, J.G., Oppenheim, P. (1956), "On Reduction", *Philosophical Studies*, VII, 6-19.

Kuhn, T.S. (1962), *The Structure of Scientific Revolutions*, Chicago: Chicago Univ. Press.

Langmuir, I. (1921), "Chemical and Metallurgical Engineering", 24, 553 as quoted by W.B. Jensen, (1984) *Journal of Chemical Education* 61: 191-200.

McLaughlin, D.R., Bender, C.F., Schaefer, H.F. (1972), *Theoretica Chimica Acta* 25: 352.

Nagel, E. (1961), *The Structure of Science*, New York: Harcourt.

Parr, R.G. , Yang, W. (1989), *Density Functional Theory for Atoms and Molecules*, Oxford: Oxford University Press.

Pauling, L. , Wilson, E.B. (1935), *Introduction to Quantum Mechanics, with Applications to Chemistry*, New York: McGraw-Hill.

Popper, K.R. (1957) "The Aim of Science", *Ratio* 1: 24-35.

Presilla-Marquez, J.D. , Graham,W.R. (1991), *Journal of Chemical Physics* 95: 5612.

Primas, H. (1983), *Chemistry, Quantum Mechanics and Reductionism*, Berlin: Springer Verlag, 2nd ed.

Putnam, H. (1965), "How Not to Talk About Meaning", in *Boston Studies in Philosophy of Science*, vol II, eds., R.S. Cohen, M. Wartofsky, New York, Humanities Press, pp. 206-207.

Ramsey, J.L. (1990), *Philosophy of Association Proceedings* 1: 485-499.

Scerri, E.R. (1993), "Correspondence and Reduction in Chemistry", in *Correspondence, Invariance and Heuristics*, Essays in Honour of Heinz Post, eds., French, S., Kamminga, H., *Boston Studies in Philosophy of Science*, volume 148, Dordrecht: Kluwer.

Schaefer, H.F. (1986), *Science* 231: 1100.

Veltner, W, McLeod, D. (1964), *Journal of Chemical Physics* 41: 235.

Weinhold, F. (1972), *Advances in Quantum Chemistry* 6: 299.

Wilson, S. (1984), *Electron Correlation in Molecules*, Oxford: Pergamon.

Part V

REALISM AND ITS GUISES

Could Theoretical Entities Save Realism?[1]

Mohamed Elsamahi

The University of Calgary

1. Introduction

Hacking (1983) introduces an attempt to defend scientific realism on the basis of the reality of theoretical entities. This position, which is called entity realism, is based on disconnecting the reality of theoretical entities from the truth and explanatory power of theories that account for them. In this way, two problems can be avoided. First, if theories about theoretical entities are rejected, the entities themselves do not have to go with them and the realist thesis that we can have knowledge of what exists in the world can be sustained. Second, theoretical entities, which will replace theories as the grounds for the realist position, would be protected from attacks on the validity of the inference to the best explanation which underlies classical or "theory" realism. In other words, theoretical entities would be able to survive the collapse of the inference to the best explanation.

Hacking and other entity realists agree with the antirealists that the classical version of scientific realism is based on the inference to the best explanation and that this basis is unsound. Therefore, entity realists reject classical realism in favor of an experimentation-oriented realism. Other entity realists who adopt Hacking's experimental argument and existential criterion for putative entities are Cartwright (1983), Giere (1988), and Harre (1986). For them, the existential criterion for putative entities consists in the entity's ability to be experimentally manipulated. The subject of this paper is a critique of this line of defending realism.

2. An Outline of Entity Realism

The position known as entity realism is intended to avoid the devastating polemics against the assumed entailment relation between explanatory success and the truth of theories, or what is known as inference to the best explanation. Entity realists want to ground scientific realism in beliefs about the existence of theoretical entities that are attained through experimentation. They assume that if such beliefs are not strictly dependent on the truth of theories, they can survive both the antirealist attacks and theory change. So, their strategy requires experimental information about theoretical entities that is not derived from or dependent upon theories.

Entity realists assume that having a true theory about a putative entity is not necessary for believing in the reality of such an entity. For example, Hacking asserts that we can believe in the existence of electrons without having to believe in any true theory about the electron (1983, 27). He also maintains that the truth or falsity of theories about entities should not affect our beliefs in the existence of such entities. Harre (1986, 93-99) and Cartwright (1983, 92) express the same conviction. This is Hacking's brief characterization of entity realism:

> But one can believe in some entities without believing in any particular theory in which they are embedded. One can even hold that no general deep theory about the entities could possibly be true, for there is no such truth (1983, 29).

Thus, establishing an entity realism that is invulnerable to the problems of theory change and theory falsification, as well as the problems related to the inference to the best explanation, requires, as we have pointed out above, a theory-free experimental route to theoretical entities. The separateness of entity realism is unattainable otherwise. For this reason, entity realists should convince us that their experimental access to theoretical entities does not depend heavily on the theories which introduce and characterize theoretical entities. This is what Hacking's experimental argument is supposed to do. In fact, Hacking's argument is primarily designed to show that the existence of a putative entity could be demonstrated by analyzing certain experimental considerations that do not, supposedly, owe their cognitive significance to the truth of the theories which refer to such an entity.

The experimental argument is based on a tacit analogy, although Hacking does not describe it as analogical. It is clear that Hacking's assertion (1983, 262-275) that using an entity to investigate another physical object is sufficient to make it as real as ordinary tools of investigation represents an argument by analogy. This argument stresses the similarity between the use of the electron in investigating the physical properties of weak neutral currents and the use of ordinary laboratory tools in exploring other physical phenomena. The analogy underlying the argument seems to imply that the successful use of putative entities to investigate something else cannot be interpreted in a metaphorical or instrumental sense and should be understood in realistic terms. But we will deal with this point later in this paper.

Hacking's concrete example is the employment of PEGGY II guns which are designed in a manner that allows the use of some of the well-understood causal properties of electrons, namely their spin, to investigate the less understood weak neutral currents. The success of the experiment suggests to Hacking that electrons have acted as tools for gathering information about weak neutral currents. For this reason, their existence should be as certain as the existence of the laboratory tools that we see and touch, like cloud chambers and galvanometers.

Giere presents a similar experimental argument that refers to the use of protons in an experiment in nuclear physics to investigate the nuclear structure. In this experiment, the proton is used as a tool for investigating other theoretical entities in the atomic nucleus. Like Hacking, Giere concludes that the successful manipulation of protons in the experiment justifies the belief in the reality of the proton (1988, 125). This leads him to accept Hacking's ontological criterion, on which entity realism rests.

Hacking's ontological criterion is derived from his experimental argument. According to this criterion, an entity has to be capable of being experimentally manipulated in order to be real (1983, 263). For example, if the photon cannot be used in an experiment to investigate something else, it may be a mere fiction. Until photons are actually used as investigative tools they remain hypothetical, regardless of all the theoretical assertions about their existence.

The manipulation criterion shows that entity realists ascribe more cognitive significance to scientific experimentation than to scientific theorizing. It also reveals the ontological orientation of entity realism. Indeed, it is an important issue for entity realists to determine which theoretical entities exist in the world and which ones do not. The manipulation criterion is intended for this specific aim. More precisely, it is intended as a decision procedure for determining the existential status of the scientific entities introduced by physical theories.

The scheme according to which entity realists apply their manipulation criterion consists of a few steps. The first involves screening theories for putative entities. Then, the existential assertions referring to putative entities in these theories are ignored, and all the entities referred to by theories are considered hypothetical. This step could be called a suspension of judgment on the referential assertions of theories. Lastly, a search is conducted for a relevant experiment that utilizes a putative entity. A prototype for such experiments is the PEGGY II experiment. If an acceptable experiment of this type is found, the manipulated entity is considered real or existent rather than hypothetical. This is how the ontological procedure of entity realism works.

As a consequence of disconnecting the truth conditions of theories from the ontological status of theoretical entities, entity realists can be free to treat the referents of discarded theories as real and the referents of acceptable theories as unreal. This reflects the conviction of the entity realists that theories lack the capacity of deciding the ontological status of their entities. For them, only the manipulability of a putative entity can settle the relevant existential questions. What this means is that entity realists conceive scientific theories as screening procedures for possible existents, not as determinants of what actually exists in the world. They project that theories introducing theoretical entities are only supposed to provide lists of hypothetical entities which have the potential for being real, together with descriptions of their possible properties. These theoretical data will require further ontological research by entity realists to determine the final ontological status of entities. Of course, this is different from the classical realists' account, which permits theories to determine which entities are real and which are hypothetical.

3. Assessing the Experimental Argument

There are four main problems for the experimental argument and the manipulation criterion that derives from it. These problems, if not answered, can undermine the whole project of entity realism. Although our discussions of the experimental argument are focused on Hacking's version, Giere's version is susceptible to the same criticism. Both are arguments by analogy from experiments which are supposed to involve manipulating theoretical entities to discover a less understood subject in physics.

3.1. The Analogy Underlying the Experimental Argument Is Questionable

Perhaps, there is a sense in which electrons are manipulated in the experiment described by Hacking. But they are not manipulated in the literal sense. For example, ordinary experimental objects can be washed, heated, and stirred but unobservable entities can not. In the PEGGY II experiment, electrons are manipulated in an indirect sense: their spin, which is a mathematized physical property, is entered into calculations and equations required for the design and interpretation of the results of the experiment. So, the assertion that electrons are used to investigate weak neutral currents actually means that certain physical and mathematical notions which derive from the conception of electron spin are used to specify and compute certain properties of weak neutral currents. The term "manipulation" here should be understood rather metaphorically. Indeed, it would be more plausible to maintain that the physical responses and reactions of the weak neutral currents in the experiment could be well-explained by the appeal to electron spin than to maintain that these currents were literally manipulated by electrons.

It is not possible for entity realists to infer from the success of an experiment which depends on electron spin that electrons are real, since the property of spin is ascribed to some other entities like photons and neutrons which entity realists consider hypothetical. So having spin is not indicative of existence, when the entity realists' criteria for existence of an entity are taken into account. This is why it is not possible for entity realists to conclude that explaining weak neutral currents by means of electron spin entails that the electron exits. Thus, it is not possible to exclude the possibility that electron spin is a theoretical device. Such devices cannot be manipulated in any literal sense.

Now if the manipulation of electrons in the experiment is not literal, it becomes questionable whether or not the analogy holds. On one side we have electrons that are manipulated in a non-literal sense, that is, in the sense of using their physical properties in the experimental calculations and in making hypotheses about a physical phenomenon. On the other side we have ordinary laboratory tools that are literally manipulated. Hacking needs to argue that manipulating electrons in a different way from manipulating things in the world of common sense would not jeopardize his analogy. Such a supplementary argument is necessary for justifying the analogy. Moreover, since Hacking does not explore the possibility that the weak neutral currents may themselves be hypothetical, he needs to show that their ontology should not affect the conclusion of the analogical argument. More specifically, Hacking needs to argue that a putative entity that is successfully used to investigate a phenomenon should be a real entity whether this phenomenon is explained in terms of hypothetical or real entities. Without such additional arguments, Hacking's experimental argument cannot go beyond affirming the usefulness of the property of the electron spin for exploring the weak neutral currents which is insufficient for substantiating the claim about the reality of the electron.

3.2. The PEGGY II Experiment Could be Interpreted in Instrumental Terms

The second problem for the experimental argument can be expressed in the form of a question like this: would the final results of the PEGGY II experiment be exactly reproduced if the electron was only hypothetical? Contrary to what is implicit in Hacking's argument, the answer is not necessarily "no." Indeed, we have no good reason to deny that a fictitious entity with a property identical to the electron spin can lead to the same experimental results of PEGGY II. The reason is that the experiment does not rely on any ontological notion. It only relies on mathematized and abstracted models and formulations which derive from the conception of the electron spin. This is why the relevant computations and equations should not be affected by the source of information about this spin. It would not matter for these computations whether the information comes from the electron that exists only in theories or from the electron that exists in the external world.

To clarify this point let us imagine that we stopped believing in the reality of the electron ten years from now. If we, after ten years, repeat the PEGGY II experiment as was conducted before giving up our belief in the reality of the electron, we should not expect to obtain different results, since the ontology of the electron decides nothing about the equations and models that the experiment utilizes. In fact, we know from modern science that hypothetical entities are frequently used by scientists to make inferences about the unobservable structure of the world. Hacking, who considers many theoretical entities (those that are not yet manipulated to investigate something else) as hypothetical, would not deny this. So it is possible, in principle, to use information from hypothetical entities in experimentation.

For Hacking, the photon, unlike the electron, belongs to the class of hypothetical entities. But we know that the physical properties of the photon, for example its velocity and momentum, appear in many equations and models of physics. What this means is that entities do not have to be real to be used as ingredients in scientific experiments. Why, then, should hypothetical entities not be used as successfully to investigate something else?

Hacking's contention that experiments which use an entity as an investigative tool, alone, can decide the ontology of the used entity rests on an unsupported assumption. This assumption could be formulated in these words: the existence of the used entity is necessary for the success of these particular experiments. In other words, Hacking supposes that no hypothetical entity can lead to the desirable experimental results achieved by experiments like PEGGY II, even if such entities were successfully used in other kinds of experiments. The problem is that Hacking provides no argument to warrant such a tacit assumption, which is his only possible line of arguing against the instrumentalist interpretation of the PEGGY II experiment.

3.3. The Experimental Argument Relies on Theories More than Experimentation

The third problem for the experimental argument is its reliance on certain experimental notions that draw heavily on theories. Although it is difficult to conceive a theory-free experimentation, it may be possible to assume that some of the simpler experimental findings are not strongly dependent on theoretical reasoning. Hacking, himself, holds a stronger version of this opinion (1983, 164-165). However, it will be very unreasonable to include the electron spin among these simpler experimental findings. Indeed, the conception of the spin of subatomic particles, including the electron spin, is the product of some sophisticated theories about the deep structure of the world.

The elementary particle spin was first suggested by Pauli to explain a prediction by Bohr's atomic theory about the lines in the spectra of certain metals. Pauli proposed that these lines are due to a magnetic interaction between the nucleus and the moving electrons of the atom, which led to the suggestion that the electron exists in two states with the same orbital motion (Ramsey 1967, 7-66). Later, these states were interpreted as the spin of the electron around an axis and a formula was introduced to calculate the electron spin using some physical constants from quantum mechanics, including Planck's constant. Among the theoretical notions involved in the concept of electron spin are magnetic moment, angular momentum of a particle, Russell Saunders coupling, and Planck's law of the electromagnetic radiation (Brown 1967, 4-132). So the concept of electron spin was posited in light of Bohr's atomic theory and was further developed with the help of certain theories from quantum mechanics.

The heavy dependence of the PEGGY II experiment on such a theoretical concept, i.e. electron spin, justifies describing this experiment as theory-dependent. This is why Hacking's remark (1983, 270) that "the making of PEGGY II was fairly non-theoretical" would be inexact if it is intended to mean that the experiment utilizes no significant theoretical information about the electron.

Now if it is correct that both the design and the interpretation of the results of the PEGGY II experiment are permeated with thoughts from theories about the electron, we would not be justified in accepting Hacking's experimental argument as significantly non-theoretical. Even if we accept, for the sake of argument, that there could be fairly non-theoretical experimentation, Hacking's argument relies on theory-laden experimentation, as shown above. Thus, the necessary condition for the distinctiveness of entity realism from theory realism, which is the reliance on "fairly non-theoretical experimentation", is not satisfied. This means that the antirealist criticism applies to entity realism as well.

3.4. The Clash with Theory Change and Scientific Progress

The fourth problem is the incompatibility of the manipulation criterion, which is derived from the experimental argument, with theory change. It will be shown below that the acceptance of this criterion could lead to disagreements with updated scientific views.

The view of entity realism that the entities referred to by theories can only be hypothetical entails the rejection of the so-called causal reference. Traditionally, when a true theory describes and defines an entity, it is conceded that the theory refers to the entity as real. This referential component of theories, which contains the existential assertions about theoretical entities, is not approved by entity realists because it emphasizes the link between the truth of theories and the reality of their entities. Since it is essential for their position to deny such a link, entity realists propose that causal reference is only sufficient for conferring a hypothetical status on putative entities. The disapproval of causal reference is explicitly asserted by Hacking (1983, 28-29), Giere (1988, 93), and Harre (1986, 65-70). As noted above, the manipulation criterion plays for entity realism the role that causal reference plays for classical realism.

Although the rejection or the acceptance of reference is primarily a debate about semantics, it can lead to serious implications for philosophy of science, particularly with regard to theory change. Indeed, the significance of reference for the philosophy of science becomes clear and evident only when the referents of theories, that is, their theoretical entities, are changed in the process of scientific progress. In this process, new theories are accepted together with their entities in the place of older theories. The referents of discarded theories go with them while new referents emerge as real. This process of entity replacement constitutes an important feature of theory change (or scientific progress).

Now would entity realists respond to such changes in science by altering their ontological decisions which are based on the manipulability criterion? In other words, would they revise the status of an entity that once acted as an investigative tool and was classified by the manipulation criterion as real if scientists, in response to theory change, declare such an entity obsolete? Apparently they would not.

In fact, entity realists are convinced that when it comes to ontological matters, their manipulation criterion is the ultimate guide. The truth conditions of theories, and consequently theory change, cannot provide the final word on the existential status of entities. In other words, causal reference and theory truth, according to entity realists, are incapable of resolving the ultimate ontological questions. Therefore, their official position will dictate the choice of adhering to the ontological decisions made by the manipulation criterion. Consequently, they should have no reason for revising the status of an entity in response to new developments in the theories of science. This implies that they consider the manipulation criterion irrevisable in the face of any change in the truth conditions of theories.

For the same reason, nothing should stop entity realists from testing the referents of discarded theories for possible existence by means of their manipulation criterion, since the falsity of a theory does not entail that its entities are fictitious. It follows that, for entity realism, it is legitimate to test an entity like the phlogiston for manipulability, i.e. for the possibility of existence in the world, regardless of the fact that phlogiston theories are obsolete.

Obviously, the manipulation criterion could end in conflicts between entity realism and science in case of theory change. Entity realists may be led by their criterion to disagree with scientists on some of the new developments in the way theoretical entities are conceived. This implies that entity realists are not obligated to accept every progress in science apart from newly introduced manipulation experiments. Put differently, entity realists are committed to progress in experimental but not theoretical science. But this position rests on the assumption that there can be non-theoretical experimental science, an assumption that entity realists need to defend.

One way to avoid the clash with scientific progress would be to deal with the manipulation criterion as revisable and with its ontological decisions as only provisional. This could only be done, it seems, by restricting the application of this criterion to the referents of accepted theories and denying its previous decisions on the existential status of the referents of the theories that become falsified. However, the manipulation criterion, in this case, will inevitably shrink into a selective verifying procedure that could, at most, ascribe more credibility to certain referents of true theories. Put differently, the criterion turns to an unnecessary verifying supplement that emphasizes the existential assertions of theories with regard to certain entities. But such a verifying supplement would consolidate, rather than undermine, the role of reference in theories. So if the manipulation criterion becomes revisable, the ontological status of entities will depend on the truth conditions of theories. Consequently, the criterion loses its significance as an ontological procedure while the truth conditions of theories regain the grasp on the existential decisions regarding theoretical entities. This precisely means the collapse of entity realism.

What our discussion reveals, so far, is that entity realism can survive as an autonomous position only if its ontological criterion is made irrevisable and overriding over science with respect to ontological issues. It follows that if entity realism is autonomous, it will not be required to comply with the existential decisions of science or with theory change. In other words, it will have no firm commitment to accept all scientific progress. For example, if science happens to replace the electron with a new entity, entity realists, having already confirmed the existence of electrons by the manipulation criterion, would have no compelling reason to agree with new science on discarding the electron.

By avoiding an unconditional commitment to the progress of theoretical science, entity realism implies a division of scientific progress into significant (related to manipulation experiments) and insignificant (theoretical). But the rationality of using a philosophical criterion, i.e. the manipulation criterion, to judge the significance of scientific progress needs to be defended. This means that entity realism needs to argue that it is rational to divide science into theoretical and experimental and to split scientific progress into significant, to be accepted, and insignificant, to be ignored. The latter distinction implies that we are not compelled to derive our beliefs about theoretical entities from the latest theories but from the latest manipulation experiments. This position is difficult to defend, since it leads us to the odd consequence that our beliefs about scientific entities need not be based on updated science. Moreover, if it happens that in the following fifty years science introduces a number of new theories but no new manipulation-type experiments, entity realists would not be required to pay any attention to such progress. So the problem of entity realism with the progress of science is largely about the justifiability of using the manipulation criterion as the determinant of the significance or triviality of scientific progress. If this attitude to scientific progress turns out to be indefensible, entity realism would be implausible and anti-progressive.

Note

[1] I would like to thank Marc Ereshefsky for very helpful comments.

References

Brown, W. (1967), "Magnetic Materials", in E.U. Condon and H. Odishaw, (eds.), *Handbook of Physics*. New York: McGraw Hill Book Co., Part 4, pp. 129-145.

Cartwright, N. (1983), *How the Laws of Physics Lie*. Oxford: Clarendon Press.

Giere, R. (1988), *Explaining Science: A Cognitive Approach*. Chicago: Chicago University Press.

Hacking, I. (1983), *Representing and Intervening: Introductory Topics in the Philosophy of Natural Science*. Cambridge: Cambridge University Press.

Harre, R. (1986), *Varieties of Realism: A Rationale for the Natural Sciences*. Oxford: Basil Blackwell.

Ramsey, N. (1967) "Hyperfine Structure and Atomic Bean Methods", E. U. Condon and H. Odishaw, (eds.), *Handbook of Physics*. New York: McGraw Hill Book Co., Part 7, pp. 66-71.

Realism, Convergence, and Additivity[1]

Cory Juhl
University of Texas—Austin

Kevin T. Kelly
Carnegie-Mellon University

1. Introduction

In one version of the debate between scepticism and scientific realism, the sceptic denies and the realist affirms the ability of science to home in on the unknown truth. Germane to this debate is the familiar claim that probabilistic updating by Bayesian conditionalization will almost surely arrive at the truth so long as one does not "close the door" on a hypothesis by assigning it probability zero. In this paper, we observe, on behalf of the sceptic, that (1) there exists a coherent agent who in fact *fails* on an *uncountable* set of possible worlds to find the truth about a simple, universal hypothesis H even though he "keeps the door open" concerning H until H is logically refuted and even though there is a trivial, logically driven method that converges to the truth about H no matter what; that (2) nonetheless, this agent "almost surely" arrives at the truth in the usual, Bayesian sense because the uncountable set of circumstances in which he fails is assigned probability zero by the agent; and that (3) the guarantee that the set of all circumstances in which the agent fails has probability zero hinges on the axiom of countable additivity.

Countable additivity has sometimes been billed as a technical convenience in the theory of probability. But in light of its invocation to license disregard for uncountably many failures to find the truth, it is actually a powerful axiom for scientific realism that should be subjected to the greatest possible philosophical scrutiny.

1. To Err is only Bayesian

Consider Fred, who watches a coin-flipping apparatus. For all Fred knows *a priori*, any sequence of heads and tails may possibly arise in the limit. Fred is interested in whether or not there will ever occur a run of two consecutive heads, the first of which occurs at an odd flip. H is the hypothesis that such a pair will *never* occur. H is a universal hypothesis (i.e. for *each* odd stage, no run of two heads will be observed starting at that stage).

Since H is universal, there is a trivial, Popperian method that Fred could use to converge to the truth value of H in the limit: conjecture the truth value 1 (i.e. "true") until H is logically refuted by the data. Conjecture 0 (i.e. "false") thereafter.

But Fred does not use the obvious method just described. He is a coherent personalist who updates by conditionalization. Consider the following two probability models of what might be going on in the flipping device. On model P_-, the flipper produces independent trials using a fair coin. For each finite data sequence e, the probability that the next flip will land heads is given by:

$$P_-(\text{heads}| e) = 0.5.$$

On model P_+, the flipper produces a fair flip unless a head on the next flip would refute H. In that case, a two-tailed coin is substituted for the fair coin inside the machine prior to the flip. For each finite data sequence e, the probability that the next flip will land heads is given by:

$$P_+(\text{heads}| e) = \begin{cases} 0 \text{ if length(e) is odd and the last entry in e is heads} \\ 0.5 \text{ otherwise.} \end{cases}$$

Fred's initial, joint probability measure P is the 50-50 mixture of P_+ and P_-. In other words, $P(S) = .5P_+(S) + .5P_-(S)$. By standard means, it can be shown that these constraints uniquely determine a probability measure over the Borel sets of infinite binary sequences. Fred adopts that measure and hence is sychronically coherent. Moreover, it is easy to verify that $1 > P(H| e) > 0$ if H is logically consistent with e, so P "keeps the door open" with respect to H and $\neg$H.

Fred now proceeds to update his measure by conditionalization. For each finite data sequence e, $P(e) > 0$, since $P_-(e) > 0$. Therefore, conditionalization can proceed according to Bayes' theorem without any concern about zeros in the denominator P(e). It is not hard to verify that P(h| e) never changes unless a head is observed in an odd position. So on any data stream in which heads eventually stop appearing in odd positions, Fred's probability for H is eventually frozen at some value less that 1, and on any data stream on which no head ever occurs in an odd position, Fred never alters his *a priori* probability assignment at all. There are evidently uncountably many such cases (all possible infinite, binary sequences can be fit into the even positions), so Fred's posterior probability for H fails to approach the truth on uncountably many possible data streams, even though the trivial Popperian method is guaranteed to stabilize to the truth value of H on *every possible* data stream.

This result raises the question whether some other Bayesian does better than Fred concerning H. Indeed, this is the case. Fred's measure fails to decrement the probability of H when a head is observed on an odd position, even though this outcome brings H "closer" to refutation. Cheryl employs a measure Q that "notices" when this happens. Q is the 50-50 mixture of the fair coin model P_+ together with a measure Q_- that "anticipates" refutation:[2]

$$Q_-(e) = \begin{cases} 0 \text{ if e refutes h} \\ 1/3^{(n+1)/2} \text{ if n is odd and e ends with 0} \\ 2/3^{(n+1)/2} \text{ if n is odd and e ends with 1} \\ 1/3^{n/2} \text{ otherwise.} \end{cases}$$

It is not hard to show that *no matter how* the data comes in, Q eventually stabilizes to 0 if H is false and Q gets ever closer to 1 if H is true. So some Bayesians are *logically guaranteed* to arrive at the truth in the limit whereas others (who "leave the door open" concerning the hypothesis) are not.[3]

3. Logical Reliability

Suppose that the *data stream* ε received "in the limit" will consist of an infinite sequence of 1's and 0's emanating from some system. For all we know, the actual data stream we are seeing is drawn from some set K, which represents our *a priori* empirical *background knowledge*. A non-probabilistic, *empirical hypothesis* H may be identified with the set of all data streams of which it is true. So both K and H are sets of data streams.

In general, an *inductive method* may be viewed as a function α(H, e) that takes a hypothesis H and a finite data sequence e as inputs and that produces a real number in the [0, 1] interval as its "conjecture" concerning H on the data provided. So in particular, if P is a probability measure on the set of all infinite sequences of 1's and 0's such that for each finite data sequence e, P(e) > 0, then $α_P(H, e) = P(H| e)$ is an inductive method.[4] But inductive methods also include functions that do not result from conditioning a probability measure.

Convergence can be understood in various senses. Consider an infinite data stream ε, a hypothesis H and a method α. α *stabilizes* to 1 [0] on H, ε just in case after reading some finite, initial segment of ε, α conjectures only 1's [0's] while investigating H. More leniently, α *approaches* 1 [0] in the limit on H, ε just in case α's conjectures get ever closer to 1 [0] as α reads more and more of ε while investigating H. Stabilization to 1 [0] implies approach to 1 [0], but not conversely.

Inductive success can also be taken in various senses. α *verifies* [*refutes*] H *in the limit* on ε if and only if α *stabilizes* to 1 [0] exactly when H is true [false] of ε. Similarly, α *verifies* [*refutes*] H *gradually* if and only if α *approaches* 1 [0] exactly when H is true [false] of ε. Verification and refutation are "one-sided" success criteria. Decision requires "two-sided" success: α *decides* H *in the limit* [*gradually*] on ε if and only if α verifies and refutes H in the limit [gradually] on ε.

Reliability is the ability to *succeed* over a wide range of possible circumstances. *Logical reliability* demands success on *every* data stream in K. Thus, we may say that α verifies H with certainty given K just in case α verifies H with certainty on each ε in K, and similarly for all the other notions of success. Recall the hypothesis H which says that there will never be two consecutive heads (1's) starting at an odd position. The obvious Popperian method decides H in the limit given vacuous background knowledge. Updating by Cheryl's measure refutes H in the limit but does not verify H in the limit and hence gradually decides H but does not decide H in the limit. Updating by Fred's measure gradually refutes H but does not gradually verify H (and hence does not gradually decide H or verify, refute, or decide H in the limit). So from the point of view of logical reliability, each method may be placed in a detailed *scale of logical reliability*, depending on the sense of limiting success according to which it succeeds concerning H over all of K.

Instead of asking whether a *given* method is reliable for H and for K, we can ask whether or not there *exists* a reliable method for H, K. Say that H is *verifiable in the limit* given K just in case some a verifies H in the limit given K; and similarly for the other notions of reliability. It may now be shown that the different senses in which reliability can be achieved for H, K are ordered by the following, proper implications from bottom to top:

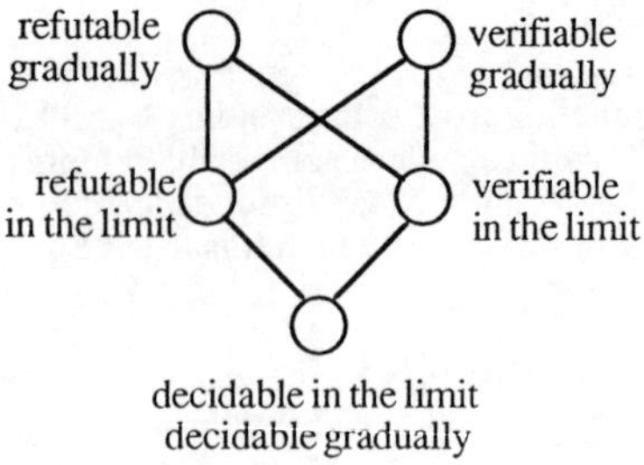

```
            refutable  ○────────○  verifiable
            gradually   ╲      ╱   gradually
                         ╲    ╱
            refutable  ○──╳───○  verifiable
            in the limit  ╱  ╲   in the limit
                         ╱    ╲
                        ○
               decidable in the limit
                decidable gradually
```

Figure 1

The reason for the equivalence of decidability in the limit and gradual decidability is that a gradual decider can be "simulated" by a limiting decider who conjectures 1 if the gradual decider conjectures a value greater than 0.5 and who conjectures 0 otherwise.

Just as methods can be ranked with respect to a problem according to the sense in which they are reliable, inductive problems (pairs of form (H, K)) can be ranked according to the senses of reliability in which they may be solved. For example, the hypothesis H_{fin}, which says that only finitely many 0's will ever be observed, is verifiable[5] but not refutable[6] in the limit, and hence is gradually refutable and gradually verifiable but not decidable gradually or in the limit. So the above diagram may be thought of as a logical scale of *underdetermination*, if underdetermination of H relative to K is taken in the plausible sense of the impossibility of a reliable inductive method for H given K.

4. Probabilistic Reliability

Probabilistic reliability, like logical reliability, is a matter of success over a wide range of possible data streams. But unlike logical probability, the probabilistic version requires success only over a set of data streams of probability 1. Despite the fact that allowing a zero probability of error seems to concede nothing, it has impressive consequences:

Proposition 4.1: Let P be a probability measure on the infinite Boolean sequences such that for each finite sequence e, $P(e) > 0$. Let H be measurable with respect to P. Let $\alpha_P(H, e) = P(H| e)$. Then

$P(\alpha_P$ gradually decides H$) = 1$.

Proof: The proposition is a special case of Halmos(1974).[7]

In other words, so long as P assigns a well-defined probability to H, P assigns a unit probability to its own ability to gradually decide H. It follows that:

Corollary 4.2: for each such probability measure P and for each H measurable with respect to P, H is decidable in the limit with probability 1 in P.

α_P is not a limiting decider with probability 1 in P, but the method α'_P that conjectures 1 if $P(H| e) > 0.5$ and that conjectures 0 otherwise is a limiting decider with probability 1 in P. Since the P-measurable hypotheses include all Borel hypotheses[8], all Borel hypotheses are decidable in the limit with unit probability in P. Since the Borel hypotheses include virtually all non-probabilistic, empirical hypotheses that would arise in practice, this result plays very much to the scientific realist's advantage. It says that despite the fact that any possible method might possibly be fooled for eternity, for most questions we may neglect all of the possible worlds in which this happens.

Another way of understanding corollary 4.2 is that every measurable hypothesis H can be *approximated* by a hypothesis H' that is decidable in the limit such that the measure of the "mismatch" H Δ H' = (H - H') $\cup$ (H - H') is 0. In other words, every measurable hypothesis is "almost" decidable in the limit.

5. Countable Additivity

The preceding results on probabilistic reliability are all proved using the axiom of *countable additivity*, which states that for each countable partition $\Pi = \{G_1, ..., G_n, ...\}$ of a given set S, the probability of S is the sum of the probabilities of the cells of the partition:

$$P(S) = \sum_{i=0}^{\infty} P(G_i).$$

Finite additivity is similar, except that P is required to be a *finite* partition of S.

Countable additivity was introduced by Kolmogorov as a "technical expedience" and many probability theorists have followed suit with lukewarm apologies, of which the following is typical:

> The general condition of countable additivity is a further restriction... — a restriction without which modern probability theory could not function. It is a tenable point of view that our intuition demands infinite additivity just as much as finite additivity. At any rate, however, infinite additivity does not contradict our intuitive ideas, and the theory built on it is sufficiently far developed to assert that the assumption is justified by its success. (Halmos 1970, 187)

Strictly speaking, a technical convenience is a principle that facilitates proofs of interesting results that could have been proven in a more roundabout way without it. Countable additivity is then no mere technical convenience because the convergence and approximation theorems of the preceding section can fail without it:

Proposition 5.1: There is a finitely but not countably additive measure P and a hypothesis H that is verifiable in the limit in the logical sense such that *no possible method* a (conditionalization or otherwise) refutes H in the limit with probability 1 in P.[9] Hence, no method decides H gradually or in the limit with probability 1 in P.

Proof sketch (details are in [Kelly 94]): Recall H_{fin}, which states that the data stream will eventually stabilize to 1. It has already been observed that H_{fin} is verifiable but not refutable in the limit in the logical sense. We now define a finitely additive measure as follows. First, observe that H_{fin} is countable. Assign non-zero probabilities to each singleton subset of H_{fin} so that all of these probabilities sum to 1/2, and let $P(H_{fin}) = 1/2$. So far, nothing violates countable additivity. Next, we place probability 1/2 nowhere in particular but arbitrarily close to the "edge" of S so that any superset R of H_{fin} that is refutable in the limit catches this extra probability and hence has probability 1. It can be shown that this assumption is consistent with finite additivity.

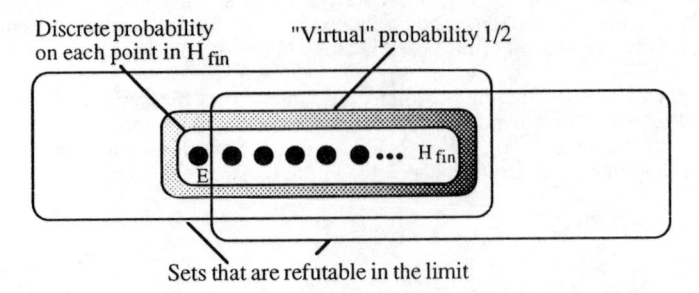

Figure 2

Let α be an inductive method. α refutes some set in the limit, namely, the set R of all data streams on which α stabilizes to 0. Case 1: suppose $H_{fin} \subseteq R$. Then $P(R) = 1$. Since a fails to refute R in the limit on each data stream in R - S, α fails with probability 1/2. Case 2: suppose H_{fin} is not a subset of R. Then α fails to stabilize to 0 on some data stream in H_{fin}, so α fails with some probability greater than 0 since each point in H_{fin} carries nonzero probability.

So the sweeping, positive results of the preceding section fail when countable additivity is dropped. Inasmuch as those results seem to banish logical underdetermination and the sceptical doubts it gives rise to, countable additivity is anything but a mere technical convenience. It is an epistemological axiom of the first importance, and should therefore be held up to the most careful philosophical scrutiny.

In fact, the principle has not held up so well against such scrutiny as it has received. As DeFinetti has pointed out, countable additivity is false on the frequentist interpretation of probability, since in the sequence 0, 1, 2, 3, ..., n, ... each number occurs with limiting relative frequency 0 but the limiting relative frequency of some number occurring is 1.[10] On the personalist side, both DeFinetti and Savage have rejected countable additivity as a general norm of coherence. High on the list of DeFinetti's reasons for rejecting the principle is that it forces a "bias" in one's probabilities on a countable partition.

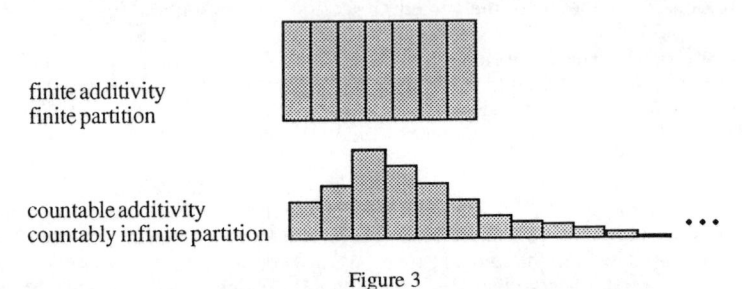

Figure 3

Suppose we are given a countable partition into events E_i, and let us put ourselves into the subjectivistic position. An individual wishes to evaluate the pi = $P(E_i)$:

he is free to choose them as he pleases, except that, if he wants to be coherent, he must be careful not to inadvertently violate the conditions of coherence. Someone tells him that in order to be coherent he can choose the pi in any way he likes, so long as the sum = 1 (it is the same thing as in the finite case, anyway!).

The same thing?!!! You must be joking, the other will answer. In the finite case, this condition allowed me to choose the probabilities to be all equal, or slightly different, or very different; in short, I could express any opinion whatsoever. Here, on the other hand, the content of my judgments enter into the pircture: I am allowed to express them only if they are unbalanced.... Otherwise, even if I think they are equally probable... I am obliged to pick 'at random' a convergent series which, however I choose it, is in absolute contrast to what I think. If not, you call me incoherent! In leaving the finite domain, is it I who has ceased to understand anything, or is it you who has gone mad? (De Finetti 1990, 123)

In fact, it is easy to see that this objectional "bias" is the very reason why countable additivity is such a potent anti-sceptical force. Consider the universal hypothesis H_0 that only 0 will ever be observed. The ancient sceptical argument against inductive generalization tells us that as soon as we declare that H_0 is true, the very next datum might be a 0. But now observe that we can partition the space of all 0-1 data streams as follows:

$$\Pi = \{H_0\} \cup \{F_e : e \text{ is a finite string of 0's followed by a 1}\}$$

where F_e denotes the set of all infinite data streams that extend finite sequence e. F_e is the proposition that H_0 will be refuted at stage length(e). Since P is a countable partition, countable additivity forces us to bias our degrees of belief over the F_e, so that most of our probability is used up on H_0, together with a finite union of F_e's, so the infinitely many remaining F_e's share only a negligible probability. In other words, the stages of inquiry constitute a countable partition through time so that countable additivity forces us to neglect the probability of H_0 being refuted in the indefinite future.

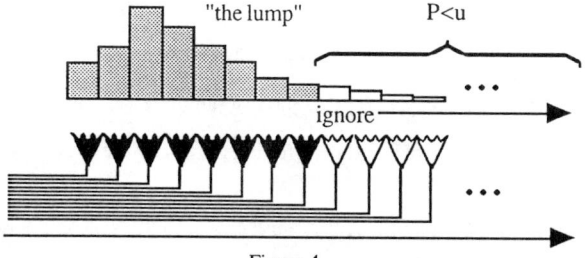

Figure 4

On the other hand, if we drop countable additivity, then the sceptical argument can be revived. For then we can let any co-finite union of F_e's carry probability 1/2 and any finite union of F_e's carry probability 0, leaving probability 1/2 for H_0. Then the probability of H0 being refuted by a given time remains constantly 1/2, no matter which time is chosen. There is nothing subtle about this. Countable additivity amounts to the flat denial that one should be concerned about being fooled in the indefinite future, which is precisely the basis of the ancient, sceptical argument against inductive generalization. It is therefore not surprising that countable additivity should lead to realist conclusions. In a similar manner, a two-phased invocation of countable additivity yields the approximation result stated in corollary 4.2 above, and proposition 5.1 shows that this argument can also be overturned when countable additivity is dropped.

6. Conclusion

One construal of the difference between realists and their critics is that the former view science as a reliable means for finding the truth and the latter do not. In this regard, probabilistic reliability suggests a realistic perspective whereas logical reliability casts a sceptical shadow. But we have seen that the realistic force of probability theory depends on the axiom of countable additivity, without which its strong convergence theorems can fail, not just for Bayesian conditionalization, but for all possible methods. For this reason, countable additivity should be at the focus of philosophical debates between realists and their critics. It is no mere technicality. Nor is it "justified by its fruits", insofar as its fruits are precisely the realist conclusions under dispute. It is a powerful realist principle, and should be scrutinized as such.

Notes

[1] Cory Juhl's contribution to this work was funded by a grant from the University Research Institute at the University of Texas at Austin.

[2] This measure was suggested to us by T. Seidenfeld.

[3] An interesting side issue: it would seem that the property of being guaranteed to find the truth on every possible data stream might serve as a constraint on a personalist's choice of an initial probability measure. Incidentally, the "obvious" measure in which a uniform density is placed over Bernoulli parameters in the unit interval assigns prior probability zero to our hypothesis (by the strong law of large numbers) and hence fails to converge to the truth whenever H is true. This measure fails even to "keep the door open" concerning H.

[4] Conditioning can be extended to cases in which the evidence has zero probability, but we may neglect the complexities that arise in such cases here.

[5] Consider the method that just repeats the last datum observed.

[6] A demon can can be constructed who fools every possible inductive method on at least one possible data stream.

[7] For an extended discussion of results of this sort, cf. (Schervish and Seidenfeld 1990).

[8] The class of Borel hypotheses is defined in (Moschavakis 1980).

[9] It is known that conditioning must behave strangely on some countable partition when countable additivity is violated (Schervish, et. al., 1984). The preceding result shows that there exist cases in which no possible inductive method succeeds with probability 1.

[10] Ibid, p. 123. Regarding statistical practice, De Finetti relates: So far as I know, however, none of them has ever taken this observation into account, let alone disputed it; clearly it has been overlooked, although it seems to me I have repeated it on many occasions.

References

De Finetti, B. (1990), *Theory of Probability*, vols. 1 and 2, New York: Wiley.

Halmos, P.R. (1974), *Measure Theory*, New York: Springer.

Kelly, K. (1993), "Learning Theory and Descriptive Set Theory", forthcoming, *Logic and Computation*.

_____. (1994), *The Logic of Reliable Inquiry*, forthcoming, Oxford University Press.

Moschovakis, Y.N. (1980), *Descriptive Set Theory*, New York: North-Holland.

Royden, H. L. (1988), *Real Analysis*, New York: Macmillan.

Savage, Leonard J. (1972), *The Foundations of Statistics*, New York: Dover.

Schervish, J., T. Seidenfeld and J. Kadane (1984), "The Extent of Non-Conglomerability of Finitely-Additive Probabilities", *Z. Wahrscheinlichkeitstheorie verw. Gebeite* 66.

Schervish, J., and Seindenfeld, T. (1990), "An Approach to Consensus and Certainty with Increasing Evidence", *Journal of Statistical Planning and Inference* 25: 401-414.

Austere Realism and the Worldly Assumptions of Inferential Statistics[1]

J.D. Trout

Loyola University of Chicago

1. Introduction

Inferential statistical tests—such as analysis of variance, t-tests, chi-square and Wilcoxin signed ranks—now constitute a principal class of methods for the testing of scientific hypotheses. In particular, inferential statistics (when properly applied) are normally understood, for better or worse, as warranting a theoretical (typically causal) inference from an observed sample to an unobserved part of the population. These methods have been applied with great effect in domains such as population genetics, mechanics, cognitive, perceptual and social psychology, economics, and sociology, and it is not often that scientists in these fields eschew causal explanation in favor of the statement of brute correlations among properties. However, the appropriateness of a particular statistical test, and the reliability of the inference from sample to population, are subject to the application of certain statistical principles and concepts.

In this paper I will consider the role of one statistical concept (statistical power) and two statistical principles or assumptions (homogeneity of variance and the independence of random error), in the reliable application of selected statistical methods. Indeed, it is a truism repeatedly found, and often demonstrated, in statistics texts, that the results of particular tests are reliable only if they satisfy certain assumptions. So, for example, the distributions to which a parametric test such as the ANOVA is applied must have equal variance, if we are to legitimately infer from sample characteristics to the population characteristics.[2] But the conformity of statistical tests to these concepts and assumptions entails at least the following modest or austere realist commitment:

(C) the populations under study have a stable theoretical or unobserved structure (call this property T, a propensity or disposition) that metaphysically grounds the observed values and permits replication and generalization; the objects therefore have a fixed value independent of our efforts to measure them.

(C) provides the best explanation for the correlation between the joint use of statistical assumptions and statistical tests, on the one hand, and methodological success on the other. The claim that (C) provides the best explanation, however, depends on the following naturalistic constraint on explanation:

(E) Philosophers should not treat as inexplicable or basic those correlational facts that scientists themselves do not treat as irreducible.

Without (C), the methodological value of such assumptions and concepts would be an inexplicable or brute fact of scientific methodology. Without (E), the philosopher of science is able to design fanciful rational reconstructions of scientific practice; such reconstructions were popular in the logical empiricist tradition, guided by *a priori* principles of rationality that often designated as irrational typical and successful scientific practices. Along the way, I will consider possible empiricist interpretations of the reliability of these principles and assumptions. Let us first turn to the considerations that weigh in favor of (E).

2. Austere Realism: Evidence from Statistical Practice

There is a handful of statistical practices, entrenched in the physical, behavioral, and social sciences alike, which depend on assumptions about the nature of the unobserved world. Without these assumptions, familiar statistical practices would be without a rationale. In what follows, I will describe several such assumptions embedded in good methodological practice. By "good methodological practice" I mean practice that displays vigilance concerning the standard virtues of experimental design, such as sensitivity and power. If we acknowledge that these principles and concepts are central to good methodological practice, then traditional empiricists cannot both deny that the appropriate domains have property T and still hold that we are sometimes justified in making inductions from samples to populations, even when those inductions concern (unobserved) observables.

The naturalistic constraint on explanation (E) mentioned in section 1 in part expresses a desire for deeper explanations. This plea for explanatory depth, however, is not new. Leibniz appealed to it when voicing his criticisms of Newton's purely mathematical description of the relation between the Earth and the Sun. Newton's equations, Leibniz said, do not *explain* how the sun and Earth attract one another through space. Instead, they describe an observed relationship that, if left without explanation, must be deemed a "perpetual miracle". In a more contemporary context, Jonathan Vogel notes that "Where explanation is concerned, more is better, if you get something for it." (1990, 659) This should not be taken to imply that an explanation is inadequate until all of the facts on which the explanation depends are themselves explained. It is the governing theory that tells us when the explanation has gone far enough. Also, reigning theory tells us when any further detail would be extraneous or would yield a misleading picture of the sensitivity of the system. A naturalistic philosophy of science avoids rational reconstruction. Accordingly, the explanatory standards of our best sciences settle the question of the depth at which an explanation should terminate.

Some pleas for deeper explanation range over ordinary and scientific contexts. The sparest arguments for realism have always begun with modest explanatory presuppositions seemingly shared by all. In its austerity, the present contention resembles the conclusion of a clever argument for the existence of unobserved structure, proposed by Paul Humphreys. I quote him at length:

> Consider an experimental situation S in which regularity R has been isolated, one in which a single observed factor A is uniformly associated with a second observed factor E; i.e., E regularly appears whenever A is present. [footnote deleted] Then introduce a third factor B which, in S, in the absence of A, is uniformly associated with a factor F. Now suppose that we claimed that a straightforward Humean regularity was sufficient, in the simple situation we have described (together with certain additional features such as temporal succession — what these are does not matter

here), to identify A as a cause of E and B as a cause of F. Suppose further that neither E nor F is observed when both A and B are present and that the situation is completely deterministic. Now ask what happened to E. Why is it not present when B appears together with A? Now, as I mentioned earlier, it is possible for someone to deny that an explanation of this fact is called for. In such a view, there are three brute facts: situations with only A also have E present; situations with only B have F present; and situations with A and B have neither E nor F. I assume, in contrast, that the burden of proof is always on those who deny that an explanation exists for a given fact. And the case we have in mind should be taken to be the most routine, everyday kind of situation, with no exotic quantum effects. (1989, 58-9)

Humphreys's argument shows that there is an austere, minimal realist position one might adopt, and that the explanatory considerations in its favor are quite modest.[3] The natural explanation for the absence of E when A and B are present is that B prevents A from causing E, where event B is itself unobservable. This minimal realism justifies its ontological commitment by appeal to relatively enduring, unobserved structures,[4] but does so without advancing ambitious claims concerning the approximate truth of theories or providing detailed descriptions of specific theoretical properties. This austere realism is achieved rather by appeal to ordinary, explanatory demands. Recall that the principal question raised by experimental situation S is why E is not present when B appears together with A? If one rejects appeals to unobserved causal factors preventing E as epistemically illicit, then the observable correlations stated above must be regarded as explanatorily basic or irreducible; they just occur, in virtue of nothing knowable.

Similarly, evidence for (C) can be found not just in the experimental settings of the sort described above, but also in the success of those statistical concepts and principles that depend for their justification on the population's possession of property T. Below, the relevant principles and concepts are the independence of random error, statistical power, and homogeneity of variance, but others could serve to exemplify property T just as nicely, such as the unbiasedness and efficiency of an estimator.

The Independence of Random Error. Random fluctuations in the behavior of objects in study populations are thought to be tractable to statistical methods. A process is random if each of the possible outcomes (or values) has an equal probability of occurring. Random error is not a threat to design validity precisely because random error is *unsystematic*; that is, it is not the result of *bias*. It is in this sense that random errors are *independent*. Two events (or processes, properties, states, etc.) are statistically independent if a change in the probability of the one event has no effect (either positively or negatively) on the probability of the other event.

Typical statistics texts mention that no two events are perfectly independent; appearances to the contrary derive from the idealization most closely realized in games of chance in which outcomes are equipossible (Humphreys 1985). The correlation between two variables never completely reaches zero or, in null hypothesis testing, the null hypothesis is always, strictly speaking, false. This fact depends on a conception of causal relations according to which they are promiscuous and far-reaching. Causal relations breed statistical nonindependence.

The principle of the independence of random error concerns, among other issues, the bias-diluting or bias-reducing effects of random error. On the one hand, the choice of research problems, samples, controls, etc., is guided by our best theories. On the other hand, the independence of each random effect on each subject in a sample (or on each measurement of an instrument) reduces the likelihood of patterned results; that is what makes certain patterns in the data, when they do recur under diverse tests, particularly striking and theoretically interesting. This latter principle depends on the assumption of the canceling effects of error, an assumption stated (and sometimes, even justified) in any standard statistics text. For sheer clarity and simplicity, it is hard to improve upon Guy's statement of the assumption, fashioned over 150 years ago: "[T]he errors necessarily existing in our observations and experiments (the consequence of the imperfection of our senses, or of our instruments) neutralize each other, and leave the actual value of the object or objects observed"(1839, 32). [5]

What must the population be like if reliable theoretical judgments can be made based upon the principle of the independence of random error? The assumption is that random error (observed and unobserved) is of opposing value and equal magnitude, and so it is assumed to have a value independently of measuring it. Moreover, the independence of random error concerns the inference from a sample to a population because the assumption is that the error structure is the same in both the observed and unobserved parts of the population. The population's possession of property T therefore offers the most plausible explanation for the holding of the principle of the independence of random error.

Power and Sensitivity. After all the time and effort expended to design and run an experiment, the researcher wants to be confident (at least about factors in the design over which the experimenter has control) that the experiment is in fact sensitive to the dependent variable so that an effect will reach significance if the latter is present. The power of a test, then, is the probability of correctly rejecting a false null. (So, power = $1-\beta$) Power is affected by a number of factors: (1) the probability of a Type I error, in which we reject the null hypothesis when it is true (2) the true alternative hypothesis, (3) sample size, and (4) the specific test to be used.

There are a number of ways that the concept of statistical power can be loosely illustrated. Using a test with low power (say, due to small sample size), is like trying to catch fish of various sizes with a large mesh net: You won't catch many, but the ones you do catch will be large. On the other hand, you will miss many smaller fish (commit many misses, or accept the null hypothesis when it is false). It is, indeed, due to the strict nonindependence of events (along with considerations of power) that we can construct "overly" sensitive tests. By increasing the sample size, we increase the probability that the postulated effect will be detected if it is there. And where the sample size is extremely large, it is highly probable that arbitrarily selected variables will yield correlations that reach significance (for more on what has been called the "crud factor", see Meehl 1990).

What must the populations be like if we are to suppose that selected factors such as sample size and size of uncontrolled error affect sensitivity and power? If we are to honor (E) rather than simply insist that an equation holds or an observed correlation obtains, we must suppose that there is some discriminable, stable property of each object such that, by introducing it into the sample, the test increases in statistical power.

Homogeneity of Variance. When comparing two populations, we are attempting to estimate the value of the same quantity in both populations; otherwise, it would make no sense to infer that a change in the value of the quantity on the treated sample is due to the introduction of the independent variable. In tests that depend on variance, we estimate the magnitude of that influence by a ratio that conforms to the appropriate distribution (t, chi-square, etc.). Where σ^2 is variance, homogeneity of variance is indicated by the relation

$$\sigma_1^2 = \sigma_2^2 = \sigma^2$$

in which the subscripts indicate the particular population.

There are parametric tests (such as the t-test and ANOVA) and nonparametric ones (such as the Wilcoxin). Parametric tests make three demands on the data they apply to: (1) the data must be drawn from a *normal* population, (2) the populations (when it appears that there are more than one) must have *equal variances*, and (3) the variable of interest must be measured on an *interval* scale.

To see how the concept of homogeneity of variance plays a crucial methodological role in deciding what test to run, we must first explain the notion of variance. We determine the variance of a population by first calculating the deviation score $(x - \bar{x})$ (the difference between each score and the sample mean). After summing the squared deviation scores, we divide the sum by the number of scores. The result is the variance of the population. Now, if tests of dispersion (such as ANOVA) are to show that, with respect to a certain variable, some set of scores was drawn a different population—and thus there is a significant difference between them—the two populations must have roughly equal variances. Otherwise, the appearance that the two sets of scores belong to different populations could be an artifact of the inequality of variance rather than of performance differences with respect to the test variable *despite* the populations' being otherwise the same.

In order to determine homogeneity, a F-ratio test is run. The F ratio represents the relation between the two estimates of variance: the variance between the means of the groups studied to the variance *within* the groups studied. This relation is expressed as follows (S^2 is the symbol for variance):

$$F = \frac{S_b^2}{S_w^2}$$

For a t-test, we calculate the F-ratio as a preliminary to a test that compares the means of two groups. The t-test makes an assumption of equal variance; that is, that the two groups are drawn from populations with equal variances. There is a rationale for the prior F test, for we want to make sure that any substantial difference in the means is a consequence not of initial differences in population variance, but in mean performance upon the introduction of the treatment.

Now, what properties must such populations possess if it is permissible to infer the suitability of a parametric test from the outcome of the F-ratio test? Minimally, it must be the case that the populations have enduring properties in virtue of which they can be systematically distinguished. And because the distribution is replicable—that is, the distribution is the result of a process that would produce a relevantly similar distribution under repetitions—the unobserved properties must be stable enough to permit such replication. By (C), this feature is best explained in terms of the population's possession of property T.

3. Explanation and Irreducible Facts: Empiricist and Otherwise Deflationary Reactions

There is a remarkably durable, empiricist conception of theory-testing according to which the epistemic interpretation of these statistical notions is exhausted by their observational content; their meaning can be defined, and their use given a rationale, solely in terms of their observational content. Indeed, population parameters are often treated by statisticians as certain sorts of mathematical fictions, as the result of indefinite or (approaching) infinite samplings (for a brief review, see Suppes, 82-83; for a discussion from the perspective of statistical inference, see Barnett 1982). Traditional empiricists have been loathe to rely on idealizations (see Trout 1995).

Nevertheless, the values that provide an apparent basis for knowledge possess features which violate traditional empiricist conceptions of knowledge. In the case of knowledge borne from statistical inference, the standard empiricist understanding of statistical assumptions takes a form of the frequency interpretation: The "real" values are simply idealizations concerning the observed values yielded under an indefinite (or infinite) number of samplings or potentially infinite sequences of trials (for the standard view, see Mises 1957). The difficulties with a frequency interpretation of probabilistic claims are well known (see, for example, Hacking 1965). Consider the .5 probability a fair coin has to come up heads. Because the frequency interpretation defines 'probability' in terms of observed frequency, no probability (of coming up heads or of coming up tails) can be assigned to an unflipped coin. The most natural explanation for the .5 distribution is that the coin has a *propensity* or *tendency* to produce a distribution of .5 heads. Let us concede that explanatory appeal to such a propensity may be vacuous when it is invoked to account for only a single (type of) observed effect, because the propensity gets framed *in terms of* the observed effect: The coin yields a .5 heads distribution because it has the propensity to yield a .5 heads distribution. Such explanations, however, are not vacuous as long as the propensities are manifested in diverse ways, and thus can be independently characterized.[6] So the propensity said to explain a fair coin's .5 heads distribution may explain other observed effects as well, for example, other effects concerning its center of gravity. On the frequency interpretation, by contrast, the fact (and the correctness of our expectation) that a fair coin will yield a .5 heads distribution is not explained in terms of an unobserved, independently specifiable disposition or propensity. Rather, that fact is basic or irreducible, not explicable in terms of any deeper causal fact. The approximation of sample values to population values in the estimation of parameters such as variance appears to be a similarly irreducible fact for the empiricist, while it is explicable to the realist in terms of a propensity or tendency of the objects to produce certain observed values.[7]

Explanations cite particular factors manifested in a variety of ways in observation. Without an explanation in terms of theoretical entities or laws, these various observable manifestations appear to be unconnected, and so one class of observational data does not provide inductive support for claims about the relations to other classes of observational data that represent other manifestations of putatively theoretical objects. For example, in cognitive psychology, prototype studies in the 1970s revealed just such a robust phenomenon. When asked to rate how representative an object is (e.g., robin, chicken, etc.) of a certain class (e.g., bird), subjects' performance was the same on both ranking and reaction time tasks (Rosch and Mervis 1975; Mervis and Rosch 1981). The convergent results of these two different test methods are taken by psychologists to indicate that the prototype effect represents a real (non-artifactual) feature of mental organization. Statistical methods were used in both sets of studies representing both measures.

There are a number of empiricist reinterpretations of the argument for C (the claim that some populations have property T). One might claim that I have not distinguished between objects which fall outside of the range of our sensory and perceptual powers, and those that we *could* observe with the unaided senses but are simply not properly situated at this time. The empiricist might claim that the argument for C has no force, since one ought not treat with epistemic parity claims about unobserv*ed* and unobserv*able* phenomena. However, the argument I have presented trades on the *success* with which these statistical concepts and assumptions have been implemented. This empiricist response, I believe, cannot bear the weight it places on this distinction. If the empiricist is to preserve the claim that induction is *ever* successful, the claim must range at least over unobserved observables. Meager as this power might be, without T this reliability of induction over unobserved observables is mysterious. The empiricist might contend that such instrumental reliability needs no explanation, but such a rejoinder runs afoul of (E).

On the other hand, one might deny that the sciences that employ statistical methods are successful, but that view has never been defended, and for good reason; these principles have been used to draw conclusions concerning observables, proving reliable in circumstances in which population values are already known. This is a common point among realists. Clark Glymour states that bootstrap principles "are principles we use in our science to draw conclusions about the observable as well as about the unobservable. If such principles are abandoned *tout court*, the result will not be a simple scientific antirealism about the unobservable; it will be an unsimple skepticism." (1985, 116) Michael Devitt states that the fundamental issue separating the realist and empiricist "is selective scepticism; epistemic discrimination against unobservables; unobservables rights." (1991, 147) The sweeping consequences of selective skepticism have been noted elsewhere. Richard Boyd replies to attacks on realist applications of abduction by pointing out that "the empiricist who rejects abductive inferences is probably unable to avoid—in any philosophically plausible way—the conclusion that the inductive inferences which scientists make about observables are unjustified." (1983, 217) If we want an explanation for the reliability of these statistical principles and concepts *at all*, we must suppose that we have at least modest theoretical knowledge.

4. Conclusion

The argument I advance for an austere realist interpretation of statistical practice depends on similar explanatory considerations, and makes a specific suggestion concerning the explanatory item: a propensity or a dispositional property invoked to account for the (observed) results of statistical applications. But for the modern philosophical vocabulary and specific measures of empiricism, there is nothing new in this general picture, and in holding it, I am in sound statistical company. Lagrange, Laplace, and Gauss, all had confidence that the populations worth studying had parameters with real values,[8] and inaccuracy or error arose not from the absence of a true value, but from both the nature of the true causes—which could be stochastic—and from our ignorance. Given this confidence in the stability of an unobserved world, it was also quite common for these 18th and 19th century figures to believe that under repeated samplings statistical methodology will bring our beliefs into conformity with the world.

The virtue of the present argument—in addition to its illustration of the austere realism represented in the reliability of statistical principles and concepts—is its demonstration that the empiricist's selective skepticism is abetted by (indeed, may depend upon) the empiricist's tolerance for brute facts or irreducible correlations. This tolerance is unnatural once philosophical standards of explanation are assessed by those of science, that is, by (E). Likewise, we should not blithely regard as inexplicable, or in need of no explanation, the correlation of the use of statistical methods and the improvements in diverse fields incident upon the introduction of these methods. More specifically, we should not take as irreducible the coincident use of statistical principles and the reliability of inferences from samples to population characteristics. In light of the general epistemic reliability of these statistical, quantitative methods—in the social and behavioral sciences as well as biology, chemistry and physics (for the latter, see Eadie et. al., 1971)—we should want to understand why they work when they do. Once achieved, this understanding leads to realism. Therefore, to the extent that one adopts the naturalistic conception of explanation, one will reject as unduly mysterious the empiricist account of the reliability of selected applications of statistical methodology, treating as a brute fact a correlation that the realist and scientist alike would explain.[9]

Notes

[1] For comments on this paper, thanks are owed to Paul Moser.

[2] I believe my argument works for many other statistical methods in addition, for example, regression. For a nice philosophical introduction to regression, see (Woodward 1988).

[3] The position I defend has affinities to that found in Humphreys (1989), as well as to Devitt's (1991) "Weak Realism" and Almeder's (1991) "Blind Realism", though my evidence derives from the successful use of statistical concepts and principles. Other realist arguments for modest knowledge of unobserved structure can be seen in Hausman (1983, 1986). On the basis of other, perhaps more specific and detailed evidence, a stronger version of realism may be warranted.

[4] Here "real" means "taxonomic in science", "natural kind" or "an isolable object that can act as a cause", rather than the sort of nonexplanatory construct that permits the "average man" fallacy.

[5] For another clear statement of this assumption, see Humphreys (1989, 48). Not all error need be observed; to suppose so is to conflate, in verificationist fashion, the concept of error with the experiential grounds for identifying error. One qualification: When the "errors necessarily existing in our observations" are large, they may conflict and thus cancel out, leaving us with a statistical test that is less powerful, and consequently less sensitive to the variable under investigation.

[6] I hold that such propensity-explanations are not vacuous even if no such independent specification can be found. Minimally, citing a propensity as a cause informs us that the effect is, in general, nonaccidental.

[7] Although I will only discuss the frequency interpretation of these principles and concepts, Bayesians have something to account for as well; they must explain the stability of their subjective estimates.

[8] The sense in which a population has a "real value" is, as presented in this paper, a far weaker sense than one might have thought some scientific realists are committed to. The above methodological concepts and principles do not presuppose that these values always license causal inference, that they are permanent (rather than just stable), or that they are exact. The fact that these assumptions play a central role in making discriminating judgments about appropriate design is insufficient to support all of the epistemological claims of scientific realism—in particular, the claim that we can have detailed knowledge of the specific nature of those unobservables.

[9] The argument present in this paper is developed in greater detail in a chapter of a forthcoming book on the philosophical foundations of quantitative methods in the social and behavioral sciences (Trout, forthcoming).

References

Almeder, R. (1991), *Blind Realism*. Lanham, MD: Rowman and Littlefield.

Barnett, V. (1982), *Comparative Statistical Inference*, 2nd ed. New York: John Wiley & Sons.

Boyd, R. (1983), "The Current Status of Scientific Realism", *Erkenntnis* 19: 45-90, reprinted in R. Boyd, P. Gasper, and J.D. Trout, (eds.) (1991), *The Philosophy of Science*. Cambridge, MA: MIT Press/Bradford. (Page references are to the latter.)

Cohen, J. (1988), *Statistical Power Analysis for the Behavioral Sciences*, 2nd ed. . Hillsdale, NJ: Erlbaum.

Devitt, M. (1991), *Realism and Truth*, 2nd ed. London: Blackwell.

Eadie, W., Drijard, D., James, F., Roos, M. and Sadoulet, B. (1971), *Statistical Methods in Experimental Physics*. London: North-Holland Publishing Co.

Glymour, C. (1984), "Explanation and Realism". In J. Leplin, ed., *Scientific Realism* (pp.173-192). Berkeley: University of California Press. Reprinted in P. M. Churchland and C. A. Hooker, eds., *Images of Science*, Chicago: University of Chicago Press, pp.99-117.

Guy, W. (1839), "On the Value of the Numerical Method as Applied to Science, but Especially to Physiology and Medicine", *Proceedings of the Royal Statistical Society* A 2: 25-47.

Hacking, I. (1965), *The Logic of Statistical Inference*. Cambridge, ENG: Cambridge University Press.

Hausman, D. (1983), "Are There Causal Relations Among Dependent Variables?" *Philosophy of Science* 50: 58-81.

Hausman, D. (1986), "Causation and Experimentation", *American Philosophical Quarterly* 23: 143-154.

Humphreys, P. (1985), "Why Propensities Cannot Be Probabilities", *The Philosophical Review* 94: 557-570.

_____. (1989), *The Chances of Explanation*. Princeton: Princeton University Press.

Meehl, P. (1990), "Appraising and Amending Theories: The Strategy of Lakatosian Defense and Two Principles that Warrant It", *Psychological Inquiry* 1: 108-141.

Mervis, C.B., and Rosch, E. (1981), "Categorization of Natural Objects", *Annual Review of Psychology* 32: 89-115.

Mises, L. von. (1957), *Probability, Statistics and Truth*, 2nd rev. ed. New York: Dover Publications (reprinted in 1981).

Rosch, E., and Mervis, C.B. (1975), "Family Resemblances: Studies in the Internal Structure of Categories", *Cognitive Psychology* 7: 573-605.

Suppes, P. (1984), *Probabilistic Metaphysics*. London: Blackwell.

Trout, J.D. (1995), "Measurement", in W.H. Newton-Smith (ed.), *A Companion to the Philosophy of Science*, London: Blackwell.

_ _ _ _ _ . (forthcoming), *Measuring the Intentional World* (unpublished book manuscript).

Vogel, J. (1990), "Cartesian Skepticism and Inference to the Best Explanation", *The Journal of Philosophy* 90: 658-666.

Woodward, J. (1988), "Understanding Regression", in A. Fine and J. Leplin (eds.), *PSA 1988, volume 1*, Lansing, MI: Philosophy of Science Association, pp.255-269.

Retrieving the Point of the Realism-Instrumentalism Debate: Mach vs. Planck on Science Education Policy

Steve Fuller

University of Durham

The realism-instrumentalism debate (RID) is often seen as the core debate in philosophy of science, yet increasing numbers of philosophers have followed Fine's (1984) lead in questioning the point of this debate. Fine's largely unchallenged rendition of RID makes it is easy to see why. RID appears doomed to stalemate. According to Fine, realists and instrumentalists are both trying to account for the string of progressive episodes in the history of science. The two sides are said to agree on what those episodes are and that they constitute progress. However, it seems that every realist story of one such episode can be matched by an instrumentalist one—and vice versa. As Fine sees it, any such story is merely an attempt to extract surplus philosophical value from the historical labors of scientists. It only adds a misleading air of inevitability to their original efforts. These labors can be more simply and adequately captured by examining the norms implicitly governing the scientists' actual practices, which, if one insists on a philosophical label, may be called the "Natural Ontological Attitude" (NOA). Admittedly, the content of NOA varies over time and place, but then the philosophical supposition that the history of science would deliver an overarching logic of inquiry was false from the start.

Philosophers (e.g. Rouse 1987) who embrace the recent turn to "historicism," "contextualism," and "practice" in science studies take their lead from Fine in critiquing this Whiggish exploitation of the historical record. However, I believe that RID deserves a new lease on life. This paper, an exercise in applied social epistemology (Fuller 1993), aims to recover some of the cultural import that was originally invested in this debate, one which motivated scientists in the decades before World War I to engage in highly polemical public exchanges while in the midst of an unprecedented period of research activity. In this context, explaining the success of science was only a means to other ends, not an end in itself.

Ironically, Fine's "surplus value" account of RID appears impressive just as long as we suppose that the context surrounding the debate is the same now as it was when versions of the realist and instrumentalist positions were first articulated a little over a century ago. One would expect, then, that the debate was originally conducted by professional philosophers in technical journals. But, of course, this was not the case. On the contrary, the prototypes of these positions—often better illustrated and more en-

---

PSA 1994, Volume 1, pp. 200-208
Copyright © 1994 by the Philosophy of Science Association

gagingly expressed—were to be found in the popular writings and lectures of the leading scientists of the day (cf. Kockelmans 1968). Professional philosophers, insofar as they followed these matters at all, did so as kibitzers, not as major players (e.g. Cassirer 1923). Philosophers began to dictate the terms of the debate only with the rise of Logical Positivism. This transition arguably began in Schlick (1974, 94-101), a hybrid account of RID written by the student of Planck who went on to occupy Mach's chair in Vienna. However, our analysis here will be confined to the differences between Schlick's two illustrious precursors.

Historians have been much quicker than philosophers to pick up on why RID has mattered to scientists (Laudan 1993). When, say, Boltzmann and Ostwald, or Planck and Mach, argued about the existence of atoms, they were not merely trying to second-guess what empirical research would eventually show; rather, they were trying to influence the direction that such research should take and the way it should be evaluated. Like judges, they would appeal to history to show precedent for alternative science policies. "Realism" and "instrumentalism" were thus comparable to schools of legal interpretation. They captured alternative strategies for extending the methods, findings, and world-view of the physical sciences into new domains. These new domains included not only the research practices of the newer academic disciplines, such as the biological and social sciences, but also, as we shall see here, educational practices at the university and sub-university level.

Although scientists nowadays rarely polemicize against each other about the ends of science, realist and instrumentalist arguments in this original sense continue to appear in their popular works, as when Weinberg (1993) defends the Supercollider on realist grounds, for its putative contribution to explanatory unity in physics. Philosophers are prone to dismiss these arguments as unsophisticated and self-serving, and hence feel no need to contest them in the public forum. Unwittingly, such philosophical disdain only serves to lower the standards of public science policy debate, as science popularizations come to be seen as either authoritative accounts of how science works or dubious cases of special pleading. However, both unappetizing options may be avoided by recalling some of the public contexts that have called for adopting a "realist" or "instrumentalist" stance.

1. Mach vs. Planck: The Aims of Science Education

Between the Franco-Prussian War and World War I (1870-1914), great strides were taken to modernize the educational systems of the imperial powers, Britain, France, and the newly consolidated German Reich—the last generally regarded, even by the Kaiser, as the economically most powerful yet socially most backward nation in Europe (Albisetti 1983). Reform moved on two orthogonal fronts: on the one hand, educational opportunities were expanded for the populace, while, on the other hand, educational credentials were introduced to differentiate and stratify occupations (Mueller et al. 1987). Today, movement on both fronts is taken as a defining feature of modernizing societies. Indeed, sometimes (as in the work of Niklas Luhmann and other systems-theorists) "growth through specialization" enjoys an axiomatic status, bolstered by a strong organismic analogy and little empirical argument. However, the relationship between the expansion and specialization of education was much more controversial at the end of the last century.

When relatively few people received formal schooling, education served as a vehicle for social mobility, enabling both the lower classes to improve their social standing and the upper classes to move easily between careers. Not surprisingly, democratic theorists called for universal formal education as a means of completely equalizing employment opportunities. Yet, many educational reformers feared that, left unchecked, the mobility of the newly educated masses would lead to social disorder. They wanted to reinvent within the "liberalized" educational system the sorts of discriminations that had traditionally restricted access to education. In practice, this meant predicating employment on credentials, the acquisition of which required the student to undergo a course of study whose content would be controlled by the relevant academic specialists.

In Germany, much of the public debate over these matters centered on the implications of introducing the natural sciences into the secondary schools, the *Gymnasien*, the curricula of which had remained uniformly humanistic throughout the nineteenth century. Most parties agreed that some form of (natural) science education should be made available in at least some of these schools—but in what form, and to what end? Answers to these questions turned on what was taken to be the distinctive epistemic contribution of the natural sciences, and its relevance to the "modern" German citizen who may not pursue scientific research as a career, but whose continued support would be needed for science to continue at its current pace.

The terms of RID start to emerge at this point. I will let the statements of Ernst Mach and Max Planck stand for representative "instrumentalist" and "realist" attitudes. They engaged in a sustained and highly personal debate about "the ends of science" that attracted considerable public attention from 1908 to 1913. Broadly speaking, Mach's instrumentalism drove him to see mass empowerment and scientific credentialism as incompatible goals for education, whereas Planck's realism led him to endorse credentialism as a necessary complement to a rapidly expanding educational system.

The cases of Mach and Planck are especially illuminating because they lived their respective ideologies. Mach championed the cause of adult education in the Austrian Parliament and authored a series of middle school science textbooks and college science textbooks for nonscience majors. Through these texts, many of the leading scientific thinkers of this century—including Einstein, Heisenberg, Carnap, Popper, and Wittgenstein—were introduced to Machian instrumentalism. Planck, on the other hand, administered many of the institutions responsible for shaping the professional identity and public voice of the German natural science community both before and after World War I. These included the German Physics Society, Berlin University, the corporate-sponsored national research institutes known as the Kaiser-Wilhelm-Gesellschaften, as well as several international scientific unions.

Unless otherwise indicated, the account that follows is drawn from Blackmore (1972, 204-27), Heilbron (1986, 47-69), and Blackmore (1992, 127-50).

Mach famously located the value of science in its ability to economize on thought: Phenomena that once could only be handled with great mental and physical effort—because they were thought to be disparate in nature—now could be epitomized in a single mathematical equation or set of formulas. In most general terms, science is an abstract labor-saving device that facilitates the satisfaction of human needs, thereby freeing up time for people to pursue other things. Mach did not hold the practice of academic science to be itself an especially interesting or ennobling pursuit. He went so far as to ridicule the practicing scientist's taste for the odd and the exceptional that was often pedagogically dignified under the rubric of "curiosity."

Given these views, Mach not surprisingly concluded that secondary schools could do justice to science without requiring an inordinate amount of specialized science education. Indeed, beyond the point of showing science's historical tendency toward economizing on effort, science education may even become self-defeating. The last thing Mach wanted was to replace the time students wasted on mastering humanistic arcana with their mastering scientific arcana. In fact, he wanted to reduce the amount of time students spent in school generally.

Here we begin to see the role that an instrumentalist orientation toward the history of science played in Mach's pedagogy. For what follows, let me distinguish its role in the education of nonscientists, which has been admired by positivists and genetic epistemologists alike (Matthews 1991), from its role in the education of scientists, which Feyerabend (1979, 195-205) has identified as the inspiration for his own project. Planck's position will then be constructed as a response to each version.

In the case of nonscientists, Mach's pedagogy followed directly from his instrumentalism. Take the "atomic hypothesis" that was then being pursued by Planck and many of his colleagues in physics. Should it be taught as part of general science education? Mach reasoned as follows: If atomism turns out to be true, then its hypotheses will be reducible to equations with specifiable applications; in that case, the appeal to atoms will be superfluous. However, if atomism turns out to be false, then it will be better for it not to have been taught at all, rather than than to risk legitimating whatever metaphorical inferences people were prone to draw from the nature of atoms.

Mach's caution about overextending the authority of science was characteristic of the Austro-German brand of positivism that would come to dominate Anglo-American philosophy of science. Its political cues were taken from Mill rather than Comte. A more familiar expression of this concern may be found in Max Weber's call for "value-neutral" science, which is often misconstrued as simply a defense of pure research, when in fact its primary aim was to persuade educators not to preempt their students' value choices by passing off their own speculations and prejudices as though they had the status of empirical truths (Proctor 1991, 134-54). Such was the perceived power of scientific rhetoric that Mach himself wanted to purge "mass" and "force" from science pedagogy, as he believed that the circulation of these terms from classical mechanics through German Idealism to *Realpolitik*—now translated as "ego" and "will"—were responsible for the sense of inevitability that politicians increasingly attached to international conflict (Blackmore 1972, 232-236). Here, then, are the ironic origins of the idea of language therapy that would later preoccupy Logical Positivism.

However much Mach's views on science education might lead students to respect the accomplishments of science, they were not designed to encourage students to enter science or even to adopt a scientific mindset—at least one that bore the stamp of the physics community. Such was the nature of Planck's opposition to Mach. It reflected Planck's view that science imparted an increasingly coherent world-picture, much like a Kuhnian "paradigm," whose strictures could deepen the understanding and formalize the practice of virtually any field. Speaking in terms of credentials, science education "added value" to other forms of technical training. For example, engineers became better engineers by mastering some of the problem-set of physics. Planck's focus, then, was on teaching students how scientists construct and solve well-formed problems, thereby enabling them to acquire what we would now call "exemplars" and "disciplinary matrices" that can be used for shaping their own practices. This was more than a matter of applying theory-neutral equations as one pleased. It required that students be sufficiently committed to some overarching model of physical reality—such as atomism—to think through its implications for some experimentally testable cases.

Now, a standard objection of realists to instrumentalists comes to life: Without a theoretical framework to suggest objects beyond phenomena that have already been economically saved, what motivation would there be for continuing to do science? In practical terms: If you do not convey some of the research orientation of professional scientists in general science education, how do you then expect to recruit the next generation of scientists and to sustain public support for cutting-edge scientific research? For Mach science was a fit subject for general education as long as students could easily assimilate it into their normal lives, whereas Planck wanted students to become acquainted with the more demanding qualities of science that contributed to its distinctive place in modern culture. Moreover, contrary to the labor-saving image of science that Mach promoted, Planck believed that as physics approached unification, each additional increment of knowledge would require increased effort without necessarily issuing in any direct practical benefits. A public accustomed to seeing science as an economizing tool could well become discouraged by such prospects.

When it came to the professional education of scientists, Mach called for a "critical-historical" approach, as immortalized in his 1883 work, *The Science of Mechanics* (Mach 1960). "Critical-historical" is an expression taken from a brand of Enlightenment theology championed, in the nineteenth century, by the "Young Hegelian" David Friedrich Strauss (Gregory 1992). The theme that runs through this tradition is that Christianity has been mystified by ecclesiastical attempts to suppress the historicity of Jesus. A properly historicized Jesus would, among other things, show that Jesus was imperfect (but hence humanly approachable), that his teachings should be seen as universalizable (and hence not the property of a particular religion), and that the historical record has been repeatedly compromised in order to maintain Church authority. Mach's innovation was to model the history of science on this vision of the history of Christianity.

The "universal message" of science that Mach believed has always faced institutional resistance is that science progresses only as an economical response to outstanding human needs. In this respect, Mach endorsed the prevailing image of Galileo as someone who succeeded by not letting theology obscure his ability to confront nature directly. However, Mach was provocative in continuing this line of argument into the present, suggesting that the institutional structure of physics itself—especially its insistence on a uniform theoretical orientation to research—impeded scientific progress. As evidence, Mach highlighted fundamental objections to Newtonian mechanics which remain just as potent as when they were first made nearly two centuries earlier, but which had been suppressed from the professional training of physicists. The most famous of these objections pertained to the existence of absolute space and time, the aether, atoms, and even mass itself.

Planck was particularly incensed by Mach's call for a critical-historical approach to professional science education. He truly believed that it would destabilize the scientific community by renewing the credibility of defunct research programs—good examples of which could be found among Mach's students who pursued experimental *Naturphilosophie* in order to follow up Goethe's phenomenological critique of Newtonian mechanics. Mach was willing to risk destabilizing the scientific community because these defunct programs typically drew on "folk science"—what we would now call "indigenous knowledge"—the revival of which was crucial for democratic empowerment. The polemics over this particular issue enabled both Planck and Mach to indulge in a religious rhetoric that confirmed Mach's worst suspicions that he was a heretic fighting against a rapidly forming clerisy in physics.

One indicator of Mach's resistance to this emerging orthodoxy was his distinctive sense of the "unity of science," which was modeled on Fechner's laws of psychophysics. By contrast, Planck cited Boltzmann's statistical unification of thermodynamics and mechanics as evidence of impending unity. In other words, Mach envisaged unity in terms of translation principles between types of phenomena that are treated as ontological equals, such as physical stimulus and psychic response, but not as reduction principles that purport to exhaustively explain one type in terms of a "deeper" type to which only cutting-edge physicists and their emulators had epistemic access. Not surprisingly, Mach found many allies across the sciences, but they were nearly all *outside* physics.

For his part, Planck argued that the epistemic distinctiveness of physics lay in its ability to reach closure on an ever wider body of observations by a larger number of observers, all encompassed under a single unifying theory. This theory may not make sense to someone without the proper training, but then the validity of such a theory would not be checked by such anthropocentric means, but by its ability to deliver the same results to any inquirer anywhere, even on Mars. Any critique that failed to respect this fundamental aspiration did not deserve the title of "science." Planck's appeal to the "independent" and "invariant" character of ultimate reality was certainly a familiar one, yet his argument implicitly conceded to Mach that universal convergence on this reality presupposed the elimination of individuality from the creative process by the enforcement of a uniform research orientation. We see here a vivid admission of the paradoxical interdependence of scientific realism and the theory-ladenness of observation.

2. RID and the Politics of Science: Then and Now

A common assumption of RID is that science is, in some significant sense, autonomous from the rest of society. At the very least, the success of science can be explained without referring to the societies that have supported it. However, once RID is fought in the arena of educational policy, such autonomy becomes difficult to maintain: How can science justify its autonomy, while at the same time meriting inclusion, if not privilege, in the key processes of social reproduction? From the Mach-Planck debate, two kinds of answers can be discerned which, over the course of this century, have proven influential as they have unsatisfactory. The realist appeals to science as an exceptionally rational world-picture whose adoption promises to add disciplined thought to any line of work. The instrumentalist portrays science as an economizing tool that can assist everyone in pursuing their ends without imposing any value orientation of its own. In practice, these alternative visions of "autonomy" have amounted to the dilemma of "use or be used."

When positions are developed dialectically, it is common for each side to scrupulously avoid the interlocutor's shortcomings, yet to remain unwittingly blind to one's own deficiencies. For Mach, the biggest threat posed by Planck's realist educational policy was clearly indoctrination. This fear led Mach to conceptualize the cultural significance of the natural sciences in terms that closely conformed to the liberal doctrine of "academic freedom" as applied to both researcher and student. Unfortunately, this doctrine was originally designed with the humanities as the center of the educational system. The range of applications of natural scientific knowledge was wider and potentially more dangerous than those of humanistic knowledge. Thus, to deny science its own value orientation was to license indirectly the appropriation of scientific knowledge for any purpose, including destructive ones, as World War I was soon to demonstrate. For every Mach who resolutely refused to involve his scientific expertise in the war effort, there were plenty of Machians, especially among the chemists (including the "pacifist" Wilhelm Ostwald), who "freely" enrolled in the Kaiser's cause (Johnson 1990, 180-183).

Unlike Mach, Planck did not believe that the problem of scientific autonomy would be solved by a sharp separation of science and values, for that would only make science captive to whomever has the power to use it. Science had to be socially recognized as its own value orientation, alongside yet noncompetitive with the state, religion, and industry. An elite functionary for most of his career, Planck was alive to corporatist tendencies in the modern nation-state that eluded Mach's democratic liberalism. Thus, Planck organized scientists in ways that enabled them to take collective control of the direction and application of their work, a strategy that included insinuating a distinctive natural scientific perspective throughout the educational system.

The success of this perspective is reflected in the ascendency of such concepts as "mental tendency," "problem set," and even "gestalt," first in experimental psychology, and later in pedagogy, and ultimately in Kuhn's (1977) account of paradigm acquisition. The seminal studies associated with these concepts frequently had subjects solve simple puzzles in classical mechanics, tasks which left the impression that the thought-patterns of physicists were especially self-conscious versions of thinking *per se* (cf. Humphrey 1951). This was a far cry from the simple capacity for precise observation and critical judgment that Mach wished to carry over from science to general education.

Nevertheless, as the religious rhetoric of the Mach-Planck debate brings out, the realist strategy placed science—especially an advanced science like physics—in an awkward political position of its own. For, if the ends of science are not merely distinct, but increasingly divergent, from other societal ends, then students will need to be given early exposure to the scientific world-picture, in order to be attracted, or at least rendered sympathetic, to scientific careers. Thus, increased control of the curriculum would seem to be necessary for continued control of the research agenda.

At least three philosophical lessons of contemporary import may be drawn from the Mach-Planck version of RID:

First, the pedagogical debate over the role of "science" in general and specialized education is still with us, and the legacies of Mach and Planck live on. Their differences are perhaps most acutely felt in the teaching of mathematics. Is it better pedagogy to make students adept at using machines that will simplify their computational tasks, or should students continue to acquire computational skills first-hand, including a conceptual framework—such as number theory—that epistemically grounds those skills? Mach or Planck? What the realist takes to be essential to the intellectual process, the instrumentalist takes to be an unnecessary inconvenience.

Second, the old mutual suspicions linger. What is to be feared more: a closed science that has been reduced to an off-the-shelf technology (Planck's fear) or the mass indoctrination of a scientific theory that affords it an extrascientific significance that it does not deserve (Mach's fear)? The Technological Menace or The Ideological Menace? While the former nightmare has become vivid enough with science's increasing relations with the state and industry, the latter appears interestingly enacted in the Creationist Controversy. Despite Mach's own strong predilection for Darwinism, Feyerabend (1979) may be right in supposing that a consistently Machian sensibility would want to challenge the hegemony that evolutionary theory enjoys in the public teaching of biology.

Third, and speaking in a more reflexive vein, the argument of this paper may be read as an instance of Mach's own critical-historical approach, which itself has been suppressed in recent discussions of the role of history in the philosophy of science. I take this suppression to be another indication of the long-term triumph of Planck's vision of science over Mach's. Planck's attitude toward history is rehearsed whenever we appear forced to choose between a "science textbook" history that depicts the past purely in terms of its anticipation of the present and a "professional historian's" history that depicts the past purely in its own terms without any reference to the present. Kuhn (1970) canonized this false dilemma—between Whiggism and relativism—when he argued, on the one hand, that normal science cannot proceed without scientists having an "Orwellian" sense of their own history, and on the other, that the past cannot be properly understood if the historian has a stake in how the events turn out (cf. Fuller 1992).

Missing from this convenient division of scientific and historical labor is the idea that the very consignment of certain events to "the past"—the fact that, in Mach's day, certain criticisms of Newtonian mechanics were considered "merely historical"—may reveal the implicit limits of contemporary scientific modes of thought. Thus, *tertium datur*: History can be used not only to legitimate the present and to recover the past, but also to alter the future by reintroducing silenced voices from the past into present-day concerns. This is more than simply telling the stories of scientists "warts and all" (*pace* Brush 1974), but rather an attempt to renegotiate the disciplinary boundary between "history of science" and "science proper." Such critical-historical work regularly occurs in the social sciences. Not surprisingly, the major historians of these fields are themselves typically regarded as field practitioners. By contrast, a historian of physics is not usually said to make a direct contribution to physics. And this is exactly how Planck and Kuhn would want it, for such a division of labor enables the pursuit of normal science. But is this how things *should* be?

References

Albisetti, J. (1983), *Secondary School Reform in Imperial Germany*. Princeton: Princeton.

Blackmore, J. (1972), *Ernst Mach*. Berkeley: California.

_ _ _ _ _ _ _ . (Ed.) (1992), *Ernst Mach—A Deeper Look*. Dordrecht: Kluwer.

Brush, S. (1974), "Should the History of Science Be Rated X?" *Science* 183: 1164-1183.

Cassirer, E. ([1910] 1923), *Substance and Function*. La Salle: Open Court.

Feyerabend, P. (1979), *Science in a Free Society*. London: Verso.

Fine, A. (1984), "The Natural Ontological Attitude." In J. Leplin (ed.), *Scientific Realism*. Berkeley: California, pp. 83-107.

Fuller, S. (1992), "Being There with Thomas Kuhn: A Parable for Postmodern Times." *History and Theory* 31: 241-275.

_ _ _ _ _. (1993), *Philosophy, Rhetoric, and the End of Knowledge: The Coming of Science & Technology Studies*. Madison: Wisconsin.

Gregory, F. (1992), "Theologians, Science, and Theories of Truth in Nineteenth Century Germany." In M.J. Nye et al. (eds.) *The Invention of Physical Science*. Dordrecht: Kluwer, pp. 81-96.

Heilbron, J. (1986), *The Dilemmas of an Upright Man: Max Planck as Spokesman for German Science*. Berkeley: California.

Humphrey, G. (1951), *Thinking*. London: Methuen.

Johnson, J. (1990), *The Kaiser's Chemists*. Chapel Hill: North Carolina.

Kockelmans, J. (Ed.) (1968), *Philosophy of Science: The Historical Background*. New York: Free Press.

Kuhn, T. ([1962] 1970), *The Structure of Scientific Revolutions*. Chicago: Chicago.

_ _ _ _ _. (1977), *The Essential Tension*. Chicago: Chicago.

Laudan, R. (1993), "Histories of Science and Their Uses: A Reviewto 1913." *History of Science* 31: 1-34.

Mach, E. ([1883] 1960), *The Science of Mechanics*. LaSalle: Open Court.

Matthews, M. (1991). "Ernst Mach and Contemporary Science Education Reforms." In M. Matthews (ed.), *History, Philosophy, and Science Teaching* New York: Teacher's College, pp. 9-18.

Mueller, D., Ringer, F., Simon, B. (Eds.) (1987), *The Rise of the Modern Educational System: Structural Change and Social Reproduction 1870-1920*. Cambridge: Cambridge.

Proctor, R. (1991), *Value-Free Science?* Cambridge: Harvard.

Rouse, J. (1987), *Knowledge and Power*. Ithaca: Cornell.

Schlick, M. ([1925] 1974), *The General Theory of Knowledge*. Vienna: Springer Verlag.

Weinberg, S. (1993), *Dreams of a Final Theory*. New York: Pantheon.

Part VI

QUANTUM MECHANICS AND COSMOLOGY

On the Paradoxical Aspects of New Quantum Experiments[1]

Lev Vaidman

Tel-Aviv University

1. Interaction-Free Measurements

I shall discuss two recently suggested quantum experiments. These experiments lead to paradoxical situations. I will argue that in the framework of a particular Many-Worlds Interpretation of quantum theory the paradoxes do not arise.

The first experiment is the "interaction-free measurement" (Elitzur and Vaidman 1993). The experimental group at Innsbruck headed by Prof. Zeilinger is working now on its realization. Before describing the experiment let me discuss a general question: "How do we know that there is an object in a region of space?"

The simplest case is when the object itself causes some physical changes outside the region:

i) The object is charged, so there is a field outside. We can measure this field.
ii) The object yields a potential outside. Aharonov and Bohm have taught us that potential even without field leads to a measurable effect.
iii) The objects radiates photons or other particles.

If the region is empty, then, of course, none of (i)-(iii) will occur. However, even if there is something in the region, it might happen that there will be no field, no potential, and no radiation outside.[2] The object may be neutral. If (i)-(iii) did not occur, how can we know that the object is in the region? There is one simple case when getting information without "touching" the object is possible. Suppose we know that an object is located in one of two boxes. If we open one of them, it might be empty. Surely we have not touched the object; nevertheless, we know now that the object is in the second box. Quantum mechanics allows a subtle variation of this case. We can (in principle) prepare a two-particle system in a correlated state such as the Einstein-Podolsky-Rosen (EPR) state. Then, measuring the location of one particle yields the location of the other even though we have not touched it. In both cases we had prior information about the object. Here we consider the situation in which we do not have any prior information.

In order to explain the meaning of a measurement without "touching", assume that the object which we are trying to locate in a certain region is a supersensitive detector, one with 100% efficiency for all kinds of particles. Nowadays experimentalists can create a single-photon state. Every time such a photon reaches the detector, the detector clicks. Applying standard logic — if A implies B, then negation of B implies negation of A— we can claim that if the detector did not click, then no particles reached it.

Locating such a detector without triggering it leads to a paradox: we obtain information about a region of space from which nothing came out. Indeed, we assumed that the detector itself sends nothing out, and since it did not click, no particle from the outside visited the region.

Let us now present Elitzur-Vaidman's solution of this task. We call it an "interaction-free" measurement. I have to mention from the outset that the measurement is not always successful. At least in half (on average) of the cases we do touch the object, and if the object is a super-detector it will click in all these cases. Still we have finite chance for success: in up to 50% of the cases the detector does not click but we are 100% sure that something is inside the region.[3]

Our method employs the Mach-Zehnder interferometer used in classical optics. In principle, it can work with any particle. The particle reaches the first beam splitter which has transmission coefficient 1/2. The transmitted and reflected parts of the particle's wave are then reflected by the mirrors and finally reunite at another, similar beam splitter (Fig. 1a). Two detectors collect the particles after they pass through the second beam splitter. We can arrange the positions of the beam splitters and the mirrors such that, due to destructive interference, no particles are detected by one of the detectors, say D_2, and all are detected by D_1. We place the interferometer in such a way that one of the routes of the particle passes through the place where the super-detector might be (Fig. 1b). We send a single particle through the system. If the interferometer is empty, than detector D_1 clicks. But if the super-detector blocks one arm of the interferometer than there are three possible outcomes of this measurement:
i) super-detector clicks, ii) detector D_1 clicks, iii) detector D_2 clicks. The probability for the first case is 1/2. In the second case (for which the probability is 1/4), the measurement does not succeed either. The particle could have reached D_1 in both cases: when the super-detector is, and when it is not there. Finally, in the third case, when the detector D_2 clicks (the probability for which is 1/4), we have achieved our goal: we know that the super-detector is inside the interferomete and it did not click. A slight modification allows us to find the detector without triggering it with the probability arbitrarily close to 50% (Elitzur and Vaidman 1993).

Note that we have succeeded to find out that a region of space *is not* empty without any particle passing through it. But we cannot find out that the region *is* empty without passing a particle through the region. Indeed, we know that the region is empty when, after passing very many photons through the interferometer, the detector D_2 remains silent. In this case we have no reason to claim that the photons have not passed through the observed region.

In this example we have presented the following paradox: nothing inside the region had any influence on the world outside, nevertheless, we obtain certain information about what is inside.

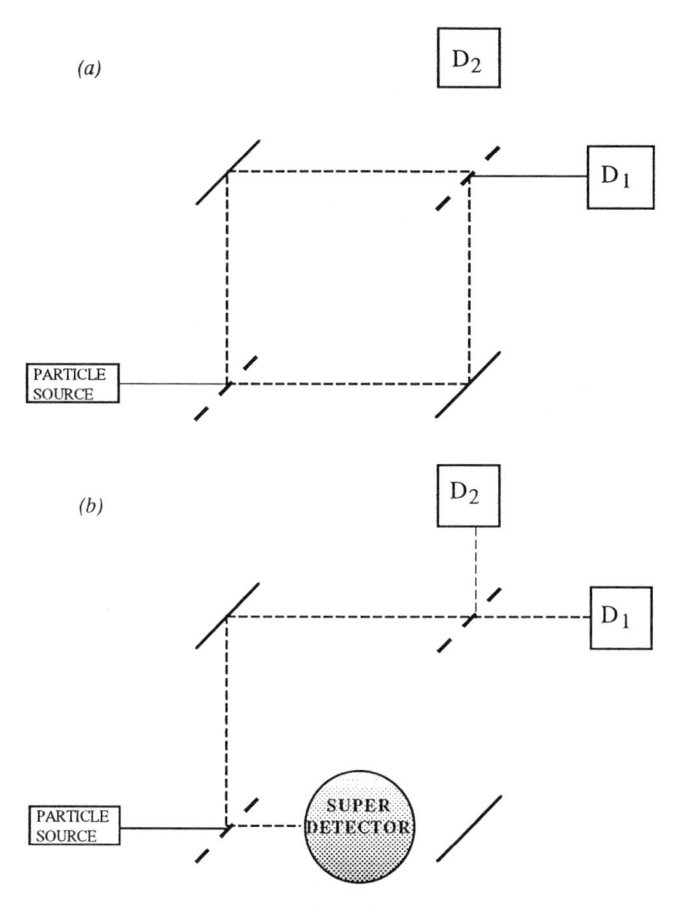

Figure 1.
a). If there is no any object inside the interferometer, D_2 never clicks.
b). When D_2 clicks after sending just one particle we know that the super-detector is inside the interferometer while it did not count any particle.

2. Teleportation of Quantum States

Let us describe now the second paradox. We will discuss a recent proposal of Bennett *et al.* (1993) for teleportation of an unknown quantum state. They found a method for transmitting an unknown quantum state of a spin-1/2 particle to a distant spin-1/2 particle without actually moving the particle from one place to another. Recently, I have found a way to generalize teleportation procedure to systems with continuous variables (Vaidman 1994).

The word "teleportation" recalls the heroes of Star Trek entering a transmitting cabin in their starship Enterprise: in a few seconds they disappear and immediately appear on a distant starship. It sounds, however, like a bad science fiction: too many laws are broken in this picture. For instance, the center of mass of a closed system should not move. But, when the heroes move far away to another starship, the center of mass of the closed system consisting of the two starships does move. However, quantum theory teaches us that the essential feature of an object is not the *matter* of which it is made, but its form. Indeed, all objects are "made" out of identical elementary particles, and what distinguishes one object from another is the state of these particles. Thus, we can consider a different kind of machine for teleportation. The receiving teleportation chamber is not empty before the transmission, but it contains elementary particles in a number equal to the number of particles of the object to be transmitted. Then, the transmission results in building the Star Trek heroes out of these particles, while the heroes in the cabin on Enterprise revert to an unstructured set of elementary particles.

One might be tempted to build the heroes in several locations, i.e., to produce several copies. This is certainly not a teleportation. But the unitarity of quantum theory prevents this possibility. It is impossible to clone an unknown quantum state. It is also impossible to identify an unknown quantum state without significantly changing it. Therefore, the only option quantum mechanics leaves, is destruction of the heroes in one place and their creation in another; i.e., teleportation of an unknown quantum state of about 10^{26} particles to an identical set of particles located in a distant region. Clearly, this is not a feasible project today, but the teleportation of a quantum state of one particle is a subject of serious consideration of many experimentalists.

In order to teleport a quantum state Ψ from one place to another we need a "quantum channel". This is an additional system of two completely correlated particles, one located in a place from where we want to teleport the state and the other is in the remote location. The first step of the teleportation procedure is a particular local quantum measurement performed on two particles: the particle in the state Ψ and the adjacent particle of the correlated pair. In the case of teleportation of a spin state of a spin-1/2 particle, this measurement has four possible outcomes; let us label them as numbers from 0 to 3. The result is transmitted to the remote location. The next step of teleportation is a "rotation" of the state of the second particle. There are four cases: If the transmitted number is zero, no rotation is required, the teleportation has been completed. In all other cases we have to rotate the state by an angle π around the x, y, or z axis, in accordance with the outcomes 1, 2, or 3. After the appropriate rotation the state of the second particle is Ψ.

The procedure for teleportation of a quantum state of a continuous variable $\Psi(q)$ is very similar. We use the original EPR pair of correlated particles as a "quantum channel". We perform a local measurement on a system which includes a particle in a state $\Psi(q)$ and one member of the EPR pair. After the local measurement, which yields two numbers a and b, the state of the second particle is $e^{ibq}\Psi(q+a)$, i.e., the original state shifted in q and the conjugate momentum p. In the next step we transmit the numbers a and b by phone. The last step is a back shift of a in q and b in p, thus bringing the second particle of the pair to the state $\Psi(q)$.

I see a paradoxical situation in this example as follows. The special theory of relativity tells us that nothing can move faster than light. Any massive body cannot move faster than light, and no information can be sent faster than light. However, the phase velocity of waves, for example, can move faster than light. As I see it, the spirit of the theory of relativity is that nothing which has "direct physical meaning by itself" should move faster than light. Conversely, only things which cannot move faster than light can have any physical meaning.

All objects are "made" out of *identical* particles, so the particles cannot characterize the object. The essence of an object, then, is in its quantum state. But in the process of teleportation of a quantum state the only part which cannot be sent faster than light is a small amount of classical information which we have to transmit by phone. Therefore, it seems that we have to admit that the essence of a quantum state is in this classical information. The spin-state of a spin-1/2 particle is described by a direction in a three-dimensional space, i.e., by two continuous variables, the two angles. Nevertheless, in teleporting this state we have to transmit only a number from 0 to 3, just two bits of information! For the case of a state with continuous eigenvalue the description is a continuum of complex numbers, while we have to transmit only two numbers.

Let me summarize the logic of the argument. The theory of relativity teaches us that we cannot move anything essential faster than light. In the process of teleportation of a quantum state, say of a spin-1/2 particle, the limitation on the velocity of transmission is on just two bits of classical information. Therefore, the essence of the quantum state of a spin is only the two bits. This however, is in conflict with the fact that the description of the state requires two real numbers.

The first step of the teleportation is closely connected to a paradoxical situation which is so much discussed that there is a consensus of agreement not to agree about its interpretations. Consider the EPR-Bohm pair of (anti)correlated spin-1/2 particles. Each particle has no quantum state by itself, each one is in a "mixed" state. After the measurement of a spin component of particle 1, the EPR-Bohm state collapses into a product state. Immediately after the measurement each particle is in a pure state. One can consider the measurement of the spin of particle 1 as a creation of its spin state and immediate teleportation of the (flipped) state to the second particle (without destruction of the state of the first particle). In this case there is no need for sending any classical information. This kind of instantaneous teleportation clearly contradicts the spirit of the theory of relativity. Let me note that there is no contradiction with the letter of Einstein's theory: there is no way of sending information faster than light using collapse of the quantum wave function.

3. A Many-Worlds Interpretation of Quantum Theory

I want to argue that the paradoxes presented above are resolved, or at least appear less paradoxical in the framework of the Many-Worlds Interpretation (MWI) (Everett 1957). Actually, the MWI itself has several interpretations, which are conceptually different. I (Vaidman 1993) take a view in which we have one physical universe which incorporates many (subjective) worlds. The physical universe is described by one wave function, which evolves deterministically according to the Schrodinger equation. This wave function, at any time, can be decomposed into a superposition of many states, each corresponding to a different story (i.e., history up to the considered time). One of the stories is the world as *you*, the reader of this paper, know it. What we perceive is just a small part of what *is* in the universe. The laws of physics relate to the whole universe, and it is not surprising that consideration of only a part of it leads to paradoxical situations. Considering all worlds together, the physical universe, resolves the paradoxes.

Let us turn now to the example of the interaction-free measurement. In the framework of the MWI it is not true that we got information about the region without anything being there. The photon which we sent into the interferometer was there, but — in another world. In our experiment three worlds (three different stories) appear: i) the super-detector clicks, ii) detector D_1 clicks, iii) detector D_2 clicks. Obtaining information in the world (iii) without any object being in the region became possible because in the world (i) a photon was in that region and it triggered the super-detector.

Now we can understand why we could not get information that the region is empty without a photon being there. In this case there is no other world, except the one we are aware of, so obtaining information about the region without being there is on the level of the whole universe. Our physical intuition correctly tells us that such situation is impossible.

Let us turn now to the example of teleportation *per se*. The first step of teleportation of a state Ψ, the local measurement, splits the world according to the possible outcomes of the local measurement. After completing the teleportation, we obtain in the remote location the state Ψ in *all* worlds. Therefore, in the whole universe (assuming that before the measurement there was just one world in the universe), the state has been transmitted from one particle to another. In this case the MWI does not tell us that the paradoxical situation does not exist on the physical level of the universe as it did in the previous example. The teleportation works not only in a world, it works in the whole universe! The MWI provides another explanation. Let us discuss it now.

In the framework of the MWI the teleportation procedure does not move the quantum state: the state was, in some sense, in the remote location from the beginning. Indeed the correlated pair, which is the necessary item for teleportation, incorporates all possible quantum states of the remote particle, and, in particular, the state Ψ which has to be teleported. The local measurement of the teleportation procedure splits the world in which this experiment has been done into worlds in all of which the state of the remote particle differs from the state Ψ by some known transformation. The number of such worlds is smaller than the necessary information for defining the state Ψ. This explains why the information which has to be transmitted for teleportation of a quantum state—the information which world is it (what tranformation has to be done)—is much smaller than the information which is needed for creation of such a state. For example, for the case of a spin-1/2 particle there are only 4 different worlds, so in order to teleport the state we have to transmit just 2 bits.

The above examples illustrate how the MWI clarifies the issue of nonlocality in quantum mechanics. The quantum state of the universe, and its components which correspond to various worlds, are nonlocal. Interactions governed by the Hamiltonian evolution change *locally* the state of the universe. If there are no splittings of worlds, the states corresponding to the worlds are changed locally too. Measurements are interactions which split the worlds into more worlds make the stories diverge. These new worlds differ one from another not only in the region of the measurement, but also in other places. This happens due to the correlations incorporated in the quantum state before the measurement. Thus, in the created worlds we obtain, effectively, nonlocal changes, while there is no nonlocal action on the physical level of the universe.

Notes

[1] It is a pleasure to thank Yakir Aharonov and Philip Pearle for stimulating discussions. The research was supported in part by grant 425/93-1 of the Basic Research Foundation (administered by the Israel Academy of Sciences and Humanities). Send reprint requests to the author, Physics Department, Tel-Aviv University, Tel-Aviv 69978 ISRAEL; e-mail: vaidman@ccsg.tau.ac.il.

[2] The object must have mass, and therefore it must create a gravitational field. But let us assume that the field is too weak to detect

[3] Very recently Kwiat *et al.* have found a new scheme for interaction-free measurements which employs the quantum Zeno effect. Their method allows detection of

the super-detector, without triggering it, with probability which is arbitrary close, but not equal to 100%. See also Vaidman (forthcoming).

References

Bennett, C.H.; Brassard, G.; Crepeau, C.; Jozsa, R.; Peres, A.; and Wootters, W.K. (1993), "Teleporting of Unknown Quantum State via Dual Classical and Einstein-Podolsky-Rosen Channels", *Physical Review Letters* 70: 1895-1898.

Everett, H. (1957), "Relative State Formulation of Quantum Mechanics", *Review of Modern Physics* 29: 454-462.

Elitzur, A. and Vaidman, L. (1993), "Quantum Mechanical Interaction-Free Measurements", *Foundation of Physics* 23: 987-997.

Kwiat, P.; Weinfurter, H.; Herzog, T.; Zeilinger, A.; and Kasevich, M. (forthcoming), in "Symposium on Modern Physics" Helsinki, June 1994.

Vaidman, L. (1993), "On Schizophrenic Experiences of the Neutron or why We should Believe in the Many-Worlds Interpretation of Quantum Theory", Tel-Aviv University preprint, TAUP 2058-93.

_____. (1994), "Teleportation of Quantum States", *Physical Review* A 49: 1473-1476.

_____.. (forthcoming), "On the Realization of Interacton-Free Measurements", *Quantum Optics*.

The Bohmian Model of Quantum Cosmology

Craig Callender and Robert Weingard

Rutgers University

1. Introduction

Philosophers of science have not paid much attention to recent developments in quantum cosmology. This fact is surprising, since quantum cosmology is replete with conceptual issues involving (e.g.) the fundamental nature of time and space, the interpretation of quantum mechanics, and the ultimate meaning of probability. One notable exception, Quentin Smith, has recently examined the Hartle-Hawking (1983) proposal. Trying to make sense of the view, he resorts to an instrumentalist picture, which treats the proposal as merely a heuristic device for the algorithm responsible for predictions. While we do not examine Smith's account here, we would like to contrast it with the model presented in this note, in which a fully realistic interpretation of quantum cosmology is developed.

Recently there has been a resurgence of interest in the de Broglie-Bohm causal interpretation of quantum mechanics. The merits of this interpretation regarding non-relativistic quantum mechanics are extolled elsewhere, and shall not be repeated here (see Albert 1993, Bell 1987, Bohm and Hiley 1993, Durr et al 1992). The present essay concerns the relationship between Bohmian mechanics and recent problems in quantum cosmology. We argue that when cosmological factors are considered, the de Broglie-Bohm interpretation remains the only satisfactory interpretation of quantum theory. This assertion is advanced with a Bohmian resolution of (one aspect of) the so-called problem of time in quantum cosmology. Moreover, the preceding is accomplished without having to split worlds, multiply minds, or ever worry about observers collapsing wavefunctions.

2. Bohmian mechanics

Nonrelativistic Bohmian mechanics is characterized by two basic equations of motion. One governs the wave function $\Psi = A\exp[iS(x)]$, the other the particles postulated by the theory. It is convenient to rewrite the Schrodinger equation as a modified Hamilton-Jacobi equation

$$dS/dt + (\nabla S)^2/2m + V + Q = 0 \qquad (1)$$

where Q, the so-called quantum potential, is given by

$$Q = -h^2 \nabla^2 R / 2mR.$$

(1) continues to be the equation of motion for the wave function. As in ordinary Hamilton-Jacobi theory, probability is conserved for particles satisfying (1), provided the particles have momentum $p = \nabla S$. This feature produces an equation for the particles:

$$mdx/dt = -\nabla(V) - \nabla(Q). \tag{2}$$

On the assumption that the probability density equals $|\Psi|^2$, Bohmian mechanics reproduces the results of ordinary quantum mechanics. The theory is deterministic, and provides a conceptually clear account of quantum phenomena; in particular, it does not suffer from a problem of measurement (see Bohm and Hiley 1993).

3. The Problem(s) of Time in Quantum Cosmology

The 'problem of time' in canonical quantum gravity (QG) and quantum cosmology seems not to refer to a particular problem, but to an ill-defined set of related problems. Time in quantum mechanics is essentially the immutable, external time of Newton, whereas general relativity treats it as an arbitrary parameter. This incompatibility creates a multitude of difficulties, e.g. the factor-ordering problem, the problem of observables, the Hilbert space problem (see Kuchar 1992 for an excellent review). No one of these is properly singled out as *the* problem of time.

Even so, from the perspective of cosmology one problem is especially vexing. In QG the Hamiltonian for the universe is identically zero. Vanishing Hamiltonians are not generally problematic, for time variables can usually be physically identified through a system's interaction with other systems. But when the entire universe is the subject, the vanishing of the Hamiltonian does present a serious difficulty.

At the very least, the vanishing of the Hamiltonian apparently spoils the desired interpretation of the principal equation of QG, the Wheeler-DeWitt equation

$$H\Psi_{wd} = [1/2 G_{abcd} p^{ab} p^{cd} - |h|1/2 R + H_{matter}] \Psi_{wd} = 0. \tag{3}$$

Here G is the DeWitt metric, representing the intrinsic geometry of a point in 'superspace.' Superspace is the configuration space consisting of the set of equivalence classes of Riemannian metrics on (usually compact) spatial 3-geometries. p^{ab} is the canonical momenta conjugate to h_{ab}, $|h|$ the volume element of the 3-space h_{ab} and R its scalar curvature. The picture sought in QG is one in which a 3-geometry evolves in superspace along an arbitrary time parameter τ. In the naive interpretation, Ψ is intended to provide the probability amplitude for a particular 3-space obtaining at a time parameter τ. Since H = 0, the wave function of the universe is independent of time. Hence the 3-space does not evolve in time. If the 3-space is considered analogous to a particle, it is in an eigenstate of zero energy, whose state cannot be affected by any measurement. Not only doesn't the universe described by (3) expand, then, but contrary to our experience, it appears utterly static. One often hears the possibly exaggerated claim that time 'disappears' in QG.

4. Interpretations of Quantum Mechanics

If the situation in cosmology is used as a yardstick by which interpretations of quantum theory are measured, the de Broglie-Bohm model seems uniquely fit. First, unlike the orthodox interpretation and its variants, it requires no external measuring apparatus to reduce superposed wave functions. Since cosmology is concerned with the wave function of the entire universe, it makes little sense to speak of external measurements collapsing the wave function. Even on 'collapse' versions not requiring measurement, such as Ghirardi et al (1986), to speak of the collapse of the universe's wave function is awkward at best, given the aforementioned static situation. Second, the main reason to quantize gravity is that Einstein's field equations look inconsistent, since they set an ordinary function of spacetime points equal to a quantity depending on quantum operators. Since in Bohm's theory the operator-formalism emerges as only a phenomenal (measurement) description of the underlying physics, this prima facie conflict doesn't signify any problem deeper than that. Third, a little-mentioned point is that in cosmology most 'predictions' are really retrodictions. Bohmian mechanics, in contrast to most interpretations, allows us to know more about the past than the future. Integrating the velocity field equation backwards in time allows for accurate retrodictions to be made. And this is the case even if the initial wave function is a superposition of states.[1]

For these reasons, Bohm's interpretation seems well suited for cosmological application. It is also preferable to the interpretations presently used in quantum cosmology, versions of the many-worlds and many-minds interpretations (see Albert and Loewer 1988 for discussion of both and references). In our opinion, though explanatorily useful, these two views are utterly fantastic. It is simply incredible that quantum probability distributions are given by the 'trajectories' of the continuous infinity of minds associated with each observer, or by the splitting of worlds or 'relative states' (whatever they may be). That so many have swallowed such notions is remarkable.

Because the many-worlds interpretation is so popular, a few remarks should particularly be directed its way. (1) Decoherence may solve the so-called 'basis problem,' but it does not explain in what sense many-worlds describes a probabilistic theory. Each measurement outcome in our universe has probability one, since the entire universe corresponds to one ray in Hilbert space. (2) The meaning of the wave function is muddled in many-worlds. The range of quantum mechanics is ambiguous: is it a theory of our world, or of the continuous infinity of worlds? (3) Unlike in Bohm's theory, wherein the classical limit could not be any clearer ($Q \longrightarrow 0$), the classical limit in most presentations of many-worlds is mysterious.[2] Further critical discussion of these issues can be found in Albert and Loewer (1988) and Bohm and Hiley (1993). When these problems are contrasted to the clarity of Bohmian mechanics, and its resolution of the problem of time (in section 7), we believe a compelling case is made for the superiority of the causal interpretation.

Finally, it has been asserted that many-worlds and/or many-minds can solve the problem of time in quantum cosmology (Squires and Collins 1993). We wish to dispute this claim. It is true that projection operators can be introduced which do not commute with the Hamiltonian, thereby producing non-trivial time evolution. But the *physical meaning* of these projection operators needs to be clarified. Why is the wave function projected, and what selects the associated eigenvalues? It is not surprising that advocates of these theories gloss these questions, for good answers are not forthcoming. If consciousness is involved in the answers, for instance, then the universe did not start evolving until conscious minds 'entered' the universe. The only way to make sense of this result is to claim the universe began with conscious minds in it, and that the appearance of prior development is illusory. Surely cosmologists can't be happy with that. If 'something else' is involved in the answers, we wish to know what it is. Until then, the purely formal application of projection operators to regain time is not physically justified.

5. The Klein-Gordon Analogy

Since QG is so poorly understood, one usually works with simple models which are described by equations sharing significant features with the Wheeler-DeWitt equation. The model used here is the relativistic particle described by the Klein-Gordon (KG) equation. To mirror the Wheeler-DeWitt equation (3) the Hamiltonian of the KG equation is turned into an operator, and imposed as a restriction on the space of physical states:

$$H\Psi_{kg} = [\tfrac{1}{2}G^{ab}P_aP_b + \tfrac{1}{2}MV]\Psi_{kg} = 0. \qquad (4)$$

The analogy with (3) is a good one: the position plays the role of the intrinsic geometry, the background metric the role of the DeWitt metric, and the potential the role of the scalar curvature. Also, the kinetic portion of each Hamiltonian is indefinite (the geometrodynamical potential is also indefinite, whereas the relativistic particle's is positive definite). In fact, the relationship is better than an analogy: when (3) is restricted to two degrees of freedom, it *is* (4).

Notoriously, the KG equation suffers from a serious problem. Its inner product $<\Psi_1|\Psi_2>$ is not positive definite, and therefore cannot be used to define a probability. A single relativistic particle has no suitable Hilbert space.[3] The traditional response to this problem is to rewrite the theory as a field theory on Fock space. States of arbitrary particle number are allowed, and the wave function is second quantized. In second quantization, the 'wave function' $\Psi(x)$ becomes a field operator $\hat{\phi}(x)$, which acts on the Hilbert space of states. The wave function of the field, however, is a functional of the field configuration $\omega(\Psi(x))$ — in that representation.

In QG the states Ψ are functionals of the metric $\Psi(h(x))$. If we were to 'third' quantize, then the Ψ becomes an operator $\hat{\phi}(h(x))$ on states, and the wave function a functional of the field $\Psi(h(x))$ — $\omega(\psi(h(x)))$.

6. Bohmian Third Quantization

Even if quantizing $\Psi(h(x))$ to obtain the operator valued field $\hat{\phi}(h(x))$ in superspace solved the Hilbert space problem, we still have the problem of the interpretation of quantum mechanics. So we would like to apply Bohm's theory to $\Psi(h(x))$ as well. Since Bohmian particle mechanics has been extended to bosonic field theory, it is possible to investigate a Bohmian version of third quantized QG.

Bohmian field theory is formulated in terms of a super-wave $\Psi[\phi(x)]$ over a full field configuration. The characteristic feature of Bohmian mechanics, the quantum potential, has a field theoretic analogue derived from the Schrodinger equation

$$Q[\phi(x)] = -\tfrac{1}{2}h^2 \int d^3x\, (\partial^2 R[\phi]/\partial\phi^2(x,t))/R[\phi].$$

$Q[\phi(x)]$ modifies the KG equation, and guarantees that Bohm's causal field agrees with predictions made by conventional quantum field theory. In Bohmian field theory $\Psi[\phi(x)]$ is interpreted as the probability amplitude for finding field values $\phi(x)$ when the system is in state Ψ. Further details about this approach can be found in Bohm and Hiley (1993), Bell (1987) and Huggett and Weingard (1994).

In Bohmian third quantization, $\Psi(h(x))$ becomes a 'physical' field, the beables of the theory, on analogy with $\phi(x)$ when Bohm's theory is applied to the scalar field. In the case of the Bohmian scalar field, there are 0-particle, 1-particle ... n-particle field configurations, but these are just φ-field configurations that interact with localized measuring devices *as if* 0,1,...n 'localized' discrete entities (with continuous trajectories) were present. But there are no such entities present, there are only φ-field configurations.

What about the similiar situation for ψ(h(x)) on the analogy between φ(x) and ψ(h(x))? The configuration space for ψ(h(x)), analogous to x for φ(x), is the space — superspace — of 3-metrics h(x). The space of h(x) doesn't, presumably, have the ontological status of the configuration space x, but Bohm's theory does assign a definite value to ψ for each 3-metric h(x) (given suitable boundary conditions). Let's emphasize, superspace is a configuration space, relative to which values of ψ are defined; its points are not the values of actual universes, any more than the x's in the φ(x) are actual values of the positions of the particles. Now, just as in the scalar field, if the wave function is appropriate, then ψ(h(x)) will be a 0,1,...n universe field configuration. But if the analogy with the scalar field is good, and the field is the 'beable,' then there aren't any 'universes.' Consequently, the theory is scarcely comprehensible.

If third quantization solved the problem of time, it would warrant further speculation about its meaning. However, there is reason to think the problem survives third quantization, for second quantization itself is a casualty of the problem. As Kuchar (1992) has emphasized, quantum field theory on a dynamical manifold suffers from its inability to define a one-particle Hilbert space. Quantum field theory on a flat background is well-defined because the background admits the relevant isometries for construction of a one-particle Hilbert space (spanned by the positive energy solutions of the KG equation). This space is subsequently used to construct the Fock space. On a dynamic background, Hilbert spaces for stationary pasts and futures can be designed; however, it is not clear whether they can be built for the dynamical region (see fn. 3). Hence, when applied to dynamic backgrounds, second quantization itself apparently suffers from the problem of time. Consequently, it is unlikely that third quantization, which merely carries out the second quantization procedure, is going to do any better.

7. The Reappearance of Time in Bohmian Cosmology

Let's start again. This time we'll be less ambitious, and approach the Wheeler-DeWitt equation from a naive Bohmian perspective.

Consider a wave function where $\Psi = A\exp[iS]$. To keep matters simple, let Ψ be a function over just two variables, χ, the radius of the universe, and θ, a spatially constant scalar field. Substitute Ψ into equation (4). Separated into real and imaginary parts, (4) becomes

$$n^{\mu\nu}\partial_\mu S \partial_\nu S A - V - Q = 0 \tag{5}$$

$$A^{-1}n^{\mu\nu}\partial_\mu(A^\nu\partial_\nu S) = 0. \tag{6}$$

(5) is the Hamilton-Jacobi equation modified by the quantum potential (with E = 0). (6) is a continuity equation. The quantum potential takes the form

$$Q = -(\partial_\chi \partial^\chi A - \partial_\theta \partial^\theta A).$$

Continuing as in regular Bohmian mechanics, trajectories for the dynamical variables are obtained.

Therefore,

$$d\chi/dt = dS_\chi(\chi,\theta); \quad d\theta/dt = dS_\theta(\chi,\theta). \tag{7}$$

Dynamics are obtained with respect to a time parameter t. (We discuss the nature of this time in section 9). Contrary to the static situation described at the outset, a Bohmian approach to cosmology admits nontrivial evolution of the dynamical variables. The importance of the trajectories can not be underestimated. They allow for both dynamics and a clear interpretation. The former makes the theory physically viable, the latter makes it comprehensible (unlike most other quantum cosmological schemes). Since the temporal evolution of cosmological parameters can be predicted (and retrodicted) it amounts to precisely the picture wanted in cosmology.

Given such success from a relatively modest point, the reader may have the feeling that we have pulled a rabbit out of a hat. Where do the time and dynamics originate? The answer is that they arise from the laws of Bohmian mechanics. Unlike most interpretations of quantum mechanics, Bohmian mechanics is explicitly a theory with two fundamental equations of motion. One governs Ψ, the other the dynamical variables of the theory. In particle mechanics, the variable is position, in field theory the full field configuration, and now in cosmology, variables such as χ,θ. In short, the 'magic' of the result ultimately stems from the interpretation's recognition that two equations of motion are needed. Indeed, from a Bohmian perspective, it is not terribly surprising that quantum cosmology cannot describe a dynamic universe when only one of these equations is used.

8. Discussion

The first item to notice is that the calculation essentially has been done before, in the WKB interpretation of Ψ (see Halliwell 1991). The difference between the approaches lies in the drastically different interpretations of the formal result. The WKB interpretation tries to extract a probabilistic interpretation of Ψ, but only when $\Psi = A \exp[iS]$. If $\Psi = A\exp[iS]$ then the conserved current $j^A = |A|^2 \nabla S$ can provide probabilities for observing classical trajectories. The interpretation states that if $\Psi = A\exp[iS]$ then there is a particular probability associated with measuring specific values of the dynamical variables. The interpretation suffers, however, because it is valid over only a very limited range, and generally, there is no reason to suppose the probability density will be nonnegative.

In our interpretation all wavefunctions permit the derivation of real deterministic trajectories. It is therefore clear and globally applicable. We do not extract probabilities from the KG equation (so it doesn't suffer from a Hilbert space problem). However, the Schrodinger equation still provides probabilistic results for quantum mechanics, thus emphasizing quantum mechanics' status as a measurement formalism. As in Bohmian mechanics these probabilities merely reflect our ignorance, not the underlying reality. That they are not objectively probabilistic is a significant virtue, in our opinion, for we wonder what it means for the entire universe to have a certain chance. Given the negligible formal difference between the WKB and causal interpretations, then, and the tremendous difference in clarity and application, we believe proponents of the WKB approach would do well to embrace our model.

The wave function employed is section 7 is complex, as in the Vilenkin (1989) model. If the wave function were real, like the Hartle-Hawking (1983) wave function, the dynamical variables would be at rest. As pointed out by Squires (1992), since the world is not static, there apparently exists an incompatibility between the Hartle-Hawking proposal and the Bohmian model. What Squires ignores is the fact that Hartle-Hawking interpret the wave function only in the classical region of minisuperspace, where it looks like the sum of two WKB solutions. Hartle-Hawking interpret this as two non-interfering descriptions of our universe. Though we wonder whether breaking the superposition is justified, if it is (as has been claimed), their proposal is compatible with the present approach: their wave function corresponds to two non-interfering sets of trajectories. If it is unjustified, then the Hawking-Hartle conjecture is physically uninteresting, and our proposal's apparent incompatibility with it does not bother us. The more general issue of real wave functions in the causal theory is rather complicated, and shall not be discussed here.

The present model should be distinguished from another recent suggestion in the Bohmian spirit, namely, Pitowsky (1991). Pitowsky essentially adds a quantum correction term to the classical gravitational field equations, as Bohm adds a quantum potential to the Hamilton-Jacobi equation. It is in that sense that the proposal is Bohmian. Unlike in QG (and thus the present proposal), Pitowsky does not quantize the metric field; that is, the wave function is still a function of x, not of h(x). The geometry is affected only through the new quantum input into the momentum-energy tensor. Consequently, Pitowsky's project is more in line with the semiclassical approach to QG, in which the matter but not the metric fields are quantized and described by Schrodinger's equation. As such, we expect it to suffer from similar difficulties, e.g., the well-known problem posed by superpositions (see Kuchar 1992). Because he has not addressed the issue of how time and dynamics arise, it is hard to say exactly how the problem of time will manifest itself for his proposal.[4] Anyway, since there are reasons for believing that the semiclassical approach is unsatisfactory, Pitowsky's approach, though interesting, is not especially promising.

9. Time in a Bohmian Universe

The time parameter labeling the trajectories in Bohmian cosmology is a theoretical posit. Like Newton's absolute time, the time in Bohmian cosmology is most naturally viewed as an unobservable, physical time, arising from the basic laws. We stress that this claim is only speculation. Clearly, a successful integration of quantum mechanics and general relativity, for instance, would demand reevaluation. Nevertheless, the picture described is the most straightforward one.

Consider the status of time in Newtonian mechanics. Newton believed time was a real relation, not be confused with its sensible measure. He also presumed it to be unique and empirically determined. The reason is that from an independent knowledge of the true forces of a moving body, a single time measure consistent with his mechanics emerges. Change this time, and what was once (for instance) a freely moving body with no forces acting on it becomes an accelerating/decelerating body apparently suffering the imposition of forces. Since Newtonian theory already tells us what the forces are, e.g., in simple physical situations, the true measure of time can be detected. Since it follows from the laws of physics, there is a clear sense in which the unique time is empirically determined. While this reasoning may be epistemically circular (for the selection of forces may originate from the selection of time, or vice versa), it is nonvicious, for it is probably endemic to the practice of theorizing.

The situation seems similar in Bohmian cosmology. Here time also looks to be a theoretical posit. The t in equations (7) cannot be directly measured; yet from a knowledge of the true 'forces' acting on the cosmological variables, t is uniquely determined (that is, if the laws of Bohmian cosmology are correct). If this is correct, Bohmian cosmology is not generally covariant. The laws define a preferred time. We would like to make two remarks on this consequence. First, though a preferred time may be upsetting in QG, it is not from the perspective of quantum cosmology. In cosmology, the goal is to watch various quantities evolve with respect to cosmic time. Bohmian cosmology allows for precisely this. If it does not apply to the infinite number of time parametrizations (most of which are pathological) compatible with general relativity, is much of value lost to cosmology?

Second, we should squarely face the fact that this consequence may be inevitable anyway. The so-called problem of functional evolution seriously threatens the covariance of general relativity (Kuchar 1988, 227). The problem stems from the possibility that the commutators between the constraint operators may not vanish:

$[H(x), H'(x)] \neq 0$.

If that obtains, a state's evolution from an initial hypersurface to a final one might depend on the foliation connecting the two hypersurfaces. One state on the initial hypersurface might be developed into two nonequivalent states on the later (earlier) hypersurface. This problem is often overlooked because it obtains only in models that are infinitely dimensional. Nevertheless, since the difficulty is equivalent to the notorious factor-ordering problem, it should be taken quite seriously. Fixing the space-time's foliation solves the problem, of course, but only at the expense of covariance. Critics of the non-covariance of Bohmian cosmology should thus bear in mind the possibility that their favorite theory may someday share the same fate.

Finally, a crucial question (perhaps *the* crucial question) confronting all quantum cosmological schemes remains to be addressed. The question is: is the time variable posited in quantum cosmology identical to the one referred to in the rest of physics, e.g., in Schrodinger's equation? Assuming there is at most one physical time, the question asks whether quantum cosmology and the rest of physics are mutually consistent.

So far as we can detect, the answer for Bohmian cosmology is uncertain, but quite promising. First, the time function found within the region of configuration space where the WKB approximation is valid is the same as that in the time-dependent Schrodinger equation (see e.g. Banks 1985). Due to the formal similarity between WKB and the present endeavor, then, the time function posited in section 7 is the same as the one found in the Schrodinger equation. Whether it is the same with a different wave function is unknown. It has been suggested that the concept of time is semiclassical, and breaks down outside the WKB region. If this is the case, then the Bohmian time is the 'right' time; if it is not so, and there are meaningful time functions outside the WKB region, then it is an open question whether they are all equivalent.

Second, the cosmic time posited by Bohmian cosmology is at least the right kind of entity to be identified with the time of dynamics. This feature is one happy consequence of Bohmian cosmology's implicit rejection of general relativity. Time in Bohmian cosmology seems to refer to real temporal relations, of the sort intimately associated with mechanics. It is not the arbitrary parameter found in general relativity. Prima facie, Bohmian time is more plausibly identified with the time of dynamics than with that found in many other proposals.

Notes

[1] Consider a universe at t=0 which is in a superposition between two cosmological properties, $\Psi = |a1\rangle + |a2\rangle$. It is now at t = present found to be in state (say) $|a2\rangle$.

Then we know with certainty that any measurement of Ψ between t=0 and t=present would have yielded |a2>. This fact goes unexplained in the orthodox interpretation. But in Bohm's theory, the reason is clear: the universe exists independently of the wave function, and was in state |a2> the entire duration. See Aharonov and Albert (1987).

[2]Gell-Mann and Hartle (1989) obtain a classical limit with their coarse-grained projection operators, though there remains the question of whether these can be defined without violating quantum mechanics (see Bohm and Hiley 1993, ch.14).

[3]If the space allows a timelike Killing vector field and a non-negative potential, the inner product will be positive. Unfortunately, neither condition is likely to be satisfied in QG. Theorems due to Kuchar make it plausible that QG does not admit the relevant isometries for a timelike Killing vector field. Additionally, the QG potential can be both negative and positive (and of course, the solutions cannot be restricted to the positive energy ones, for these correspond to the physically significant contraction of the 3-volume).

[4]If he introduces time like others in the semiclassical approach, he must isolate some classical degrees of freedom to play the role of time. Such a split is notoriously difficult (see Unruh and Wald 1989). If he tries to manage without time, say with a path integral calculating the transition amplitudes between states (as he indicates he might, p.349), then the integral typically depends on choice of foliation, and additionally leads to violations of the Hamiltonian constraints; see Kuchar (1992) for details.

References

Aharonov, Y. and Albert, D. (1987), "The Issue of Retrodiction in Bohm's Theory", in *Quantum Implications*, B.J. Hiley and David Peat (eds.). New York: Routledge, 224-226.

Albert, D. (1993), *Quantum Mechanics and Experience*. Cambridge: Harvard University Press.

_____ . and Loewer, B. (1988), "Interpreting the Many Worlds Interpretation", *Synthese* 77: 195-213.

Banks, T. (1985), "TCP, Quantum Gravity, The Cosmological Constant and All That ...", *Nuclear Physics* B249: 332-360.

Bell, J. (1987), *Speakable and Unspeakable in Quantum Mechanics*. Cambridge: Cambridge University Press.

Bohm, D. and Hiley, B.J. (1993) *The Undivided Universe*. NY: Routledge.

Durr, D., Goldstein, S., and Zanghi, N. (1992) "Quantum Equilibrium and the Origin of Absolute Uncertainty", *Journal of Statistical Physics* 67: 843-907.

Gell-Mann, M. and Hartle, J. (1989), "Quantum Mechanics in the Light of Quantum Cosmology", in *Proceedings of the 3rd International Symposium on the Foundations of Quantum Mechanics*. Kobyashi, S. (ed.). Tokyo: Physical Society of Japan.

Ghiradi, G., Rimini, A. and Weber, T. (1986), *Physical Review* D34: 470.

Halliwell, J. (1990), "Introductory Lectures on Quantum Cosmology", in *Quantum Cosmology and Baby Universes* Proc. Jerusalem Winter School, Coleman, S. et al (eds.). New Jersey: World Scientific.

Hartle, J. and Hawking, S. (1983), "Wave Function of the Universe", *Physical Review* D28: 2960-2975.

Huggett, N. and Weingard, R. (forthcoming), "Interpretations of Quantum Field Theory", *Philosophy of Science*.

Kuchar. K. (1992), "Time and Interpretations of Quantum Gravity" in *General Relativity and Relativistic Astrophysics*, Kunstatter, K. et al (eds.). New Jersey: World Scientific, 211-314.

Pitowsky, I. (1991), "Bohm's Quantum Potentials and Quantum Gravity", *Foundations of Physics* 21: 343-352.

Smith, Q. (forthcoming) "The Physical Meaning of Hawking's Quantum Cosmology".

Squires, E. (1992), "An Apparent Conflict Between the de Broglie-Bohm Model and Orthodoxy in Quantum Cosmology", *Foundations of Physics Letters* 5 71-75.

Squires, E. and Collins, P. (1993) "Time in a Quantum Universe", *Foundations of Physics* 5: 913-921.

Unruh, W. and Wald, R. (1989), "Time and the Interpretation of Canonical Quantum Gravity", *Physical Review* D40: 2598-2614.

Vilenkin, A. (1989), *Physical Review* D39: 1116.

Should we Believe in the Big Bang?:
A Critique of the Integrity of Modern Cosmology

Graeme Rhook and Mark Zangari

La Trobe University,

1. Introduction

Although the relativistic, hot big bang (RHBB) model is generally regarded as "spectacularly successful: In short, it provides a reliable accounting of the history of the Universe from about 0.01 sec after the [big] bang until today, some 15 billion years later" (Turner 1992, 1), it has recently faced serious attacks from a number of physicists who cite long lists of (they claim) critical anomalies (e.g. Arp, *et.al*. 1990, Burbidge, 1992, Arp and van Flandern, 1992, Lerner, 1993a, Narlikar 1993). In this paper, we shall not attempt to adjudicate on such disputes, nor do we wish to take a position on the normative criteria that should govern science. Rather, our primary concern is to highlight the tensions between the normative criteria accepted by proponents of the RHBB (or demanded of rival theories) and the actual progress of the big bang program. Our critique focuses on four central aspects of the theory: 1) the interpretation of cosmic red shifts, 2) the cosmic background radiation, 3) the inflation hypothesis and 4) the dark matter program. The interpretation of the red shift, rather than providing an incontrovertible basis for the theory, has been called into question by observational evidence and we shall query the adequacy of the response to this challenge offered by supporters of the big bang. In sections 2.1 and 2.2, we shall examine the status of the RHBB's claim that its explanation of the microwave background is clearly superior to alternative accounts. Because of a serious anomaly, this claim currently stands only in conjunction with a hypothesis, referred to as 'inflation'. However, inflation has come under considerable attack, both with regard to its poor empirical standing and because of its violation of the normative criteria supposedly accepted by its adherents. In section 2.3, we briefly outline the inflation hypothesis and suggest why, despite its problems, it has received considerable support from within the RHBB program. Finally, we shall argue in section 3 that the well known claim that the majority of the matter in the universe is invisible, or "dark" has spawned a research program that has the potential to insulate itself almost completely from empirical falsification. In the light of this analysis, we suggest that the big bang program, rather than being a highly corroborated and satisfactory account of the large scale universe, has a probability of ultimate success that does not justify its current dominance of contemporary cosmology.

2. The Interpretation of Cosmic Red Shifts

It is generally claimed that the RHBB model receives strong observational support from a phenomenon known as the *Hubble expansion*. This refers to the systematic red-shift in the electromagnetic spectra of distant astronomical objects that varies with their distance from our galaxy. This red-shift/distance relation was first accurately described by Edwin Hubble (1929) and the phenomenon is now interpreted as basically a Doppler effect, implying that distant objects are receding from the Milky Way at a rate proportional to their distance (see e.g. Weinberg 1972, 415-18). The constant of proportionality relating the recessional velocity to distance is now referred to as the 'Hubble constant' (H_0) and the effect is generally believed to be evidence that the universe is currently expanding.

However, the theory is already confronted by evidence that is in conflict with this, possibly its most central claim and, as we shall see, the response offered by supporters of the big bang is indicative of an approach often used to protect the theory from anomalous data. Contrary to the theory's assertion that cosmic red shifts are the result only of recessional velocities, there is evidence that nearby objects appear to have 'intrinsic redshifts' independent of their velocity relative to the earth (Arp *et.al.*, 1990, Arp and van Flandern, 1992). Furthermore, disproportionately many groups of objects with markedly different redshifts (which means they should be at vastly different radial distances from our galaxy) appear to be close together in the sky, some even possessing luminous connections. Neither of these overly concern big bang cosmologists, although the typical strategy they employ to fend off the possible threat makes the theory particularly insensitive to empirical anomalies.

Arp and Van Flandern accuse the big bang supporters of unjustifiably ignoring these examples of contrary evidence which were "visible but no one saw them because they we not supposed to be there" (1992, 267). However, despite Arp and Van Flandern's claim, some supporters of the big bang have acknowledged that serious anomalies in the red-shift/distance relation would indeed "show that the fundamental kinematics of the big bang are seriously incomplete" (Peebles, *et.al.* 1991, 775). Peebles *et.al.* go on to concede that the discovery of objects with a large blue-shift would constitute a serious anomaly for the RHBB model (*loc.cit.*). But it is not clear that such an observation is at all probable, even if the standard interpretation of the red shift were clearly false. Furthermore this particular example of setting a quite severe criterion for what constitutes a serious anomaly is by no means atypical. Consider the big bang supporters' more general claim regarding anomalous structures:

> ...they would be a problem for the Big Bang model itself only if it were shown that there is no plausible way to account for these structures within the relativistic expanding world model. (ibid, 769)

This attitude underlies the sort of perception about the degree of success enjoyed by the RHBB that is in evidence in Schramm's claim that "So far none of the Big Bang's predictions has failed. Of course, we still have unsolved problems..." (1993, 31). Such claims contain irreducibly subjective judgements about what constitutes a *serious* anomaly, as opposed to a failed prediction or a mere "problem". Furthermore, in the case of supporters of the big bang, the criterion for what constitutes a plausible hypothesis to solve "problems" is stretched by their perception of the relative implausibility of the RHBB being false. As we shall discuss in sections 2.3 and 3, this allows the survival of research programs that appear to have a low *prima facie* likelihood of success.

3. Cosmic Background Radiation

Alpher and Herman (1948) in their calculation of light element production during a hot, early phase of the universe, predicted that the present universe should be filled with electromagnetic radiation remaining from that era, but greatly cooled due to the expansion (for details, see Alpher and Herman in Balbinot 1990, 129-157). In the light of this prediction, the subsequent discovery of the (now measured) 2.7° microwave background by Penzias and Wilson (1965) generally tipped the balance of opinion in favour of the then struggling big bang theory and, ever since, has been cited as decisive evidence in its favour. However, we shall argue that the RHBB's explanation of the background radiation is not clearly superior to potential rivals on three counts. Firstly, there are plausible rival theories available; secondly, the RHBB's prediction of the angular distribution of the background is in conflict with the dynamics of the expanding RHBB universe, i.e. the theory suffers from a *prima facie* inconsistency and finally, the big bang's currently favoured response to this problem, known as the *inflation* hypothesis, has attracted serious criticism.

3.1. A Rival Account of the Cosmic Background Radiation

Because the existence of a background of microwave radiation was predicted as a consequence of the big bang, its account, unlike that of rivals, was granted immunity against accusations of being *ad hoc*. Competing theories were then forced into constructing *post hoc* explanations for the radiation which did not carry the force of being prior predictions, and which themselves lay open to charges of being *ad hoc*. However, *ceteris paribus*, this distinction alone does not seem to be sufficient grounds for clearly favouring either theory. It is interesting to speculate that Hoyle in his (1946), and others, were already considering the mechanism which contributes material to a rival account of the background radiation (i.e. the supernova model). One may wonder what the accepted cosmology might be today if the thermalised scattering of starlight by supernova ejecta had been investigated then, resulting in the prediction of isotropic electromagnetic radiation at a temperature of 2.8°K (see Hoyle in Balbinot *et.al.* 224).

The cosmic background radiation according to Alpher and Herman's (1948) prediction possessed the following properties: a temperature of about 5°K, a thermal (or Planck) energy distribution, and constant angular distribution, i.e., almost perfect isotropy in the cosmic rest frame. These three parameters provide at least a minimum criterion for the adequacy of any alternative account. The most prominent rival to the big bang's explanation of the microwave background is based on the thermalisation of starlight by fine particles that may be present in interstellar and intergalactic space. As alluded to above, Hoyle (in Balbinot *et.al.* 224) suggests that condensed iron filaments produced from the vaporised cores of supernovae might provide the medium for the thermalising pro cess, and thus explain the thermal properties of the background radiation. Its temperature can be simply derived by assuming that the energy emitted in the production of the helium observed in the universe is thermalised, and the calculation yields 2.8°K—almost precisely the measured value (although assumptions about the efficiency of the absorption of the radiation may introduce a further factor of about 2.). The isotropy is assured by the mean scattering time for the photons being short relative to the age of the universe. On the strength of this model alone, the RHBB cannot claim to provide an unrivalled explanation for the cosmic background radiation.

The RHBB's claim to clear superiority is further eroded by confirmation of the rival theory's novel prediction that radiation from distant sources should be attenuated at radio frequencies due to absorption by the intergalactic medium (Lerner 1993b).

Furthermore, the rival theory seems to be at an advantage in one other important respect. The microwave background is explained by appealing to well understood physical processes currently occurring in the universe, or which occurred at times when the state of the universe was not radically different to the present. By contrast, the big bang both *explains* the microwave background as a "relic" of an age when the universe was qualitatively different to its present state, and treats the radiation as *evidence* for the existence of this state. Unlike its rival which appeals to occurrent process, the big bang relies on positing processes which occurred only in the past and for which there is little direct evidence, other than the phenomena to be explained. As we shall now argue, the standing of the RHBB relative to its rival(s) is weakened further both by the presence of inconsistent requirements within the standard big bang theory, and by the nature of the amendment made in order to recover the theory's consistency.

3.2. Consistency of the RHBB Account of the Background Radiation's Isotropy

The predicted and observed isotropy of the microwave background, while presenting the theory with one of its greatest successes, also poses one of its greatest problems. The isotropy is explained by proposing that different regions of the early universe interact d to reach thermal equilibrium, thus smearing out any local anisotropies.[1] This interaction required that the various regions were causally accessible to each other (i.e. that they lay within each others' *particle horizons*—see Zel'dovich and Novikov 1983, pp. 35-41). However, this is in direct conflict with the kinematics of the standard model. Because the "radius" of the universe (i.e. the scale factor) was much larger than the particle horizon in the early universe, by the time the photons currently detected in the cosmic background radiation were emitted (about 100,000 years after the big bang), the universe contained about 10^6 causally unconnected regions (Linde 1990, 26), making it impossible for all parts of the plasma from which the radiation was supposedly emitted to be in thermal equilibrium. This difficulty has come to be known as the horizon problem. Thus, while the energy distribution and approximate temperature (up to a factor of about 2) of the background radiation agree with the RHBB predictions, its extreme isotropy (to at least 10^{-4}) is highly problematic.

As it stands, without some appropriate amendment, the standard big bang model, rather than receiving unqualified support from the confirmation of its prediction of the existence of the cosmic background radiation, is inconsistent with its observed properties. While the horizon problem has long been recognised as pressing by proponents of the RHBB (e.g. Weinberg, 1972), a solution to this problem emerged from a hypothesis proposed by Alan Guth (1981) which has come to be known as *inflation*. This has spawned a large research program in its own right, although as we will discuss in the next section, one that is also faced with problems no less serious than those it is supposed to solve. If inflation proves unacceptable, and we shall argue that there are good reasons to believe this is the case, then the standard big bang model is left with major, unresolved inconsistencies.

3.3. Inflation

Apart from the horizon problem, the big bang theory faces another difficulty known as the *flatness problem*. The current expansion rate and density of the universe define a critical density ρ_c for which the universe is "flat". Following convention, we introduce the parameter Ω as the ratio of the actual density ρ to the critical density. That is, $\Omega = \rho/\rho_c$ so $\Omega=1$ corresponds to the case where the actual density of the universe is precisely the critical, "flat" value. The evolution of the RHBB universe is extremely sensitive to the value of Ω during its earliest stages and is highly unstable if Ω differs from 1 by any appreciable amount. For if $\Omega > 1$, even by a very small amount, the universe re-collapses almost immediately after the big bang; if $\Omega<1$, the expansion is so rapid that significant accumulation of matter would be impossible. It has been shown that for the universe of the present density to be expanding at the rate H_0 some 10-15 billion years after the big bang, Ω could not differ from 1 by more than 1 part in 10^{59} at the Planck time ($t_P \approx 10^{-43}$ seconds after the big bang) (Linde 1990, p.24). Fine tuning to an accuracy of sixty decimal places cannot be satisfactorily explained as serendipity, so the big bang is faced with the challenge of explaining why the universe was "created" flat, that is, with Ω so close to 1.

Inflation attempts to solve both the flatness and horizon problems by appealing to grand-unified theories (GUT) in particle physics. The standard model is extended to include a very short period of super-rapid exponential growth during the first 10^{-35} seconds. The scale factor could increase by a factor of 10^{30}, thus explaining the apparent flatness and, in addition, requiring that $\Omega=1$ still be the case in the present universe. This also provides a solution to the horizon problem since the observable universe "grew" out of a region tiny enough to have reached equilibrium prior to the growth. While the details of the various inflationary models are not our concern here (see Linde 1990), the spawning of a huge inflation research program highlights the degree to which cosmologists find it plausible to introduce more and more speculative physics in the face of challenges to the big bang, rather than seriously pursuing alternative solutions that rely on existing physics but lie outside the RHBB framework.

The inflation hypothesis has been accused of introducing more mysterious problems than it solves and of being "metaphysics", rather than physics because of its lack of testable claims (Rothman 1989, pp.29-50). However, the "metaphysics" accusation is not completely justified because inflation is falsifiable, due to the fact that GUTs do make at least one specific, testable claim: that protons are unstable and eventually decay. However, all experiments testing this have yet to detect a single proton decay event. In all likelihood it seems the proton is stable and the GUTs on which inflation rests are probably wrong (Lerner 1993a). Given its *prima facie* small likelihood of success, a natural question emerges regarding why the inflation program itself, like the universe it describes, underwent such rapid growth in the 1980's. While we can only speculate, it seems that a significant impetus for the work came from a synergy between the two branches physics concerned with phenomena at the opposite extremes of scale: particle physics and cosmology. The collaboration between the two from the late 1940's to 1970's produced the nucleosynthesis calculations that are now regarded as another firm pillar of evidence for the big bang (Wagoner in Balbinot *et.al.* 1990, 159-185). Modern particle physics, whose theories describe properties manifested only at energies that are unachievable in any foreseeable experiment, finds almost unbounded energies in the very early universe. Thus, the RHBB provides a "cosmic laboratory". In return, cosmology utilises specimens from the "particle zoo" to solve its own problems. Inflation is one such solution, and we will shortly encounter another in the dark matter hypothesis.

We would summarise the relative standings of the RHBB and its rivals in respect of the microwave background radiation in the following way: the RHBB model must invoke inflation (or some other mechanism) to fully account for the properties of the cosmic background radiation, while a rival explanation requires the presence of particulate material that has not, as yet, been directly detected nor ruled out. Although supporters of the big bang argue that their model empirically out-performs its rivals in accounting for the background radiation (Peebles *et.al.* 1991), they generally support this by appealing only to the energy spectrum. While it may be true that the closest fitting frequency distribution belongs to the standard model, this still fails to provide satisfactory accounts of the temperature and isotropy. Adding inflation to the model to resolve the isotropy problem removes any obvious empirical advantage that the big bang may claim, given the difficulties with the inflation hypothesis mentioned earlier. Just as rivals seek empirical justification of a mechanism to account for the radiation's thermal spectrum (while easily accounting for its isotropy and temperature), the big bang must find an empirically acceptable explanation for the isotropy. Given the relative likelihoods of finding evidence for an interstellar absorptive medium and a GUT-based inflationary period, the justification for claiming that the cosmic bac kground radiation unequivocally supports the RHBB ahead of other models seems thin.

4. Dark Matter

Apart from the Hubble expansion and the cosmic background radiation, the big bang receives considerable observational support from a third piece of evidence: the relative abundances of light nuclei, supposedly synthesized during the first 1,000 seconds of the universe. Superficially, this appears to have met with great success, at least in there being little qualitative disagreement with observation (Boesgaard and Steigman, 1985, although Lerner 1993a challenges this in some detail). However, this success has, paradoxically, turned out to also present the theory with its greatest challenge. Primordial nucleosynthesis calculations place limits on the density of baryonic matter ρ_B in the universe. Expressing this as a fraction of the critical density, $\Omega_B \equiv \rho_B/\rho_c$, Boesgaard and Steigman (1985) specify that $0.011 < \Omega_B < 0.19$. While the observed value $\Omega_B \approx 0.019$ lies well within this range providing the RHBB with one of its greatest successes, big bang cosmologists nonetheless enthusiastically embrace models with $\Omega=1$. According to Turner (1992, 19), favoured models of structure formation with $\Omega \neq 1$ are inconsistent with the isotropy of the cosmic background radiation and recall from section 2.3 that the model with inflation also predicts that $\Omega=1$.

Rather than regarding the prediction that $\Omega=1$ as an anomaly pointing to defects in the inflation or structure formation hypotheses, big bang cosmologists use it to make arguably one of the boldest predictions in the history of science: that 98% of the matter in the universe must be "missing" or "hidden" or "dark", and that at least 80% of this matter is not the ordinary baryonic, matter of which atomic nuclei consist, but something quite new.

This bold prediction of the RHBB could in principle be experimentally confirmed, and indeed its confirmation would provide quite dramatic support for the big bang theory. On the other hand if the prediction was refuted then the big bang theory would face quite grave difficulties. As we shall see, the problem for refuting the prediction is in constructing, even in principle, an experiment or series of experiments that could do so.

The big bang theory shares, with respect to this particular problem, some similarities with the Ptolemaic program. A particular Ptolemaic theory, consisting of a configuration of the appropriate geometric figures, could easily be shown to be inconsistent with the data. However once a particular theory is refuted the program enables another to be posited and this process might continue indefinitely. Consequently the Ptolemaic program itself is not so easily refuted by the failure of experiments to confirm a particular theory within the program. The prediction and search for the nonbaryonic dark matter shares this feature.

The dark matter hypothesis is not recent, dating back to at least Sandage (1956). Yet despite its relative longevity, the existence of dark matter has not yet been confirmed. The constraints provided by nucleosynthesis and the theories dealing with the formation of structures of cosmological scales are quite demanding. Given that baryons cannot make up the missing mass, more exotic particles have been offered as candidates, although none of these have been actually detected, and no candidate has won anything like general acceptance. The comments of Schramm and Riordan (1991, 201), prominent supporters, indicate the current state of the program.

> Virtually every theorist has his or her own pet theory... Because very little is known about the properties of dark matter, other than the fact that it exists, there is plenty of room for theoretical speculation right now.

What is of particular interest is the attitude of the supporters of the RHBB to the unresolved difficulties posed by the fact that a viable dark matter candidate has failed to emerge. Consider, for example, the assessment of another of its prominent supporters (Stephen Hawking in Riordan and Schramm 1991, viii):

> So there are two possibilities; either our understanding of the very early Universe is completely wrong, or there is some other form of matter in the Universe that we have failed to detect. The second possibility seems more likely, but the required amount of missing "dark' matter is enormous; it is about a hundred times the matter we can directly observe.... The quest is on for finding 99% of the Universe.

Hawking's assuredness is by no means universal but it is not unfair to say that it is characteristic of supporters of the RHBB. However it is far from clear that the consequences of this attitude for modern cosmology are at all benign. The potential dangers are illustrated in some of the theoretical responses that proponents are prepared to countenance. Consider for example Schramm's and Riordan's description of a theory generated by the program: (1991, 200)

> A final dark horse candidate is what has become known as "shadow matter". In any Theory of Everything, which purports to unify gravity with the other three fundamental forces at times before 10^{-43} second, there can arise a parallel universe (commingled with our own) that has completely different forces and particles... It is a shadow universe that occupies the very same physical space as the familiar Universe but has no normal interaction with it other than through the force of gravity. We can imagine that the particles of shadow matter might form shadow atoms and molecules. There could be shadow rocks and plants, even shadow people, planets, stars, and galaxies that would pass right through our own almost completely unnoticed... the only influence this shadow universe would have upon our own world be through the force of gravity.

Contrast this with Schramm's expressed attitude regarding the criteria that an adequate cosmological model must meet (1993, 31):

> A cosmological theory must explain these three pillars in quantitative detail, using theoretical models that agree as far as possible with known laboratory physics.

If the big bang theory turns out to be valid then the search for dark matter may very well be viewed in retrospect as a program that succeeded brilliantly in the face of not insignificant experimental and theoretical obstacles. On the other hand, if the big bang theory is false, or more particularly if its requirement that $\Omega=1$ is false, then it is not clear what sort of experiments would be capable of terminating the program. The program has already moved away from considering the particles that have currently been detected and is looking toward particle physics or other theoretical programs in the hope that they might provide appropriate candidates. However if a candidate provided proves problematic, either because its particular characteristics do not conform to the requirements of the big bang theory itself or because the theory on which the candidate is based loses support, then appeal can be made to other speculative theories or alternatively it might be decided that the program must await further theoretical, experimental and technological developments before its adherents can hope to have their predictions confirmed.

If dark matter is not detected by a particular experiment then it is possible to argue that dark matter is in fact more weakly interacting, darker and more transparent than was thought. The shadow matter hypothesis illustrates the limit of this strategy. The shadow matter hypothesis ensures, *ex hypothesi*, that the dark matter cannot be detected except by its gravitational effects and 'explains' why any experiment which attempted to detect it other than by its gravitational effects would fail. Presumably the theory about the distribution and configuration of the shadow matter can be modified so that it fits the observations about the actual world. Consequently it is not immediately obvious how this particular hypothesis might, even in principle, be falsified. In short, the dark matter program is characterised by the fact that, although it is in principle possible to confirm the program's prediction should it be valid, it is not possible to falsify the program should it be wrong.

5. Conclusion

The dark matter program is proceeding in a way which ensures that it is insulated from empirical falsification, while the success of the dark matter program is crucial for the viability of the inflation hypothesis. In turn, the viability of the inflation hypothesis is necessary in order to resolve an internal inconsistency within the theory. The conflation of this series of dependencies provides grounds for rejecting the claim that the big bang theory enjoys a clear superiority over rival accounts of the characteristics of the cosmic background radiation. In addition, supporters of the RHBB insulate the theory from alternative interpretations of the red shift by invoking severe criteria for determining what constitutes a serious anomaly. In the light of these criticisms we suggest that the big bang theory's dominance of contemporary cosmology is not justified nor does it deserved to be considered as a highly corroborated and satisfactory account of the evolution of the large scale universe. While we consider that the big bang is the most developed theory presently available, this alone is not sufficient grounds for the degree of support it currently enjoys especially in the light of the features we have described.

Notes

[1](Linde 1990, 26, estimates that the alternative—i.e. the various regions remained isolated and so never reached equilibrium, b ut somehow underwent identical local evolution—has a probability of at most 10-24)

References

Alfvén, H. (Feb. 1971), "Plasma Physics Applied to Cosmology", *Physics Today* 24: 28-33.

Alpher, R.A. and Herman, R. (1948), "Evolution of the Universe", *Nature* 162: 774.

Arp, H.C. and van Flandern, T. (1992), "The case against the big bang", *Physics Letters* A 164: 263-273.

_ _ _ _ _ ., Burbidge, G., Hoyle, F., Narlikar, J.V. and Wickramasinghe, N.C. (1990), "The extragalactic Universe: an alternative view", *Nature* 346: 807-812.

Balbinot, R., Bergia, S., Bertotti, B., and Messina, A. (eds) (1990), *Modern Cosmology in Retrospect*, Cambridge: Cambridge University Press.

Boesgaard, A.M. and Steigman, G. (1985), "Big Bang Nucleosynthesis: Theories and Observations", *Annual Review of Astronomy and Astrophysics* 23: 319-78.

Burbidge, G. (1992), "Why Only One Big Bang?", *Scientific American* 266: 96.

Grünbaum, A. (1990), "Pseudo-Creation of the Big Bang", *Nature*, 344: 821-822.

Guth, A. (1981), "Inflationary Universe: A Possible Solution to the Horizon and Flatness Problems", *Physical Review* D23: 347-356.

Hoyle, F. (1946), "The Synthesis of the Elements from Hydrogen", *Monthly Notices of the Royal Astronomical Society* 106: 343-383.

Hubble, E.P. (1929), "A relation between distance and radial velocity among extragalactic nebulae", *Proceedings of the National Academy of Science U.S.* 15: 517-554.

Lerner, E.J. (1993a), "The Case Against the Big Bang", *Progress in New Cosmologies*, Plenum Press.

_ _ _ _ _ _ . (1993b), "Confirmation of Radio Absorption by the Intergalactic Medium", *Astrophysics and Space Science* 207: 17-26.

Linde, A. (1990), *Particle Physics and Inflationary Cosmology*, Switzerland: Harwood Publishers.

Maddox, J. (1989), "Down with the Big Bang", *Nature* 340: 425.

Narlikar, J. (1992), "The Concepts of 'Beginning' and 'Creation' in Cosmology", *Philosophy of Science* 59: 361-371.

_ _ _ _ _ _ . (1993), "Challenge for the Big Bang", *New Scientist* 138: 27-31.

Peebles, P.J.E., Schramm, D.N., Turner, E.L. and Kron R.G. (1991), "The case for the relativistic hot Big Bang cosmology", *Nature* 35 2: 769-776.

Penzias A.A. and Wilson R.W. (1965),"A Measurement of Excess Antenna Temperature at 4080 MHz", *Astrophysical Journal* 142: 419.

Riordan, M. and Schramm, D. (1991), *The Shadows of Creation: Dark Matter and the Structure of the Universe*, New York: Freeman.

Rothman, T. (1989), *Science à la Mode*, Princeton: Princeton University Press.

Sandage, A.R. (1956), "The Red Shift", *Scientific American* 195, No.3: 170-200.

Schramm, D. (1993), "The big bang strikes back", *New Scientist* 138: 31.

Turner, M.S. (1992) *Inflation after COBE: Lectures on Inflationary Cosmology*, Preprint FERMILAB-Conf-92/313-A.

Weinberg, S. (1972), *Gravitation and Cosmology: Principles and Applications of the General Theory of Relativity*, New York: John Wiley & Sons Inc.

Zel'dovich, Y.B. and Novikov, I.D. (1983), *Relativistic Astrophysics Vol. 2: The Structure and Evolution of the Universe*, Chicago: University of Chicago Press.

Part VII

STATISTICS AND EXPERIMENTAL REASONING

On the Nature of Bayesian Convergence[1]

James Hawthorne

University of Oklahoma

1. Introduction

Bayesians assess the inductive support for theoretical hypotheses on the basis of two sorts of factors, one fairly objective, the other highly subjective. The objective factor consists of the *likelihoods* or *direct inference probabilities* that theoretical hypotheses specify for evidential events. This is the means by which evidence affects inductive support. The subjective factor consists of the *prior probabilities* assigned to the various competing hypotheses. For a Bayesian agent the *prior* probability of a hypothesis represents how plausible the agent considers the hypothesis to be *before* the impact of evidence is considered, and Bayesian agents may radically differ in their initial plausibility assessments. Bayes' formula combines likelihoods with an agent's prior probabilities to produce the agent's *posterior probability* for each hypothesis. *Posterior* probabilities represent how plausible an agent considers the hypothesis to be *after* the evidence is taken into account. Thus, to the extent that the values of subjective prior probabilities continue to affect the values of posterior probabilities as evidence accumulates, Bayesian agents will continue to differ regarding how plausible they consider theoretical hypotheses to be. If Bayesian induction is to yield either an objective assessment or intersubjective agreement among agents regarding the inductive support for hypotheses, then the evidence must somehow produce a *convergence to agreement* among the posterior probabilities of different Bayesian agents in spite of their initial disagreements about the plausibilities of the hypotheses. Hence, advocates of Bayesian induction have investigated the circumstances under which evidence might "wash out" the effects of subjective priors and bring agents into agreement on posterior probabilities. Bayesian convergence theorems establish conditions under which accumulating evidence can induce agents to converge to agreement.

In this paper I will describe some important features of Bayesian convergence, including some features that have not been widely recognized by Bayesian logicians. I will discuss three sorts of Bayesian convergence results. The first shows how the objectivity of simple inductions, which assess the likelihoods of individual events, depends on the objectivity of posterior probabilities of general theoretical hypotheses. This sort of convergence result shows the pivotal role of theoretical hypotheses in *systematizing* simple inductive inferences in a manner that bolsters their objectivity.

PSA 1994, Volume 1, pp. 241-249
Copyright © 1994 by the Philosophy of Science Association

The second convergence result reveals that (except in very special circumstances) evidence can induce Bayesian probability functions to *converge to agreement* on the posterior probabilities for theoretical hypotheses *only if* the convergence is to 0 (refutation) or 1 (confirmation). The third result establishes general conditions under which the evidence *will very probably* compel posterior probabilities of theoretical hypotheses to converge to 0 or 1.

2. Simple Induction

Simple induction is a form of inference in which an agent infers the probability that some particular outcome e will result from some initial state of affairs c on the basis of some sequence of evidential events. Simple inductions are inferences about specific "occurrent events" rather than general hypotheses or theories. Let '$(c_1 \cdot c_2 \cdot \ldots \cdot c_n)$' represent a conjunction of descriptions of the initial states, initial conditions, or experimental arrangements for a series of past observations, and let their respective outcomes form the conjunction '$(e_1 \cdot e_2 \cdot \ldots \cdot e_n)$'. I will use '$c^n$' to abbreviate the conjunction '$(c_1 \cdot c_2 \cdot \ldots \cdot c_n)$' of descriptions of n initial states, and 'e^n' to abbreviate the conjunction of the n descriptions of respective outcomes. Agents may also employ some relatively uncontroversial background knowledge b as a premise in simple inductive inferences. Background knowledge (including auxiliary hypotheses) represented by 'b' will typically include relatively uncontroversial claims about how pieces of familiar instrumentation work and about methods and conditions under which various kinds of observable phenomena may be reliably detected or measured.

The evidence for a simple induction may be a sequence of events that are very similar to the event whose probability is to be inferred—as when we infer the probability that the next toss of a particular bent coin will come up heads from a sequence of outcomes of previous tosses of the same coin. In such cases simple induction is typically called *induction by enumeration*. But, under the term *simple induction* I mean to include a much broader class of inferences than those that merely employ enumerations of similar cases. In general the evidence for simple inductions may be a very diverse collection of previous events. The evidence might include descriptions of how other objects of various shapes have tumbled, and how they have bounced on different surfaces. Indeed, the simple inductive evidence for an event could include all of the evidence that one might normally employ in the confirmation of the kinds of theoretical hypotheses that would be relevant to the individual event in question.

Let α and β be two Bayesian agents whose respective probability functions are P_α and P_β. For any sentences A and B, $P_\alpha[A \mid B]$ represents α's conditional probability for A given B, her assessment of how probable (how plausible, how likely to be true) A would be if B were true (and similarly for $P_\beta[A \mid B]$). For r a real number between 0 and 1, '$P_\alpha[e \mid c \cdot c^n \cdot e^n \cdot b] = r$' represents the *simple inductive probability* of e on $(c \cdot c^n \cdot e^n \cdot b)$ for agent α. I.e. if the premises in the conjunction '$(c \cdot c^n \cdot e^n \cdot b)$' represent α's total relevant evidence regarding e, then α's rational degree of confidence in e, the plausibility of e for α, should be r.

Even when agents possess the same total evidence, they may strongly disagree on the simple inductive probability of an outcome e of condition c—i.e. the probability values r and s for the simple inductive inferences $P_\alpha[e \mid c \cdot c^n \cdot e^n \cdot b] = r$ and $P_\beta[e \mid c \cdot c^n \cdot e^n \cdot b] = s$ may differ greatly. How might agents be induced to come to agreement about the probability of e as the evidence accumulates? If the evidence could bring the agents to concur on the truth of some general theoretical hypothesis that states deterministic or stochastic laws governing situations like c, then they would come to agree on how likely e is. Each agent would invoke the theoretical hypothesis, together with c and b, and arrive at the same conclusion about the likelihood that e is true. Indeed, one of the primary reasons that agents seek to confirm a theoretical hypothesis is so that it may be used as a kind of objective inference ticket that *systematizes simple inductions*. The role of theoretical hypotheses in the systematization of simple inductions is central to the scientific enterprize. We will see precisely how inductive systematization works in a Bayesian context after first exploring the relationship between theoretical hypotheses and descriptions of individual events from a Bayesian perspective.

3. Theoretical Hypotheses and Likelihoods

Let $H = \{h_1, h_2, ...\}$ be a class of competing theoretical hypotheses that bear on whether event e will result from condition c. H may contain an infinite number of alternatives, and two Bayesian agents α and β may disagree widely on how plausible they are. But suppose that α and β are like-minded enough that they consider only the theoretical hypotheses in H to have some non-zero degree of initial plausibility. Thus, for each agent the sum of prior probabilities of hypotheses in H is 1 (i.e. $\sum_j P_\alpha[h_j|b] = 1 = \sum_j P_\beta[h_j|b]$). And since the hypotheses in H are alternatives, any distinct pair of them, h_i and h_j, should be incompatible (i.e. $P_\alpha[h_i \cdot h_j|b] = 0$, and similarly for β).

The hypotheses in H may be deterministic or statistical, and they may be extremely broad theories or their scope may be quite narrow. But whatever their scope, Bayesians typically suppose that the principal epistemic role of theoretical hypotheses is to underwrite *relatively objective* probabilities for individual events. Taken together with initial conditions c and background knowledge b, each theoretical hypothesis h_i should provide a fairly unambiguous indication of the probability that e will occur (if h_i is true). This is one of the main reasons that we construct theoretical hypotheses in the first place. Bayesians usually call these probabilities *likelihoods* or *direct inference probabilities*. Likelihoods take the form $P_\alpha[e|c \cdot h_i \cdot b] = r$, and Bayesian agents are supposed to generally agree on their values.

Logicist Bayesians usually call likelihoods *direct inference* probabilities. Logicists think of likelihoods as objective, logical relationships, as an extension of the logical entailment relation to a form of probabilistic entailment that can logically link stochastic object-language sentences (e.g. about propensities) to descriptions of individual chance events. By contrast, *Personalist Bayesians* do not think of likelihoods as *logically determinate*, but they also usually consider likelihoods to have a "more objective" status than other probabilities. Personalists often maintain that relative to theoretical hypotheses there will usually be a high degree of intersubjective agreement among Bayesian agents on the likelihoods of events. Indeed, the main rational for using Bayes' Theorem to *calculate posterior probabilities* of theoretical hypotheses is that the likelihoods are generally considered by Bayesians to be more objective, or more subject to intersubjective agreement than the posterior probabilities they are used to calculate. Henceforth I will suppose that agents agree on likelihoods for events relative to hypotheses in H. As a result we may drop the subscripts 'α' and 'β' from likelihoods, and for each hypothesis h_i we write $P_\alpha[e|c \cdot h_i \cdot b] = P_\beta[e|c \cdot h_i \cdot b] = P[e|c \cdot h_i \cdot b]$.

There is another important facet to the kind of objectivity that theoretical hypotheses should afford likelihoods of events. Consider the probability of event e relative to both $(c \cdot h_i \cdot b)$ *and* to the previous evidence $(c^n \cdot e^n)$—i.e. consider $P_\alpha[e|(c \cdot h_i \cdot b) \cdot (c^n \cdot e^n)]$. When $(c \cdot h_i \cdot b)$ furnishes a direct inference probability $P[e|c \cdot h_i \cdot b] = r$, the old evidence $(c^n \cdot e^n)$ should become irrelevant to the probability of e (relative to $(c \cdot h_i \cdot b)$). The idea is that the old evidence plays it's role through the inductive support it provides for h_i. But given that h_i (together with $(c \cdot b)$) *is* true, the old evidence should be screened off from influence on the probability of e. Thus, we may reasonably suppose that the following *independence condition* holds for each Bayesian agent α: $P_\alpha[e|(c \cdot h_i \cdot b) \cdot (c^n \cdot e^n)]$ = $P[e|c \cdot h_i \cdot b]$. If a hypothesis failed to *screen off* predicted events from previous evidence in this way, then each time an agent appealed to the hypothesis to predict an event she would have to employ a vast collection of previous observational and experimental data as initial conditions. This would largely undermine one of the chief reasons for formulating theoretical hypotheses in the first place. Imagine what it would be like if to compute a future location for a planet we not only had to appeal to a gravitational theory and a few observations of the planets' past locations, but also had to employ in the computation all of the data from the experiments and observations that went to confirm the theory of gravitation.

Finally, another plausible *independence condition* will be used in the next section. The initial condition statement c that combines with a hypothesis to determine the likelihood of an outcome should not by itself function as evidence for or against the hypothesis. It should only become relevant to the support of a hypothesis when it is conjoined with its associated outcome e. That is, although adding (c·e) to any previous evidence might count as additional evidence for h_i, when the outcome e is still in question, c should not by itself be relevant to the posterior probability of h_i. Thus, for each agent α we should have $P_\alpha[h_i|c \cdot c^n \cdot e^n \cdot b] = P_\alpha[h_i|c^n \cdot e^n \cdot b]$. (Equivalently, relative to $(c^n \cdot e^n \cdot b)$, the initial condition c should be no more likely to have occurred if h_i is true than if it is false, i.e. $P\alpha[c|h_i \cdot (c^n \cdot e^n \cdot b)] = P\alpha[c|\neg h_i \cdot (c^n \cdot e^n \cdot b)]$.)

We are now ready to see how agents may be induced to come to agreement about the probability of an event e as the evidence accumulates. The main idea is simple. If the accumulating evidence comes to strongly support the same hypothesis in H for all agents, then that hypothesis may be used as a premise that generates a likelihood for e on which all agents agree. Thus, inductive confirmation of hypotheses can systematize simple inductive inferences. In the next section I will spell out the details of this strategy from a Bayesian perspective.

4. Bayesian Convergence for Simple Inductions

Under the conditions described in the previous section the convergence to agreement of simple inductions for Bayesian agents reduces to their convergence to agreement on the posterior probabilities of the theoretical hypotheses in H. To see this notice that for any Bayesian agent α:

(I) $P_\alpha[e|c \cdot c^n \cdot e^n \cdot b]$ = $\Sigma_j P_\alpha[e|(c \cdot h_j \cdot b) \cdot (c^n \cdot e^n)] \times P_\alpha[h_j|c \cdot c^n \cdot e^n \cdot b]$
= $\Sigma_j P[e|c \cdot h_j \cdot b] \times P_\alpha[h_j|c^n \cdot e^n \cdot b]$.

The first line of this equality is a theorem of probability theory (since H is a set of mutually incompatible hypotheses to which Bayesian agents assign probabilities that sum to 1). The second line follows directly from the first line and the two independence conditions described in the previous section.

Equation (I) exhibits the connection between simple inductions, likelihoods, and Bayesian theory confirmation. It suggests two different sorts of convergence for simple inductions. The first is the kind that comes from highly confirming a hypothesis in H. If the posterior probability that α assigns hypothesis h_i approaches 1 as the evidence increases (i.e. if $P_\alpha[h_i|c^n \cdot e^n \cdot b] \longrightarrow 1$, as n increases), then the simple induction probabilities for α will approach the likelihoods that h_i specifies (i.e. $P_\alpha[e|c^n \cdot e^n \cdot b] \longrightarrow P[e|c \cdot h_i \cdot b]$). Call the convergence to 1 of agent α's posterior probabilities for a hypothesis (and the convergence to 0 of its alternatives) *Type 1 Hypothesis Convergence* for α.

Call the convergence of α's simple inductive probabilities to the likelihoods specified by a hypothesis *Type 1 Simple Inductive Convergence* for α. Then the first convergence result that flows from equation (I) says that, for each Bayesian agent:

Type 1 Hypothesis Convergence implies *Type 1 Simple Inductive Convergence.*

Equation (I) also suggests a second sort of Bayesian convergence, a variety of convergence to agreement *among* agents. Call the convergence to agreement of agents α and β on their posterior probabilities for all hypotheses in H *Type 2 Hypothesis Convergence* for α and β (i.e. for each h_j, $|P_\alpha[h_j|c^n \cdot e^n \cdot b] - P_\beta[h_j|c^n \cdot e^n \cdot b]| \longrightarrow 0$). Call the convergence to agreement by α and β on the probabilities for simple inductions *Type 2 Simple Inductive Convergence* for α and β (i.e. $|P_\alpha[e|c^n \cdot e^n \cdot b] - P_\beta[e|c^n \cdot e^n \cdot b]| \longrightarrow 0$). Then the second sort of convergence result that flows from equation (I) is that, for each pair of agents:

Type 2 Hypothesis Convergence implies *Type 2 Simple Inductive Convergence.*

Each of the two types of *simple inductive convergence* lends a kind of objectivity to simple Bayesian inductions. Each shows how the Bayesian evaluation of hypotheses leads to a systematization of simple inductions by reducing the convergence problem for simple inductions to a convergence problem for theoretical hypotheses. Indeed, inductive systematization turns out to be even more regimented than one might have expected. For, somewhat surprisingly:

Type 2 Hypothesis Convergence implies *Type 1 Hypothesis Convergence.*

That is, if evidence can bring a pair of Bayesian agents who possess even *moderately diverse* prior probabilities for theoretical hypotheses (in a sense to be made precise in the next section) into agreement about posterior probabilities, then they must come to agree that the posterior probability of one particular hypothesis, h_i, approaches 1 (and posteriors of its competitors approach 0).

In the next section I will explain the connection between the two types of hypothesis convergence in more detail. Its implications for the nature of simple induction are striking. If evidence induces agreement among Bayesian agents on the probability of an event (*Type 2 Simple inductive Convergence*) by causing the agents to converge to agreement on the posterior probabilities of hypotheses (*Type 2 Hypothesis Convergence*), then it may do so *only* by raising the posterior probability of one hypothesis towards 1 for each Bayesian agent (*Type 1 Hypothesis Convergence*), thus forcing the simple inductions of all agents towards agreement with the (direct inference) likelihoods specified by that hypothesis (*Type 1 Simple Induction Convergence*).

5. Likelihood Ratios and Bayesian Convergence for Theories

Bayesian induction regarding theoretical hypotheses essentially depends on *likelihood ratios*. Likelihood ratios are ratios of direct inference probabilities for competing hypotheses and will be abbreviated by '$LR[e^n|j/i]$', where by definition:

$$LR[e^n|j/i] = P[e^n|c^n \cdot h_j \cdot b] / P[e^n|c^n \cdot h_i \cdot b].$$

Likelihood ratios measure how many times more (or less) likely the evidence would be according to one hypothesis as compared to another.

The central role of likelihood ratios in Bayesian induction becomes apparent when we consider the ratio of posterior probabilities for a pair of hypotheses. The ratio of their posterior probabilities equals the product of their likelihood ratio with the ratio of their prior probabilities:

(II) $\quad P_\alpha[h_i|e^n \cdot c^n \cdot b] / P_\alpha[h_i|e^n \cdot c^n \cdot b] = LR[e^n|j/i] \times (P_\alpha[h_j|b] / P_\alpha[h_i|b])$.

This equality is a theorem of probability theory (provided that the probability of the initial conditions is the same relative to each hypotheses— i.e. $P_\alpha[c^n|h_i \cdot b] = P_\beta[c^n|h_j \cdot b]$). For simplicity I will assume that this proviso holds; a much weaker assumption would suffice, see (Hawthorne, forthcoming), but would complicate the exposition unnecessarily.

The relationship between ratios of posterior probabilities and likelihood ratios expressed by equation (II) is really all there is to Bayesian induction for theoretical hypotheses. The *absolute* probability of a hypothesis comes directly from the sum of these ratios. To see this, first consider the *odds*, Ω_α, against a hypothesis h_i relative to evidence, defined as follows:

(III) $\quad \Omega_\alpha[\neg h_i|e^n \cdot c^n \cdot b] = P_\alpha[\neg h_i|e^n \cdot c^n \cdot b] / P_\alpha[h_i|e^n \cdot c^n \cdot b]$
$\quad\quad\quad\quad\quad\quad\quad\quad = \sum_{j \neq i} P_\alpha[h_j|e^n \cdot c^n \cdot b] / P_\alpha[h_i|e^n \cdot c^n \cdot b]$.

The odds against a hypotheses is the sum of the relative plausibilities for its competitors, the sum of instances of equation (II). Equations (II) and (III), then, imply the following relationship between the odds against a hypothesis and likelihood ratios:

(IV) $\quad \Omega_\alpha[\neg h_i|e^n \cdot c^n \cdot b] = \sum_{j \neq i} LR[e^n|j/i] \times (P_\alpha[h_j|b] / P_\alpha[h_i|b])$.

The probability of a hypothesis on evidence is related to the odds against the hypothesis by the following formula:

(V) $\quad P_\alpha[h_i|e^n \cdot c^n \cdot b] = 1 / (1 + \Omega_\alpha[\neg h_i|e^n \cdot c^n \cdot b])$.

Taken together, equations (IV) and (V) express a form of Bayes's theorem in terms of the odds against a hypothesis; this formulation makes the role of the likelihood ratios more perspicuous than the more usual form of the theorem, which is:

(VI) $\quad P_\alpha[h_i|e^n \cdot c^n \cdot b] = P[e^n|c^n \cdot h_i \cdot b] \times P_\alpha[h_i|b] / \sum_j P[e^n|c^n \cdot h_j \cdot b] \times P_\alpha[h_j|b]$.

If h_i comes to make the evidence negligibly likely in comparison to some alternative, h_j—i.e. if $LR[e^n|i/j]$ converges to 0—then the inverse likelihood ratio, $LR[e^n|j/i]$, blows up to infinity, and the odds against h_i blow up with it, by equation (IV). Hence, by equation (V), the probability of h_i goes to 0. On the other hand, if *every* alternative to h_i makes the evidence negligibly likely in comparison to h_i—i.e. if for every alternative h_j, $LR[e^n|j/i]$ converges to 0—then by equation (IV) the odds against h_i converge to 0. When this happens, equation (V) says that the probability of h_i converges to 1.

Now, suppose the accumulating evidence does not drive the probability of h_i to either 0 or 1. Then for some alternative hypothesis h_j, the likelihood ratios $LR[e^n|j/i]$ will neither blow up nor converge to 0. If these likelihood ratios do not go to extremes, then equation (II) implies that the ratio of posterior probabilities of h_j to h_i will remain under the influence of their prior probabilities. Thus, if the prior probabilities of agents α and β diverge radically for h_i and h_j, then so must their posterior probabilities. Indeed, if the likelihood of evidence relative to h_i agrees with the likelihood relative to h_j, then $LR[e^n|j/i] = 1$, and the evidence yields *no change* in the ratio of their posterior probabilities from the ratio of their priors. So, when the evidence fails to take the likelihood ratios to extremes, the initial plausibility assessments will continue to significantly affect the posterior probabilities of hypotheses.

In light of the central role played by likelihood ratios in Bayesian induction the following theorem should not be too surprising. See (Hawthorne, forthcoming) for details and a proof.

Theorem. *Non-Zero Convergence is Convergence to One.*
Let h_i be some hypothesis in H, and suppose that the following conditions are satisfied:
  i) for agent α there is a number r such that, for all n, $P_\alpha[h_i|e^n \cdot c^n \cdot b] \geq r > 0$;
  ii) there is another agent β who's probability function P_β *modestly differs* with P_α on the prior plausibilities for hypotheses in the sense that there is some fraction q between 0 and 1 such that, for every h_j in H other than h_i,
  $P_\beta[h_j|b] \leq q \times P_\alpha[h_j|b]$;
  iii) $\lim_n | P_\alpha[h_i|e^n \cdot c^n \cdot b] - P_\beta[h_i|e^n \cdot c^n \cdot b] | = 0$.
Then, for every h_j in H other than h_i, $\lim_n LR[e^n|j/i] = 0$; and $\lim_n P_\alpha[h_i|e^n \cdot c^n \cdot b] = 1$.

A *modest difference* between two agents α and β (as expressed in condition (ii)) simply means that for all alternatives to a hypothesis h_i in H, agent β's prior probabilities for the alternatives are at least slightly below (e.g. less than 99.99% of) the respective prior probabilities that α assigns them. The only hypothesis to which β assigns a higher prior probability than α is h_i. The theorem implies that if a community of Bayesian agents is diverse enough to contain even one pair of agents who *modestly differ* on a hypothesis h_i (which doesn't acquire an arbitrarily low posterior probability), then these two agents can *converge to agreement* about the posterior probabilities of hypotheses *only if* the whole community comes to agree that the evidence increasing confirms h_i and increasingly refutes its competitors (since, for all alternatives h_j, $\lim_n LR[e^n|j/i] = 0$ *for every agent*). (Alternatively, if h_i acquires an arbitrarily low posterior probability for some agent α, then for some h_j, $LR[e^n|j/i]$ must get arbitrarily large, by equations (V) and (IV); so h_i acquires an arbitrarily low posterior probability for *all* agents.)

It takes just one *modestly differing* pair of agents in the community for the theorem to apply. There are, of course, special classes of probability functions, representing highly restricted communities of Bayesian agents, for which convergence short of 0 and 1 may occur. Suppose, for example, that all agents in a community agree on the prior probabilities for most hypotheses, but disagree on priors of a few. If all of the hypotheses on which they initially disagree become increasingly refuted by the evidence, then everyone in the community will converge on common values other than 0 or 1 for posterior probabilities of the unrefuted hypotheses. However, if the community contains even one agent who's probability function *modestly differs* from another's (on a hypothesis who's posterior remains above some $r > 0$), then convergence to agreement implies that every hypothesis converges to 0 or 1 *for all agents in the community*.

Finally, notice that evidence can only distinguish between hypotheses in H that disagree on the likelihoods of at least some possible events. The influence of the prior probabilities of empirically equivalent hypotheses cannot be washed out unless each is refuted relative to some other empirically distinct hypothesis. If empirically equivalent alternatives to the true hypothesis are among the alternatives in H, then at best the evidence can refute their empirically distinct competitors, and bring the disjunction of the true hypothesis with its empirical equivalents in H to converge to 1. The relative sizes of posterior probabilities of the true hypothesis and its empirical equivalents remains equal to the relative sizes of their priors (by equation (II)).

6. The Likelihood of Obtaining Refuting Evidence

Thus far I have argued that Bayesian convergence ultimately reduces to the convergence of posterior probabilities of theoretical hypotheses to either 0 or 1. Equations (IV) and (V) together show that a hypothesis h_i can become highly refuted only if at least one alternative h_j makes the evidence much more likely than does h_i, so that $LR[e_n|j/i]$ blows up. And h_j may in turn become refuted relative to some other hypothesis. If, however, h_i is to become highly confirmed, it can only do so only by driving $LR[e_n|j/i]$ to 0 for all alternative hypotheses h_j in H. Thus, the crucial question becomes: Is there any reason to think that accumulating evidence will cause empirically distinct alternatives of the true hypothesis to become *increasingly refuted* relative to the true hypothesis via likelihood ratios? I will briefly describe a third kind of Bayesian convergence theorem that establishes that if two hypotheses are empirically distinct, then a sufficiently long sequence of evidence can almost certainly do the job.

L.J. Savage's Bayesian convergence theorem (1972, 46-50) is just such a result. It says that if the accumulating evidence consists of a sequence of independent, identically distributed events (i.e. if the evidence is drawn from repetitions of the same kind of observation or experiment, like repeated tosses of the same coin), then false alternative hypotheses will *almost certainly* become *highly refuted*, and the true hypothesis (or its disjunction with empirically equivalent competitors) will become *highly confirmed*.

Hesse (1975) and Earman (1992) argue convincingly that Savage's theorem presupposes conditions on the evidence that are unrealistic for most real cases of scientific theory testing. In particular, Savage's assumption that the evidence is a sequence of independent, identically distributed events is generally not satisfied. Hesse and Earman also point out that Savage's theorem puts no bounds on the rate at which convergence takes place; for all we know a billion observations would not be enough to bring about any noticeable degree of convergence.

There is a generalized version of Savage's theorem, (in Hawthorne, forthcoming) that avoids the main objections raised by Hesse and Earman. In this version of the theorem the evidence need not consist of identically distributed events, nor is it required to consist of independent events (although the independence of evidential events *relative to a theory* is a perfectly reasonable assumption in scientific contexts). This version also provides bounds on the rate of Bayesian convergence that explicitly depend on a quantitative information-theoretic measure of the *quality of the evidence*. I do not have space to go deeply into the details of the theorem, but I will briefly describe its main features.

Suppose that h_i is a true hypothesis in H (although, of course, the agents are unaware of its truth) and let h_j be one of its empirically distinct competitors. Also suppose that the background claims b are true and that a sequence of initial states or observational conditions c^n holds. Let E^n be the set of all possible outcome sequences that might result from c^n. That is, each member of E^n is a conjunction e^n that describes a possible sequence of outcomes that may result from the conditions described by c^n.

Hypothesis h_i will assign a higher likelihood to some of the outcome sequences described in E^n than does the competitor h_j, and it will assign a lower likelihood to others. This is just what it means for h_i and h_j to be empirically distinct. Consider the following subset of possible outcome sequences in E^n: $S^n(m) = \{e^n \mid e^n \in E^n$ and $LR[e^n|j/i] < 1/2^m\}$. Each outcome sequence in the set $S^n(m)$ is one for which the likelihood according to h_i is a factor of more than 2^m (e.g. 2^{100}) times larger than the likelihood h_j specifies for it.

Let '$V[S^n(m)]$' denote the *disjunction* of all of the possible outcome sequences in $S^n(m)$. '$V[S^n(m)]$' asserts that, for the first n observations c^n, one of the outcome sequences will occur that makes the likelihood ratio $LR[e^n|j/i]$ less than $1/2^m$. If '$V[S^n(m)]$' *is true* for some very large value of m (e.g. m = 100), then, an e^n *does occur* that makes $LR[e^n|j/i]$ extremely small (e.g. less that $1/2^{100}$). Generally this will make h_j highly unlikely on the evidence (relative to h_i).

How likely is it that '$V[S^n(m)]$' is true? The true hypotheses h_i answers this question with the following *direct inference probability*:

$$P[\ V[S^n(m)]\ |\ c^n \cdot h_i \cdot b] \geq 1 - (1/n) \times \underline{VQI}^n[i/j|i] / (\underline{EQI}^n[i/j|i] - (m/n))^2.$$

This probability will converge to 1 as n increases, provided only that h_j differs from h_i (at least slightly) about the likelihoods of some possible outcome sequences. (The terms $\underline{EQI}^n[i/j|i]$ and $\underline{VQI}^n[i/j|i]$ are information-theoretic measures of the extent of disagreement between h_i and h_j about the likelihoods of the various possible outcomes. They are, respectively, a measure of the *mean* and *variance* of the *expected values* of $\log(LR[e^n|i/j])$ in the set E^n. If h_j differs even slightly from h_i regarding the likelihoods of some possible outcome sequences in E^n, for increasing values of n, then the ratios ($\underline{VQI}^n[i/j|i] / \underline{EQI}^n[i/j|i]^2$) will be bounded above. It follows immediately that for any chosen value of m, $P[\ V[S^n(m)]\ |\ c^n \cdot h_i \cdot b]$ goes to 1 as the amount of evidence, n, increases. See (Hawthorne, forthcoming) for details.

Thus, if h_i is true, then for each empirically distinct alternative hypothesis h_j, the likelihood ratio $LR[e^n|j/i]$ will almost surely go to 0. We saw earlier that this kind of convergence brings with it a more general convergence, the convergence of simple inductive inferences to the values of direct inference probabilities. Therefore, if Bayesian agents discover a true hypothesis and test empirically distinct competitors against it in a contest of likelihood ratios, they will almost surely come to agree on the inductive support for theories and on the simple inductive probabilities of individual events.

Notes

[1] I wish to thank Chris Swoyer for numerous valuable comments and suggestions.

References

Earman, J. (1992), *Bayes or Bust?: A Critical Examination of Bayesian Confirmation Theory*. Cambridge: MIT Press.

Hawthorne, J. (forthcoming), "Bayesian Induction Is Eliminative Induction", *Philosophical Topics*, v. 21, no. 1.

Hesse, M. (1975), "Bayesian Methods and the Initial Probability of Theories", in G. Maxwell and R. Anderson, (eds.), *Induction, Probability, and Confirmation*, Minnesota Studies in the Philosophy of Science, vol. 6. Minneapolis: U. of Minnesota Press, pp. 50-105.

Savage, L. (1972), *The Foundations of Statistics*. New York: Dover.

The Extent of Dilation of Sets of Probabilities and the Asymptotics of Robust Bayesian Inference[1]

Timothy Herron, Teddy Seidenfeld, and Larry Wasserman

Carnegie-Mellon University

1. Overview

We discuss two general issues concerning diverging sets of Bayesian (conditional) probabilities—divergence of "posteriors"— that can result with increasing evidence. Consider a set $\mathcal{P}$ of probabilities typically, but not always, based on a set of Bayesian "priors." Incorporating sets of probabilities, rather than relying on a single probability, is a useful way to provide a rigorous mathematical framework for studying sensitivity and robustness in Classical and Bayesian inference. See: Berger (1984, 1985, 1990); Lavine (1991); Huber and Strassen (1973); Walley (1991); and Wasserman and Kadane (1990). Also, sets of probabilities arise in group decision problems. See: Levi (1982); and Seidenfeld, Kadane, and Schervish (1989). Third, sets of probabilities are one consequence of weakening traditional axioms for uncertainty. See: Good (1952); Smith (1961); Kyburg (1961); Levi (1974); Fishburn (1986); Seidenfeld, Schervish, and Kadane (1990); and Walley (1991).

Fix E, an event of interest, and X, a random variable to be observed. With respect to a set $\mathcal{P}$, when the set of conditional probabilities for E, given X, strictly contains the set of unconditional probabilities for E, for each possible outcome $X = x$, call this phenomenon *dilation* of the set of probabilities (Seidenfeld and Wasserman 1993). Thus, *dilation* contrasts with the asymptotic merging of posterior probabilities reported by Savage (1954) and by Blackwell and Dubins (1962), which we discuss briefly in section §2.

For a wide variety of models for Robust Bayesian inference the extent to which X dilates E is related to a model specific index of how far key elements of $\mathcal{P}$ are from a distribution that makes X and E independent. Some sets $\mathcal{P}$ use a class of priors generated by a "neighborhood" of a focal distribution P. These include: the ε-contamination class of priors, the Total Variation class of priors, and symmetric neighborhoods of a prior. The extent to which X dilates E in these sets is related to a model specific index of how far P is from a distribution that makes X and E independent. In other sets $\mathcal{P}$, e.g., in the Frechet class, and models given by lower and upper probabilities for atoms, the extent of dilation may be indexed by departures from independence of the probabilities that are extreme points of (the convex closure of) $\mathcal{P}$ rather than by referring to some obvious focal member of $\mathcal{P}$. In section §3, we discuss this connection between independence and indices for the extent of dilation.

In section §4, we consider phenomena related to asymptotic dilation. At a fixed confidence level, $(1-\alpha)$, Classical interval estimates (based on an m.l.e., $\hat{\theta}$), $A_n = [\hat{\theta} - a_n, \hat{\theta} + a_n]$, have length $O(n^{-1/2})$ (for a sample of size n). Of course, the confidence level correctly reports the (prior) probability that $\theta \in A_n$ for each $P \in \mathcal{P}$, $P(A_n) = 1-\alpha$, independent of the prior for θ. However, as shown by Pericchi and Walley (1991), if an ε-contamination class is used for the prior on the parameter, there is asymptotic (posterior) dilation for the A_n, given the data $(x_1, ..., x_n)$. That is, the asymptotic *lower* posterior probability P_* $(= \inf_{P \in \mathcal{P}})$ for A_n is 0,

$$\lim_{n \to \infty} P_*(A_n \mid x_1, ..., x_n) = 0. \qquad \text{(a.s.)}$$

By contrast, if the intervals A'_n are chosen with length $O(\sqrt{\log(n)/n})$, then there is no asymptotic dilation. This is explained by using H.Jeffreys' (1967, §5) theory of Bayesian hypothesis testing. In section §4, we discuss how the class of priors and the asymptotic rate of dilation for Bayesian (posterior) and Classical interval estimates are related. First, however, we summarize two familiar results about the merging of conditional probabilities since these are in sharp contrast with the effects of dilation.

2. The merging of conditional probabilities with increasing shared evidence

As a backdrop to the discussion of dilation, we begin by pointing to two well known results about the asymptotic merging of Bayesian posterior probabilities.

2.1 Savage (1954, §3.6) provides an (almost everywhere) approach simultaneously to consensus and to certainty among a few Bayesian investigators, provided:

(1) they investigate finitely many statistical hypotheses $\Theta = \{\theta_1, ..., \theta_k\}$
(2) they use Bayes' rule to update probabilities about Θ given a growing sequence shared data $\{x_1, ...\}$. These data are identically, independently distributed (i.i.d.) given θ (where the Bayesians agree on the statistical model parametrized by Θ).
(3) they have prior agreement about *null* events. Specifically (given condition 2), there is agreement about which parameter values have positive prior probability.

By a simple application of the strong law of large numbers, Savage concludes that, almost surely, the agents' posterior probabilities will converge to 1 for the true value of Θ. Asymptotically, with probability 1, they achieve consensus and certainty about Θ.

2.2 Blackwell and Dubins (1962) give an impressive generalization about consensus without using either "i" of Savage's i.i.d. condition (2). Theirs is a standard martingale convergence result which we summarize next.

Consider a denumerable sequence of sets X_i ($i = 1,...$) with associated σ-fields $\mathcal{B}_i$. Form the infinite Cartesian product $X = X_1 \otimes ...$ of sequences $(x_1, x_2, ...) = x \in X$, where $x_i \in X_i$. That is, each x_i is an atom of its algebra $\mathcal{B}_i$. Let the measurable sets in X (the events) be the elements of of the σ-algebra $\mathcal{B}$ generated by the set of measurable rectangles. Define the spaces of histories $(H_n, \mathcal{H}_n)$ and futures $(F_n, \mathcal{F}_n)$ where $H_n = X_1 \otimes ... \otimes X_n$, $\mathcal{H}_n = \mathcal{B}_1 \otimes ... \otimes \mathcal{B}_n$, and where $F_n = X_{n+1} \otimes ...$ and $\mathcal{F}_n = \mathcal{B}_{n+1} \otimes ...$.

Blackwell and Dubins' argument requires that P is a *predictive*, σ-additive probability on the measure space (X, $\mathcal{B}$). (That P is *predictive* means that there exist conditional probability *distributions* of events given past events, $P^n(\bullet \mid \mathcal{H}_n)$.) Consider a probability Q which is in agreement with P about events of measure 0 in $\mathcal{B}$: $\forall E \in \mathcal{B}$, P(E) = 0 iff Q(E) = 0. That is, P and Q are mutually absolutely continuous [**m.a.c.**]. Then Q, too, is σ-additive and predictive if P is, with conditional distributions $Q^n(\mathcal{F}_n \mid \mathcal{H}_n)$.

Blackwell & Dubins (1962) prove there is almost certain asymptotic consensus between the conditional probabilities P^n and Q^n.

Theorem 1. For each P^n there is (a version of) Q^n so that, almost surely, the distance between them vanishes with increasing histories: $\lim_{n \to \infty} \rho(P^n, Q^n) = 0$ [a.e. P or Q], where ρ is the uniform distance (total variation) metric between distributions. (That is, with μ and ν defined on the same measure space (M, $\mathcal{M}$), ρ(μ,ν) is the least upper bound over events $E \in \mathcal{M}$ of $\mid \mu(E) - \nu(E) \mid$.)

Thus, the powerful assumption that P and Q are mutually absolutely continuous (Savage's condition 3) is what drives the merging of the two families of conditional probabilities P^n and Q^n.

3. Dilation and short run divergence of posterior probabilities.

Throughout this section, let $\mathcal{P}$ be a (convex) set of probabilities on a (finite) algebra $\mathcal{A}$. For a useful contrast with Savage-styled, or Blackwell-Dubins-styled asymptotic consensus, the following discussion focuses on the short run dynamics of upper and lower conditional probabilities in Robust Bayesian models.

For an event E, denote by $P_*(E)$ the "lower" probability of E: $\inf_\mathcal{P} \{P(E)\}$ and denote by $P^*(E)$ the "upper" probability of E: $\sup_\mathcal{P} \{P(E)\}$. Let $(b_1, ..., b_n)$ be a (finite) partition generated by an observable B.

Definition. The set of conditional probabilities $\{P(E \mid b_i)\}$ **dilate** if
$$P_*(E \mid b_i) < P_*(E) \leq P^*(E) < P^*(E \mid b_i) \quad (i = 1, ..., n).$$

That is, dilation occurs provided that, for each event b_i in a partition B, the conditional probabilities for an event E, given b_i, properly include the unconditional probabilities for E.

Here is an illustration of dilation.

Heuristic Example. Suppose A is a highly "uncertain" event with respect to the set $\mathcal{P}$. That is, $P^*(A) - P_*(A) \approx 1$. Let {H,T} indicate the flip of a fair coin whose outcomes are independent of A. That is, P(A,H) = P(A)/2 for each $P \in \mathcal{P}$. Define event E by, E = {(A,H), (A^c,T)}.

It follows, simply, that P(E) = .5 for each $P \in \mathcal{P}$. (E is *pivotal* for A.) But then,
$$0 \approx P_*(E \mid H) < P_*(E) = P^*(E) < P^*(E \mid H) \approx 1$$
and
$$0 \approx P_*(E \mid T) < P_*(E) = P^*(E) < P^*(E \mid T) \approx 1.$$

Thus, regardless of how the coin lands, conditional probability for event E dilates to a large interval, from a determinate unconditional probability of .5. Also, this example mimics Ellsberg's (1961) "paradox," where the mixture of two uncertain events has a determinate probability.

3.1 Dilation and Independence.

The next two theorems on existence of dilation serve to motivate using indices of departures from independence to gauge the extent of dilation. They appear in (Seidenfeld and Wasserman 1993).

Independence is sufficient for dilation.

Let Q be a convex set of probabilities on algebra $\mathcal{A}$ and suppose we have access to a "fair" coin which may be flipped repeatedly: algebra C. Assume the coin flips are independent and, with respect to Q, also independent of events in $\mathcal{A}$. Let P be the resulting convex set of probabilities on $\mathcal{A} \times C$. (This condition is similar to, e.g., DeGroot's assumption of an extraneous continuous random variable, and is similar to the "fineness" assumptions in the theories of Savage, Ramsey, Jeffrey, etc.)

Theorem 2: If Q is not a singleton, there is a 2×2 table of the form $(E, E^c) \times (H, T)$ where both:

$$P_*(E \mid H) < P_*(E) = .5 = P^*(E) < P^*(E \mid H)$$
$$P_*(E \mid T) < P_*(E) = .5 = P^*(E) < P^*(E \mid T).$$

That is, then dilation occurs.

Independence is necessary for dilation.

Let P be a convex set of probabilities on algebra $\mathcal{A}$. The next result is formulated for subalgebras of 4 atoms: (p_1, p_2, p_3, p_4)

The case of 2×2 tables.

	b_1	b_2
A_1	p_1	p_2
A_2	p_3	p_4

Define the quantity $S_P(A_1, b_1) = P(A_1, b_1) / P(A_1)P(b_1) = p_1/(p_1+p_2)(p_1+p_3)$, and we stipulate that $S_P(A_1, b_1) = 1$ if $P(A_1)P(b_1) = 0$. Thus, $S_P(A_1, b_1) = 1$ iff A and B are independent under P and "S_P" is an index of dependence between events.

Lemma 1: If P displays dilation in this sub-algebra, then

$$\inf\nolimits_P \{S_P(A_1, b_1)\} < 1 < \sup\nolimits_P \{S_P(A_1, b_1)\}.$$

Theorem 3: If P displays dilation in this sub-algebra, then there exists $P^\# \in P$ such

$S_P\#(A_1,b_1) = 1$.

Thus, independence is also necessary for dilation.

3.2 The extent of dilation

We begin by reviewing some results that obtain for the ε-contaminated model (Seidenfeld and Wasserman 1993). Given probability P and $1 > \varepsilon > 0$, define the convex set $\mathcal{P}_\varepsilon(P) = \{(1-\varepsilon)P + \varepsilon Q$: Q an arbitrary probability}. This model is popular in studies of Bayesian Robustness. (See Huber 1973 and 1981; Berger 1984.)

Lemma 2. In the ε-contaminated model, dilation occurs in algebra $\mathcal{A}$ iff it occurs in some 2x2 subalgebra of $\mathcal{A}$.

So without loss of generality, the next result is formulated for 2x2 tables using the notation of Lemma 1.

Theorem 4: $\mathcal{P}_\varepsilon(P)$ experiences dilation if and only if
case 1: $S_P(A_1,b_1) > 1$
$\varepsilon > [S_P(A_1,b_1) - 1] \bullet \max\{ P(A_1)/P(A_2) ; P(b_1)/P(b_2) \}$
case 2: $S_P(A_1,b_1) < 1$
$\varepsilon > [1 - S_P(A_1,b_1)] \bullet \max\{ 1 ; P(A_1)P(b_1)/P(A_2)P(b_2) \}$
and case 3: $S_P(A_1,b_1) = 1$
P is internal to the simplex of all distributions.

Thus, dilation occurs in the ε-contaminated model if and only if the focal distribution, P, is close enough (in the tetrahedron of distributions on four atoms) to the saddle-shaped surface of distributions which make A and B independent. Here, S_P provides one relevant index of the proximity of the focal distribution P to the surface of independence.

Definition: For $B \subset \mathcal{B}$ (*B* not necessarily a binary outcome) define the *extent of dilation*
by $\Delta(A, B) = \min_{b \in B} [P^*(A|b) - P^*(A) + (P_*(A) - P_*(A|b))]$.

For the ε-contamination model we have

Theorem 5: $\Delta(A, B) = \min_{b \in B} [\varepsilon(1-\varepsilon)P(b^c) \div \varepsilon+(1-\varepsilon)P(b)]$

In this model, $\Delta(A, B)$ does not depend upon the event A. Moreover, the extent of dilation is maximized when $\varepsilon = \sqrt{P(b_\Delta)} \div (1+\sqrt{P(b_\Delta)})$, where $b_\Delta \in B$ achieves the minimum for $\Delta(A, B)$.

Similar findings obtain for *total variation neighborhoods*. Given a probability P and $1 > \varepsilon > 0$, define the convex set $\mathcal{U}_\varepsilon(P) = \{Q: \rho(P, Q) \le \varepsilon\}$. Thus $\mathcal{U}_\varepsilon(P)$ is the uniform distance (total variation) neighborhood of P, corresponding to the metric of Blackwell-Dubins' consensus. As before, consider dilation in 2x2 tables. Define a second index of association: $d_P(A,B) = P(AB) - P(A)P(B)$

(Informal version of) **Theorem 6:** $\mathcal{U}_\varepsilon(P)$ experiences dilation if and only if P is sufficiently close to the surface of independence, as indexed by d_P.

The extent of dilation for the total variation model also may be expressed in terms of the d_P- index, though there are annoying cases depending upon whether the set $\mathcal{U}_\varepsilon(P)$ is truncated by the simplex of all distributions.

Whereas, in the previous two models, each of the sets $\mathcal{P}_\varepsilon(P)$ and $\mathcal{U}_\varepsilon(P)$ has a single distribution that serves as its natural focal point, some sets of probabilities are created through constraints on extreme points directly. For example, consider a model where $\mathcal{P}$ is defined by the lower and upper probabilities on the *atoms* of the algebra $\mathcal{A}$. In section §2 of "Divisive Conditioning" (Herron, Seidenfeld, and Wasserman 1993 - hereafter referred to as DC), these sets are called **ALUP** models. For convenience, take the algebra to be finite with atoms $a_{i,j}$ ($i = 1,2$; $j = 1, ..., n$) and where $A = \cup_j a_{1,j}$ and $b_j = \{a_{1,j}, a_{2,j}\}$. For each atom $a_{i,j}$, denote the lower and upper probability bounds achieved within the (closed set) $\mathcal{P}$ by $\beta_{i,j}$ and $\gamma_{i,j}$, respectively. Likewise, for an event E let β_E and γ_E, denote the values $P_*(E)$ and $P^*(E)$. We discuss dilation of the event A given the outcome of random quantity $B = \{b_1, ..., b_n\}$.

Dilation conditions for an ALUP model are easy to express in terms of extreme values within $\mathcal{P}$.

Theorem 7. (i) $P^*(A) < P^*(A|b_j)$ *iff* $\gamma_{1,j} \beta_A c - \gamma_A \beta_{2,j} > 0$

and (ii) $P_*(A) > P_*(A|b_j)$ *iff* $\gamma_{2,j} \beta_A - \gamma_A c \beta_{1,j} > 0$.

Next, given events E and F and a probability P, define the (covariance-) index

$$\delta_P(E,F) = P(EF)P(E^c F^c) - P(E^c F)P(EF^c).$$

Within ALUP models, the extent of dilation for A given $B = b_j$ is provided by the $\delta_P(A,b_j)$ (covariance-) index. Given an event E, use the notation $\{P_*(E)\}$ and $\{P^*(E)\}$ for denoting, respectively, the set of probabilities within $\mathcal{P}$ that achieve the lower and upper probability bounds for event E. Specifically: let $P_{1,j}$ be a probability such that $P_{1,j} \in \{P^*(A)\} \cap \{P^*(a_{1,j})\} \cap \{P_*(a_{2,j})\}$, and let $P_{2,j}$ be a probability such that, $P_{2,j} \in \{P_*(A)\} \cap \{P_*(a_{1,j})\} \cap \{P^*(a_{2,j})\}$. (Existence of $P_{1,j}$ and $P_{2,j}$ are demonstrated in §2 of **DC**.) Then a simple calculation shows:

Theorem 8. $\Delta(A, B) = \min_j [\delta_{P1,j}(A,b_j)P_{2,j}(b_j) - \delta_{P2,j}(A,b_j)P_{1,j}(b_j)]$.

Thus, as with the ε-contamination and total variation models, the extent of dilation in ALUP models also is a function of an index of probabilistic independence between the events in question.

Observe that the ε-contamination models are a special case of the ALUP models: they correspond to ALUP models obtained by specifying the lower probabilities for the atoms and letting the upper probabilities be as large as possible consistent with these constraints on lower probabilities. Then the extent of dilation for a set $\mathcal{P}_\varepsilon(P)$ of probabilities may be reported either by attending to the S_P-index for the focal distribution of the set (as in Theorem 5), or by attending to the δ_P-index for the extreme points of the set (as in Theorem 8).

4. Asymptotic Dilation for Classical and Bayesian interval estimates

In an interesting essay, L.Pericchi and P.Walley (1991, pp. 14-16), calculate the upper and lower probabilities of familiar Normal confidence interval estimates under an ε-contaminated model for the "prior" of the unknown Normal mean. Specifically, they consider data $x = (x_1,..., x_n)$ which are i.i.d. $N(\theta,\sigma^2)$ for an unknown mean θ and known variance σ^2. The "prior" class $\mathcal{P}_\varepsilon(P_0)$ is an ε-contaminated set $\{(1-\varepsilon)P_0 + \varepsilon Q\}$, where P_0 is a conjugate Normal distribution, $N(\mu,v^2)$, and Q is arbitrary. Note that pairs of elements of $\mathcal{P}_\varepsilon(P_0)$ are *not* all mutually absolutely continuous since Q ranges over one-point distributions that concentrate mass at different values of θ. Hence, Theorem 1 does not apply.

For $\varepsilon = 0$, $\mathcal{P}_\varepsilon(P_0)$ is the singleton Bayes' (conjugate) prior, P_0. Then the Bayes' posterior for θ, $P_0(\theta|x)$, is a Normal $N(\mu',\tau^2)$; where $\tau^2 = (v^{-2} + n\sigma^{-2})^{-1}$, $\mu' = \tau^2[(\mu/v^2) + (n\hat{x}/\sigma^2)]$, and where $\hat{x}$ is the sample average (of x). The standard 95% confidence interval for θ is $A_n = [\hat{x} \pm 1.96\sigma/n^{.5}]$. Under the Bayes' prior P_0 (for $\varepsilon = 0$), the Bayes' posterior of A_n, $P_0(A_n | x)$, depends upon the data, x. When n is large enough that τ^2 is approximately equal to σ^2/n, i.e., when $\sigma/\sqrt{n}^{.5}$ is sufficiently small, then $P_0(A_n | x)$ is close to .95. Otherwise, $P_0(A_n | x)$ may fall to very low values. Thus, asymptotically, the Bayes' posterior for A_n approximates the usual confidence level. However, under the ε-contaminated model $\mathcal{P}_\varepsilon(P_0)$ (for $\varepsilon > 0$), Pericchi and Walley show that, with increasing sample size n, $P^n_*(A_n) \to 0$ while $P^{n*}(A_n) \to 1$. That is, in terms of dilation, the sequence of standard confidence intervals estimates (each at the same fixed confidence level) dilate their unconditional probability or coverage level.

What sequence of confidence levels avoids dilation? That is, if it is required that $P^n_*(A'_n) \geq .95$, how should the intervals, A'_n, grow as a function of n? Pericchi and Walley (1991, 16) report that the sequence of intervals $A'_n = [\hat{x} \pm \zeta_n\sigma/n^{.5}]$ has a posterior probability which is bounded below, e.g., $P^n_*(A'_n) \geq .95$, provided that ζ_n increases at the rate $(\log n)^{.5}$. They call intervals whose lower posterior probability is bounded above some constant, "credible" intervals.

A connection exists between this the rate of growth for ζ_n that makes A'_n credible, due to Walley and Pericchi, and an old but important result due to Sir Harold Jeffreys (1967, 248). The connection to Jeffreys' theory offers another interpretation for the lower posterior probabilities $P^n_*(A'_n)$ arising from the ε-contaminated class.

Adapt Jeffreys' Bayesian hypothesis testing, as follows. Consider a (simple) "null" hypothesis, H_0: $\theta = \theta_0$, against the (composite) alternative H_0^c: $\theta \neq \theta_0$. Let the prior ratio $P(H_0) / P(H_0^c)$ be specified as γ:$(1-\gamma)$. (Jeffreys uses $\gamma = .5$.) Given H_0, the x_i are i.i.d. $N(\theta_0,\sigma^2)$. Given H_0^c, let the parameter θ be distributed as $N(\mu,v^2)$. Then, when the data make $|\hat{x}-\theta_0|$ large relative to $\sigma/n^{.5}$ the posterior ratio $P(H_0|x) / P(H_0^c|x)$ is smaller than the prior ratio, and when $|\hat{x}-\theta_0|$ is small relative to $\sigma/n^{.5}$ the posterior odds favor the null hypothesis. But to maintain a constant posterior odds ratio with increasing sample size *rather than being constant —as a fixed significance level would entail—* the quantity $|\hat{x}-\theta_0| / (\sigma/n^{.5})$ has *to grow* at the rate $(\log n)^{.5}$ though, of course, the difference $|\hat{x}-\theta_0|$ approaches 0.

In other words, Jeffreys' analysis reveals that, from a Bayesian point of view, the posterior odds for the usual two-sided hypothesis test of H_0 versus the alternative H_0^c depends upon *both* the observed type$_1$ error (or significance level), α and the sample size, n. At a fixed significance level, e.g. at observed significance $\alpha = .05$, larger samples yield ever higher (in fact, unbounded) posterior odds in favor of H_0. To keep posterior odds constant as sample size grows, the observed significance level must decrease towards 0.

It is well known that Classical confidence intervals can be obtained by inverting a family of hypothesis tests, generated by varying the "null" hypothesis. That is, the interval $A_n = [\hat{x} \pm 1.96\sigma/n^{.5}]$, with confidence 95%, corresponds to the family of unrejected null hypotheses: each value θ belonging to the interval is a null hypothesis that is not rejected on a standard two-sided test at significance level $\alpha = .05$.

rejected null hypotheses: each value θ belonging to the interval is a null hypothesis that is not rejected on a standard two-sided test at significance level α = .05. Consider a family of Jeffreys' hypothesis tests obtained by varying the "null" through the parameter space and, corresponding to each null hypothesis, varying the prior probability which puts mass γ on H_0. Say that a value of θ, θ = $θ_0$, is rejected when its posterior probability falls below a threshold, e.g., when P(H_0|x) < .05 for the Jeffreys' prior P(θ = $θ_0$) = γ. The class of probabilities obtained by varying the null hypothesis forms an ε-contaminated model: {(1-γ)P(θ|H_0^c) + γQ}, with extreme points (for Q) corresponding to all the one-point "null" hypotheses.

Define the interval B_n of null hypotheses, with sample size n, where each survives rejection under Jeffreys' tests. The B_n are the intervals $A'_n = [\bar{x} \pm \zeta_n \sigma/n^{.5}]$ of Pericchi and Walley's analysis, reported above. What Pericchi and Walley observe, expressed in terms of the required rate of growth of ζ_n for credible intervals (intervals that have a fixed lower posterior probability with respect to the class $\mathcal{P}_\varepsilon(P_0)$) is exactly the result Jeffreys reports about the shrinking α-levels in hypothesis tests in order that posterior probabilities for the "null" be constant, regardless of sample size. In short, credible intervals from the ε-contaminated model $\mathcal{P}_\varepsilon(P_0)$ are the result of inverting on a family of Jeffreys' hypothesis tests that use a fixed lower bound on posterior odds to form the rejection region of the test.

We conclude our discussion of asymptotic dilation by relating the length of an interval estimate of a parameter θ to the shape of a (symmetic) class of priors for θ. Consider interval estimation of a normal mean, 0 < θ < 1. (This restriction to θ = (0,1) is for mathematical convenience.) We use a prior (symmetric) family S_α of rearrangements of the density $p_\alpha(θ) = (1-α)θ^{-α}$, for 0 < α < 1. Let the interval estimate of θ be $A_n = [\hat{θ} - a_n, \hat{θ} + a_n]$. For constants C > 0 and d, write $a_n = \{n^{-1}(C+d\log n)\}^{1/2}$.

Theorem 9. For the S_α model, there is asymptotic dilation of A_n if and only if d < α.

(The proof is given in §7 of **DC**.)

5. Summary

In contrast with Savage's, and Blackwell and Dubins' well known results about the merging of Bayesian posterior probabilities given sufficient shared evidence, in this paper we reported two aspects of the contrary case, which we call dilation of sets of probabilities. Let $\mathcal{P}$ be a set of probabilities. The quantity X **dilates** the probabilities for an event E provided the set of conditional probabilities for E, given X = x, properly contains the set of unconditional probabilities for E, for each possible outcome of X. Thus, when X dilates E and probabilities are updated by Bayes' rule, the revised opinions about E given X diverge for certain.

In section §3 we indicated how, for several classes of probabilities used in Robust Bayesian inference, the extent of dilation may be gauged by an index of how far away key elements of $\mathcal{P}$ are from distributions that make X and E independent. In section §4 we discussed how ordinary Classical confidence intervals (at a fixed confidence level) experience asymptotic dilation as a function of sample size. Relative to the choice of the class of priors, we explain how to adjust the length of the intervals to avoid asymptotic dilation.

This inquiry, we think, points out two new ways in which Classical and Robust Bayesian statistical inference may be related to each other. We hope to continue our

Note

[1] Timothy Herron and Teddy Seidenfeld were supported by NSF grant SES-9208942. Larry Wasserman was supported by NSF Grant DMS-90005858, and NIH grant R01-CA54852-01.

References

Berger, J.O. (1984), "The Robust Bayesian Viewpoint", with discussion, in J.B. Kadane, (ed.), *Robustness of Bayesian Analysis* Amsterdam: North-Holland, 63-144.

_____. (1985), *Statistical Decision Theory*, (2nd edition). N.Y.: Springer-Verlag.

_____. (1990), "Robust Bayesian analysis: sensitivity to the prior", *J. Stat. Planning & Inference* 25: 303-328.

Blackwell, D. & Dubins, L. (1962), "Merging of opinions with increasing information", *Ann. Math. Stat.* 33: 882-887.

DeRobertis, L. and Hartigan, J. A. (1981), "Bayesian inference using intervals of measures", *Annals of Statistics* 9: 235-244.

Ellsberg, D. (1961), "Risk, Ambiguity, and the Savage Axioms", *Quart. J. Econ.* 75: 643-669.

Fishburn, P.C. (1986), "The Axioms of Subjective Probability", *Statistical Science* 1: 335-358.

Good, I.J. (1952), "Rational Decisions", *J. Royal Stat. Soc.* B, 14: 107-114.

Herron, T., Seidenfeld, T., and Wasserman, L. (1993), "Divisive Conditioning", Tech.Report #585, Dept. of Statistics, Carnegie Mellon University. Pgh., PA 15213.

Huber, P.J. (1973), "The use of Choquet capacities in statistics", *Bull. Inst. Int. Stat.* 45:181-191.

Huber, P.J. (1981), *Robust Statistics,* New York: Wiley.

Huber, P.J. and Strassen, V. (1973), "Minimax tests and the Neyman-Pearson lemma for capacities", *Annals of Statistics* 1: 241-263.

Jeffreys, H. (1967), *Theory of Probability* (3rd ed.), Oxford: Oxford University Press.

Kyburg, H.E. (1961), *Probability and Logic of Rational Belief*, Middleton, CT: Wesleyan Univ. Press.

Lavine, M. (1991), "Sensitivity in Bayesian statistics: the prior and the likelihood", *Journal of American Statistics Association* 86: 396-399.

Levi, I. (1974), "On indeterminate probabilities", *Journal of Philosophy* 71: 391-418.

____ . (1982), "Conflict and Social Agency", *Journal of Philosophy* 79: 231-247.

Pericchi, L.R. and Walley, P. (1991), "Robust Bayesian Credible Intervals and Prior Ignorance", *Int. Stat. Review* 58: 1-23.

Savage, L.J. (1954), *The Foundations of Statistics*, New York: Wiley.

Seidenfeld, T., Kadane, J.B., and Schervish, M.J. (1989), "On the shared preferences of two Bayesian decision makers", *Journal of Philosophy* 86: 225-244.

Seidenfeld, T., Schervish, M.J., and Kadane, J.B. (1990), "Decisions without Ordering", in W.Sieg (ed.), *Acting and Reflecting*, Dordrecht: Kluwer Academic, pp. 143-170.

Seidenfeld, T. and Wasserman, L. (1993), "Dilation for sets of probabilities", *Annals of Statistics*. 21: 1139-1154.

Smith, C.A.B. (1961), "Consistency in statistical inference and decisions", *J. Royal Stat. Soc. B.* 23: 1-25.

Walley, P. (1991), *Statistical Reasoning with Imprecise Probabilities*, London: Chapman Hall.

Wasserman, L. and Kadane, J.B. (1990), "Bayes' theorem for Choquet capacities", *Annals of Statistics* 18: 1328-1339.

In Search of a Pointless Decision Principle[1]

Prasanta S. Bandyopadhayay

University of Rochester

1. Introduction

"Maximizing expected utility (MEU)" is one assumption of (strict) Bayesian decision theory [Savage 1972]. According to the principle of MEU, in a given decision situation, the decision maker should choose one of the alternatives with maximal expected utility [For an excellent discussion of decision theory, see Jeffrey 1990]. However, MEU as the foundation of Bayesian decision theory has been under attack. One counterexample that seems to dispute MEU as a rationality principle is offered by Daniel Ellsberg[Ellsberg 1961]. While discussing what has been dubbed as the Ellsberg paradox, I will consider briefly three decision principles, each of which is different from the MEU principle. These three principles are, (i) Henry Kyburg's Principle, (ii) Peter Gardenfors and Nihls-Eric Sahlin's Maximin criterion (MMEU) [Sahlin 1985, Gardenfors and Sahlin 1988, Sahlin 1993] and finally, (iii) the "weak-dominance principle(WDP)" [Despite Isaac Levi's warning against the term "weak-dominance" of a possible confusion with the similar kind of dominance notion, I stick to this term because of the lack of a better substitute, in private communication] that draws its inspiration from Kyburg's theory which I will endorse [Kyburg 1983a, Kyburg 1983b, Kyburg 1990, Kyburg 1994]. The purpose of this paper is to answer four things: (a) Why are the strict Bayesians who espouse MEU as the standard of rationality really threatened by a paradoxical conclusion? (b) How can an improvement on strict Bayesianism have a better handle on issues like the Ellsberg problem and some other decision problems? (c) How does my decision principle address the Ellsberg problem and some others? (d) What are the ordering properties of the Principle that I advocate and what are its implications?

2. The Ellsberg paradox

Suppose an urn contains 30 red balls and 60 black or yellow balls. The proportion of black balls to the yellow balls in the urn (represented as b) is unknown. One ball is to be drawn from the urn. There are two decision situations, A and B. Each situation has two alternatives. The two alternatives for A are a_1 and a_2. If I choose a_1, I will get $100 if a red ball is drawn and I will receive $0 otherwise. If I choose a_2, then I will get $100 if a black ball is drawn, otherwise I won't receive anything. Under the above circumstances,

I am told that B is the second situation. B, like A, has two alternatives, a_3 and a_4. If I choose a_3, I will get $100 if a red or yellow ball is drawn, otherwise I will receive $0. If I choose a_4, I will get $100 if a black or yellow ball is drawn, otherwise I will receive $0. The two decision situations above can be represented as follows:

States

Acts A	s_1 1/3 Red	s_2 Black ——— 2/3 ———	s_3 Yellow
a_1	$100	$0	$0
a_2	$0	$100	$0

States

Acts B	1/3 Red	Black ——— 2/3 ———	Yellow
a_3	$100	$0	$100
a_4	$0	$100	$100

MEU does not tell us which course of action one should take in either situation. The expected value of a_1 is the closed interval [100/3, 100/3] and the expected value of a_2 is the closed interval [0, 200/3] provided we represent the uncertainty about the proportion of black balls in the urn which could be anything from 0 to 2/3. In decision situation A, most of the people prefer a_1 to a_2. Likewise, in decision situation B, we cannot say anything about our preference regarding a_3 and a_4 if we stick to MEU. If we represent the uncertainty involved in the proportion of black balls which could be any number from 0 to 2/3, then in B, we get the following intervals. In B, the expected utility calculation yields the closed interval [100/3, 300/3] for choice a_3. We get the closed interval [200/3, 200/3] for a_4 from the expected utility calculation. Again, MEU does not help us. In decision situation B most of the people prefer a_4 to a_3. These two preferences, namely, a_1 over a_2 and a_4 over a_3, are not only intuitive, but also agree, as we have cited, with most people's choices. Why is this decision problem called a paradox?

This decision situation is called a paradox because according to the Bayesian decision theory, one should choose a_4 if and only if one chooses a_2. For any given value of b that represents the uncertainty of the proportion of black balls in the urn, the expected utility of a_2 exceeds that of a_1 if and only if the expected utility of a_4 exceeds that of a_3. This equivalence holds for a Bayesian because he adheres to another Bayesian principle that Leonard Savage calls the "sure thing principle". According to the sure thing principle, the choice between two alternatives must be unaffected by the value of outcomes corresponding to states for which both alternatives have the same payoff. In the matrix above, the outcome a_1 and a_2 are the same for S_3 (i.e., yellow), and the outcome a_3 and a_4 are also the same for S_3. Also, a_1 has the same outcome as a_3, except for S_3 (i.e., yellow); whereas a_2 has the same outcome as a_4 except for S_3. In a situation like this, the sure thing principle requires that a_1 is preferred to a_2 if and only if a_3 is likewise preferred to a_4. Thus, if we take recourse to the standard point valued approach to the MEU principle, then for the Bayesian the rational agent should prefer a_1 to a_2 if and only if the agent should prefer a_3 to a_4, or equivalently

(i) $100 b > 100/3$ if and only if $(1-b) 100 > 200/3$ for all values of b.

Most people however prefer a_1 to a_2 and a_4 to a_3, or equivalently:

(ii) $100/3 > 100 b$ if and only if $200/3 > (1-b)100$
for all values of b in the interval [0, 2/3].

There is *no* b that makes (ii) true (except 1/3). On the other hand, if b= 1/3, then the instance of (ii), i.e., $100 > 100$ if and only if $200 > 200$, is true because both the sides of the equivalence are false. Under this special case when b=1/3, the expected utilities of all of the acts are equal. Then the decision maker should be indifferent between a_1 and a_2 and between a_3 and a_4. However, when b=1/3 one cannot both strictly prefer a_1 to a_2 and a_3 to a_4. Thus, the sure thing principle is violated. In what follows, I will discuss three decision principles to see how they address the Ellsberg problem and some others.

3. Kyburg's decision principle

Kyburg provides a decision theory based on probability intervals. According to Kyburg, probability represents a necessary relation between a set of sentences regarded as accepted and a given sentence. For him, this relation does not determine a real number representing the probability of the sentence S as being true relative to knowledge corpus K. Kyburg characterizes the probability of a sentence S relative to a set of sentences K, as the interval [p, q] if and only if the following conditions are met:

(I) S is known in K to be equivalent to a sentence of the form "*a* is an element of set *b*".

(II) "*a* is an element of *c*" is a sentence in K.

(III) The proportion of *c*'s that are *b*'s is known in K in the interval [p, q].

(IV) Relative to K, *a* is a random member of *c* with respect to *b*.

Kyburg provides his central decision rule which he calls the *Principle III*.

The decision maker ought to reject any choice a_i for which there is an act a_j whose minimum expected utility exceeds the maximum expected utility of a_i.

Consider Kyburg's decision rule in the case of the Ellsberg paradox. In A, for example the maximum expected utility of a_1 is 100/3 which is not less than the minimum utility of a_2, which is 0. On the other hand, for B, the maximum expected utility of a_3 is 300/3, which is not less than the minimum expected utility of a_4 which is 200/3. Therefore, for Kyburg, all four acts in the Ellsberg problem are perfectly rational. Though Kyburg's decision rule does not provide a unique decision among these four alternatives, it may often eliminate decisions in other situations. Unfortunately, in the Ellsberg paradox, no action can be eliminated by his central decision rule. Kyburg adds a further decision principle that one should maximize one's minimum gain. By virtue of this principle, a_1 in A and a_4 in B come out to be rational. The additional principle is called the *Minimax* principle. Kyburg thinks that in many cases his decision rule along with the minimax principle, provides good decisions.

4. Gardenfors and Sahlin's decision rule

Gardenfors and Sahlin propose a decision theory that provides a model for an agent's beliefs about the states of world in a decision situation. Their theory, like Kyburg's, is motivated by the thought that the strict Bayesians are too restrictive when they assume a definite probability for each proposition. Unlike Kyburg's theory which is based on probability intervals, the theory assumes that the agent's beliefs can be represented by a set of P of probability distributions, the set of epistemically possible distributions which consist of those measures which are consistent with beliefs the agent has. The set of probability distributions is constrained by a second order measure of epistemic reliability. According to the theory proposed by Gardenfors and Sahlin, the agent exploits those measures of P that are epistemically reliable, i.e. a subset of P/E$_1$ of P used when making a decision. The theory provides a two-step rule for reaching a decision. First, the expected utility of each choice coupled with each probability distribution P in P/E is calculated and the minimal expected utility of each alternative is determined. Then the largest minimal expected utility among the alternative is chosen. The decision rule Gardenfors and Sahlin provide is the following:

(MMEU) The alternative with the maximum of the minimum expected utility ought to be chosen.

For the Ellsberg paradox, we find that the minimum of the expected utility of a_1 and a_2 are 100/3 and 0 respectively and the minimum of the expected utility of a_3 and a_4 are 100/3 and 200/3 respectively. According to MMEU, we have to choose the largest among the minimum of the expected utilities. a_1 and a_4 become the rational choices according to the MMEU. Interestingly, this matches with the intuition of most people.

5. The third decision rule

I put forward a decision principle within the framework of Kyburg's theory. Like Kyburg's theory, my theory is based on probability intervals. The probability of a proposition S being true as an element of our knowledge corpus K belongs to the interval [p, q]. If a rational agent has complete knowledge about an event, its associated probability interval would be reduced to a point, whereas if his information is incomplete, the associated probability interval would be wider. If a rational agent is completely ignorant of the event, then his complete ignorance can be represented by the interval from 0 to 1.

Since I am proposing a decision principle it is worthwhile to provide a connection between the degrees of belief and utilities with actions. I assume that the utility function U is a real-valued function defined over sentences. Then the outcomes of decisions can be regarded as certain propositions of coming true. In my theory, as in Kyburg's, since probabilities are intervals, expected utilities have to be construed as intervals. The expected utility of a sentence S is the interval consisting of the utility of S's being true multiplied by the lower probability of S, and the utility of S's being true multiplied by the upper probability of S. If the probability of S is the closed interval [p, q], and utility for S is U, the expected utility of S is [Up, Uq].

Since the expected utilities are intervals, it does not make sense to maximize intervals in the same way which we maximize expected utilities for points. Although we cannot maximize an interval in the usual way, Kyburg suggests a principle that seems to be rational in this context. That is, one ought not to choose an action whose maximum expected utility is less than the minimum expected utility of some other action. In decision situations, Kyburg's principle eliminates many acts as irrational. I want to strengthen Kyburg's theory by adding further conditions to his theory, and including Kyburg's central principle. I call my decision principle the W*eak dominance principle*:

For all closed intervals I and J, if min(I) < min (J) and max(I) < max(J), and either min(I) < min(J) or max(I) < max(J), then one must choose the act corresponding to J.

The strategy of the paper is to argue that neither do I agree with strict Bayesianism nor with Gardenfors and Sahlin type theory nor with Kyburg's theory in all its nuances. (I) Recall the Ellsberg problem. In this problem, the intuitive decision is to prefer a_1 to a_2 in decision situation A and a_4 to a_3 in decision situation B. In A, the expected utility a_1 is [100/3, 100/3] whereas, the expected utility of a_2 is [0, 200/3]. In B, expected utility of a_3 is [100/3,300/3] and expected utility of a_4 is [200/3, 200/3]. In the Ellsberg paradox, Kyburg's Principle and WDP cannot eliminate any of the actions as irrational. Kyburg suggests that if one uses the *Minimax principle*, then he can choose a_1 in A and a_4 in B. This decision agrees with the intuitions of most people. I will argue that although the *Minimax* principle yields a unique and intuitive solution in the Ellsberg paradox, the *Minimax* principle cannot provide a general and intuitive solution to all decision situations.

Consider an Ellsberg type decision situation with the same probability distributions for red, black and yellow balls as in the Ellsberg paradox. The only difference between the former and the latter is the utility matrix

States

Acts	1/3		2/3
	Red	black	Yellow
a_1	1	0	0
a_2	0	300	0

Based on probability intervals, the expected utility of a_1 is [1/3, 1/3] and that of a_2 is [0, 600/3]. According to the *Minimax* principle, a_1 is the correct decision, although the majority of the people will recommend a_2 as the correct choice. In the Ellsberg paradox, our intuition matches with the choice demanded by the *Minimax* principle, whereas in this Ellsberg type situation, most people would not agree with the decision provided by the latter principle. Our intuition varies from one case to the next. Under these two situations, our decisions are made based on the different psychological factors affecting our relevant decisions. In the original Ellsberg paradox, despite the uncertainty about the proportion of black balls to yellow balls (it can be anything between zero to two thirds), my analysis of the situation indicates that most of the people prefer a_1 to a_2 since the expected utility of a_1 lies at the center of the closed interval of a_2. On the other hand, in the Ellsberg type situation just presented, the expected utility of a_1 is the closed interval that lies more close to zero of the closed interval of a_2 than to 600/3 of the same interval. So in the latter situation, a_2 seems to be the rational decision for most people.

Even in the Ellsberg type situation, there may be some people who choose a_1 to be the rational decision. A hungry person may opt for a_1 in the Ellsberg type situation. For him, buying a hamburger for that price rather than choosing a_2 for a possible larger fortune seems to be rational. However, those who are relatively well off and whose main concern is not buying hamburgers and other cheap things may opt for a_2. For them, gambling for a large fortune is much more attractive, hence rational. According to the *Minimax* principle, any agent ought to choose a_1 in the Ellsberg type decision situation. On my account, whether the agent will choose a_1 or a_2 depends on the psychological factors affecting his decision. If the agent is risk averse, he may choose a_1. On the other hand, if the agent is risk prone, he may opt for a_2. So, in the Ellsberg type decision situation, though the *Minimax* principle demands a_1 to be the correct decision, this may not be the right decision in a given decision situation for a particular agent. So, one should not apply the *Minimax* principle expecting to get both a unique and intuitive decision in all decision situations.

(II) I will provide a utility matrix associated with two sets of probability distributions(e.g., p_1 and p_2) to see how my decision criterion fairs amidst the two other decision rules. For the sake of the discussion, I denote the utility matrix with two probability distributions by M.

	States		Probability Distributions
Acts	s_1	s_2	$p_1(s_1)=.4$
a_1	1	-1	$p_1(s_2)=.6$
a_2	-1	1	$p_2(s_1)=p_2(s_2)=.5$
a_3	0	0	

Expected utility(a_1)= [-.2, 0]
Expected utility(a_2)=[0,.2]
Expected utility(a_3)=[0,0]

According to Kyburg, any of the choices are perfectly rational since we have no way of determining that any of the options are irrational based on his principle[For Levi too, all options are both E-admissible and S-admissible. Therefore, all options are feasible. For lack of space, I have to leave out Levi's well-articulated decision theory from the paper. For Levi's theory, see Levi 1974, Levi 1988]. For Gardenfors and Sahlin a_2 and a_3 are the only feasible options. The MMEU criterion rejects a_1 as the rational choice because it is not the largest of the minimum of expected values. My decision rule, by contrast, demands a_2 to be chosen as rational because that option is no worse than its alternatives under any circumstances and it is at least better than any alternative under some circumstances. So, a_2 seems to be the natural choice among all other alternatives and my decision principle alone agrees with the choice.

6. Ordering relationship of the decision rule

Recall WDP: For all closed intervals I and J, if min(I) < min (J) and max(I) < max(J), and either min(I) < min(J) or max(I) < max(J), then one must choose the act corresponding to J. The principle induces a strict partial ordering, i.e. it is irreflexive and transitive on the intervals to which it is applied. The principle induces irreflexive ordering on intervals since the interval I cannot be better than itself. Therefore, that the ordering relation is irreflexive follows from the definition of <. The principle produces an ordering on intervals which is also transitive. Consider the first conjunct of the antecedent. That conjunct consists of two conjuncts, namely min(I) < min(J) and max(I) < max(J). For all I, J, and K, if they have a transitive relation, then if I< J and J< K, then by transitivity, I necessarily < K. Suppose I< J, then min(I) < min(J). Since J < K, min(J) < min(K).Therefore, min(I)< min(K). By the similar argument, max(I) < max(J). Consider the other limb of the conjunction, min(I) < min(J) or max(I) < max(J).

Take its first disjunct, i.e. min(I) < min(J). For all I, J and K, if min(I) < min(J), as it has been shown that min(J) < min(K), hence, the claim min(I) < min(K) is proved. Take the rest of the disjunct, i.e. max(I) < max (J). For all I, J, and K, if max(I) < max(J) and max(J) < max (K), then show that max (I) < max(K). We have shown before that max(J) < max(K). Therefore, max(I) < max(K). Q.E.D. One interesting ordering relation the weak-dominance principle produces on intervals is that it forms a lattice. A strict partially ordered set is called a *lattice* if and only if for any two elements a and b, there is a least upper bound and a greatest lowest bound for a and b.

In the given framework we can characterize strict preference and indifference relations between two acts. We rewrite WDP and divide it into two parts.

(i) Given that the two intervals are not disjoint, if their intervals have the same maximum, choose the act corresponding to the interval with higher minimum.

(ii) Given that the two intervals are not disjoint, if their intervals have the same minimum, choose the act corresponding to the interval with higher maximum.

The two conditions (i and ii) give *strict preference orderings* over acts. The agent is *indifferent* between two acts iff both max(I) = max(J) and min(I) = min(J). In other words, selection of one act over the other, in this situation, is irrelevant to our principle. There are many acts for which neither the agent has strict preference nor is he indifferent between them. We will return to these acts which are *incomparable* to one another.

When an agent is indifferent between two acts, one of the proposed conditions to eliminate all acts except one is to consider whether the choice the rule recommends remains invariant even after the application of the condition called the *mixture condition*. The intuition behind this condition is that if the agent is indifferent between two acts, then the agent will be indifferent between them and the third act of tossing a fair coin and doing the first act on heads and the second on tails. Suppose the utilities of both a_1 and a_2 are [1,0]. Then, if we add the "mixed" act of flipping a fair coin and doing a_1 on heads and a_2 on tails, we get the following matrix.

	s_1	s_2
a_1	0	1
a_2	1	0
a_3	1/2	1/2

The values for a_3 are calculated by considering that its utility under either state is the expected utility of a bet on a fair coin that pays 0 and 1.

We are here no longer indifferent among three acts after the application of mixture condition, though pairwise, we may be indifferent to one another. For an agent, a_1 and a_2 are still indifferent, but neither a_1 is indifferent to a_3 nor is a_2 to a_3. Among three acts, a_3 has the higher minimum, i.e. 1/2, whereas both a_1 and a_2 have higher maximum than a_3. Our rule does not recommend any action being rational in this situation. We call these acts, a_1, a_2 and a_3, incomparable leaving open the possibility that the act which an agent will choose depends upon his or her mental make up. In fact, in the spectrum between two extreme acts (same minimum, but different maximum, and same maximum, but different minimum), most acts fall into the category of being incomparable. Suppose a_1 is [1, 10,000] and a_2 is [100, 120]. Assuming that a_1

and a_2 represent expected utilities of both acts, the *optimistic reasoner* may choose a_1 arguing that since the true value must lie in between the closed interval, why not take the risk? If we hit the Jack-pot, then we may get 10,000. So, the optimistic reasoner chooses a_1. The *pessimistic reasoner*, on the other hand, may opt for a_2. He may contend that if we choose a_1 and unfortunately a_1 is not the true state of nature, then he may loose a huge amount of money. Then, why not go for a_2 which lies in between the interval [100, 120]? Our principle does not provide us a unique solution to these cases. This does not represent any drawback on the part of the principle rather it shows that our principle is realistic in capturing the true state of nature.

7. Relationship among the three decision rules

I will consider whether any decision principle is entailed by any other principle by providing an intuitive justification for the connection among different principles. I will frequently refer to Sahlin's utility matrix with two probability distributions as M* which is as follows:

Acts	States s_1	s_2	Probability distributions
a_1	12	-10	$p_1(s_1) = .04$, $p_1(s_2) = .06$;
a_2	-9	11	$p_2(s_1) = .06$, $p_2(s_2) = .04$;
a_3	0	0	

Expected utility $(a_1) = [-1.2, 3.2]$
Expected utility $(a_2) = [-1, 3]$
Expected utility $(a_3) = [0, 0]$

For Kyburg's principle, all acts are rational [Levi, however, thinks a_1 and a_2 to be E-admissible and only a_2 to be S-admissible]. According to the MMEU, a_3 is the correct choice. For the sake of the discussion, I will reformulate some of the above decision principles in the following manner:

(1) If two intervals are disjoint, reject the one with lower maximum.

(2) Given that the two intervals, namely I and J are not disjoint, then if max(J) > max(I), but it is not the case that min(J) is < min (I), then choose the act corresponding to J.

(3) Choose the one with highest minimum.

Principle 1 is equivalent to Kyburg's principle. Principle 2 represents WDP. Gardenfors and Sahlin accept principle 3, that is, the *Minimax* principle. Let's begin with rule 1. Rule 1 implies neither rule 2 nor rule 3. Principle 1 is a special case of the principle 2[needless to rehearse the obvious steps of the proof]. For example, where I=[2,3] and J = [4,4], both rules reject I as a rational act. Consider now J' = [3,4] in addition to I. Rule 1 rejects neither of them, whereas rule 2 rejects I as a rational act. Principle 1 does not imply principle 3. If we consider M*, then we find that all acts are rational from the standpoint of Kyburg's principle (i.e., principle 1). But, neither act a_1 nor act a_2 is recommended by principle 3.

Consider rule 2. Rule 2 implies rule 1, for the reason given above. Clearly, rule 2 does not imply rule 3. Consider M. In M, rule 2 recommends a_2 as the rational choice whereas rule 3 accepts both a_2 and a_3 as rational choices. Consider rule 3. Rule 3 does not entail rule 1. Consider M*. According to rule 3, a_3 is the correct choice. It shows a clear violation of rule 1 since all acts are rational according to rule 1.Nor does principle 3 entail principle 2 either. Suppose c_1 and c_2 are two acts with the closed intervals, [.2, .5] and [.2, .6]. According to principle 2, we ought to choose c_2, whereas for principle 3, we cannot reject either of them. In short, principle 2 entails 1 whereas for the rest of them the entailment relationship is not that straightforward.

8. Concluding remarks

Based on the weak-dominance principle, which rests on the interval notion of probability, I argued for both: (i) why should not one expect to reach a unique option from two different sets of options in connection with the Ellsberg paradox and (ii) that, at least under one circumstance, the principle gives better results than the rest of the principles discussed above. I showed that the principle is strict partially ordered and also discussed that it forms a lattice.

Note

[1] The author acknowledges his debt to Prasun Basu, John G. Bennett, Abhijit Dasgupta, Deepnarayan Gupta, Jack Hall, Henry Kyburg Jr., Isaac Levi, Patrick Maher, Michael Mathias, Abhaya Nayak, Nihls-Eric Sahlin, Paul Weirich for their suggestions and comments on earlier versions of the paper and to Peter G. Found for suggesting to him the title of the paper. John G. Bennett deserves special mention for helping him in thinking through the difficult areas of decision theory.

References

Ellsberg, D. (1961), "Risk, Ambiguity, and the Savage Axioms" *Quarterly Journal of Economics,* 75, 528-557. Also, in Gardenfors and Sahlin (eds).

Gardenfors, P.. and Sahlin, N. (1988), "Unreliable Probabilities, Risk Taking, and Decision Making" in Gardenfors and Sahlin (eds). *Decision, Probability, and Utility,* Cambridge: Cambridge University Press.

Kyburg, H.E., Jr. (1983a), *Epistemology and Inference,* Minneappolis: University of Minnesota Press.

_____. (1983b). "Rational Belief" in *The Behavioral and Brain Sciences,* Vol 6: No 2, 231-273.

_____. (1990). *Science and Reason,* Oxford: Oxford University Press.

_____. (1994)."Believing on the Basis of the Evidence" in *Computational Intelligence,* Vol 10; No 1.

Levi, I. (1974), "On Indeterminate Probabilities", in *Journal of Philosophy* 71: 391-418.

_____ . (1988), *Hard Choices,* Cambridge: Cambridge University Press.

Jeffrey, R. (1990), *The Logic of Decision,* Chicago: University of Chicago Press.

Sahlin, N. (1985), "Three decision rules for generalized probability representations" in *The Behavioral and Brain Sciences*, Vol 8, No 4: 751-753.

_____. (1993), "On Higher Order Beliefs" in *Philosophy of Probability* (ed) J.P. Dubucs, Holland: Kluwer Academy.

Savage, L. (1972). *The Foundations of Statistics*, New York: Dover Publications, Inc.

The New Experimentalism, Topical Hypotheses, and Learning from Error[1]

Deborah G. Mayo

Virginia Polytechnic Institute and State University

1. Introduction: The New Experimentalists

Following a period during which philosophers of science focused on theory to the near exclusion of experiment, a number of philosophers, historians and sociologists of science have, in one way or another, turned their attention to experimentation, instrumentation, and laboratory practices.[2] Considerable work in philosophy of science of the last decade reflects this surge of interest in experiment, as promoted by Ackermann, Cartwright, Franklin, Galison, Giere, Hacking and others. Where has this movement taken us and where do we still have to go?

In asking this question, my focus is on that subset of the experimentalist movement whose members, following Ackermann (1989), I dub the "New Experimentalists". Although their agendas differ, members of this group share the core thesis that aspects of experiment might offer an important, though largely untapped, resource for addressing key problems in philosophy of science. In particular, their hope is to find ways to steer a path between the old logical empiricism, where observations were deemed relatively unproblematic, and the more pessimistic post-Kuhnians, who take the failure of logical empiricist models of appraisal as leading to underdetermination and holistic theory change, if not to denying outright the role of evidence in constraining appraisal. To steer this path it is suggested that we clear away the obstacles created by old-style accounts of how observation provides a basis for appraisal (via confirmation theory or inductive logic) and repave the way with an account rooted in the actual procedures for arriving at experimental data and experimental knowledge.

Why is it thought that turning to these experimental practices will offer up new pathways for grappling with philosophical problems about evidence and inference? The answer, as I see it, can be summed up with Ian Hacking's apt slogan: "experiment may have a life of its own" (1983, 160).

There are three main senses in which the life of experiment may be independent of theories and theorizing, and each corresponds to an important theme brought out by the New Experimentalist work. First, the claim of an independent life for experiment, the one initially emphasized by Hacking (1983), asserts that the *aims* of experimental inquiry may be quite independent of testing, confirming or filling out some theory. Instead, actual experimental inquiries focus on manifold local tasks: checking instruments, ruling out extraneous factors, getting accuracy estimates, distinguishing real effect from artifact, and estimating the effects of background factors.

The second reading of the slogan asserts that experimental data may be *justified* independently of theory, that experimental evidence need not be theory-laden in any way that invalidates its role in grounding experimental arguments. "A philosophy of experimental science", insists Hacking, "cannot allow theory-dominated philosophy to make the very concept of observation become suspect." (1983, 185)

A third reading of the slogan asserts that experimental knowledge may be retained despite changes of theory. Says Galison, "experimental conclusions have a stubbornness not easily canceled by theory change." (1987, 259) This suggests that experimental knowledge may serve not only in adjudicating between rival theories, but also as a basis for progress in science.[3]

In exploring these three themes the New Experimentalists have opened up a new and promising avenue for grappling with key challenges currently facing philosophers of science. Less clear is whether the new attention being accorded experiment has paid off in advancing solutions to these problems. Nor is it clear that they have demarcated a program for working out a philosophy or epistemology of experiment. For sure, they have given us an important start: their experimental narratives offer a rich source of illustrations of how experiment lives its own life apart from high-level theories and theorizing. But something more general and more systematic seems to be needed to show how this independence is achieved and how it gets us around the problems of evidence and of inference in so-called theory dominated philosophies. My aim in this paper is to suggest why the New Experimentalism has come up short and propose a way to remedy this. I will illustrate a portion of my proposal utilizing Galison's (1987) interesting experimental narrative on neutral currents. All references to Galison will be to this work.

2. Getting Small: Topical Hypotheses and the Local Discrimination of Error

To begin, I suggest we pursue seriously the first reading of the slogan, "experiment has a life of its own". Galison states it clearly:

> [E]xperimentalists' real concern is not with global changes of world view. In the laboratory the scientist wants to find local methods to eliminate or at least quantify backgrounds, to understand where the signal is being lost, and to correct systematic errors. (245)

For Galison, the question "How do experiments end?" (as in the title of his book) asks "When do experimentalists stake their claim on the reality of an effect? When do they assert that [it]...is more than an artifact of the apparatus or environment?" (4) The answer, in a nutshell, is only after having sufficiently well ruled out or subtracted out various backgrounds. Accordingly, a central experimental task is investigating and debating claims about backgrounds.

More recently, Hacking refers to the kind of claims experiment investigates as "topical hypotheses"—like topical creams—in contrast to deeply penetrating theories. Hacking claims:

> It is a virtue of recent philosophy of science that it has increasingly come to acknowledge that most of the intellectual work of the theoretical sciences is con-

ducted at [the level of *topical* hypotheses] rather than in the rarefied gas of systematic theory. (Hacking 1992, 45)

To their credit, the New Experimentalists have been the leaders in this recognition. At the same time I think this points to the reason the New Experimentalists have come up short. The reason, as I see it, is that the experimental practices that have the most to offer in understanding these local tasks are still largely untapped. These are the activities involved in experimental design, experimental modeling, and data analysis—activities which, in practice, receive structure from statistical methods and arguments.

This is not to say the experimental narratives do not include the use of statistical methods. In fact, their narratives are chock full of specific applications of statistical techniques, e.g., techniques of data analysis, significance tests, confidence interval estimates, and other methods from what I propose to call *standard error statistics*.[4] What has not been done is explain how these methods are used to accomplish reliably the local tasks of arriving at data, learning about backgrounds, and so on.

In rejecting old-style accounts of confirmation as the wrong way to go, the New Experimentalists seem dubious about the value of utilizing statistical ideas to construct a general account of experimental inference. Theories of confirmation, inductive inference, and testing, were born in a theory-dominated philosophy of science, and this is what they wish to move away from. The complexities and context dependencies of actual experimental practice just seem recalcitrant to the kind of uniform treatment dreamt of by philosophers of induction. And since it is felt that overlooking these complexities is precisely what led to many of the problems that the New Experimentalists hope to resolve, it is natural to find them skeptical of the value of general inference accounts. Ironically, where there is an attempt to employ formal statistical ideas to give an overarching structure to experiment, some New Experimentalists revert back to the theory-dominated philosophies of confirmation, testing, and decision, particularly Bayesian philosophies (e.g., Franklin 1986, 1990).

The central position of what may be called "theory-dominated" philosophies of confirmation or testing is that the task of a theory of statistics begins with data or evidence already in hand, and seeks to provide some uniform rule (akin to deductive logic) to relate evidence (or evidence statements) to any theory, hypothesis, or decision of interest. Most commonly, the rule is to operate by providing some quantitative measure of support, confirmation, credibility or probability to hypotheses. Examples are the inductive logics of Carnap and of subjective Bayesians.

Galison is right to doubt that it is productive to search for "an after-the-fact reconstruction based on an inductive logic" (3). Such accounts, at their best, serve to reconstruct scientific inferences after-the-fact, rather than capture the methods actually used, though I will not argue this here. Where the New Experimentalists shortchange themselves is in playing down the use of local statistical methods at the experimental level—the very level they exhort us to focus on.

Those philosophers of statistics who have entered the experimentalist discussions (e.g., Howson and Urbach 1989) have encouraged this downplaying of the methods from standard error statistics. Embracing the theory-dominated philosophy of subjective Bayesian confirmation theory, Howson and Urbach reject standard error statistics as inappropriate, and regard its widespread use in experimental practice as unwarranted. Now it is true that the conglomeration of local tools comprising standard error statistics looks inadequate from the perspective of the aims of theory-dominated confirmation theory, because they do not provide a uniform quantitative measure of the bearing of evidence on hypotheses. But when it comes to the New Experimentalist aims, exactly the reverse is the case. Standard error statistics provide just the tools needed for investigating the topical hypotheses in experimental learning.

After all, if what we want are tools for discriminating signals from noise, ruling out artifacts, distinguishing backgrounds, and so on, then we really need tools for doing that. And these tools must be applicable with the kind of information scientists actually tend to have.[5] The conglomeration of methods and models from standard error statistics is the place to look for forward-looking procedures to obtain data in the first place, and which are apt even with only vague preliminary questions in hand. As such, these tools can provide the needed structure to the practices given a central place by the New Experimentalists.

3. Arguing From Error

Rather than approach the statistical tools in their formal setting, I shall begin right off with how I think they are used in experimental learning. Their aim, as I see them, is to direct experimental activities so as to allow us to give experimental arguments. The arguments follow a pattern of what might be called *an argument from error* or *learning from error*. The overarching structure of the argument is guided by the following thesis:

> It is learned that an error is absent when (and only to the extent that) a procedure of inquiry (which may include several tests) with a high probability of detecting the error if it existed, nevertheless failed to do so.

Such a procedure of inquiry, we can say, is one with a high capability of severely probing for errors—we may call it a *reliable (or highly severe) error probe*. According to the above thesis, we can argue that an error is absent if it fails to be detected by a highly reliable error probe.

Alternatively, the argument from error can be described in terms of a test of a hypothesis, H, that a given error is absent. The evidence indicates the correctness of hypothesis H, when H passes a severe test—one with a high probability of failing H, if H is false. An analogous argument can also be given to infer the presence of an error.[6]

Standard error statistics provides tools for reliable error probes that are robust across different scientific domains, with very minimal assumptions. The New Experimentalist offerings reveal (whether intended or not) the function and rationale of these statistical tools from the perspective of actual experimental practice—the very understanding missing from theory-dominated perspectives on scientific inference. Standard statistical tools, thus understood, can return the favor to the New Experimentalist program. Its already well-worked-out models and methods, I believe, provide the needed general framework for pursuing the different ways in which experiment lives a life of its own.

Here I shall focus on a first step, corresponding to the first construal of our slogan. This first step is to utilize the New Experimentalist narratives, together with this thesis about arguing from error, to understand the role of error statistics in distinguishing genuine effects from artifacts.

4. Distinguishing Effects From Artifacts: Galison and Neutral Currents

Galison's (1987) work is especially congenial. I shall follow a portion of his discussion of the discovery of neutral currents. Although by the end of the 1960s,

Galison tells us, the "collective wisdom" was that there were no neutral currents (164, 174), soon after (from 1971-1974) "photographs...that at first appeared to be mere curiosities came to be seen as powerful evidence for" their existence. (135) I am just going to focus on one particular analysis for which Galison provides detailed data. Abstracted from the whole story, this part will obviously not give an understanding of either the theory at stake or the sociological context. But it is sufficient to bring out the answer to Galison's key question: "[H]ow did the experimentalists themselves come to believe that neutral currents existed? What persuaded them that they were looking at a real effect and not at an artifact of the machine or the environment?" (136)

Here are the bare bones of the experimental analysis: Neutral currents are described as those neutrino events without muons. Experimental outcomes are described as muonless or muonful events, and the recorded result is the ratio of the number of muonless and muonful events. (This ratio is an example of what is meant by a statistic—a function of the outcome.) The main thing is that the more muonless events recorded, the more the result favors neutral currents. The worry is that recorded muonless events are due, not to neutral currents, but to inadequacies of the detection apparatus.

Experiments were conducted by a collaboration of researchers from Harvard, Wisconsin, Pennsylvania, and Fermilab, the HWPF group. They recorded 54 muonless events and 56 muonful events giving a ratio of 54/56. The question is: Does this provide evidence of the existence of neutral currents?

> For Rubbia [from Harvard] there was no question about the statistical significance of the effect. ...Rubbia emphasized that 'the important question in my opinion is whether neutral currents exist or not... The evidence we have is a 6-standard-deviation-effect.' (Galison, 220)

The "important question" revolved around the question of the statistical significance of the effect. I will refer to it as the *significant question*. Galison puts it this way:

> Given the assumption that the pre-Glashow-Weinberg-Salam theory of weak interactions is valid (no neutral currents), then what is the probability that HWPF would have an experiment with as many recorded muonless events as they did? (220)

Three points need to be addressed: How might the probability in the significant question be interpreted? Why would one want to know it? and, How might one get it? While the answers to these questions are found to be problematic from the point of view of theory-dominated accounts of inference, this is not the case were one to adopt the point of view of the New Experimentalism. I will consider each in turn.

(i) *Interpreting the significant question*

What is being asked when one asks for the probability that HWPF would have an experiment with as many recorded muonless events as they did, given no neutral currents? The question, in statistical language, is: How (statistically) significant is the number of recorded excess muonless events? Here I want to explain the significant question informally.

The experimental result, recall, was the recorded ratio of muonless to muonful events, namely, 54/56. The significant question, then, is: What is the probability that HWPF would get as many as (or more than) 54 muonless events, given there are no neutral currents? One way to cash out what is wanted is this: How often, in a series of experiments such as the one done by HWPF, would as many muonless events be expected to occur, given there are no neutral currents?

But there is only this one experimental result, not a series of experiments. True, the series of experiments here is a kind of hypothetical construct. What we need to get at is why it is perceived as so useful to introduce this hypothetical construct into the data analysis.

(ii) What is the value of answering the significant question?

The quick answer is that it is an effective way to distinguish real effect from artifacts. Were the experiment so well controlled that the only reason for failing to detect a muon is that the event is a genuine muonless one, then artifacts would not be a problem and this statistical construct would not be needed. But artifacts are a problem. From the start a good deal of attention focused on the backgrounds that might fake neutral currents. (Galison,177) A major problem was escaping muons. "From the beginning of the HWPF neutral-current search, the principal worry was that a muon could escape detection in the muon spectrometer by exiting at a wide angle. The event would therefore look like a neutral-current event in which no muon was ever produced." (Galison, 217)

The problem, then, is to rule out a certain error: construing as a genuine muonless event one where the muon simply never made it to the spectrometer, and thus went undetected. To relate this problem to the significant question, let us introduce some abbreviations. If we let hypothesis H be

H: neutral currents are responsible for (at least some of) the results

then, within this piece of data analysis, the falsity of H is the artifact explanation:

H is false (the artifact explanation): recorded muonless events are due, not to neutral currents, but to wide-angle muons escaping detection.

Our significant question becomes:

What is the probability of a ratio (of muonless to muonful events) as great as 54/56, given that H is false?

The answer is the *significance probability* or *significance level* of the result.

Returning to the relevance of knowing this probability (the significance level), suppose it were found to be high. That is, suppose as many or even more muonless events would occur frequently, say more often than not, even if H is false (and it is simply an artifact). What is being supposed is that a result, as or even more favorable to H than the HWPF result, is fairly common due, not to neutral currents, but to wide angle muons escaping detection. Were that so, the HWPF result clearly does *not* provide grounds to rule out wide-angle muons as the source (the artifact explanation). Were one to proceed by taking such a result as grounds for ruling out the artifact explanation, one would be wrong more often than not. That is, the probability of correctly detecting the artifact explanation (not-H) would be less than .5. The procedure would be an unreliable error probe. Since high significance level means low reliability, results are not taken to indicate H unless the significance probability is low.

Suppose now that the significance probability is very low, say 0.01 or 0.001. This means that it is extremely improbable for so many muonless events to result, if H were false and the HWPF researchers were really only observing the result of muons escaping. Since escaping muons could practically never be responsible for so many muonless events, their occurrence in the experiment is taken as good grounds for rejecting the artifact explanation. That is because, following an argument from error, the procedure is a highly reliable probe of the artifact explanation. This was the case in the HPWF experiment, although the probability in that case was considerably smaller. But how do you get the significance probability?

(iii) How is the significant question answered?

The reasoning I just described does not require a precise value of the significance probability. It is enough to know that it is or is not very low—that the procedure is or is not fairly reliable. But how does one arrive at even a ballpark figure? The answer comes from the use of various standard statistical analyses, but to apply them (even qualitatively) requires information about how the artifact in question could be responsible for certain experimental results. Statistical analyses are rather magical, but they do not come from thin air. They send the researcher back for domain-specific information. Let us see what the HWPF did.

The data used in the HWPF paper is as follows: (Galison, 220)

Visible muon events	56
No visible muon events	54
Calculated muonless events	24
Excess	30
Statistical significant deviation	5.1

The first two entries just record the HWPF result. What about the third entry, the calculated number of muonless events? This refers to the number calculated or expected to occur because of escaping muons. This calculation comes from separate work deliberately carried out to find out how an event can wind up being recorded "muonless", not because no muon was produced (as would be the case in neutral currents), but because the muon never made it to the detection instrument.

The group from Harvard, for example, created a computer simulation to model statistically how muons could escape detection by the spectrometer by exiting at a wide angle. This is an example of what is called a "Monte Carlo" program.

> By comparing the number of muons expected not to reach the muon spectrometer with the number of measured muonless events, they could determine if there was a statistically significant excess of neutral candidates. (Galison, 217)

In short, the Monte Carlo simulation afforded a way (not the only way) to answer the significant question.

The reason probability arises in this part of the analysis is not because the hypothesis about neutral currents is a statistical one, much less because it quantifies credibility in H or in not-H. Probabilistic considerations are deliberately *introduced* into the data analysis because they offer a way to model the expected effect of the artifact (escaping muons). Statistical considerations, we might call them "manipulations on paper" (or on computer), afford a way to subtract out background factors that cannot literally be controlled for. In several places, Galison brings out what I have in mind:

> In a sense the computer simulation allows the experimentalist to see, at least through the eye of the central processor, what would happen if a larger spark chamber were on the floor, if a shield were thicker, or if the multiton concrete walls were removed.

The Monte Carlo program can do even more. It can simulate situations that *could never exist in nature*. ... One part of the Gargamelle demonstration functioned this way: suppose the world had only charged-current neutrino interactions. How many neutral-current *candidates* would there be? (265)

It was calculated that 24 muonless events would be expected in the HWPF experiment due to escaping muons. Next, Galison explains, "they wanted to know how likely it was that the observed ratio of muonless to muon-ful events (54/56) would fall within the statistical spread of the calculated ratio (24/56), due entirely to wide-angle muons." (220) The difference between the ratio observed and the ratio expected (due to the artifact) is 54/56 - 24/56 = 0.536. How improbable is such a difference even if the HWPF experiment *were* being done on a process where the artifact explanation is true (i.e., where recorded muonless events were due to escaping muons)? This is "the significant question" again, and finally we can answer it.

The simulation lets us model the relevant features of what it would be like were the HWPF study actually experimenting on a process where the artifact explanation is true. It tells us it would be like experimenting on a process that generates ratios (of m events to m-less events) where the average (and the most likely) ratio is 24/56. (This corresponds to the hypothetical sequence of experiments we spoke of.) The statistical model tells us how probable different observed ratios are, given the average ratio is 24/56. In other words, the statistical model tells us what it would be like to experiment on a process where the artifact explanation is true; namely, certain outcomes (observed ratios) would occur with certain probabilities. (Most experiments would yield ratios close to the average (24/56); the vast majority would be within two standard deviations of it.)

Putting an observed difference between recorded and expected ratios in standard deviation units allows one to use a chart to read off the corresponding probability. The standard deviation (generally estimated) gives just that—a standard unit of deviation that allows the same standard scale to be used with lots of different problems in different scientific domains. Any difference exceeding two or more standard deviation units corresponds to one that is improbably large (occurring less than 2% of the time).

Approximating the standard deviation of the observed ratio shows the observed difference to be 5.1 standard deviations.[7] This is so improbable as to be off the charts; so, clearly, by significance test reasoning, the observed difference indicates that the artifact explanation is untenable. It is practically impossible for so many muonless events to have been recorded, were they due to the artifact of wide angle muons. The procedure is a highly reliable artifact probe.[8]

This is just one small part of a series of experimental arguments that took years to build up. Each involved this kind of statistical data analysis to distinguish real effects or signals from artifacts, to estimate the maximum effect of different backgrounds, and to rule out key errors *piece-meal*. They are put together to form the experimental arguments that showed the experiment could end.

5. Conclusion

The New Experimentalists are right to insist on the centrality of the tasks of distinguishing and subtracting out backgrounds, quite apart from the aim of testing high-level theories. They are also right to suppose that experimental practices offer especially powerful tools for these local tasks. While their experimental narratives offer a rich source of illustrations, something more general is needed to understand how experimental practices accomplish these tasks. In this paper I have showed how a standard error statistical tool (significance tests) together with an experimental narrative, can serve to articulate the procedure for distinguishing artifacts in an important class of cases. The next step or set of steps would be to explore how a handful of standard (or canonical) statistical models permit analogous arguments from error to be substantiated across a wide spectrum of experimental inquiries. These, still mostly untapped, tools, I believe, are the key to advancing solutions to the problems about evidence and inference that the New Experimentalist movement set for itself.

Notes

[1] This research was supported by an NSF award in Studies in Science, Technology and Society. I gratefully acknowledge that support.

[2] A collection of this work may be found in Achinstein and Hannaway (1985). For a good selection of interdisciplinary contributions, see Gooding, Pinch, and Schaffer (1989).

[3] Giere (1988) and Hacking (1983) have especially stressed how this sort of progress is indicated when an entity or process becomes so well understood that it can be used to investigate other objects and processes.

[4] I use this label rather than the labels often given to specific components of this methodology, e.g., Fisherian tests, Neyman-Pearson or Orthodox statistics, because the latter are associated with certain inference philosophies that do not necessarily reflect the uses of these methods in experimental practice.

[5] In contrast, to get a Bayesian inference going, an agent requires a prior probability assignment to an exhaustive set of hypotheses, among other things.

[6] I discuss severe tests in Mayo (1991). A full discussion of arguing from error, and a development of the corresponding error statistics approach occurs in Mayo (forthcoming).

[7] The standard deviation is estimated using the recorded result and a standard statistical model. It equals

$$\left(\frac{24}{56}\right)\sqrt{\frac{1}{24}+\frac{1}{56}} = 0.105. \text{ (Galison 1987, 220-221)}$$

[8] Galison points out that a different analysis of the HWPF data resulted in a different level of significance—still highly significant. The error statistics approach does not mandate one best analysis—several are used to check and supplement one another.

References

Achinstein, P. and Hannaway, O. (eds.) (1985), *Observation, Experiment and Hypothesis in Modern Physical Science*. Cambridge, MA: MIT Press.

Ackermann, R. (1985), *Data, Instruments, and Theory.* Princeton: Princeton University Press.

————————. (1989), "The New Experimentalism", *The British Journal for the Philosophy of Science* 40: 185-190.

Cartwright, N. (1983), *How the Laws of Physics Lie.* Oxford: Clarendon Press.

Franklin, A. (1986), *The Neglect of Experiment.* Cambridge: Cambridge University Press.

——————. (1990), *Experiment, Right or Wrong.* Cambridge: Cambridge University Press.

Galison, P. (1987), *How Experiments End.* Chicago: University of Chicago Press.

Giere, R. (1988), *Explaining Science.* Chicago: University of Chicago Press.

Gooding, D., Pinch, T., and Schaffer, S. (eds.) (1989), *The Uses of Experiment: Studies in the Natural Sciences.* Cambridge: Cambridge University Press.

Hacking, I. (1983), *Representing and Intervening.* Cambridge: Cambridge University Press.

——————. (1992), "The Self-vindication of the Laboratory Sciences", in A. Pickering (ed.), *Science as Practice and Culture.* Chicago: The University of Chicago Press, pp. 29-64.

Howson, C. and Urbach, P. (1989), *Scientific Reasoning: The Bayesian Approach.* La Salle: Open Court.

Mayo, D. (1991), "Novel Evidence and Severe Tests", *Philosophy of Science* 58: 523-552.

—————. (forthcoming), *Error and the Growth of Experimental Knowledge.* Chicago: The University of Chicago Press.

Of Nulls and Norms[1]

Peter Godfrey-Smith

Stanford University

1. Introduction

When the Neyman-Pearson approach to the testing of statistical hypotheses was introduced in the 1930's and 1940's it was presented with an accompanying philosophy of science. Jerzy Neyman held a strong form of pragmatism.[2] He sought not just to model scientific decision on practical decision, but in a sense to *reduce* it to practical decision. Neyman-Pearson techniques are used to compare hypotheses in the light of experimental data, and they contain an asymmetry, in that one hypothesis—the "null"—typically gets the benefit of the doubt. For Neyman, the choice of which hypothesis gets this benefit is governed by behavioral considerations. When presented as an account of testing in science this claim was met with vigorous resistance.

Though the philosophy was controversial, the methods became standard. Philosophers and statiscans have produced alternative interpretations of many distinctive features of these tests. But they are not the only ones who have had to give a new rationale for Neyman-Pearson methods. These methods are taught with textbooks, which generally have to say something about which hypotheses should receive the benefit of the doubt.

My aim here is to compare (i) what philosophers and other foundational thinkers say, and (ii) what textbooks say, about the choice of null hypotheses and the control of error rates in testing. In the future I hope to extend this work to include (iii) what scientists actually do. This paper is a preliminary report.

2. Outline of the Problem

Suppose there are two alternative hypotheses H_0 and H_1 describing an unknown parameter of a system. An observable variable X is related probabilistically to H_0 and H_1, and functions $P(X|H_0)$ and $P(X|H_1)$ describe these relations. You intend to accept H_0 or H_1 according to the observed value of X. You must first decide which values of X, if you observe them, should prompt you to accept H_0 and which should prompt you to reject H_0 for H_1. We will call H_0 the "null hypothesis," and the set of values of X which will lead to the rejection of H_0 is the "critical region," symbolized "C."

Neyman-Pearson techniques control the risks associated with inference from incomplete data by controlling the rates of two different kinds of errors —the famous "Type I" and "Type II" errors. Given $P(X|H_0)$ and $P(X|H_1)$ it is possible to define, for any given critical region, the likelihoods of various right and wrong decisions. First, we can define the likelihood that an observation will fall into the critical region, prompting the rejection of H_0, even though H_0 is true. This is a Type I error, and the likelihood of a Type I error (given H_0) is called the "size" of the test, symbolised "α." That is, $\alpha = \Pr(X \in C|H_0)$. A good critical region makes Type I errors unlikely. One way to do this is to make the critical region very small, but this will also tend to make the chance of rejecting H_0 small when H_1 is true. A failure to reject H_0 when H_1 is true is a "Type II error." The likelihood of a Type II error (given H_1) is symbolized "ß." So $ß = \Pr(X \notin C|H_1)$. The "power" of the test is (1-ß). To some extent power can be maximized while keeping α constant—this is a central aim of Neyman-Pearson methods. Also, both errors can be reduced by increasing the sample size. But after fixing the sample size and the general location of the critical region, the two error rates must be traded off. Making Type I errors rarer requires making Type II errors more common, and vice versa.

The Type I error rate, α, is usually set at some standard value such as 0.05, 0.01 or 0.001. There are no widely recognized standard values for ß. Instead ß is usually allowed to vary, within broad limits, as a consequence of the choice of α. Acceptable values of ß can easily be around 0.3 or 0.4.[3] It is rarely asked that ß be as low as α. The null hypothesis receives the benefit of the doubt because the chance of rejecting this hypothesis falsely is rigidly controlled and kept low, while the chance of falsely rejecting the other hypothesis is less tightly controlled and generally higher. This is why the decision to regard a hypothesis as the "null" is epistemically important. Some writers do not use the term "null," but this does not remove the problem if they retain the quantitative asymmetry between α and ß.

I have described the simplest case, where the choice is between two definite hypotheses. H_0 might be "Black snakes and brown snakes are the same average length" and H_1 could be "The average length of black snakes is a foot longer than brown." Often the choice is between H_0 and a *range* of alternative H_i's, a "compound" alternative. We might suspect that black snakes are longer but have no view on how much longer they are. Then, given α, the hypothesis that brown snakes are six inches longer will have one value for ß and the hypothesis that they are two feet longer will have a smaller ß. A test will have high power for some alternatives but low power for others. This makes things more complicated but does not remove the asymmetry. There are ways in which power *could* be controlled rigidly even in a situation like this, but as a matter of fact α is generally set strictly and the power, within broad limits, is left to take care of itself.[4]

3. Four Views of the Asymmetry

I will describe three ways in which this asymmetry might be justified. A fourth view aims to downplay the asymmetry rather than justify it.

3.1 The Pragmatic Justification

Firstly there is Neyman's own view, found most clearly in his 1950 textbook. For Neyman, a statistical test is used when we must make a two-way behavioral choice. To "accept H_0" is to decide to act a certain way. H_0 is "true" if the world is in a state such that the action associated with H_0 is better than the alternative action. To accept H_0 we need not believe that the world is in such a state. It is just a decision to act as if the world is that way (1950, 258-259). In decisions like this, Neyman says, one type of wrong behavioral decision will typically be more serious than the other. Whichever mistake is more serious is regarded as the Type I error (Neyman 1950, 263; 1942, 304).

As Neyman-Pearson techniques became accepted, Neyman's philosophy prospered in some circles as well. Abraham Wald developed a non-Bayesian decision theory in which Neyman's "inductive behavior" concept is central (Wald 1950, 10). A well-known textbook presentation from this period, Cramér 1951, also resembles Neyman's account. Around this time the methods and the philosophy were often presented as a package (Neyman 1942, 293, 301).

There was also a chorus of dissent. In the forefront was R.A. Fisher, who mounted vigorous and sometimes personal attacks on the rival school (1956). A basic theme of both Fisher's attack and some early philosophical treatments is that decisions about hypotheses in science are, and should be, made without knowledge of the behavioral consequences of accepting hypotheses, and hence without knowledge of which mistake is more practically serious (Fisher 1956, 102-103, Jeffrey 1956, 242). A large literature has addressed this issue (Mayo 1992).

3.2 The Doxastic Justification

Suppose Neyman's philosophy is discarded but we seek to retain the methods. One alternative view of the α/β asymmetry explains it in terms of an asymmetry in the *attitudes* the researcher takes to the different hypotheses (Levi 1962). On this view, when an observation in the critical region occurs the researcher rejects H_0. But when an observation falls outside the critical region the researcher merely suspends judgement. So a Type I error is a false belief, but a Type II error is qualitatively different and less serious—a "regrettable suspension" perhaps. On this account any hypothesis can take the role of a null.[5]

3.3 The Semantic Justification

Another way to understand the asymmetry is to claim that some hypotheses are "natural nulls," in relation to their alternatives. They have this status in virtue of their content. The term "null" does of course have definite connotations—a hypothesis that there is no effect, or nothing going on. If one takes a pragmatic or doxastic attitude to the asymmetry these connotations are misleading. Indeed, Neyman did not like the term (1942, 304), which was associated with Fisher (1935). Others have expressed reservations (Kendall and Stuart 1961), but the term "null" has become common in discussions of Neyman-Pearson methods. In some ways this term is an import from a rival approach.

If hypotheses of "no effect" are nulls, then the asymmetry between α and β operates as the wielder of Occam's razor. The more serious error is multiplying effects beyond necessity, rather than not recognising enough effects. The asymmetry establishes a bias in favor of the simpler hypothesis.[6]

3.4 The Deflationary View

Finally, it is possible to deny that the asymmetry is as important as it appears. This I take to be part of the position advanced by Alan Birnbaum (1977). Birnbaum claims there are two interpretations of Neyman-Pearson tests, an *evidential* and a *behavioral* interpetation. The behavioral interpretation is associated with what I called the "pragmatic" view of the asymmetry. Birnbaum claims the behavioral interpretation need not be taken as a literal description of testing in science, though in other domains it may be applied literally. The behavioral interpretation is applied to science in a "heuristic or hypothetical" way, as a means of illustrating some abstract concepts (1977, 32-33).

This is not yet an answer to our question. Even if some aspects of a hypothetical behavioral decision are useful in illustrating Neyman-Pearson techniques, why should the *asymmetry* in error probabilities carry over to the evidential application of the tests? Birnbaum gives no general answer. Within the evidential interpretation he does allow background information to make some pieces of information more valuable than others (1977, 38-39). I think the key to Birnbaum's view of the asymmetry lies in his formulation of the principle behind the evidential interpretation of the tests:

(Conf): A concept of statistical evidence is not plausible unless it finds "strong evidence for H_1 against H_0" with small probability (α) when H_0 is true, and with much larger probability (1-ß) when H_1 is true. (1977, 24. Some symbols changed)

This view has a "two-way" characteristic. Whether a test delivers good evidence for H_1 depends not just on the size but also on the power of the test. *Both* error probabilities are relevant in assessing *each* direction of evidential relevance. (On another view, when a test is significant this is good evidence against H_0 whether or not the test had a high power.) For Birnbaum the typically higher value of ß, compared to α, is not of great importance as long as (1-ß) remains "much larger" than α. How much? If (1-ß) is 0.5 this is a fairly low power test, and even then (1-ß) will be at least 10 times larger than α, for standard values of α.

Because Birnbaum takes both error probabilities to be relevant to all evaluations of evidence, his evidential interpretation of the tests reduces the importance of the asymmetry. Precise values of α and ß may be adjusted according to our epistemic interests, but the evidential function of the test is served as long as there is a certain basic relationship between α and ß.

Before moving on we must note the possibility of positions which combine some of these alternatives. So there are *simple* and *combination* positions. There are two kinds of (consistent) combination positions. Firstly, it might be maintained that the choice of a null can be governed by either one principle or another. For example, when the practical consequences of decisions are known these might determine the choice pragmatically, and otherwise the researcher might be compelled to make the assignment semantically. This is a *disjunctive* combination position. Birnbaum probably had a disjunctive combination view, with pragmatic and deflationary components.

Secondly, it might be held that one or more of the criteria will coincide in their judgements. Perhaps wrongly rejecting the simpler hypothesis is more practically serious than wrongly rejecting the alternative. This is a *conjunctive* combination position. I understand Giere 1977 to propose this conjunctive view for one type of experiment—testing "simple causal hypotheses" (1977, 57). In the case of testing theories, Giere's view is probably deflationary (1977, 61).

4. A Sample of Textbooks

We will now look at how the asymmetry is handled in some statistics textbooks. Textbooks are interesting here as they occupy an intermediate position between the theory and practice of science. They are normative but practical, as they are designed to train and perhaps indoctrinate those entering a field (Kuhn 1962). I will discuss a sample of 12 textbooks.[7]

My presentation will be informal, in part because so far the situation is turning out to be complicated. It is difficult to place many texts into the categories given above. Several of my assignments are tentative. This can be taken to reflect on the descriptive value of the categories, or to reflect on the conceptual rigor of the texts, or both. I view it as reflecting on both. Also, most texts which can be readily categorised fall into a combination position.

The single most common pattern is a combination of a pragmatic view with some other. It is often unclear what the other view is, and whether the combination is disjunctive or conjunctive. In fact, some books say the asymmetry has a pragmatic justification some of the time, but not all the time, and nothing is said about what should happen the rest of the time. I will start with an illustration of this pattern.

Sokal and Rohlf 1981 is a widely used statistics textbook for biologists, which I have several times heard referred to as "the Bible." Here is the core of their discussion of the asymmetry.

> In most applications, scientists would wish to keep both of these [types of] errors small, since they do not wish to reject a null hypothesis when it is true, nor do they wish to accept it when another hypothesis is correct.
>
> However, we should note that there are special applications, often non-scientific, in which one type of error would be less serious than the other, and our strategy of testing or method of procedure would obviously take this into account. (1981, 163)

From the first sentence it appears that Sokal and Rohlf think there should be no asymmetry. A doxastic view is not supported, and they do not specify the semantic properties of a null. The second sentence seems to say that the *only* situations where the errors are of unequal importance are "special." Here the pragmatic view holds, but these cases are distinguished from the usual scientific cases. Sokal and Rohlf's examples, however, feature levels of α at or below 0.05 (a level they have reservations about), and levels of ß over 0.5 (1981, 163). So I interpret Sokal and Rohlf as having a pragmatic view combined disjunctively with something which is probably not doxastic, but which is not clear.

Another book with a combination including a pragmatic view and something which is not doxastic is Weiss and Hassett 1982. Their presentation is even structurally split. In the main chapter on hypothesis testing the framework is not pragmatic, and a doxastic view of the asymmetry is also probably ruled out. They permit the experimenter, for example, to "accept the null as being a reasonable hypothesis" after a non-significant result (1982, 264).[8] Then at the end of the book there is a separate chapter on "Planning a Study," where the framework is pragmatic and α and ß are determined by the costs of errors and the interests of the experimenter (1982, 559).

In contrast, Rice 1988 has a disjunctive combination view which accepts the pragmatic and semantic criteria, and clearly indicates their different roles (1988, 274). The null can be simpler, *or* it can be more practically serious to falsely deny it. It is not implied that these criteria coincide.

Iman and Connover 1983 also have a combination view, which may be like Rice's, but might be conjunctive. Their discussion begins with light bulb testing examples which suggest a pragmatic view, and they acknowledge different penalties for different mistakes. But then they state that a null "is often, but not always, a version of the statement 'Any observed change or difference is due to chance variability'" (1983, 208), and is usually "worded in a way that reflects the status quo" (1983, 209). They give examples of typical nulls, including claims that the defendent is innocent, the new bulb burns as long as the established one, and so on.

Since Iman and Conover seem to recognise differences between error costs as a typical feature of tests, but also think a null generally has a certain content, they may think the semantic and the pragmatic ways of choosing a null coincide in general. The "status quo" hypothesis will tend to also be the one whose false rejection is more practically serious. So they may have a conjunctive view.

A view which more clearly combines the semantic and pragmatic, and views the criteria as coinciding, is presented in Crosby 1977, an introductory text for psychologists.

> Researchers usually feel that the consequences of making a Type I error are more serious than the consequences of a Type II error. If the null hypothesis is rejected, the researcher may publish the results in a journal and the results might be reported by others in textbooks or in newspaper or magazine articles. Researchers don't want to mislead others or risk damaging their reputations by publishing results that aren't there in the population....
> In contrast to the consequences of publishing false results, the consequences of a Type II error are not seen as being very serious. (Crosby 1977, 152)

Nulls are assumed to have a characteristic content which is unpublishable if supported. Only rejections of the null, which find some new effect, are publishable. A behavioral asymmetry is a *consequence* of an asymmetry in content. This may describe a significant portion of practice in psychology, where the semantic view of nulls is common. It does not apply to all fields though. Experiments which support a null hypothesis are often publishable.[8]

Cohen 1977 may present the same view as Crosby. Cohen 1977 is a technical manual for psychologists, focused specifically on power. Cohen says Type I errors are more serious than Type II because "failure to find is less serious than finding something that is not there" (1977, 56). This can be read in two ways. It may be same view as Crosby's, where "finding" implies reporting the find. However, it might say that when a Type II error occurs the researcher does not falsely believe that an effect or relationship is absent; she merely believes that she has not found an effect, and remains in a state of suspension of judgement. This would express a doxastic view.

Many combination views accept or suggest a semantic criterion (Rice, Iman and Connover, Crosby, Cohen). This is not universal though. Some texts explicitly caution against it.

Freund 1984 is an introductory text with a combination of the pragmatic criterion and perhaps the doxastic. The first detailed illustration given is a pragmatic case concerning the organization of airport car-parks. At one point this case is described as "typical" (1984, 282), although it is also admitted that it is rarely possible to determine "cash payoffs." Freund then rejects the semantic view. He says that though the term "null" has apparent connotations, nowadays the term can be used for "any hypothesis set up primarily to see whether it can be rejected." When a result is nonsignificant, as long as action is not forced upon the researcher, there is a choice between accepting the null and "reserving judgement," according to the circumstances (1984, 285-87). Thus we have the pragmatic criterion and the doxastic, considered as alternatives.

Another book which rejects a strong semantic criterion is Snedecor and Cochran 1980. They mention the dictionary meaning and connotations of the term "null," but say that in modern usage the term can mean any hypothesis being tested (1980, 64). Their positive view is hard to classify. It may be partly doxastic.

On the other hand, the only texts I looked at which clearly endorsed a *single* view of the asymmetry endorsed a semantic one. These were two recent texts, Larsen and Marx 1990 and Freeman et al 1990. Larsen and Marx define a null hypothesis as a "statement about the value of a parameter that reflects either the status quo or the *absence* of any special effect associated with the treatment being investigated" (1990, 377. See also Freeman et al 1990, 432). These books have broken from the pragmatic view completely. Lest a pattern be perceived too readily, the most recent text of all, Mendenhall and Sincich 1992, presents a pragmatic view (perhaps also semantic). This book is directed at engineers as well as scientists—perhaps a pragmatic holdover is more likely there.

The last book I will mention is older, but relevant firstly because it is still used, and secondly because its age supports a point I will make later. This is the massive, famous and generally philosophically savvy work by Kendall and Stuart (1961). Kendall and Stuart acknowledge that in some cases practical costs of mistakes can play a role in setting α and β. But they do not take this to be typical, and their treatment of cases in which behavior is not directly relevant is uncertain and nimble-footed. They dislike the term "null," and say that mere "convention or convenience" can bestow the benefit of a low α on a hypothesis (1961, 172). So the semantic view is rejected. They may have a doxastic view, but it is expressed with strain. They introduce the expressions "accept" and "reject" with the disclaimer that these terms

> are not intended to imply that any hypothesis is ever finally accepted or rejected in science. If the reader cannot overcome his philosophical dislike of these admittedly inapposite expressions, he will perhaps agree to regard them as code words, "reject" standing for "decide that the observations are unfavourable to" and "accept" for the opposite. (1961, 163)

This is hard to interpret. The expression "finally" in the first sentence makes it unclear whether they think there are problems with ordinary, revisable acceptance of a hypothesis. Secondly, if the opposite of "decide that the observations are unfavourable to" is "not decide that the observations are unfavourable to," then this passage asserts a doxastic view. If the opposite is "decide that the observations are not unfavourable to," then there is no doxastic asymmetry. I am not sure what they mean.

5. Conclusion

I have looked at 12 textbooks. Most have a combination view, and many do not give much guidance on the relation between criteria they recognize. It often appears to me that the conversation has gone roughly as follows:

Neyman: "Evaluating a hypothesis in science is *just* like deciding whether to buy a batch of light bulbs, so the error asymmetries are the same"

{Loud chorus of protests, scattered cries of "Bolshevism!"[10] etc.}

Textbooks: "Evaluating a hypothesis in science is *of course* different from making decisions about light bulbs, in part because mistakes in science have no definite practical consequences. But the error asymmetries are (for some reason) just the same."

Some of this ambiguity might be taken to support Birnbaum's view. Birnbaum said the "behavioral interpretation" of Neyman-Pearson testing has heuristic and illustrative value. As Birnbaum might predict, it is common for texts to use pragmatic cases to introduce the methods, but then claim that these cases are not typical of science. As I said earlier though, this does not solve the problem. Why should the asymmetry carry over to non-behavioral applications? The part of Birnbaum's account which answers this question, in a delationary way, is not supported by the texts I looked at.

I would like to propose a view which is in some ways the opposite of Birnbaum's. Birnbaum saw the pragmatic cases as serving a heuristic function, a helpful role. I suspect they often do the opposite. Neyman-Pearson methods were introduced to science largely by those who really accepted a pragmatic, behaviorist picture of testing, such as Neyman and Wald. As the methods became established in textbooks in the 1950's and 1960's they carried this philosophical baggage with them. It was always incumbent on those who wanted to keep the methods but lose the philosophy to provide an alternative rationale for features like the α/β asymmetry. In my view, both the semantic and doxastic criteria provide defensible alternative accounts, which need to be recognized for what they are and explored further. What I think we observe in some textbooks is a kind of philosophical inertia. The pragmatic picture is used to give intuitive meaning to the machinery of the tests, and it is used to *avoid* giving a clear alternative explanation of *why* β is routinely allowed to be five or ten times larger than α.

This uncertainty is especially visible in Kendall and Stuart *1961*, which was written in the period when Neyman-Pearson methods were becoming standard but the foundational debate was underway. This work shows signs of a tension between the appeal of the existing pragmatic way of thinking about testing, and the substantial problems associated with this view of science. The reader may object that Sokal and Rohlf *1981* is as indecisive as Kendall and Stuart *1961*. That is true. I do not want to claim a historical trend from as small and ambiguous a data set as this. A systematic examination of books from different decades is needed.

In the practice of science itself there is evidence of widespread employment of the semantic criterion (note 9) and, sometimes, the doxastic criterion. In fact, there may be a trend over recent decades towards an increasingly strong semantic interpretation of the asymmetry. The semantic criterion is especially interesting philosophically, as its use is Occamist. When there is a general principle that the null is the hypothesis of no effect or no relationship, the α/β asymmetry lends special protection to the more parsimonious view of the world. Further, this practice puts a rough *figure* on the extent to which the parsimonious hypothesis should be favored. Cohen (1977) proposes that a maximum β of 0.2 would be a good standard, when α is set at 0.05. This would set the difference between the two errors at a factor of four, which Cohen regards as a rough estimate of their relative epistemic seriousness. In existing practice a larger difference is often tolerated. In contexts in which the semantic criterion is used, can these facts about practice be viewed as a *measure* of the extent to which an Occamist bias is sanctioned in various disciplines? Could we model the effects of this type of Occamism, practiced to varying degrees? What is the rough degree of Occamist bias we would recommend, for a given field or problem, at reflective equilibrium?

6. Summary

Text	Pragmatic	Doxastic	Semantic	Deflationary
Kendall & Stuart 1961	Yes	? (Disj)		
Cohen 1977	?	?	? (Conj)	
Crosby 1977	Yes		Yes (Conj)	
Snedecor & Cochran 1980		?		
Sokal & Rohlf 1981	Yes		?	?
Weiss & Hasset 1982	Yes			
Iman & Conover 1983	Yes		Yes (Conj?)	
Freund 1984	Yes	Yes (Disj)		
Rice 1988	Yes		Yes (Disj)	
Freeman et al 1990			Yes	
Larsen & Marx 1990			Yes	
Mendenhall & Sincich 1992	Yes		?	

Notes

[1] Thanks to Stephen Downes and Isaac Levi for comments on earlier drafts.

[2] This is rarely visible in the original papers Neyman wrote with E.S. Pearson (Neyman and Pearson 1967). That would be no surprise if, as Mayo (1992) has argued, Egon Pearson did not agree with Neyman's views, and with what is often called the "Neyman-Pearson" philosophy of statistics.

[3] Source: textbook examples and asking biologists.

[4] We could demand that a test detect a difference of 10% between the two populations with a power of 0.9, for instance.

[5] If the situation is epistemically benign enough for ß to be comparable to α, I assume for Levi we are free to take the same attitude to a significant and a nonsignificant result. Just to confuse the issue, Neyman (1942, 303) uses language tantalizingly close to Levi's.

[6] It does this as long as ß does not fall *below* α. See Kendall and Stuart 1961 and Mayo 1985 for dangers here—a test can acquire a pathologically high power.

[7] Some textbooks I have used for other reasons. Others are used in courses at Stanford, and others were chosen by taking a random armful from the Stanford Mathematics and Computer Science Library. I do not discuss a small number of books which say nothing at all about the asymmetry, but all the books which say something about it are discussed.

[8] They also say when a result is nonsignificant "the null hypothesis cannot be rejected" (1982, 264), which does suggest a doxastic view.

[9] I intend to look at this in a longer study. For example, within two debates in biology there is substantial bi-partisan agreement about the null hypothesis. The neutral theory of molecular evolution is accepted by many on both sides of the debate to be an appropriate null (Crow 1987), but data which support it are publishable. Beatty (1987) disagrees with this view of neutralism, but has a very restricted view of the roles nulls play. Also, in the debate about extra-terrestrial causes of mass extinctions there is bipartisan support for the view that the absence of an extra-terrestrial cause is the null (Raup 1987).

[10] See Mayo 1992.

References

Beatty, J. (1987), Natural selection and the Null Hypothesis. In J. Dupre (ed.) *The Latest on the Best: Essays on Evolution and Optimality.* Cambridge, MA: MIT Press.

Birnbaum, A. (1977), The Neyman-Pearson Theory as Decision Theory, and as Inference Theory; with a Criticism of the Lindley-Savage Argument for Bayesian Theory. *Synthese* 36: 19-49.

Cohen, J. (1977), *Statistical Power Analysis for the Behavioral Sciences* (revised edition). New York: Academic Press.

Cosby, P.C. (1977), *Methods in Behavioral Research* (3rd edition). Palo Alto: Mayfield.

Cramér, H. (1951), *Mathematical Methods of Statistics.* Princeton: Princeton University Press.

Crow, J. (1987), Neutral Models in Molecular Evolution. In Nitecki and Hoffman 1987.

Fisher, R.A. (1935), *The Design of Experiments.* Edinburgh: Oliver and Boyd.

_____. (1956), *Statistical Methods and Scientific Inference.* Edinburgh: Oliver and Boyd.

Freeman, D. R. Pisani, R. Purves and A. Adhikari (1990), *Statistics* (3nd edition). New York: Norton.

Freund, J.E. (1984), *Modern Elementary Statistics* (6th edition). Englewood Cliffs: Prentice Hall.

Giere, R.N. (1977), Testing Versus Informational Models of Statistical Inference. In R.G. Colodny (ed.) *Logic, Laws and Life. Some Philosophical Complications.* Pittsburgh: University of Pittsburgh Press,

Iman, R.L. and W.J. Conover (1983), *A Modern Approach to Statistics.* New York: John Wiley.

Kendall M.G. and A. Stuart (1961), *The Advanced Theory of Statistics. Volume II: Inference and Relationship*. New York: Hafner.

Larsen, R.J. and M.L. Marx (1990), *Statistics*. Englewood Cliffs: Prentice Hall.

Levi, I. (1962), On the Seriousness of Mistakes. Reprinted in I. Levi *Decisions and Revisions*. Cambridge: Cambridge University Press, 1984.

Mayo, D.G. (1985), Behavioristic, Evidentialist, and Learning Models of Statistical Testing. *Philosophy of Science* 52: 493-516.

_____. (1992), Did Pearson Reject the Neyman-Pearson Philosophy of Statistics? *Synthese* 90: 233-262.

Mendenhall, W. and T. Sincich (1992), *Statistics for Engineering and the Sciences*. (3rd edition) San Francisco: Dellen.

Neyman, J. (1942), Basic Ideas and Some Recent Results of the Theory of Testing Statistical Hypotheses. *Journal of the Royal Statistical Society* 105: 292-327.

_____. (1950), *First Course in Probability and Statistics*. New York: Henry Holt.

Neyman, J. and E.S. Pearson (1967) *Joint Statistical Papers*. Berkeley and Los Angeles: University of California Press.

Nitecki, M.H. and A. Hoffman (1987), *Neutral Models in Biology*. Oxford: Oxford University Press.

Raup, D. (1987), Neutral Models in Paleobiology. In Nitecki and Hoffman 1987.

Rice, J.A. (1988), *Mathematical Statistics and Data Analysis*. Belmont: Wadsworth.

Snedecor, G.W. and W.G. Cochran. (1980), *Statistical Methods* (7th edition). Ames: Iowa State University Press.

Sokal, R.R. and F.J. Rohlf (1981), *Biometry*. (2nd edition). New York: Freeman.

Wald, A. (1950), *Statistical Decision Functions*. New York: Wiley.

Weiss, N. and M. Hassett (1982), *Introductory Statistics*. Reading: Addison-Wesley.

Part VIII

HISTORICAL CASE STUDIES AND METHODOLOGY

Experiment, Speculation and Law: Faraday's Analysis of Arago's Wheel[1]

Friedrich Steinle

Georg-August-Universität, Göttingen

1. Introduction

This paper deals with the mutual relation of speculative considerations, experimental activity, and the attempt to establish "laws of nature." I shall not give a general analysis, but instead study a historical example. It is taken from the work of Michael Faraday - one of the most original and most successful experimenters.

I shall start by giving a rough sketch of Faraday's view, thereby making clear the particular question on which my paper concentrates. In part three, I analyze an episode of Faraday's actual work, and conclude in part four with a rough analysis of the type of explanation given by Faraday.

2. Law and speculation

Faraday repeatedly emphasizes the necessity of distinguishing "that knowledge which consists of assumption, by which I mean theory and hypothesis, from that which is the knowledge of facts and laws; never raising the former to the dignity or authority of the latter, nor confusing the latter more than is inevitable with the former." (1844, 285-286)

It is important to note that Faraday uses the term "theory" here—as mostly—in a rather narrow and specific sense. Examples are Ampère's theory of molecular currents, the one-or-two-fluid-theories of electricity, the various theories of heat (1854, 481), the corpuscular and the wave-theory of light (1852, 408), etc. What he seems to have in mind are assumptions which aim at explaining sensible experience and phenomena by means of causes lying beyond the reach of experience. In his view such theories have an intrinsically speculative character; indeed he often uses the terms "theory" and "speculation" interchangeable[2].

Chemistry, the field in which Faraday was originally trained, may provide a nice illustration. Besides "theoretical" endeavours such as the atomic theories of Dalton or Boscovich, we find in Faraday's time diverse attempts at classifying chemical substances according to their chemical properties, of distinguishing "elementary" from "composite" ones, and of systematizing the "elementary" ones. These attempts were mainly founded on a large body of empirical knowledge about the chemical behaviour of substances, and did not necessarily involve any considerations of atoms, intermolecular forces, or the like. The classifying systems aimed at in this way are not "theories" in Faraday's sense.

The outlines of the relation Faraday finds between theoretical speculations, and considerations of facts and laws have been analyzed in some detail by Cantor (1991, ch.8). I shall only mention the main point, thereby leaving untouched the problems involved in the very concepts of facts and laws. For Faraday, only facts and laws are reliable and lasting elements of science. In this respect, theories are of considerably lower dignity and authority. Nevertheless, they are seen to play an important and indispensable role in science in that

> ... not only are they useful in rendering the vague idea more clear for the time, giving it something like a definite shape, that it may be submitted to experiment and calculation; but they lead on, by deduction and correction, to the discovery of new phenomena, and so cause an increase and advance of real physical truth, which, unlike the hypothesis that led to it, becomes fundamental knowledge not subject to change. (1852, 408)

The point can perhaps best be illustrated by the metaphor, used by Cantor (1991, 211), of a scaffolding which is indispensable for the process of erecting the building of science, but is removed when the building is finished and does not form an integral constituent of it.

Given this view, however, the question arises of how the relation between theories and facts/laws influences or even determines Faraday's everyday experimental work. Faraday claims that, although the discovery of new phenomena may essentially be driven by theoretical considerations, they no longer play any role in the resulting system of facts and laws. This entails the assumption that, in between these two stages, theories must have been somehow left aside. Can we identify such a process in Faraday's work? Regarding experimentation, we must adress a further question: if theories are to be put away, what then are the driving forces and guidelines directing his experimental activity?

Faraday's experimental procedures have been studied intensely from various aspects within the last decade (Cantor 1991, Crawford 1985, Romo/Doncel 1994, Tweney 1985 and 1992, among others). It is in particular David Gooding who has, in many studies (1985, 1990, 1992, for example), provided impressive examples of Faraday's experimentation being often independent of particular theories. The above mentioned questions, however, seem to me still to be open.—My own account takes as a starting point Faraday's methodological view. I try to make out the degree in which his claims actually correspond to his actual practice. I include, moreover, the point that the structure of his explanations of phenomena differs significantly from those usually given at his time.

As a result of having analyzed some periods of Faraday's laboratory work, I propose the thesis that a process of leaving theories aside can indeed be traced in Faraday's experimental work. I cannot present here the full material supporting this thesis. What I shall do, instead, is try to render it at least partially plausible by presenting one case-study in detail.

3. The case of Arago's wheel

I shall analyze Faraday's treatment of a phenomenon, known as Arago's disc or Arago's wheel. In 1825 François Arago announces his finding that a copper-disc, rotating in a horizontal plane, sets in motion a magnetic needle which is suspended horizontally above it without any mechanical contact (Arago 1825). Babbage and Herschel, in experimenting on this effect, find an "inverse" effect, in which the magnet, when rotated, puts the copper disc in motion. Although Arago claims to have found the effect in discs of a wide range of materials, as well in fluids and even gases, Babbage and Herschel claim that an "unequivocal" effect occurs only in "best conductors of electricity" (Babbage/Herschel 1825, 484-485 and 472-480).

Even more obscure than this experimental dissension is the explanation of the effect: it is not understandable in terms of magnetism, since copper is an entirely unmagnetic material, and, in the rest state, there is not the slightest indication of a force between the magnet and the copper disc. Babbage and Herschel, in order to give an explanation, speculate that the poles of the magnet "induce" contrary magnetic poles in the copper disc, hereby causing a slight attractive force (1825, 485-487). This process of induction is supposed, moreover, to take some period of time. In rotating the disc—or the magnet—the attractive force would, therefore, no longer be directed perpendicularly to the plane of the disc, but exhibit a component parallel to it, causing thereby the rotation of the magnet, respectively, the disc.

Faraday is attracted to Arago's effect quite early on (for details see Bradley 1989, 14 and Romo/Doncel 1994, 14). His first experiments are clearly driven by the idea of explaining the effect by electrical currents induced by the magnet. As to the origin of this idea, one can only speculate. It is in some degree made plausible by the above-mentioned experimental result of Babbage and Herschel. Faraday is, furthermore, strongly stimulated by Ampère's theory of molecular currents as the cause of magnetism (see Romo/Doncel 1994, 15-20). He derives from this theory an expectation of induced currents, flowing in the form of a vortex in the same direction as the supposed Amperian molecular currents in the magnetic pole[3].

This situation is essentially unchanged when, on 29 August 1831, he for the first time has an effect of electromagnetic induction before his eyes. It is not surprising to find that he quickly revives his old interest in Arago's effect (see his laboratory diary, reproduced in Martin 1932, 367, entry no. {17}. I shall give these entry-numbers in {}). He performs indeed a series of experiments on this problem, commencing sooner at the margin of his other experiments on electromagnetic induction, but culminating in some days work of highest intensity, dedicated exclusively to this point.

He starts, not surprisingly, by looking for any induced electric effects in Arago's disc. This idea, taken alone, however, is not so specific so as to lead directly to the construction of experiments. Now Faraday has learned already to distinguish between two different sorts of induction: volta-electric and magneto-electric induction, as he calls them some weeks later. Thus he attempts to make out which of these two appears to be the more promising candidate {54, 71, 72}. But, even this question decided, a possible experimental arrangement remains largely underdetermined. Faraday makes now intensive use of the vortex-theory, which predicts the form and direction of the supposed induced currents. This enables him to design a first experimental setting with a clear goal of what should be measured, at which place the contacts should be positioned, etc. {77}. When this first experiment delivers no positive result, he successively changes the arrangement in order to "optimize" it for the detection of the sought electric effect {78-80}. This "optimization procedure" is partly directed by his long experimental experience, partly by the underlying theory, partly by both factors together. It is successful in that it leads finally to an arrangement which shows "very distinct and constant" electrical effects {104}. The role of the vortex-theory in this process can be well characterized in Faraday's own words as "rendering the vague idea more clear for the time, giving it something like a definite shape, that it may be submitted to experiment and calculation", leading in this way, "by deduction and correction, to the discovery of the new phenomena ..." (already quoted above).

Shortly after this successful experiment he sets about "to examine [the] effect more minutely" {110}. This "examination" consists in a remarkable series of 46 experiments, at the end of which he has grasped a complete explanation of Arago's effect. The explanation is of the sort which Faraday claims to be independent of theories. This claim stands in sharp contrast to the degree in which his previous experiments are driven by speculative theory. If we are to take his claim seriously, we should find within this series of 46 experiments a process of leaving aside theories. Due to lack of space I shall not discuss all of these experiments but instead give the main lines and discuss in detail only some points of particular importance.

Faraday proceeds in two major stages: first he uses an apparatus with a rotating disc {110-136}, secondly one with a copper strip linearly moved {137-156}. To give an impression, I have assembled some of Faraday's numerous drawings in the *Diary*. Fig.99 shows the apparatus; the other figures here show the rotating disc viewed from the side, together with the position of the magnetic pole and the electric contacts, as well as the galvanometer with its connecting wires.

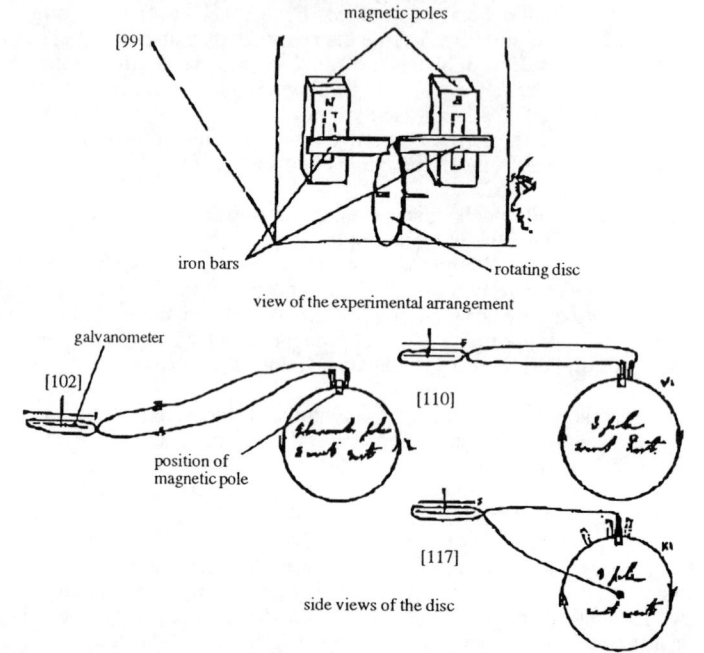

plate 1: Faraday's experiments with the rotating disc (figures of Martin 1932, 381-4)

The arrangement by which Faraday had, for the first time, detected the electric effect (fig.102) serves now as starting point: two electric contacts, as well as the magnetic pole, are arranged at the periphery of the disc in order to detect the imaginary vortex. Faraday now goes on to study the influence of the position of the two peripheral contacts and of the direction of rotation on the detected electric current ({110-114}, fig.110). His results are summarized in a rule:

115. Hence changing the direction of the motion of the plate changes the current, and also changing the conductors to the right or left reverses the current.

This rule, however, is not at all what the theory had predicted; indeed it scarcely makes any sense when thinking of vortices. Faraday gives only the comment "that the condensed imaginary bisected vortex was not so definite as supposed, if existing at all."{116}. He then analyzes the effect of only one peripheral contact, the other one becoming fixed at the axis of the disc as "the most neutral part of it" (fig.117). The results of these experiments {117, 118} show that the direction of the induced current depends on the direction of rotation only, and not on the position of the peripheral contact. What is affected by this position is only the intensity of the current, in that it decreases with increasing distance between contact and magnetic pole. This result enables Faraday to understand immediately his previous results with two peripheral contacts: Two conductors "only shewed the difference of intensity of the two currents setting into them ..."{119}.

This episode is quite characteristic for his procedure. By varying systematically certain experimental parameters and studying the corresponding changes of the experimental outcomes, he establishes empirical rules. In a further step he then establishes *relations* among these rules, in that some of them are derivable from others. Like the rules themselves, these relations do not need to make use of "theories". In the example given, the main point is some empirical knowledge about the measurement of differences among electrical effects.

Such methodological principles are applied now not casually, but quite systematically. This becomes clear by looking at the whole of Faraday's experiments with the rotating disc. In the course of this series he varies a great number of parameters of the apparatus:

- the direction of rotation of disc,
- the number and position of the peripheral contacts {110-122},
- the position of the magnetic pole {123-126},
- the number of magnetic poles {127- 128},
- the distance between magnetic pole and disc {128},
- the plane of rotation of the disc {133, 134},
- the position of the inner contact on the disc {136}.

The vortex-theory is mentioned twice further. In varying the position of the magnetic pole, Faraday tries to imagine the expectations derived from this theory. The experimental results, however, are completely incompatible with these expectations. Faraday's concluding remark, ("Hence if any vortex it must be very diffuse" {126}), can be understood as a definite "laying aside" of such considerations; at least the vortex-theory is no longer mentioned. This point deserves attention, for the question posed above becomes pressing now. If the directing force of the vortex-theory has now definitely left behind, what considerations do instead guide Faraday's experimentation? He does not make use of, or search for, another theory. There are, instead, the mentioned methodological principles which direct his activity.

There arises, however, a serious problem. Since not all experimental parameters can be varied, a selection must be made. This selection may be, possibly implicitly, yet again driven by theory, which would reveal the idea of having left theory behind as illusory. Looking to the list of parameters given above, this supposition may perhaps be made in the case of the parameter "position of the magnet." In all other cases, however, I cannot see how the vortex-theory—or any other one—should have influenced the selection. Instead the essential point is Faraday's great experimental experience in electromagnetic phenomena, which makes certain parameters appear as having a possible influence on the induction phenomenon. Although this has, of course, to do with conceptual frameworks, it does not necessarily involve theories in Faraday's sense. Faraday is well aware of this problem: when forming new concepts or even terms, he endeavours to hold them free of "theoretical" preconceptions (see Ross 1961). As Gooding (1992, 128) emphasizes, he describes new phenomena "as fully and unselectively as possible".

When Faraday then changes his apparatus, replacing the rotating disc by a linearly moved strip or plate of metal (fig.137), he makes only a very short but remarkable comment: "Then experimented with plates, as being simpler."{137}. "Simplicity", as used here by Faraday, seems to be an important factor of his considerations. I cannot analyze this concept in detail here[4]. It becomes clear, however, that "simplicity" in Faraday's use is intimately connected with a process of derivation: the "simpler" phenomena with the strip serve for Faraday as an aid to understand the more complex ones with the disc. The above-mentioned case of explaining the two-contact-rule by the one-contact-rule provides another example.

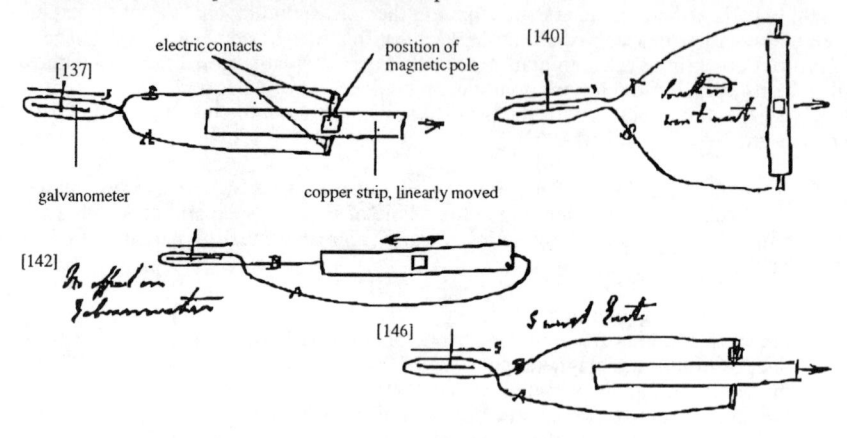

plate 2: Faraday's experiments with a linearly moved strip (figures of Martin 1932, 386-7)

With the new arrangement Faraday again varies many experimental parameters:

- the direction of motion of the strip {137- 138},
- the position of contacts at the strip {139},
- the dimensions of the strip {140, 143, 144, 155}, fig.140,
- the direction in which currents can possibly be detected, relative to the direction of motion {142}, fig.142,
- the relative motion between the contacts and the strip {141},
- the material of the strip: Cu, Fe, Pb, Sn, Zn {145, 152-154},
- the position of the magnetic pole {146}, fig.146,

- the polarity of the magnet {147, 148},
- the distance between the magnetic poles {151},

In contrast to the series with the rotating disc, there are no longer any traces of "theoretical" considerations which may have led to these variations, or, to the selection of these particular parameters.

In {155} Faraday describes an experiment in which he moves a single wire connected to the galvanometer perpendicularly to its own direction. The experiment shows that the wire, "when taken through [i.e. when moved between the poles of the magnet], produced very distinct effect". Being just a variation of the one described in {140} (see fig.140), this experiment is of high importance for Faraday. It brings out in particular clarity a general rule which in his paper reads as follows:

> All the experiments combine to prove that when a piece of metal (and the same may be true of all conducting matter) is passed either before [a] single pole, or between opposite poles of a magnet, electrical currents are produced across the metal transverse to the direction of motion...(1832, 119)

This rule plays a central role in that i) it names the essential conditions for producing induced currents; and, ii) it states clearly the transverse direction of the induced current relative to the direction of motion.

Faraday recognizes the importance of this experiment immediately: It forms effectively—besides one other experiment {156}, designed to study a special feature—the endpoint of the whole series of experiments dealing with Arago's effect. Moreover, after this day there is break of more than four weeks in Faraday's experimenting. In this time he works out his paper for the Royal Society, which already contains the complete explanation of Arago's effect.

To summarize: Faraday, in the whole series of experiments, is engaged almost exclusively in varying a multitude of parameters, in finding empirical rules, and in relating them to more general or "simple" ones. A central aim hereby is to determine which of the parameters are indispensable or essential for the analyzed effect, and then to reduce the arrangement to these essential conditions.

4. Explaining phenomena

To make clear what the result of Faraday's procedure looks like, I shall, finally, sketch the main lines of his explanation of Arago's effect as presented to the Royal Society in November/December 1831. I refer to the first version of his paper (as given in Romo/Doncel 1994), since it flowed directly out of the experimental series described[5]. (The numbers refer to the paragraphs of Faraday 1832.)

After having described Arago's effect and the state of investigation, Faraday proceeds in three stages. He describes first his experiments with the rotating disc (84-100), secondly the "simpler experiment" (101) with a linearly moved strip (101-110). Then he presents the actual explanation. He applies the general rule of the direction of the induced current in a wire to the rotating disc by imagining it as being composed of single wires as "spokes". He is well aware that such a construction is not unproblematic, since the possibility of interactions between the single spokes must be taken into account. The various experimental outcomes, however, convince him that, in the case of the rotating disc, the "spoke"-construction is well suited. He concludes that, within the rotating disc, the induced currents "will approximate towards the direction of the radii" (119).

This holds, however, only under the assumption that the currents do not completely change or even cease if the galvanometer gets disconnected. This point is not trivial, for the galvanometer forms an integral constituent of the closed circuit. Only in the later version of his paper does Faraday add a passage concerning this point. He assumes that when the galvanometer is disconnected, the currents "return in the parts of the plate on each side of and more distant from the place of the pole" (123). He gives no experimental evidence for this assumption. It is clear, however, that in principle such evidence could be gained by procedures not essentially different from those already used.

Once the point of radial currents is established, the explanation is straightforward. What is still missing is any account of the mechanical force exerted between the disc and the magnet. This is explained by a phenomenon, established by Faraday ten years earlier. Between an electric current in a wire and a nearby magnetic pole, a mechanical force is exerted whose direction is well defined; it depends on the direction of the current and the polarity of the magnet (Faraday 1821). This rule can immediately be applied to the interaction between an induced current in a wire and the magnetic pole responsible for the induction. In combination with the "spoke"-rule for the rotating disc, this leads directly to the explanation of the mechanical force between the magnet and the rotating disc as observed in Arago's effect.

One may summarize Faraday's explanation schematically (plate 3). It starts with the electrical phenomena of the disc (i) and relates these, via the phenomena of the strip (ii), to those observed in a single wire (iii). In this case, the mechanical force between currents and magnetic poles (iv) is already explained by a rule, and Faraday can combine it with the induction rule. He then goes "the way back" to the rotating disc (v, vi), making thereby understandable the mechanical phenomena occuring there.

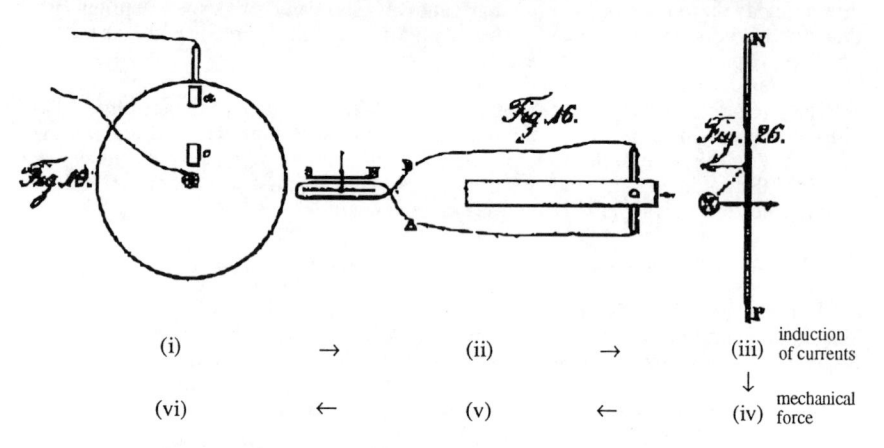

(i)	$\rightarrow$	(ii)	$\rightarrow$	(iii)	induction of currents
					$\downarrow$
(vi)	$\leftarrow$	(v)	$\leftarrow$	(iv)	mechanical force

plate 3: The main steps of Faraday's explanation of Arago's disc
(figures of Faraday 1839-55, plate I)

The main elements of which this explanation makes use are these:

- the rule of electromagnetic induction,
- the rule of the mechanical force between wires and magnets,
- the combination-rule of the disc out of single wires as "spokes",
- the assumption that the currents are present even when the galvanometer is disconnected.

None of these elements depends in any way on concepts of the nature of currents, or by what means the interaction between magnets and currents takes place, and so on, or in short: on theories in the sense described above. In contrast to the account of Babbage and Herschel no entities or processes need to be postulated which lie principally beyond the reach of experience.

Faraday's explanation of Arago's effect provides, therefore, a characteristic example of his concept of science as revealing the mutual interdependence of phenomena without involving theories. The steps of this explanation bear, moreover, a close resemblance to the main steps of his experimental proceeding. The *type* of explanation aimed at in the laboratory practice seems not to be different from that offered in his paper. In his experimental activity we can clearly trace a process of starting with theories, leaving them aside succesively, and finally ending with a system of another sort. The methodological guidelines driving this process are the principles of

- varying as many as possible parameters of the experimental apparatus,
- establishing empirical rules about the effect of these variations,
- attempting to find out which of the experimental conditions are essential for the effect sought,
- reducing the apparatus—as far as possible—to these essential conditions, thus obtaining the phenomenon as "simple" or as "pure" as possible.

To conclude I would like to add one further observation: Faraday explains the Arago-effect by classifying it—for experimental reasons—as a phenomenon of electro-magnetic induction, combined with magneto-electric mechanical interaction. To come back to the example of chemistry mentioned above, his procedure is rather like that of a chemist who endeavours to classify some substance to be a composite of other, more elementary substances, than to that of one who tries to explain the properties of the substance out of some assumed properties of atoms or molecules. As elsewhere in his work (see, for example, Gooding's observation in (1992), 138, note), Faraday's training as a chemist seems to have left its traces also in his methodology.

Notes

[1] I would like to thank Lorenz Krüger, Ryan Tweney. and Larry Holmes for critical and stimulating discussion, and the Deutsche Forschungsgemeinschaft for supporting my research.

[2] What Cantor (1991, 209) finds to be the second type of hypotheses in Faraday's work is essentially what Faraday calls theories.

[3] I shall not deal with the question of the degree to which his impression was really justified; for a discussion see Romo/Doncel 1994, 17-20

[4] I analyze this concept in another place (Steinle, forthcoming).

[5] For a detailed study of the successive alterations and augmentations of the manuscript see the thorough analysis of Romo and Doncel (1994).

References

Arago, F. (1825), *"L'action que les corps aimantés et ceux qui ne le sont pas exercent les uns sur les autres"*, *Annales de chimie et de physique* 28: 325-326.

Babbage, C. and Herschel, J.W.F. (1825), "Account of the Repetition of M.Arago's Experiments on the Magnetism Manifested by Various Substances During the Act of Rotation", *Philosophical Transactions* 115: 467-496.

Bradley, J.K. (1989), *The Early Electromagnetic Experiments of Faraday and Their Replication*. M.Phil.Diss. Univ. Cambridge.

Cantor, G.N. (1991), *Michael Faraday: Sandemanian and Scientist*. Basingstoke: Macmillan.

Crawford, E. (1985), "Learning from Experience" in D. Gooding and F.A.J.L. James (eds.), pp. 211-227.

Faraday, M. (1821), "On some new Electro-Magnetical Motions, and on the Theory of Magnetism", *Quarterly Journal of Science 12:* 74-96; quoted of (1839-55), II, pp. 127-47.

_____. (1832), "Experimental Researches in Electricity, i.e. I. On the induction of electric currents...", *Philosophical Transactions 122:* 125-62; quoted of (1839-55), series I.

_____. (1839-55), *Experimental Researches in Electricity*, 3 vols. London: Taylor.

_____. (1844), "A Speculation Touching Electric Conduction and the Nature of Matter", *Philosophical Magazine 24:* 136-44; quoted of (1839-55), II, pp. 284-239.

_____. (1852), "On the Physical Character of the Lines of Magnetic Force", *Philosophical Magazine* 3: 401-28; quoted of (1839-55), III, pp. 407-437.

_____. (1854), "Observations on Mental Education", *Lectures on Education Delivered at the Royal Institution of Great Britain*, London, pp. 39-88; quoted of: *Experimental Researches in Chemistry and Physics*. London (1859): Taylor and Francis, pp. 463-91.

Gooding, D. (1985), "In Nature's School: Faraday as an Experimentalist" in D. Gooding and F.A.J.L. James (eds.), pp. 105-135.

_____. (1990), *Experiment and the Making of Meaning*. Dordrecht: Kluwer.

_____. (1992), "Mathematics and Method in Faraday's Experiments", *Physis* 29: 121-147.

Gooding, D. and James, F.A.J.L. (eds.) (1985), *Faraday Rediscovered*. Basingstoke: Macmillan.

Hacking, I. (1983), *Representing and Intervening*. Cambridge: Cambridge University Press.

Martin, T. (ed.) (1932), *Faraday's Diary.* vol.I, London: Bell.

Romo, J. and Doncel, M. (1994): "Faraday's Initial Mistake Concerning the Direction of the Induced Currents, and the Manuscript of Series I of his 'Researches'", *Archive for History of Exact Sciences*.

Ross, S. (1961), "Faraday Consults the Scholars: The Origins of the Terms of Electrochemistry", *Notes and Records of the Royal Society of London* 16: 187-220.

_ _ _ _ . (1965), "The Search for Electromagnetic Induction 1820-1831", *Notes and Records of the Royal Society of London* 20: 184-219.

Steinle, F. (forthcoming), "Looking for a 'Simple Case': Faraday and Electromagnetic Rotation", *History of Science*.

Tweney, R. (1985), "Faraday's Discovery of Induction: A Cognitive Approach" in D. Gooding and F.A.J.L. James (eds.), pp. 189-209.

_ _ _ _ _ _. (1992). "Stopping Time: Faraday and the Scientific Creation of Perpetual Order", *Physis 29:* 149-164.

Scientists' Responses to Anomalous Data: Evidence from Psychology, History, and Philosophy of Science[1]

William F. Brewer and Clark A. Chinn

University of Illinois at Urbana-Champaign

1. Introduction

The purpose of this paper is to provide a systematic account of the role of anomaly in theory change in science. We adopt a naturalist approach to the philosophy of science (e.g., Giere 1985; Maffie 1990) and support our proposal with evidence from the history of science and from the psychology of science.

Previous discussions of the topic of anomaly in theory change have rarely used evidence from psychology. Probably this has been due to the relatively immature state of research in the psychology of science (cf. Gorman forthcoming; Gholson, Shadish, Neimeyer, and Houts 1989). However there is now enough research on the topic of belief change (cf. Chinn and Brewer 1993b) to provide a foundation for theory development in the area of responses to anomaly. Much of the research in this area uses data from undergraduates and children and not scientists (see Dunbar forthcoming, for an exception). One would prefer to have direct experimental studies of scientists, but to the degree that one can assume similar underlying reasoning processes in scientists and nonscientists (see Brewer and Samarapungavan 1991), then the data from undergraduates can provide relevant information.

The next section of this paper presents an overview of our analysis of the forms of response that individuals make when confronted with anomalous data. Then, in the core of the paper we go through each postulated form of response and provide evidence for our position from experimental psychology and from the history of science. Finally, we attempt to draw some implications of our analysis for general issues in the philosophy of science.

2. Psychological Responses to Anomaly

In idealized form we conceptualize the situation in which anomaly occurs as follows: An individual currently holds theory A. The individual then encounters data that appear to be inconsistent with theory A. The anomalous data may or may not be accompanied by theory B, which is intended to explain much of the domain of data explained by theory A, plus the new anomalous data.

What are the possible responses of the individual to the inconsistent data? We postulate that there are seven basic responses to anomalous data: (1) *ignore* the data and retain theory A; (2) *reject* the data and retain theory A; (3) *exclude* the data from the domain of theory A; (4) hold the data in *abeyance* and retain theory A; (5) *reinterpret* the data and retain theory A; (6) reinterpret the data and make *peripheral changes* to theory A; and (7) *change* theory A, possibly in favor of theory B.

We would like to argue that this is an exhaustive set of the psychologically plausible responses to anomalous data. When individuals are faced with anomalous data, they have to resolve three basic problems. (a) They have to decide whether they believe the data. If they do not accept the anomalous evidence, then the data are no longer a problem. (b) They have to decide if the anomalous data can be explained and, if so, how the data are to be explained. (c) They have to decide if the data require a change in their current theory. In Table 1 we give the seven forms of response to anomalous data and display for each the pattern of problem solutions that are associated with that form of response. We think the seven categories cover the psychologically plausible combinations of solutions to the three problems.

In the next sections of the paper we will discuss each of these forms of response to anomalous data, and for each one we will present experimental evidence from psychology and selected examples from the history of science. In our analysis of the evidence from the history of science, there are often two different forms of evidence. There are self reports by the scientist who has observed the anomalous data about his or her own response to the data. In addition, there are reports about the reception of the anomalous data by other members of the scientific community when the discovery is made public.

3. Ignoring

Ignoring anomalous data is perhaps the most extreme way of disposing of it. When an individual ignores data, he or she does not even bother to explain it away. Theory A remains intact and totally unscathed.

3.1. Psychology

There is no good evidence in the psychology literature on ignoring anomalous data. Typical studies in this area overtly present the anomalous data to the learner and do not give the learners the option to ignore the data.

3.2. History of Science

There are frequent reports of "anomalous data" in the popular press that are typically ignored by scientists. Thus, physicists ignore reports of new perpetual motion machines. Psychologists ignore claims for the existence of ESP. Astronomers ignore the predictions of astrology. Biologists ignore the reports of sightings of the Loch Ness monster.

However, it is not merely fringe ideas like ESP and astrology that are ignored by scientists. Data that are later accepted by the scientific community may be similarly ignored. For example, it is reputed that some Aristotelian philosophers refused to look through Galileo's telescope, which was revealing data anomalous to the Aristotelian world view (Fermi and Bernardini 1961, 59). Another example can be seen in the response of physicists to the finding that hot water freezes faster than cold water. This fact was known for hundreds of years in the writings of Aristotle, Descartes, and Bacon (Osborne 1979), yet after the development of thermodynamics it vanished from the scientific literature until rediscovered by a Tanzanian high school student! (Mpemba and Osborne 1969).

Table 1
Characteristics of the Seven Responses to Anomalous Data

Response Type	Characteristics of the Response		
	Are Data Believed	Are Data Explained	Is Theory Changed
Ignoring	No	No	No
Rejection	No	Yes	No
Exclusion	Yes (or maybe)	No	No
Abeyance	Yes	Not now	No
Reinterpretation	Yes	Yes	No
Peripheral Change	Yes	Yes	Yes (peripheral)
Theory Change	Yes	Yes	Yes (core)

4. Rejection

Rejecting data is similar to ignoring data in that the individual does not accept the data, and does not make any changes to theory A. The difference is that in ignoring data, the individual does not even attempt to explain the data away; in rejection, the individual has an explanation for why the data should be rejected. Individuals who reject data use a wide variety of reasons for the rejection; however, three very common forms of rejection are: (a) arguing that there was a fundamental methodological error in the way the data were obtained; (b) arguing that the data were merely a random effect; and (c) declaring the data to be fraudulent.

4.1. Psychology

Chinn and Brewer (1992, 1993a, 1994) have conducted a series of experiments investigating how undergraduates respond to anomalous data. In these experiments, students learned about the theory that the mass extinctions at the end of the Cretaceous period were caused by a meteor impact. Students received an array of evidence supporting the theory. One piece of evidence was that the KT boundary, a thin layer of clay separating Cretaceous from Tertiary sediments, contains a high concentration of iridium at many sites around the world; because iridium is rare on the earth's surface but common in meteors, the high concentration of iridium implies that a large meteor struck the earth. After reading about the meteor theory, nearly all the students reported a strong belief in the theory. Next students were given data that contradicted the meteor theory. For example, some students received data that supported the conclusion that the KT boundary was deposited over 10,000 years, which is much longer than one would expect if a meteor had struck the earth. Across these experiments, more than half of the undergraduates rejected data that contradicted the meteor theory. Some examples of their written justifications for rejection are: "I'm skeptical of his methods of obtaining his theory"; "Subtracting the year of the top and bottom rock doesn't seem to be a very accurate way of dating the [sediments]"; and "It needs additional support from other scientists to support it."

4.2. History of Science

For scientists it appears that the most common ground for rejecting data is methodological error. Holton (1978) has provided a careful analysis of the dispute between Millikan and Ehrenhaft over the nature of the charge on the electron. Each side thought the other was making methodological errors. Ehrenhaft accused Millikan of selecting his data in order to support his view that the charge was unitary, while Millikan thought that Ehrenhaft was including bad data points in his analyses, which would give results that would appear to support Ehrenhaft's position against unitary charge.

When Osborne assigned a laboratory assistant to carry out the first modern laboratory investigation of the Mpemba effect this individual reported back that his data showed that hot water did freeze before cold, "But we'll keep on repeating the experiment until we get the right result" (Mpemba and Osborne 1969, 174).

A final example of focusing on methodology can be seen in the response of Hewish and Bell to their discovery of pulsars. Hewish stated that when they realized the source was producing very regular pulses, he treated the data "with skepticism bordering on incredulity" and noted that "in 99 cases of 100 peculiar 'variable radio sources' turn out to be some kind of electrical interference—from a badly suppressed automobile ignition circuit, for example" (Hewish 1968, 25). He noted, "I could not believe that any natural source would radiate in this fashion, and immediately consulted astronomical colleagues at other observatories to inquire whether they had any equipment in operation which might possibly generate electrical interference at a fixed sidereal time" (Hewish 1975, 1081).

5. Exclusion

Another possible response to contradictory data is to place it outside the domain of the theory. In this case the individual declares that theory A is not intended to explain the data and so does not have to make a judgment about the validity of the data. When anomalous data are excluded from the domain of a theory, they obviously do not lead to any theory change.

5.1. Psychology

Karmiloff-Smith and Inhelder (1975) investigated children attempting to balance blocks on a narrow metal support. Some of the blocks had their weight evenly distributed so that they balanced at their geometric center. Other blocks had a mass of lead hidden at one end, so that they balanced far off center. At the age of 6 or 7, children developed a geometric theory of balancing: things balance in the center. Of course, the unbalanced blocks would not balance in the center, and when the children tried to make them balance in the center, the blocks kept falling. Instead of changing their geometric center hypothesis, the children declared that the uneven blocks were impossible to balance and did not worry about them further. From our perspective, it appears that the children declared those blocks to be outside the domain of the theory they were developing. In this way, they were able to preserve their theory unaltered.

5.2. History of Science

According to Thagard (1992), a case of exclusion occurred in geophysics and geology around the middle of this century. Geophysicists accepted the principle of isostasy, which asserts that a block of the earth's crust that has not been recently disturbed is in gravitational equilibrium, with equal forces pushing upward and downward. Geologists, on the other hand, were concerned with the striking similarity of land fossils on opposite sides of oceans. They explained this by postulating that land bridges periodically rose and fell between continents. The two ideas are in contradiction; isostasy forbids the rise and fall of land bridges. Thagard concludes, "The conflict between the hypotheses of isostasy and land bridges was tolerated because they belonged to different disciplines. Only geophysicists were concerned with the validity of isostasy, and only geologists were concerned with the existence of land bridges" (1992, 162). Kuhn makes a similar generalization and notes that one response to anomaly is for scientists to treat it as the "concern of another discipline" (1962, 37).

6. Abeyance

An individual need not come up with an immediate explanation for unexplained data. Individuals who hold a particular theory can place the unexplained data in abeyance, promising to deal with it later. In common with all of the earlier forms of response, placing the data in abeyance leaves the individual's initial theory unchanged. Yet with abeyance, the individual assumes that theory A will someday be articulated so that it can explain the data.

6.1. Psychology

Brewer and Chinn (1991) investigated how undergraduates respond to data that support the ideas of relativity and quantum mechanics but contradict most people's fundamental beliefs about space, time, and matter. Asked to explain the results of experiments supporting quantum mechanics, such as the double-slit experiment, one student consistently denied the quantum mechanics interpretation of the data but could not offer his own explanation. Instead, he wrote that scientists would eventually figure out how to explain the data without accepting the notions of quantum mechanics. In response to one question, he wrote, "Not sure—I'll tell you in 20 years," indicating that he was willing to hold the data in abeyance for some time.

6.2. History of Science

Conant (1951) has given a good example of abeyance in the early history of chemistry. He pointed out that phlogiston theorists knew that calcination of a metal typically increased the weight of the metal, which was inconsistent with their theory. Conant stated, "the quantitative facts of calcination seems to have been accepted by the majority of chemical experimenters in the 1770's as just one of those things which cannot be fitted in." He goes on to note that "this attitude is much more common in science than is often believed. Indeed, it is in a way a necessary attitude at certain stages of development of any concept" (1951, 183). Holton amplifies this argument and postulates a state of "suspension of disbelief" which is "the scientist's ability during the early period of theory construction and theory confirmation to hold in abeyance final judgments concerning the validity of apparent falsification of a promising hypothesis" (1978, 212).

7. Reinterpretation

The difference between reinterpreting data and rejecting data is that an individual who reinterprets data accepts the data as something that should be explained by theory A, whereas the individual who rejects data does not believe that the data should be considered. In the case of reinterpretation, supporters of theory A and theory B can agree at some level about the data, but at a theoretical level, they give different interpretations of the data. As with all of the previous forms of response to anomalous data, reinterpretation does not require a theory change.

7.1. Psychology

Earlier we described a series of experiments by Chinn and Brewer (1992, 1993a, 1994) that were directed at investigating how undergraduates respond to anomalous data after they have read about the theory that a meteor impact caused the mass extinctions at the end of the Cretaceous period. Some students in these experiments reinterpret the anomalous data as being entirely consistent with the meteor theory. For example, when some of these students encountered data indicating that the KT boundary was deposited over 10,000 years, they argued that a meteor impact would produce so much dust that it is only natural to expect the dust to take 10,000 years or more to settle. These students reinterpreted the data as being just what they would expect if the meteor theory were true.

7.2. History of Science

The decade-long history of scientists' reactions to the meteor theory of Cretaceous extinctions (Alvarez, Alvarez, Asaro, and Michel 1980) is rife with reinterpretations. According to Raup (1986), an early reaction to the iridium data was to argue that the iridium might have seeped down into the K-T boundary from layers of limestone above the K-T boundary. This explains the anomalously high concentration of iridium without the need to make any changes in existing theories of extinction, which assumed only terrestrial mechanisms.

8. Peripheral Theory Change

Another response to anomalous data is for individuals to make a minor modification in their current theory. An individual who responds in this way clearly accepts the data, but is unwilling to give up theory A. This is the first response to anomalous data that we have discussed that involves any change in an individual's initial theory.

8.1 Psychology

Vosniadou and Brewer (1992) have provided evidence for peripheral theory change in children's beliefs about the shape of the earth. They discovered that young children (ages 4-6 years) tend to have a flat-earth theory of the shape of the earth. Then, when young children are told by adults that "the earth is round," they are faced with anomalous information. Many of the children account for this anomalous information by making peripheral changes to their flat-earth view. Some of the children interpret the data from the adults to indicate that the earth is a flat disc, while other children adopt a two-earth theory in which there is a flat earth on which we live and a round earth up in space. Both of these approaches account for the anomalous information about the earth being round, but leave the basic flat-earth belief intact.

Rowell and Dawson (1983) studied eighth, ninth, and tenth graders who believed that weight determines how much water is displaced by heavier-than-water objects. They presented these children with contradictory data that showed that it is volume, not weight, that is the determining factor. About half of these students responded, not by changing to a volume theory, but by changing to a mixed weight and volume theory in which both weight and volume have something to do with displacement. Thus, under the impact of anomalous data the students did change their theories, but they clung to the core belief that weight played a role.

8.2. History of Science

The early responses to Galileo's first telescope observations provide a good example of a peripheral theory change. One of Galileo's opponents conceded that mountains on the moon were visible through the telescope. However, he argued that the mountains were embedded in a perfectly transparent crystal sphere (Drake 1980, 48). This modification accounted for the anomalous data but allowed the philosopher to retain the core theory that the moon was a perfect sphere.

Another good example of peripheral change comes from the famous conflict in the last part of the 18th century between phlogiston explanations of combustion and calcification and Lavoisier's oxygen explanation. Phlogiston theorists believed that a substance called phlogiston was contained in combustible bodies and was released into the air during burning. Lavoisier developed the theory that combustion involved consumption of oxygen from the surrounding air. Musgrave (1976) has shown that phlogiston theorists were able to give an account of the relevant data concerning combustion and calcification. But as Lavoisier accumulated evidence in favor of his oxygen theory, the phlogiston theorists were forced to make one change after another in their auxiliary assumptions. In some versions of phlogiston theory, phlogiston was heavy, in others it was imponderable, and in still others it had negative weight. Similarly, in some versions it penetrated the pores of containers, and in others containers were impervious to penetration. Thus, the phlogiston theorists could preserve their core hypothesis—that phlogiston existed and explained combustion and calcification—only by making dramatic changes in auxiliary assumptions about the properties of phlogiston (Musgrave 1976).

9. Theory Change

The strongest effect anomalous data can have on an individual is to force the individual to change to a new theory. By theory change we mean change in one or more of the individual's core beliefs (cf. Lakatos, 1970). In this form of response to contradictory information, the individual accepts the new data and explains it by changing the core beliefs of theory A or by accepting an alternate theory.

9.1 Psychology

Most psychological experiments in which some students ignore, reject, or reinterpret data or change peripheral beliefs also report that a few students do change core beliefs in response to contradictory data. In the experiment discussed earlier, Rowell and Dawson (1983) found that three of the twelve students who held a weight theory of water displacement changed to a volume theory after observing contradictory data that supported the volume theory. Using the same domain, Burbules and Linn (1988) devised an educational intervention using contradictory data that convinced more than 70% of adolescents to shift from a weight theory of displacement to a volume theory of displacement.

9.2 History of science

The history of science suggests that theory change often requires a series of empirical anomalies, which collectively appear to be better explained by an alternate theory (Kuhn 1962). The chemical revolution provides a good example of this type of theory change. After more than a decade of active experimentation, major phlogiston theorists switched one by one to become proponents of Lavoisier's oxygen theory (Musgrave 1976).

Sometimes, theory change occurs much more rapidly. Röntgen's discovery of X-rays met with some initial disbelief, but within a month the scientific community was convinced, as the basic phenomena were quickly and easily replicated (Glasser 1934). Similarly, Soddy and Rutherford's theory that radioactive decay was associated with disintegration of one element into other elements met with relatively little resistance, even though the theory violated the existing core belief that atoms were fixed and unalterable. Soddy and Rutherford published their initial paper in 1902. In 1903, Ramsay and Soddy produced strong empirical evidence showing that helium was a breakdown product from radium, which was confirmed by Pierre Curie in 1904. A few opponents raised their voices, notably Lord Kelvin, but at a meeting of the British Association for the Advancement of Science in 1903, it was clear that Lord Kelvin had very few supporters (Badash 1966). Again, strong and easily replicable anomalous data led to a relatively quick abandonment of a core belief.

10. Conclusions

We have attempted to provide a comprehensive account of the role of anomaly in theory change in science. We have adopted a naturalistic approach to this issue and therefore have provided evidence from both history of science and the psychology of science. We think this is the first time that an attempt has been made to develop a complete, systematic classification of how people respond to anomalous data. We think the consistency of the results from the history of science and from laboratory studies lends considerable verisimilitude to our analysis.

Our analysis shows that five of the seven types of responses to anomalous data do not involve theory change on the part of the scientist. This finding provides strong evidence against any simple form of a falsification thesis.

Our analysis provides strong support for the theory-laden nature of data (Hanson 1958). There is convincing evidence from both the experimental literature in psychology and from the history of science that individuals faced with data that are inconsistent with their current theory will engage in a variety of types of theory-based rejection or reinterpretation of the evidence (see Brewer and Lambert 1993 for a more detailed discussion of the psychological data). However, our analysis does not support the view that data are so theory laden that theory choice is arbitrary. It appears that data frequently play a decisive role in producing theory change, even when the data are at first ignored, rejected, or otherwise explained away.

Our analysis suggests that scientists' responses to anomalous data are not unique to scientists, but are similar to the responses of nonscientist adults and children. However, there is need for more research on this topic. For example, are the responses to anomalous data that are described in this paper restricted to individuals raised in cultures that have been strongly influenced by the scientific community, or are these responses simply aspects of more general aspects of the human cognitive apparatus? (cf. Brewer and Samarapungavan 1991). In addition, research is needed that contrasts scientists and nonscientists in order to uncover those aspects of scientific reasoning that *are* determined by scientific training.

Finally, our analysis of the forms of responses to anomalous data suggests the need for an analysis of the factors that lead a scientist to adopt one particular response rather than another when faced with anomalous data. Chinn and Brewer (1993a; 1993b) have begun an analysis of this issue in terms of issues such as the degree of entrenchment of the current theory, the availability of a plausible alternative theory, and the credibility of the data. However, this is a rich area for future research.

Note

[1] We would like to thank Robert McCauley and Paul Thagard for comments on an earlier draft of this paper.

References

Alvarez, L.W., Alvarez, W., Asaro, F. and Michel, H.V. (1980), "Extraterrestrial Cause for the Cretaceous-Tertiary Extinction", *Science* 208: 1095-1108.

Badash, L. (1966), "How the 'Newer Alchemy' Was Received", *Scientific American* (August) 215 (2): 88-95.

Brewer, W.F. and Chinn, C.A. (1991), "Entrenched Beliefs, Inconsistent Information, and Knowledge Change", *Proceedings of the 1991 International Conference on the Learning Sciences*, Charlottesville, VA: Association for the Advancement of Computing in Education, pp. 67-73.

Brewer, W.F. and Lambert, B.L. (1993), "The Theory-Ladenness of Observation: Evidence from Cognitive Psychology", *Proceedings of the Fifteenth Annual Conference of the Cognitive Science Society*, University of Colorado-Boulder, Lawrence Erlbaum, pp. 254-259.

Brewer, W.F. and Samarapungavan, A. (1991), "Children's Theories vs. Scientific Theories: Differences in Reasoning or Differences in Knowledge?" in R. R. Hoffman and D.S. Palermo (eds.), *Cognition and the Symbolic Processes: Applied and Ecological Perspectives*. Hillsdale: Lawrence Erlbaum, pp. 209-232.

Burbules, N.C. and Linn, M.C. (1988), "Response to Contradiction: Scientific Reasoning During Adolescence", *Journal of Educational Psychology* 80: 67-75.

Chinn, C.A. and Brewer, W.F. (1992), "Psychological Responses to Anomalous Data", *Proceedings of the Fourteenth Annual Conference of the Cognitive Science Society*, Indiana University, Lawrence Erlbaum, pp. 165-170.

_____. (1993a), "Factors that Influence How People Respond to Anomalous Data", *Proceedings of the Fifteenth Annual Conference of the Cognitive Science Society*, University of Colorado-Boulder, Lawrence Erlbaum, pp. 318-323.

_____. (1993b), "The Role of Anomalous Data in Knowledge Acquisition: A Theoretical Framework and Implications for Science Instruction", *Review of Educational Research* 63: 1-49.

_____. (1994), "The Role of Anomalous Data in Belief Change", Unpublished manuscript. Champaign, University of Illinois.

Conant, J.F. (1951), *Science and Common Sense*. New Haven: Yale University Press.

Dunbar, K. (forthcoming), "*In Vivo* Cognition: Knowledge Representation and Change in Real-World Scientific Laboratories", in R. J. Sternberg and J. Davidson (eds.), *Insight*, Cambridge: MIT Press.

Drake, S. (1980), *Galileo*. New York: Hill and Wang.

Fermi, L. and Bernardini, G. (1961), *Galileo and the Scientific Revolution*. Greenwich: Premier Books.

Gholson, B., Shadish, W.R., Jr., Neimeyer, R.A. and Houts, A.C. (eds.) (1989), *Psychology of Science*. Cambridge: Cambridge University Press.

Giere, R.N. (1985), "Philosophy of Science Naturalized", *Philosophy of Science* 52: 331-356.

Glasser, O. (1934), *Wilhelm Conrad Röntgen and the Early History of the Roentgen Rays*. Springfield: Charles C. Thomas.

Gorman, M.E. (forthcoming), "Psychology of Science", in W. O'Donohue and P. Kitchener (eds.), *Psychology and Philosophy: Interdisciplinary Problems and Responses*. Allyn & Bacon.

Hanson, N.R. (1958), *Patterns of Discovery*. Cambridge: Cambridge University Press.

Hewish, A. (1968), "Pulsars", *Scientific American* (October) 219 (4): 25-35.

_____. (1975), "Pulsars and High Density Physics", *Science* 188: 1079-1083.

Holton, G. (1978), "Subelectrons, Presuppositions, and the Millikan-Ehrenhaft Dispute", *Historical Studies in the Physical Sciences* 9:161-224.

Karmiloff-Smith, A. and Inhelder, B. (1975), "'If You Want to Get Ahead, Get a Theory'", *Cognition* 3: 195-212.

Kuhn, T.S. (1962), *The Structure of Scientific Revolutions*. Chicago: University of Chicago Press.

Lakatos, I. (1970), "Falsification and the Methodology of Scientific Research Programmes", in I. Lakatos and A. Musgrave (eds.), *Criticism and the Growth of Knowledge*. London: Cambridge University Press, pp. 91-196.

Maffie, J. (1990), "Recent Work on Naturalized Epistemology", *American Philosophical Quarterly* 27: 281-293.

Mpemba, E.B. and Osborne, D.G. (1969), "Cool?", *Physics Education* 4: 172-175.

Musgrave, A. (1976), "Why Did Oxygen Supplant Phlogiston? Research Programmes in the Chemical Revolution", in C. Howson (ed.), *Method and Appraisal in the Physical Sciences: The Critical Background to Modern Science, 1800-1905*. Cambridge: Cambridge University Press, pp. 181-209.

Osborne, D.G. (1979), "Mind on Ice", *Physics Education* 14: 414-417.

Raup, D.M. (1986), *The Nemesis Affair: A Story of the Death of Dinosaurs and the Ways of Science*. New York: W.W. Norton & Company.

Rowell, J.A. and Dawson, C.J. (1983), "Laboratory Counterexamples and the Growth of Understanding in Science", *European Journal of Science Education* 5: 203-215.

Thagard, P. (1992), *Conceptual Revolutions*. Princeton: Princeton University Press.

Vosniadou, S. and Brewer, W.F. (1992), "Mental Models of the Earth: A Study of Conceptual Change in Childhood", *Cognitive Psychology* 24: 535-585.

Methodology, Epistemology and Conventions: Popper's Bad Start

John Preston

University of Reading

1. Introduction

Popper's philosophy is founded on his early conceptions of the problem of demarcation, of its solution, and of the roles of epistemology and methodology. I shall give an account of these, and of his idea of the alternative, 'naturalistic' conceptions. I then hope to show that Popper's conceptions are confused.

2. The Problem of Demarcation

At the beginning of *The Logic of Scientific Discovery* Popper tells us that "the task of the logic of scientific discovery, or the logic of knowledge, [is]... to analyse the method of the empirical sciences" (Popper 1959, 27), and that its first job is "to put forward a **concept of empirical science**, in order to make linguistic usage, now somewhat uncertain, as definite as possible, and in order to draw a clear line of demarcation between science and metaphysical ideas" (1959, 38-39). In a somewhat positivistic mood, he tells us that he regards this 'problem of demarcation' as the central problem in the theory of knowledge, more fundamental, even, than the problem of induction. Finding a 'criterion of demarcation' would be finding a principled way of distinguishing between "the empirical sciences on the one hand, and mathematics and logic as well as 'metaphysical' systems on the other" (1959, 34). Although he does not explicitly draw out the implications of this remark here, we can say that any comprehensive criterion of demarcation must also give us a way of distinguishing science, not just from metaphysics, but from philosophical activities such as epistemology.

Popper recognises that he isn't the first person who has sought such a criterion. The Logical Positivists, in particular, also sought to separate scientific sheep from metaphysical goats. According to Popper, the positivists had two suggested criteria: the verifiability principle and the criterion of inductive methods. The former was an attempt to characterize the domain of meaningful statements, the latter an attempt to characterise the domain of science. Popper argues that both criteria are misconceived. He goes on to offer a diagnosis, suggesting that these positivist criteria come to grief fundamentally because they are attempts to discover a difference in the natures of empirical science and metaphysics. Positivists, he says, interpret the problem of demarcation in a **naturalistic** way; they interpret it as if it were a problem of natural science. ...[T]hey believe they have to discover a difference, existing in the nature of things, as it were, between empirical science on the one hand and metaphysics on the other. (1959, 35).

This naturalistic approach, he says, relies on 'the dogma of meaning':

[N]othing is easier than to unmask a problem as 'meaningless' or 'pseudo'. All you have to do is to fix upon a conveniently narrow meaning for 'meaning', and you will soon be bound to say of any inconvenient question that you are unable to detect any meaning in it. Moreover, if you admit as meaningful none except problems in natural science, any debate about the concept of 'meaning' will also turn out to be meaningless. The dogma of meaning, once enthroned, is elevated forever above the battle. It can no longer be attacked. It has become (in Wittgenstein's own words) 'unassailable and definitive'. (1959, 51).

Popper's alternative to naturalism is that a plausible criterion of demarcation is a "**proposal for an agreement or convention**" (1959, 37), rather than a description. Such a proposal, he holds, must be formulated and discussed against the background of some agreed purpose, the choice of which goes "beyond rational argument" (ibid.). In other words, we have to choose the epistemic ideals which we want science to aspire to: "decisions about the way in which scientific statements are to be dealt with... depend on the **aim** which we choose from among a number of possible aims" (1959, 49). We might, for example, choose to see science as aiming at a system of absolutely certain, irrevocably true statements, or instead to see it as aiming at highly falsifiable conjectures. Each would be a **proposal**: there is no question of there being 'true' or 'essential' aims of science (1959, 38). This is a very important and strongly conventionalist element in Popper's philosophy, later picked up and expanded on by Paul Feyerabend, who chooses to see the dispute between positivists and realists over the nature of scientific theory in the same terms.[1]

Popper argues that such proposals about the aim of science can only be evaluated by analyzing their logical consequences and pointing out their philosophical fertility, their power to solve philosophical problems. "My only reason", he says, "for proposing my criterion of demarcation is that it is fruitful: that a great many points can be clarified and explained with its help" (1959, 55). Its ultimate philosophical test is its assistance in detecting inconsistencies and inadequacies in other theories of knowledge, as well as its own ability to avoid inconsistencies.

3. Epistemology and 'Methodology'

According to Popper, epistemology, which he identifies with a theory of the empirical method (1959, 39), a theory of experience (ibid.), the logic of scientific discovery (1959, 49), the theory of scientific method (ibid.), a theory of theories (1959, 59), and methodology (1959, 51), is concerned with the choice of methods:

The theory of method, in so far as it goes beyond the purely logical analysis of the relations between scientific statements, is concerned with **the choice of methods** - with decisions about the way in which scientific statements are to be dealt with. These decisions will of course depend in their turn upon the **aim** which we choose from among a number of possible aims. (1959, 49).

Popper calls such a theory of the rules of scientific method a 'methodology' (1959, 52). He points out that methodological rules aren't logical rules, but conventions, "rules of the game of empirical science" (1959, 53). We should not, he argues, characterize empirical science only by the formal, logical structure of its statements, since if we do so we will be unable to separate it from metaphysics. Empirical science should be characterized by its methods. He thus aims "to establish the rules, or if you will the norms, by which the scientist is guided when he is engaged in research or in discovery" (1959, 50). Among the examples of methodological rules which he later gives are:

(1) Scientific statements are never to be regarded as conclusively verified,

(2) A well-tested hypothesis that has proved its mettle is not to be discarded without good reason,

(3) Only such statements may be introduced in science as are inter-subjectively testable (1959, 56),

and

(4) We are not to abandon the search for universal laws and for a coherent theoretical system, nor to give up our attempts to explain causally any kind of event we can describe. (1959, 61).

Each of these is a plausible attempt to capture one of the methodological rules operative in science. "In establishing these rules", he says,

> we may proceed systematically. First a supreme rule is laid down which serves as a kind of norm for deciding upon the remaining rules, and which is thus a rule of a higher type. It is the rule which says that the other rules of scientific procedure must be designed in such a way that they do not protect any statement in science against falsification. (1959, 54).

Who is doing this 'establishing'? Popper's answer is: the philosopher of science, the epistemologist, not the scientist. Epistemology itself chooses which methodological rules scientists can be said to be following. Decisions about the way in which scientific statements are to be dealt with depend upon the aim **we** choose, and what the rules of scientific method are depends on **our** attitude to science (1959, 49). This is why there can be no account of the real aims and methods of science, independent of our attitude to science.

In choosing such rules of scientific methodology, Popper proposes that we "adopt such rules as will ensure the testability of scientific statements, which is to say, their falsifiability" (1959, 49). This is his famous criterion of demarcation. According to Popper this criterion constitutes the highest-order methodological rule from which other such rules or conventions are systematically drawn (but not deduced). It is to be judged by its fruitfulness in clarifying and explaining, by the conformity of the resulting methodological rules to the scientist's intuitive idea of the aim of science. Whether these 'methodological' investigations belong to philosophy "does not really matter much" (1959, 55), he says, because "the majority of the problems of theoretical philosophy... can be re-interpreted... as problems of method" (1959, 56, the example being Occam's razor).

The alternative 'naturalistic' view held that there is a discoverable matter of fact about what a scientist's methodology is. But the naturalistic approach, Popper complains, simply cannot accommodate methodology, since it recognises only empirical statements and tautologies, and methodological rules are neither. The naturalist cannot accept that there is "a genuine theory of knowledge, an epistemology or a methodology" (1959, 51). In consequence, he must think of methodology as an empirical science which studies the actual behaviour of scientists. But, Popper says,

> ...what I call 'methodology' should not be taken for an empirical science. I do not believe that it is possible to decide, by using the methods of an empirical science, such controversial questions as whether science actually uses a principle of induction or not. And my doubts increase when I remember that what is to be called a 'science' and who is to be called a 'scientist' must always remain a matter of convention or decision. (1959, 52).

What **does** Popper call 'methodology'? He believes that an issue in 'methodology', like the issue of whether science uses a principle of induction or not, is to be decided by comparing two different systems of rules and asking whether the system which includes a principle of induction gives rise to inconsistencies or not. So his famous eschewal of induction amounts to a **decision** that induction is not needed in science, rather than a discovery that it is not in fact used (see 1959, 52-3). Naturalists fail to notice that "whenever they believe themselves to have discovered a fact, they have only proposed a convention" (1959, 53), and therefore their conventions, like the verifiability principle, turn into dogmas. So far Popper.

4. Why Epistemology isn't Methodology

I want to suggest that this account of epistemology and methodology is extensively confused. I shall do so by comparing it with the alternative, 'naturalistic' conception which Popper opposes.

Notice firstly how much more natural than Popper's own conception of the problem of demarcation is the conception he attributes to the positivists. One would, pre-theoretically, think that an accurate description of the differences between science and non-science would give one insight into those distinctions. Why, after all, can't there be an account of the differences which actually and already exist between these disciplines? Philosophy seeks understanding, and philosophers of science can surely strive to understand the already-existing differences between science and other activities. In denying that any such descriptive account is possible, does Popper perhaps mean to imply that there are no such differences (a very radical and ridiculous thesis)? Does he mean that in 'proposing' a convention we epistemologists propose such differences? Well, if 'propose' doesn't mean 'discover', then presumably it means 'create': but surely epistemologists don't have to **create** differences between philosophy, metaphysics, theology and science! How would they go about doing so? And why would they do so anyway, since the end result of their proposal would simply not be an understanding of that which they originally strove to understand? We might agree with Popper, as against the Logical Positivists, that our task as epistemologists is not to bring about the overthrow of metaphysics, but it has to be said that the non-descriptive project Popper has in mind seems confused.

What is more, Popper does not give good reasons why the descriptive project he rejects should be as disastrous as he thinks. It is not true that the Logical Positivists treated methodology as an empirical science. Wittgenstein's *Tractatus* had explicitly recognised methodological rules ('principles of nature' such as the 'law of causality'), and the positivists themselves recognised the existence of conventions and stipulations. Descriptive or naturalistic conceptions can admit the existence of methodological rules of science and philosophical descriptions, embodied in "a genuine theory of knowledge, an epistemology" (1959, 51), together covering everything Popper refers to as 'methodology', without making the mistake of thinking that epistemology is normative. The naturalistic approach can and must recognise the existence of methodological rules as well as empirical statements and tautologies. Popper has done nothing to show that a descriptive approach to the interpretation of scientific theories and to scientific methodology must issue in the conclusion that scientific theories are meaningless pseudo-science. This conjecture is formulated on the basis of one case: those positivists who became instrumentalists because of their commitment to the verification principle.[2] But to hold that laws of nature are rules is not to hold that they are meaningless.

What about what Popper called the 'dogma of meaning'? Here he seems to be suggesting that the naturalistic approach fails by virtue of being dogmatic or unfalsifiable, that it fails to satisfy his criterion of demarcation. But if the naturalist is right in thinking that a criterion of demarcation is a **description** of the difference between science and non-science, rather than a proposal, the 'dogma of meaning' objection is completely irrelevant. Even if Popper is right in his selection of falsifiability as the correct criterion of demarcation, this objection accuses a philosophical view of not meeting a condition of adequacy for a **scientific** theory. It also fails to address anyone who does not accept such a criterion of demarcation. All Popper has shown is that naturalism is incompatible with his own view. He has so far failed to motivate that view, and has done nothing to impugn the naturalistic approach.

Pace Popper, therefore, it matters very much indeed whether what he calls 'methodology' is part of philosophy or not. 'Methodological investigations' (1959, 55) might be methodological proposals, or descriptive philosophical investigations, or factual inquiries in the social science of science. But it is essential to separate these. The naturalistic approach can't be faulted for failing to make room for what Popper calls 'methodology', since that project is incoherent, as I shall now show.

The theory of knowledge is descriptive and naturalistic, not normative. Insofar as it applies to science, it aims to describe the methods scientists use, not to select such methods. Which methodological rules scientists adhere to does not depend upon the aim we (the describers) choose, or upon our attitude to science, but on the activities of scientists. There can be such a thing as an account of the real aims and methods of scientific activities (although perhaps not of 'the aim' of science, since there may not be a single such aim) independent of our attitude to science, since scientists do follow methodological rules. It is methodology, properly so-called, that is normative, since methodology consists of rules that guide scientists in their research activities. Popper is perfectly entitled to identify the theory of method as part of the theory of knowledge. But he is confused in identifying the theory of method with methodology itself. A theory of the rules of scientific method (supposing there to be such a thing) is not a methodology.

The epistemology of methodology is not an empirical science but a philosophical account of the behaviour of scientists. We must carefully separate such a philosophical description of the methodology scientists use, from their methodology itself, which is a set of rules telling them what to do and what not to do. The latter is normative, the former isn't. We can't decide issues in the epistemology of methodology, which is a descriptive discipline, in the way Popper recommends, by comparing different systems of methodological rules and asking whether systems with a particular principle give rise to inconsistencies or not. Such an investigation might give us an idea of whether a methodology is inconsistent, and thus whether it is unsuccessful. But (setting aside problems about the very identification of the rules the scientists claim to follow) it won't tell us anything about whether that is the methodology that these scientists use or not. And there are no factual issues in methodology itself!

Methodological rules are indeed conventions, but whether a particular set of such conventions is the set of rules scientists follow is a matter of fact, not a matter to be decided by a convention. The fact that they are conventions doesn't mean that there is 'no fact of the matter' as to whether they are in operation in a given case. Methodological rules have a determinate status because they are the rules scientists do (or do not) follow. Whether or not a particular scientist or scientific community follows any particular rule is a determinate matter, since there are well-known and intelligible criteria for when a person's behaviour is rule-guided, and for identifying which rule it is that is being followed. (We have to ask them, as well as studying their activities). The methodological structure of science is something there to be described, and doesn't owe its existence or character to the describers' choice of aims. If we misidentify the nature and status of methodological rules, we will of course be less likely to discern those rules and their operation. (Feyerabend, notoriously, later fails to find any such rules, and goes on to deny that there are any! We will have to ask whether these conclusions aren't the result of his initial Popperian misconceptions about methodology).

The Popperian thesis that epistemologists choose which rules scientists follow is ridiculous: there simply isn't room for us to decide which rules they follow. And what on earth, on the Popperian account, would their following a particular rule consist in? Solely in the fact that **we** see them as doing so? It's surely not for philosophers to lay down methodological rules in science. (A Cartesian conception, according to which philosophy lays down rules of scientific method, rules for discovering the truth, isn't available to Popper). Epistemology can 'establish' the rules or norms by which the scientist is guided only in the sense that it can discover these rules, not in the sense that it can set them up or legitimate them. The systematic connection between methodological rules does not, therefore, "[make] it appropriate to speak of a **theory** of method" (1959, 54) unless by this term we just mean a methodology (properly so-called). But a methodology isn't a theory of anything, any more than the rules of chess constitute a theory of chess. Empirical science can indeed be characterized partly by an hierarchical ordering of its methodological rules. But to say this is not to say that it can be characterized by a theory of method, since a set of rules is a method, not a theory of method. And the conventions that are methodological rules are not "the pronouncements of this theory" (ibid.). It is true to say that "Profound truths are not to be expected of methodology" (ibid.), but only because methodology consists of rules, not truths. Popper's own examples of the truths that constitute methodology (the inequality between degree of corroboration and probability, and his theorem on truth-content), are nothing like methodological rules (despite the fact that he did, remember, earlier make a perfectly good stab at identifying some examples of methodological rules).

The only tasks available to philosophers here are: (a) to identify, delineate, and describe the rules that any particular group of scientists really follow, (b) to criticize the rules they follow, and (c) to suggest rules that they might want to follow instead. Naturalistic philosophers will start by identifying the norms that scientists are bound by, as well as those they **regard** themselves as bound by. We might then profitably criticize the rules scientists follow on the grounds that their doing so means that they fail to attain the aims they aspire to, or on the grounds that they conflict with each other. And we could criticize the aims or ends themselves, as unsuitable, inhumane, or limited ideals.

5. Popper's Criterion of Demarcation

Popper's use of his criterion of demarcation embodies this confusion between epistemology and methodology. This critique has extensive implications which ramify throughout Popper's work. One implication is as follows. A necessary precondition for showing that there is, or is not, a single criterion of demarcation is to clarify the status such a criterion would have to have. The falsifiability criterion, and therefore falsificationism itself, has gained a spurious plausibility from having an unclear status.

A criterion of demarcation must be **either** an attempted description of the difference between science and non-scientific activities, and thus a description of part of the system of methodological rules which scientists allegedly aspire to follow, **or** a suggested methodological rule. If, on the one hand, it's the former, then it's a philosophical description rather than a proposal or convention. So it is answerable to the realities of scientific practice, can be criticised for failing to track the actual contours of science/non-science boundaries, and would itself be falsified if we found that scientists don't actually use that rule to guide their practice. Popper's criterion has sometimes been impugned on this count (by Feyerabend, for example), and the charge cannot be evaded by claiming that it's a proposal, not a description. Falsificationists like Popper and the early Feyerabend are thus refuted by the history of science. Such a 'criterion of demarcation' cannot be judged by its fruitfulness, since whether a particular such criterion is the criterion (or a criterion) operative in science is not a matter of convention. If, on the other hand, it's the latter, it is a proposal, and can be judged by its fruitfulness, but then it has no claim to be that which demarcates science from non-science (since presumably one wouldn't **propose** a methodological rule which was already in situ). What it can't be is what Popper wants it to be, a normative criterion which those who are doing epistemology decide to adopt. Insofar as one is doing epistemology, one is neither forced, nor able, to 'adopt' methodological rules, since epistemology is part of philosophy, not part of science, and methodological rules live, move, and have their being in science, not in philosophy. (Of course, one may think that there are rules of philosophical method, but this doesn't at the same time make them rules of scientific method!).

Methodological rules (properly so-called) are "constructed with the aim of ensuring the applicability of our criterion of demarcation" (1959, 54) only if the scientists who construct and use the rules have among their aims the ambition to produce highly falsifiable theories. Aims and methodology, unsurprisingly, go together in that only when one has uncovered the methodology of a particular science is one in a position to say what the aims of that science are.

This confusion between rules and descriptions is particularly striking when it comes from those who insist most vigorously on the importance of the distinction between nature and convention. In *The Open Society and its Enemies* (Popper, 1945) Popper argues that humans moved away from the primitive, magical attitude which had characterized early societies only when they came to consciousness of the distinction between the natural and the conventional. He insists on distinguishing sharply between natural laws, which describe "a strict, unvarying regularity which either in fact holds in nature or does not hold" (1945, 57) and normative laws "such rules as forbid or demand certain modes of conduct" (ibid.), and characterises the distinction as a fundamental one (1945, 58).[3]

The most sensible things Popper says about his criterion of demarcation are that we might use such a criterion in order to firm up linguistic usage, and that it is to be judged by the conformity of the resulting methodological rules to the scientist's intuitive idea of the aim of science. The rest of what he says simply doesn't chime with these more modest and wholly naturalistic ambitions.

6. The Conventionality of 'Science'?

What about Popper's idea that "what is to be called a 'science' and who is to be called a 'scientist' must always remain a matter of convention or decision" (1959,

52)? This is not just exaggerated, but untrue. We are not free to decide what is to be called 'science'. If we don't already understand roughly what is to be called 'science', how could we even begin to suggest or discern a criterion which would demarcate science from non-science?! We couldn't even start. As for the suggestion that 'definitions are dogmas', and that any definition of science would be a dogma (1959, 55), we must reply that a definition is precisely what cannot be a dogma, since definitions are not substantive, whereas dogmas must be. Furthermore, the naturalistic philosopher need not even aim for a definition of science, but may settle for a descriptive characterization, which needn't (indeed, had better not) be dogmatic. Insofar as the scientist has an "intuitive idea of the goal of his endeavours" (1959, 55), allegedly necessary to measure our newly-minted conception of empirical science against, he already has a conception of empirical science, an understanding of what science is.

But so do the rest of us. The **concepts** of science, scientists, and the scientific are 19th century inventions: the word 'scientist' was coined around 1840 by William Whewell. But they are now part of everyone's conceptual resources, and an understanding of what science is is not something vouchsafed to scientists alone. It involves being able to deploy concepts like science, art, craft, technology, engineering, etc. It involves knowing one's way around this particular semantic field of human activities (see Kuhn 1983, 567). Understanding science itself, of course, is something deeper. It involves knowing science, not merely its conceptual location. But as philosophers of science our interest is in achieving a perspicuous overview of the concept of science (among others).

7. Conclusion

Popper thinks that the aim and the methodology of science are a matter of convention. He is a conventionalist about our account of the fundamental nature of science and methodology: he thinks that there is 'no fact of the matter' about the aim of science, and no fact of the matter about which methodological rules scientists really follow. (Conventionalism about rules might be a form of rule-scepticism. But Popper accepts that the 'methodologist' or the epistemologist can adopt methodological rules, so his conventionalism can't be generalised rule-scepticism). For him, science has no 'nature'. We can see it under the aspect of different ideals, each of which will generate a different picture of the scientific enterprise.

This conventionalist point of view cannot simply rest on Popper's critique of 'essentialism'. That critique was salutary, and reminded us that science does not consist of a body of incorrigible knowledge about ahistorical essences. Anyone who accepts a distinction between the conceptual and the factual must admit that science has a nature, a conceptual essence, which can be described. But although the naturalist insists that science has a nature, he must also insist that the nature of science is not an ahistorical essence, but something that changes over time and culture. This is the sophisticated conventionalist position.

Popper's conventionalism, however, is misplaced. Rules of scientific methodology are there to be discovered by epistemologists, not stipulated or 'laid down' by them. Methodological rules are the rules scientists follow, not merely the rules we choose to see them as following.

Notes

[1] This suggestion there are no 'true' or 'essential' aims of science conflicts with Popper's later assertion that finding satisfactory explanations is the aim of science. See "The Aim of Science", in (Popper 1972).

[2] What is more, it is based on a misinterpretation of that case. For a recognition that Popper's main argument against instrumentalism is circular, see (Worrall 1982, 205).

[3] See, however, Peter Winch's critique of Popper in "Nature and Convention", in (Winch 1972).

References

Kuhn, T.S. (1983), "Rationality and Theory Choice", *Journal of Philosophy* 80: 563-570.

Popper, K.R. (1945), *The Open Society and its Enemies*. London: Routledge.

_ _ _ _ _ _ _. (1959), *The Logic of Scientific Discovery*. London: Hutchinson.

_ _ _ _ _ _ _. (1972), *Objective Knowledge: An Evolutionary Approach*. Oxford: Clarendon Press.

Winch, P. (1972), *Ethics and Action*. London: Routledge.

Worrall, J. (1982), "Scientific Realism and Scientific Change", *Philosophical Quarterly* 32: 201-231.

Sherlock Holmes, Galileo, and the Missing History of Science[1]

Neil Thomason

University of Melbourne

Much of the history of science is missing much of the time. The missing history I have in mind is not missing because of lost documents or state secrecy. Nor is it missing because it is obscure. I am talking about central episodes in the history of science and absolutely central issues in the logical structure and evidential support of those episodes. In fact, the episodes are so central and well known, and the missing history (once seen) so obvious, that one feels compelled to hold that there must be a widespread systematic bias among historians and philosophers against seeing certain explanatory patterns.

My explanation for this missing history of science is that there is a strong tendency among historians and philosophers of science toward what I will call "Psychological Predictivism" to distinguish it from "Logical Predictivism." Logical Predictivism is the position that, if an observed phenomenon provides good evidence for a hypothesis, then that hypothesis (plus unproblematic auxiliary hypotheses), predicts the phenomenon. That is, a necessary condition for an observation to support a major hypothesis is that

(Major Hypothesis + Unproblematic Auxiliary Hypotheses) —> Observation

Logical Predictivism is false, for well-known reasons. But it is an intuitively attractive position, and I would like to point out one of its strong attractions: it precludes using an implausible, *ad hoc* auxiliary hypothesis to connect observations to a major hypothesis. After all, if one can just pick and choose among auxiliary hypotheses (including quite problematic ones), one can use *ad hoc* maneuvers of a most intellectually shoddy kind to support almost any hypothesis. Psychological Predictivism is the psychological correlate of Logical Predictivism — the visceral sense that Logical Predictivism is correct. Slightly more precisely,

"Psychological Predictivism" =$_{df.}$ the tendency to believe when presented with a hypothesis that if a phenomenon is good evidence for the hypothesis, then that hypothesis (plus unproblematic auxiliary hypotheses) predicts the phenomenon.

I think that most of us, regardless of how vehemently we may reject Logical Predictivism as a claim about the logic of evidential support, have strong tendencies toward Psychological Predictivism. Here is a test for your tendency for this dread cognitive dysfunction:

> Suppose that your friend Mary makes a claim that you only partially understand and don't accept. You ask Mary, "Does your hypothesis predict phenomenon P?" and Mary says, "No. On my hypothesis, P is quite unlikely." The next day, Mary tells you, "Wonderful news — P has been observed! I'm really surprised! Still, the observation of P means that my hypothesis is almost certainly right!"

Even as I write Mary's story, my immediate gut reaction is that MARY CAN'T DO THAT!, although intellectually I have been thoroughly disabused of Logical Predictivism. She can't predict that something probably won't happen and then conclude that it strongly supports her theory when it does. Insofar as you share my reaction, you are a Psychological Predictivist, regardless of your sophisticated philosophical views on evidential support. In this reaction you are in good philosophical company, for in more sophisticated forms Logical Predictivism has influenced much philosophy of science. The impact of the views of Popper and Lakatos is a tribute to the power of Psychological Predictivism.

I believe that many people's visceral sense that Mary has contradicted herself is an important fact. It makes me suspect that when many historians of science believe that a phenomenon strongly supported a hypothesis, they will then semi-automatically decide that that hypothesis also predicted that phenomenon. I am not claiming that *all* historians and philosophers will *always* force *all* history into the Predictivist mold — my example below from Galileo shows this is clearly false. Rather, my claim is that the strength of Psychological Predictivism means that *much* history is forced into the Predictivist mold.

Below, I will discuss historians' accounts of Galileo's discovery of the phases of Venus. It is not a case where the true history is unknown. Here the facts are too blatant to have been missed. In fact, there are two traditions, one which gets the story right and one which distorts it. My interest here is the distorting tradition, its strength and its origins.

But before examining the pernicious effects of Psychological Predictivism on the histories of Galileo's discovery, let's turn to a counter-example to Logical Predictivism. Since I am at the Victorian Centre for the History and Philosophy of Science, I will examine the work of the great Victorian methodologist, Sherlock Holmes.

1. Sherlock Holmes' Method of "Double Deduction"

A Scandal in Bohemia opens with Dr. Watson dropping in at 22B Baker Street after a longish absence. Holmes makes a series of dazzling observations about details of Watson's private life, after which Watson enquires as to Holmes' method. Holmes points out that "the difference between us is that you see and I observe" and then asks the dumfounding double question:

> "How do I know that you have been getting yourself very wet lately, and that you have a most clumsy and careless servant girl?"

> "My dear Holmes,... this is too much. ... It is true that I had a country walk on Thursday and came home in a dreadful mess; but, as I have changed my clothes, I can't imagine how you deduce it. As to Mary Jane, she is incorrigi-

ble, and my wife has given her notice; but there again I fail to see how you could work it out." ...

"It is simplicity itself," said [Holmes]; "my eyes tell me that on the inside of your left shoe, just where the firelight strikes it, the leather is scored by six almost parallel cuts. Obviously they have been caused by someone who has very carelessly scraped round the edges of the sole in order to remove crusted mud from it. Hence, you see, my double deduction that you had been out in vile weather, and that you had a particularly malignant boot-slitting specimen of the London slavey."

While one may be less than enthusiastic about Holmes' social attitudes or his phrase "double deduction", one must admire his method, especially as a contrast to Logical Predictivism.

To pedantically spell out what Holmes does so eloquently: there is an observation — the six almost parallel cuts on Watson's left shoe; and there are two hypotheses, both of which were initially quite implausible, given Watson's staid newly married life: (H1) *Watson had been tramping in mud (hence getting very wet)* and (H2) *the Watsons employed a clumsy servant*. Before the observation was made, neither hypothesis (plus plausible auxiliary hypotheses) would have predicted the phenomenon of the parallel cuts. Even if one assumes that Watson had been tramping in mud (H1), it is still very unlikely that there would be parallel cuts in the shoes — if only because it is very unlikely that the Watsons would employ so clumsy a servant. And if one assumes that the Watsons did employ a clumsy servant (H2), it is very unlikely that there would be parallel cuts on the shoes — if only because it is very unlikely that Watson would have so much crusted mud on his shoe that it would be necessary to scrape it off.

Further, the two hypotheses were epistemically independent: before the observation was made, accepting one hypothesis provided very little or no reason for accepting the other. After the observation was made, both hypotheses were reasonably well supported. Finally, even the conjunction of the two hypotheses did not predict the parallel cuts on either shoe, for presumably even clumsy servants cleaning mud off shoes don't usually score the leather. Holmes' "double deduction" is highly non-Predictivist. So what is the logical structure of Holmes' *tour de force*? Here are two alternative non-Predictivist accounts: ARCHED Hypotheses and Stepped Hypotheses.

ARCHED [Advance Reinforcing Chancy Hypotheses Explaining Data] Hypotheses are named after that central architectural form, the arch. Both halves of an arch are, by themselves, ill-supported until they are united by a keystone. Together they support the keystone and in this way support each other. Great overall strength is gained by each part which otherwise, standing by itself, would tumble to the earth. Column 1 supports column 2 which in turn supports column 1. And so it is with ARCHED hypotheses. For a more detailed account of ARCHED hypotheses, including a Bayesian analysis of them, see Thomason (1994).)

In the simplest and most dramatic form, *two* ill-supported hypotheses are combined to provide an explanation for a phenomenon which otherwise is very difficult to explain. In particular, it must be a phenomenon that neither hypothesis explains by itself. In jointly explaining the phenomenon, the two otherwise ill-supported hypotheses lend considerable support to each other. Combined, but not singly, they provide a (very?) plausible explanation for the phenomenon. That is, if they are ARCHED hypotheses, one can agree with Sherlock that the cuts are good evidence that Watson had been in the mud if and only if one also agrees the cuts are good evidence that the Watsons employed a clumsy servant. To return to our ARCHED hypotheses metaphor, both the hypotheses are now well supported because *together*, although not singly, they provide a good explanation for the cuts on Watson's boot. Each is ill-supported without the other.

Stepped Hypotheses are used when the phenomenon is taken as establishing a first, hitherto problematic hypothesis. Then this hypothesis is taken as an unproblematic auxiliary hypothesis which the second hypothesis relies on to explain the phenomenon. On this account, Holmes argued in this way: lines like that must be due to scraping off mud. Therefore, Watson probably had been tramping in mud (Hypothesis 1). But who could have done such a terrible job in scraping off the mud? Not Watson or Mrs. Watson. It was a clumsy servant (Hypothesis 2). That is, the mud scraping hypothesis is taken as correct even in the absence of a well-supported account of who could have done such a lousy job of mud-scraping.

Which did Holmes use: ARCHED Hypotheses or Stepped Hypotheses? It does not matter for our purposes; I suspect that my childhood sense that Holmes' argument is so compelling comes from some compound of both. (Of course, Dr. Watson's astonished affirmation of Holmes' conclusion also helped considerably.)

Given the intuitive plausibility of ARCHED and Stepped hypotheses, one should expect to find them widespread in science. And, if Psychological Predictivism is common among historians, one should expect often to find such hypotheses incorrectly depicted as a single hypothesis (plus unproblematic, unmentioned background assumptions) predicting a phenomenon.

2. How Galileo Established Both That Venus Circles the Sun and That Venus is Opaque and Dark

Now we must turn to Galileo's telescopic discovery of the phases of Venus, both as Galileo and his contemporaries understood it and as it is now widely reported. We will examine the widespread assertion that the Copernican theory predicted that Venus has phases like the moon. Consider the particularly clear presentation in Giere's *Understanding Scientific Reasoning*. There are a series of figures illustrating the Ptolemaic and Copernican systems. One is entitled "The Ptolemaic model predicts that one cannot observe Venus fully illuminated" and another is entitled "The Copernican model predicts that one can observe Venus fully illuminated." (1991, 63 - 68).

Giere's accompanying analysis states:

Step 4. There are likewise two predictions: (1) the Ptolemaic prediction that Venus can never be seen fully illuminated...; and (2) the Copernican prediction that Venus can be seen going through a complete set of phases, including being fully illuminated... .

Giere's figures and text present a persuasive case that the Copernican model (presumably including unproblematic auxiliary hypotheses such as that light travels in straight lines) predicts that Venus has phases.

However, Copernicus, Galileo, Kepler, and their contemporaries knew and emphasized that an additional, quite problematic auxiliary hypothesis must be assumed for the Copernican hypothesis to predict the phases of Venus: Venus is an intrinsically dark, opaque body. If Venus is self-illuminated like the sun or if it is translucent like crystal or amber, Venus would always be fully illuminated — one would never see moon-like phases. Diagrams such as Giere's implicitly presuppose the now unproblematic auxiliary hypothesis that Venus is a dark, opaque body. But this hypothesis was problematic in the 16th century. The source of planetary light had been an open question since the Greeks. Because the phases of the planets are not seen by the naked eye, there had been a simmering controversy as to whether Venus is opaque, or self-illuminating, or translucent. By the middle of the fourteenth century, Albert of Saxony clearly was aware that the nature of Venus' substance could not be established by the available evidence. In his widely studied *Questiones super quatuor libros de Celo et Mundo,* Albert wrote,

> the question, do the stellar bodies other than the sun and moon receive their light from the sun can be thought as neutral; *the reasons one gives for one side can be as easily refuted as those one gives for the other side.* Therefore, for the love of Aristotle, the prince of philosophers, I will refute the six opinions formulated against Aristotle's opinion, in favor of Avicenna's opinion, and I will assert that all the stellar bodies other than the sun and moon, whether they are planets or fixed stars, received their light from the sun. (Ariew 1987, 84, my italics).

In Book I, Chapter 10, of *De Revolutionibus,* Copernicus shows he was clearly aware of this Scholastic controversy. In discussing various proposed orderings of the spheres, he writes:

> According to those who follow Plato, since they consider that all stars, being otherwise dark bodies, shine by the solar light which they receive, if they were below the Sun, on account of their short separation from it, they would be seen ony as halves, or at most as not completely round. For they generally reflect upwards, that is towards the Sun, the light which they have received, as we see in the new or waning Moon. ... On the other hand, those who place Venus and Mercury below the Sun ... do not admit that these heavenly bodies have any opacity like the moon's. On the contrary, these shine either with their own light or with the sunlight absorbed throughout their bodies.

There is no evidence that Copernicus ever accepted the opacity of the planets or believed that his hypothesis was committed to Venus having phases like the moon.

Galileo was well-acquainted with Scholastic natural science and the possibility that the planets are not dark is discussed in his *Dialogue Concerning the Two Chief World Systems.* Salviati is speaking:

> Add to these another difficulty; for *if* the body of Venus is intrinsically dark, and like the moon it shines only by illumination from the sun, which seems reasonable, then it ought to appear horned when it is beneath the sun, as the moon does when it is likewise near the sun — a phenomenon which does not make itself evident in Venus. For that reason, Copernicus declared that Venus was either luminous in itself or that its substance was such that it could drink in the solar light and transmit this through its entire thickness in order that it might look resplendent to us. In this manner Copernicus pardoned Venus its unchanging shape ... (Galileo 1632, 334, my italics)

Galileo's understanding of the situation comes through clearly in his January 1, 1611, letter to Giuliano de' Medici announcing his discovery of Venus's phases. After describing them in considerable detail, Galileo continues:

> From this marvellous experience we have a sensible and sure proof of two great suppositions which have been doubted until now by the greatest minds of the world. One is that all planets are dark by nature (the same for Mercury as for Venus). The other is that Venus must necessarilly revolve around the sun, just like Mercury and all the other planets.... (Abetti 1954, 106)

Kepler's enthusiastic letter of March 28, 1611, responding to Galileo's discovery, further emphasizes how open the issue of the source of Venus' light was:

> *Unexpected by me in any way was your observation,* for on account of the unusual brightness of Venus I believed light of its own to be inherent in it. I meditate much internally on what surface one ought to ascribe for its globe. Astonishing, unless Venus is all gold; or, as I said in my *Foundations of Astrology,* amber. (Drake 1984, 204, my italics)

To the best of my knowledge, only one of Galileo's contemporaries or predecessors thought that Copernican theory predicts the phases of Venus. This was Galileo's student, Benedetto Castelli. And his presentation of the idea indicates that the phases of Venus were not a well-known prediction of Copernican theory. Here is an extract from Castelli's letter to Galileo, December 5, 1610, proposing that Galileo investigate the phases of Venus:

> ..., *after various thoughts passed through my mind,* I finally hit on this: that being true, as I hold most true, the Copernican arrangement of the world, *Venus would have to have,* at like elongations from the Sun, sometimes a hornéd appearance and sometimes not hornéd, according as it is found beneath or beyond the Sun; ... (Drake 1984, 206, my italics)

As an example of using problematic hypotheses to explain data, Galileo's is second only to Sherlock's. Copernicus, Galileo and Kepler all understood that there were two problematic hypotheses: (1) Venus revolves around the Sun and (2) Venus is an opaque, intrinsically dark body. Neither hypothesis *by itself* makes the prediction that Venus has phases like the moon. But, given Galileo's telescopic reports of the phases of Venus, an astronomer could insist that the phases were strong evidence for one of the hitherto problematic hypotheses only if he acknowledged that the observations were strong evidence for the other as well. These two hypotheses provided such a powerful explanation for the phenomenon that, once the super-lunar reliability of the telescope was established, no one seriously doubted the truth of either hypothesis.

This is a particularly clear example of ARCHED or Stepped Hypotheses in action. I'm inclined to think it is a Stepped Hypothesis — the observation of a gibbous Venus established its darkness and opacity and observations of the entire cycle established that it goes around the Sun. In any case, whether ARCHED or Stepped Hypotheses, the major participants at the time clearly held that there were two quite distinct hypotheses involved.

Yet, despite Copernicus', Galileo's and Kepler's unambiguous words, historians often present Galileo's discoveries as a predictable (and predicted) consequence of Copernican theory. In fact, Galileo's use of the discovery of the phases of Venus to support the Copernican hypothesis is widely presented as Giere presents it: a straightforward example of a single hypothesis (Copernicus' and sometimes Brahe's) predicting a novel fact (that Venus has moon-like phases). A typical example comes from Crombie's *Augustine to Galileo* (1952, 318, also see 48 and Crombie 1959, 210):

[Galileo] also confirmed Copernicus' deduction that Venus, because of the position he held it to have inside the earth's orbit, would have phases like the moon...

Further, historians recounting Galileo's discoveries sometimes mention predictions that in fact weren't made or debates that in fact didn't happen. Some report that the absence of phases was an objection to Copernicanism. The claim is made in Dreyer's 1906 classic *History of the Planetary Systems from Thales to Kepler* (Dreyer, 414):

> Before the end of the year 1610 ... the discovery of the phases of Venus deprived the opponents of Copernicus of a favourite weapon.

Finocchiaro, author of an important commentary writes, that to assess the overall strength of the evidence regarding the Copernican hypothesis during Galileo's lifetime, "... one has to look at the counter-arguments, and there were plenty of them."

> *The appearance of the planet Venus* was the basis of another objection. For if the Copernican system were correct, then this planet should exhibit phases similar to those of the moon but with a different period; however, none were visible (before the telescope). The reason why Venus would have to show such phases stems from the fact that in the Copernican system it is the second planet ... (Finocchiaro 1989, 17 and 18, italics in original)

There appear to be two problematic claims here. First, that if the Copernican hypothesis were true, Venus "should exhibit phases similar to those of the moon" and "would have to show such phases". Second, there is apparent implication that the absence of the phases of Venus was "a counter-argument", an "objection", raised to the Copernican hypothesis before the invention of the telescope. Finocchiaro cites no one who raised this objection, I know of no one, and Kepler's and Castelli's words are strong evidence that this was not seen as an objection to Copernicanism.

Kuhn's account in *The Copernican Revolution* also adds a historical episode that apparently did not happen:

> ... the Copernicans, or at least the cosmologically more radical ones, had anticipated the sort of universe that the telescope was disclosing. *They had predicted a detail, the phases of Venus, with precision.* More importantly, they had anticipated, at least vaguely, the imperfections and the vastly increased populations of the heavens. ... There are few phrases more annoying or more effective than "I told you so." (Kuhn 1957, 222 and 224, my italics)

In Kuhn's commentary, there is no hint of what is so explicit n Copernicus and the rest—that the source of planetary light was open to serious debate. Copernicus' discussion of the possibility that the planets "shine either with their own light or with the sunlight absorbed throughout their bodies" is omitted from Kuhn's translation of the section from I:10 of *De Revolutionibus*. (1957, missing from 177)

Who are these unnamed "cosmologically more radical" Copernicans that had predicted "with precision" the phases of Venus before the telescope was invented? Certainly not that cosmological radical Kepler or anyone who had much impact on him. Both Kepler's words ("Unexpected by me in any way") and Castelli's words ("... after various thoughts passed through my mind, I finally hit on this...") indicate that the phases of Venus were not well-known predictions by "cosmologically more radical" Copernicans.

Stillman Drake's account also invents an episode:

> The other discovery ... was that Venus passes through a regular series of changes in shape precisely like those of the moon. *Copernicus had been puzzled by the apparent absence of such changes, which were required by his theory.* [12]
>
> [Ftnt 12]: *De Revolutionibus*, i, 10: "Neither do they grant that any darkness similar to that of the moon is found in the planets, but they assume that these are either self luminous or are lighted by sunlight throughout their whole bodies." Copernicus refrained from giving his own opinion on the problem. Galileo was much impressed by the fact that this apparent contradiction of the senses had not deterred Copernicus from adhering to the heliocentric system; cf. *Dialogue*, pp. 334-35. (Drake 1957, 74 - 75, my italics)

Yet, the very passage from *De Revolutionibus* that Drake quotes in his footnote shows why "such changes" were not "required" by Copernicus' theory. Further, on the cited pages of Drake's translation of the *Dialogue* we find, "In this manner Copernicus pardoned Venus its unchanging shape ..." and on the sidebar are these words: "Venus, according to Copernicus, is either luminous by itself or is of transparent material."

These are normally cautious and judicious historians who know the key texts very well. Yet, they report that the Copernican theory predicts that Venus has phases and often they do not mention that the texts discuss the possibility that Venus is self-illuminating or translucent. Further, there is a strong historical tradition of inventing additional historical facts about the "predictions" of "cosmologically more radical" Copernicans, or "counter-arguments" or Copernicus being "puzzled." Starting at least in d'Alembert's articles in Diderot's *Encyclopedie* (v. IV, p. 160 and v. XII, p. 453), it continues through Adam's Smith's 1795 *History of Astronomy* (IV. 47 and IV.49), was very widespread in the 19th century and continues to this day. (For many more examples, see Thomason (forthcoming))

Philosophers have done worse than historians; the claim that the Copernican hypothesis *per se* predicts Venus' phases is very widespread among philosophers as well as historians — as well as Giere, it is found in Russell (1961, 520), Popper (1963, 98 and 246), Lakatos and Zahar (1978, 183-4), Chalmers (1982, 71), Gardner (1983, 221), Shea (1972, 110), Wallace (1992, 203) and others. Very few primarily philosophical accounts tell the story correctly — Clavelin (1974, 200) is a noble exception.

Having in my lifetime misinterpreted more than my fair share of arguments, I certainly do not want to claim that it is easy to get the history of science right. Still, when the well-known primary texts are so clear and the scholarly misinterpretations so striking one can, with some chutzpah, propose that scholars are falling prey to the power of Psychological Predictivism.

I suspect that what has happened is that our historians and philosophers were implicitly thinking along these lines: "The Copernican hypothesis *must predict* the phases of Venus — why else would the phases have been such excellent evidence for it? Further, since the prediction is so obvious, some Copernicans *must have* made it. Since this prediction is so clearly inconsistent with naked eye observations, the absence of the phases *must have* puzzled Copernicus and his followers and *must have* been raised as an objection to Copernicanism before Galileo's telescope was invented." If some such explanation is correct, Psychological Predictivism has not only resulted in the serious distortion of a very well-known episode in the history of science; but it has also resulted in historians inventing a difficulty and a debate where there was none.

As I said above, I am *not* claiming that Psychological Predictivism has resulted in no historians getting the history right. Many clearly present Galileo's argument — Casper (1959, 200) and Cohen (1960, 81-83) are two particularly clear examples and there may be others.

> If Venus possessed light of its own, phases occasioned by reflection of sunlight could hardly be detected telescopically or in any other way. Castelli's deduction was formally invalid without explicit assumption that Venus was as dark as the moon. Galileo's critics may suppose that he would overlook that and risk his reputation with Kepler, but I should like to know from them what evidence Galileo had that Venus was dark, before he began to observe its phases. Since even the possibility was unthinkable to Kepler, it is hardly objective to attribute to Galileo its automatic assumption. (Drake 1984, 203 -204)

There are others, including Ariew's "The Phases of Venus before 1610", which freed me from Psychological Predictivism in this particular matter.

And it is clear from my e-mail that many historians and some philosophers know the correct story. I am simply claiming that many accounts, even by excellent historians and philosophers, are strikingly inconsistent with the key relevant passages in *very* well known texts. I am proposing an explanation for the tradition of incorrect accounts. And I am further proposing that when the texts are more ambiguous or the episode is not as well known, Psychological Predictivism should have a tremendous impact.

3. Why the Invisibility of ARCHED and Stepped Hypotheses Matters

When a historian comes across an intuitively good argument, there is a very powerful tendency to understand it in terms of good argument forms. Psychological Predictivism provides one constraint on what is placed on that list. As a result, the evidential relationship of ARCHED and Stepped Hypotheses is not part of the historian's tool kit; the ARCHED and Stepped metaphors have been absent from the philosophy of science. As Maslow said, when the only tool you have is a hammer, everything looks like a nail. As a result, the use of ARCHED and Stepped Hypotheses in science can often go unnoticed even when the evidence is blatant. In subtler cases, it almost certainly will not be noticed. If all this is correct, there is a distinct possibility that the evidential structure of many of the most important scientific advances is quite different from how it is standardly understood. After all, the logical structure of Galileo's explanation of the moon-like phases of Venus is astonishingly straightforward when compared to the more technically complex explanations in most science. There may be far more ARCHED hypotheses in the history of science than one can discover by reading the secondary literature.

In fact, there may be far more ARCHED or Stepped hypotheses in the history of science than one can discover by reading even the primary literature. Scientists themselves normally have a great incentive to downplay any ill-supported aspects of their hypotheses. So they may strongly tend to hide an explanation's ARCHED or Stepped characteristics. Thus, neither scientists nor historians reliably report such hypotheses. I believe that they are very common — and that they are usually missing from both the primary and the secondary literature.

Notes

[1]This is an earlier version of a much modified (Thomason (forthcoming)). I am particularly grateful to Michael Ellis, Keith Hutchison, Ross Phillips, Brian Ellis, Len O'Neill, Martin Tamny, and the Victorian Centre for the History and Philosophy of Science.

References

Abetti, Giorgio (1954), *The History of Astronomy* translated by Betty Burr Abetti. London: Sidgwick and Johnson.

Ariew, R. (1987), "The Phases of Venus Before 1610". *Studies in History and Philosophy of Science*, 18: 81 - 92.

Casper, M. (1959), *Kepler,* translated by C. Doris Hellman. New York: Abelard - Schuman.

Chalmers, A. (1982), *What is this thing called Science?,* second edition. St. Lucia: University of Queensland Press.

Clavelin, M., (1974), *The Natural Philosophy of Galileo: Essay on the Origins and Formation of Classical Mechanics,* translated by A. J. Pomerans. Cambridge: MIT Press.

Cohen, I.B. (1960), *The Birth of a New Physics.* Garden City: Doubleday and Company.

Crombie, A.C. (1952), *Augustine to Galileo.* London: Falcon Educational Books.

_____. (1959), *Augustine to Galileo,* revised edition. Hammondsworth: Penguin Books.

Diderot and D'Alembert (eds.) (1772, 1774), *Encyclopedie ou Dictionnaire Raisonneé Des Sciences, Des Artes et Des Metiers,* A Livourne, De L'Imprimerie Des Editeurs.

Drake, S. (1957), *Discoveries and Opinions of Galileo.* Garden City: Doubleday - Anchor.

Drake, S. (1984), "Galileo, Kepler and Phases of Venus", *Journal of the History of Astronomy,* XV: 198 - 208.

Dreyer, J.L.E. (1906/1953), *History of the Planetary Systems from Thales to Kepler.* Cambridge: Cambridge University Press. Reprinted as *A History of Astronomy from Thales to Kepler*. Cambridge: Dover Publications.

Finocchiaro, M. (ed. and translator) (1989), *The Galileo Affair: A Documentary History.* Berkeley and Los Angeles: University of California Press.

Galilei, G. (1632 /1967), *Dialogue Concerning the Two Chief World Systems — Ptolemaic & Copernican,* translated by Stillman Drake, second edition. Berkeley and Los Angeles: University of California Press.

Gardner, M. R. (1983), "Realism and Instrumentalism in Pre-Newtonian Astronomy" in John Earman (ed.) *Testing Scientific Theories,* Volume X in the series *Minnesota Studies in the Philosophy of Science.* Minneapolis: University of Minnesota Press.

Giere, R. (1991), *Understanding Scientific Reasoning,* third edition. Fort Worth: Holt, Rinehart and Winston.

Gingerich, O. (1986), "Galileo's Astronomy" in Wallace, W. (ed.) *Reinterpreting Galileo.* Washington, D.C.: Catholic University of America Press.

Kuhn, T. (1957), *The Copernican Revolution: Planetary Astronomy in the Development of Western Thought.* Cambridge: Harvard University Press.

Lakatos, I., and Zahar, E. (1978), "Why Copernicus's Programme Superseded Ptolemy's" in Worrall, J. and Greg Currie (eds.) *The Methodology of Scientific Research Programmes: Philosophical Papers,* Volume I. Cambridge: Cambridge University Press.

Popper, K. (1963), *Conjectures and Refutations.* New York: Harper and Row.

Russell, B. (1961), *History of Western Philosophy and Its Connection with Political and Social Circumstances from the Earliest Times to the Present Day,* new edition. London: George Allen & Unwin Ltd.

Segre, M. (1991), *In the Wake of Galileo.* New Brunswick: Rutgers University Press.

Shea, W.R. (1972), *Galileo's Intellectual Revolution.* New York: Macmillan.

Smith, A. (1795), *The Principles Which Lead and Direct Philosophical Enquiries; Illustrated by the History of Astronomy,* reprinted in Wightman, W.P.D., J. C. Bryce and I.S. Ross (eds.) (1980) *Adam Smith: Essays on Philosophical Subjects.* Oxford: Clarendon Press.

Thomason, N. (forthcoming), "1543—The Year Copernicus Didn't Predict the Phases of Venus", in Anthony Corones and Guy Freeland (eds.) *1543 and All That.* In the series, *Australasian Studies in History and Philosophy of Science.*

_____. (1994), "The Power of ARCHED Hypotheses: Feyerabend's Galileo as a Closet Rationalist," *British Journal for the Philosophy of Science,* 45: 255-264.

Wallace, W. (1992), *Galileo's Logic of Discovery and Proof,* Volume 137 in the series, *Boston Studies in the Philosophy of Science,* Dordrecht: Kluwer Academic Publishers.

Westfall, R.S. (1971), *The Construction of Modern Science.* Cambridge: Cambridge University Press.

How to Remain (Reasonably) Optimistic: Scientific Realism and the "Luminiferous Ether"

John Worrall

London School of Economics

1. Laudan on the "Non-Referring" but Empirically Successful "Optical Ether"

In the course of his forceful (1981) attack on "convergent realism", Larry Laudan attempted to turn an influential pro-realist consideration on its head. Scientific realists have wondered how a theory could enjoy the sort of empirical predictive success exhibited by presently accepted theories in the "mature" sciences, and yet be radically wrong in what it claims is going on "behind" the empirical phenomena. If nothing like the electrons and other particles postulated by current physics exists, how can the theories that there are such things have made the great range of successful predictions that they do about phenomena observed in particle accelerators and the like? Laudan directs such realists to the history of science and to a list of "once successful" theories which are "(by present lights) non-referring"—a list of successful theories which gave a "central" role to notions that, according to theories we accept *now*, do not exist (1981, 26). He supposes, apparently quite reasonably, that, whatever account of "approximate truth" the realist relies on (and, notoriously, there is as yet no generally accepted formal account on offer), that account will entail that a theory involving central terms with no referent cannot be "approximately true". (But see section 4.) He claims, then, that, in view of the history of science, this realist consideration is *self-defeating*. If, as the realist recommends, we hold that presently accepted theories (in the "mature" sciences) are approximately true, then it follows that these earlier theories *were* radically false, despite their predictive success. The empirical success of presently accepted theories can, then, hardly be used as an argument for their approximate truth.

Laudan's argument seems especially strong because it hits, not at particular realist arguments, but directly at the realist's *general*, underlying intuitions. Laudan himself later described his (1981) as having

> challenged the intuitions which motivate the realist enterprise by arguing .. that many (now discredited) scientific theories of earlier eras exhibited an impressive sort of empirical support, arguably no different in kind from that enjoyed by many contemporary physical theories. Yet we now believe that many of those earlier theories profoundly mischaracterized the way the world really is. More specifically, we now believe that there is nothing in the world which even

approximately answers to the central explanatory entities postulated by a great many successful theories of the past (1984, 157)

Laudan's argument is, however, only as strong as the quality of his historical examples of "once successful but (by present lights) nonreferring" theories. Some of the entries on his list (which he insisted could be extended "*ad nauseam*") are in fact strikingly unimpressive. He must have been working with some very loose notion of scientific "success" in order to count, for example, the vapid, truly etherial, conjectures of Hartley and LeSage even as having "enjoyed some measure of empirical success" (1981, 27). The argument needs examples of theories which were successful in the genuinely *predictive* sense. We know that *any* theory can be made to have correct empirical consequences by "writing those consequences into" it; the cases that have traditionally induced realist-inclinations in even the most hard-headed are cases of theories, designed with one set of data in mind, that have turned out to predict entirely unexpectedly some further general phenomenon. Laudan himself gives special emphasis to the one item on his list that seems unambiguously to fit this bill — Fresnel's wave theory of light and its associated elastic-solid "luminiferous ether".

It would be difficult to argue that Fresnel's theory counts as "immature" science; and *impossible* to deny that it was impressively successful predictively. Aside from the much-rehearsed case of the light spot at the centre of the shadow of a small opaque disc (for the real story of this case see my 1989), Fresnel's theory of the wave surface inside biaxial crystals turned out to predict the existence of *internal* and *external conical refraction*. Fresnel never realised that these latter predictions follow from his theory; and the phenomena themselves were not even thought of, let alone known to occur, until, after Fresnel's death, they were derived from the theory by William Rowan Hamilton (1833) and confirmed experimentally by Humphrey Lloyd (1833).

So there is at least one theory on Laudan's list that was unambiguously successful. And, so Laudan asserts, that theory "centrally" involved a notion denied any real referent by theories *subsequently* accepted in science. Within Fresnel's theory "the optical ether functioned centrally in explanations of reflection, refraction, interference, double refraction, diffraction, and polarization"(1981,27); but Maxwell's theory and then the General Theory of Relativity entail that there is just no such thing as Fresnel's elastic optical ether. According to these later theories, light "in fact" consists of vibrations of the electromagnetic field, a field which is "*sui generis*", explicitly *not* a manifestation of the contortions of some underlying material medium. (As is well known, Maxwell himself tried hard to produce a "mechanical model" for the field—that is, to explain the field in terms of some underlying material medium. But his failure in this attempt, and the failure of his contemporaries and successors, led to the acceptance of what might be called the "mature" version of Maxwell's theory— a theory that sees the field as a "primitive" part of the furniture of the universe.)

The case of Fresnel's theory of the elastic ether poses, then, *via* Laudan's argument, a sharp challenge for the realist—a *prima facie* strong historical reason to be "pessimistic" about the likely fate of *currently* accepted theories. Two realist responses to the challenge are possible: (i) deny that the ether played a "central" role in Fresnel's theorising, and (ii) deny that the ether has in fact been rejected by later science. Both of these strategies have found interesting particular instantiations in the recent literature—in Hardin and Rosenberg (1982) and in Kitcher (1992). Both accounts see the two, apparently quite different, strategies as in fact closely interrelated.

I argue (in sections 2 and 3 respectively) that neither realist strategy succeeds. Each does, however, point to important features of the logical relationship between

Fresnel's theory and the theories that later replaced it. In section 4 I argue that, when that logical relationship is properly described and its consequences fully drawn out, although this case of a theory-shift *is* inconsistent with scientific realism as normally construed, it "confirms" a view that has been called *structural realism*. This was first developed by Poincaré (and is very different from the sort of anti-realism that is usually attributed to him). In fact, Poincaré—who fully anticipated the "pessimistic induction" argument—used exactly this historical case of the shift from Fresnel's elastic ether to Maxwell's electromagnetic field theory as the chief illustration of his general view of the aim and status of theories. Structural realism encourages an *optimistic* induction from the history of theory-change in science, but an optimistic induction concerning the discovery of *mathematical structure* rather than individual ontology.

2. Did the "Luminiferous Ether" Really Play a Central Role in the Success of Fresnel's Theory?

It can't, I think, sensibly be denied that Fresnel *believed* in the the ether as a real, material medium. He refers to such a material medium explicitly (and in explicitly "realist" terms) at various points in his scientific work. For example, in his famous (1818) "Prize Memoir" on diffraction, he characterised the general problem of diffraction as follows : "Given the intensities and the relative positions of any number of systems of light waves of the same wavelength, propagating in the same direction, to determine the intensity of the vibrations resulting from the concourse of these different systems, that is to say, the velocity of the oscillations of the molecules of the ether" (248). In his (1822, 136) he stated that the properties of polarised light are simply explained on his theory "by supposing that, in light waves, the oscillations of the molecules of the ether are executed at right angles to the rays".

It doesn't however follow that the ether played a "central role" in his theory. Indeed the standard view in history of science—following Whittaker's influential treatment—is that Fresnel's account of ether-dynamics, was pure window-dressing: whatever Fresnel's own beliefs, the heuristic impetus for his theoretical work came from mathematical considerations—the mechanical-dynamical considerations attached themselves only later, only unsuccessfully and certainly without any independent empirical success. Charting Fresnel's route to his theory of the wave-surface in birefringent crystals, Whittaker remarked (1951,119):

Having ... arrived at his result by reasoning of a purely geometrical character, [Fresnel] now devised a dynamical scheme to suit it.

(A similar claim that Fresnel's ether played no "generative" role in the development of his theory of polarisation is made in Buchwald 1989.)

No realist should advocate a "realist attitude" towards *all* theoretical claims—even theoretical claims within successful theories. Some play no effective role and are, in Kitcher's (1992) terminology, "presuppositional". (Newton's assumption that the centre of mass of the universe is at absolute rest is surely a case in point.) If the standard story were correct, then the realist would be justified in claiming that, since the material ether was a "merely presuppositional posit", and since he holds no brief for such notions, the eventual rejection of that material ether represents no threat to his position.

The claim that the ether was "central" figures as a bald, unsubstantiated assertion in Laudan's paper unaccompanied by either analysis or argument. Nonetheless it is, I believe, correct or at any rate correct enough to block this escape route for the realist. I cannot argue the case fully here, but detailed analysis (forthcoming) shows that the suggestion that the material ether was an idle component in Fresnel's system is significantly misleading. Whittaker claims, for example, that "geometrical reasoning" led the way in Fresnel's development of the wave surface in birefringent crystals. However, this "geometrical reasoning" itself did not spring from nowhere, but was based on Hooke's law, Huygens's principle, the principle of superposition ("coexistence of small movements") and other assumptions of a *general* mechanical kind. Moreover, although Fresnel's extension to cover all crystals of Huygens's famous sphere/spheroid construction for the two refracted beams can be characterised mathematically as a process of putting two equations together (by introducing three parameters for Huygens's two), that process was in turn undoubtedly guided by Fresnel's "realist" belief that there could only be one light-carrying medium and the "natural" assumption that, in the general case, the coefficients of elasticity of that medium in the three orthogonal directions in space will be different. Fresnel *did* get some important heuristic mileage out of certain general mechanical-dynamical ideas concerning some sort of mechanical medium with some sort of vibrating parts.

Whittaker and others are, I believe, wrong: Fresnel did *not* first operate geometrically and only later "interpret" his theory dynamically. Rather than two separate stages, Fresnel was, I think, working all the time with a mix of basic mechanical ideas, mathematics and known experimental results. (This is surely generally true in theoretical physics.) He arrived at an account of light as a wave motion in a mechanical medium possessing certain *general* properties in accord with the known principles of dynamics. This account was dramatically successful empirically. He (and his successors) made various attempts to *strengthen* the account into a fully-fledged dynamical theory in which everything would follow from a "natural", "unified" account of the forces operating on the ether particles. These attempts were, however, immediately refuted (or if you prefer, immediately ran into some notable and stubborn anomalies).

The problems with the attempts to strengthen the theory of the wave surface into a fully-fledged dynamical theory of the ether are well-known: the most famous being the problem of the longitudinal wave. Although Fresnel continued to talk about the etherial "fluid" even after he became convinced that the oscillations of the medium are *transverse* to the direction of propagation of the wave, he was fully aware that transversality meant that the medium had to have resistance to *shear* (and this makes it, at least according to later terminology, unambiguously an elastic *solid*). "Ordinary" elastic solids are resistant *both* to shear *and* to compression, and hence transmit *both* transverse *and* longitudinal waves. Fresnel took it that his and Arago's experimental results on the interference of polarised light establish that no longitudinal component plays any role in optical effects. What happens to the longitudinal wave in the elastic solid ether? Fresnel simply hypothesised it away. He assumed that the ether must be infinitely (or "near infinitely") resistant to compression so that the longitudinal wave travels at infinite or "near infinite" velocity and so can somehow be ignored. Unfortunately hypothesising the longitudinal wave away once is not enough: even when you have got, or rather have given yourself, a purely transverse wave, it ought on mechanical principles to develop a longitudinal component again whenever it meets the boundary of a medium of a different optical density. So you have to *keep on* hypothesising away such longitudinal components. That is, you have to keep on violating the laws of mechanics.

The intractablity of this and other problems and their role in leading eventually to a replacement theory is what lies behind Whittaker's treatment (and those of Hardin and Rosenberg and of Kitcher). But this represents illicit use of hindsight: there was, of course, no reason to believe back in 1821, say, that those problems would prove intractable. There certainly was reason in 1821 (when Fresnel completed his paper on the form of the wave-surface in biaxial crystals) to regard the ether as problematic. But—as is now generally recognised following the work of Kuhn, Lakatos and others—the highest-level, most interesting parts of theoretical science are invariably problematic, when examined closely. If the realist is in the business of advocating a realist attitude only toward entirely unproblematic theories, then realism is indeed restricted in scope.

If my claims about Fresnel's theoretical work are at all correct, then the realist still faces a threat from this case, and cannot convincingly hide behind the problems that Fresnel's basic theory of the ether undoubtedly faced. The elastic-solid ether played a problematic, but nonetheless somewhat positive role, so the realist had better have *something* positive to say about it.

3. Was Fresnel Talking about Electromagnetic Waves all Along?

Hardin and Rosenberg (1982) make the, at first glance audacious, claim that the "something positive" the realist can say about the ether is that, contrary to Laudan, we still believe it to exist: the realist can justifiably view Fresnel as talking about the electromagnetic field all along when he used the term "ether". They concede (p.611) that this account of reference "severs it from the detailed beliefs of" Fresnel. Indeed, this is hardly a question of detail: Fresnel could scarcely have believed that he was referring to an entity which was first thought of only some decades after his death! Hardin and Rosenberg point out—citing the problems faced by the elastic ether —that various features of Fresnel's accounts of optical phenomena undoubtedly sit more easily with Maxwell's field theory than with his own notion of an elastic solid. And they argue (613-4) that, given this and given the "continuity of causal role" stretching from Fresnel's ether to Maxwell's field and beyond, a realist can reasonably regard Fresnel as referring to the electromagnetic field when he used the term "ether":

Looking back across the range of theories from Fresnel to Einstein, we see a constant causal role being played in all of them; that causal role we now ascribe to the electromagnetic field. One permissible strategy of the realist is to let reference follow causal role. It seems not unreasonable, then, for realists to say that 'ether' referred to the electromagnetic field all along.

But in fact, the causal role is only "constant" if we ignore certain inconvenient features of the earlier theory: for example, the elastic solid ether *ought* to have had the effect of slowing down the planets as they moved through it. If we are allowed to be similarly selective in other cases, there seems equally to be a "constant causal role" between, say, Aristotle's notion of a body's desire to be in its natural place and Einstein's notion of a body moving along a geodesic in curved space-time; or between the seventeenth century notion of a witch and twentieth century notions of sufferers from mental illnesses of various kinds. If it is "not unreasonable ... for realists to say that 'ether' referred to the electromagnetic field all along", it seems equally to fail to be unreasonable to say that "desire for its natural place" refers to "necessity to move along a geodesic" or to say that "witch" (sometimes) refers to "sufferer from certain kinds of mental illness". The causal role in explaining certain kinds of "odd" behaviour that was attributed in the seventeenth century to a person's being a witch is certainly now attributed to mental afflictions of various kinds. Laudan makes much the same point in his (1984)—taking it to be a severe embarrassment for Hardin and Rosenberg's account (whether or not it really is an embarrassment will be considered in section 4).

Philip Kitcher's (1993) treatment of this issue has a similar theme, though with interesting variations. Kitcher concedes that Fresnel's term "ether" fails to refer, but suggests (1993, 147) that Fresnel's tokens of the term "light wave" nonetheless *do* refer. Fresnel undoubtedly believed that such waves were waves in his elastic medium but, so Kitcher claims, since Fresnel's "dominant intention" was "that of talking about the wavelike features of light, *however they happen to be realized*", it is reasonable, in view of what science now tells us, to take tokens of that term (or perhaps most of them) as "genuinely refer[ring] to electromagnetic waves of high frequency" (146).

So far as the ether itself is concerned, Kitcher asks if "the schemata employed by Fresnel and other wave theorists of the early and middle nineteenth century contain *ineliminable* commitments to the ether?"; and, concluding that they do not, infers that "[t]he ether is a prime example of a presuppositional posit" (1993, 149). I have suggested that the ether was in fact a problematic, but certainly *non*-idle notion in Fresnel's approach. Fresnel got real heuristic assistance from ideas of a *general* mechanical kind. But are those ideas nonetheless "eliminable"? Can't the principle of superposition, Hooke's law and the other assumptions that Fresnel used be cut off from their original mechanical bearings and be re-clothed in the terms of some other theoretical framework?

If this is a question about logical possibilities rather than possibilities practically available to Fresnel and his contemporaries, then the answer is of course "yes". But if "ineliminability" requires that there be *no other* theory that explains the phenomena at issue then no theory, no theoretical notion, is ineliminable. In cases such as this one where we are considering a theory that was later replaced, no recourse need be had to abstract ideas about "underdetermination" in order to establish this: the replacing theory itself forms a constructive proof of the "eliminability" of the earlier one. Maxwell's theory of the electromagnetic field shows that the role that Fresnel believed could only be played by a material medium can also be played by the field. A realist committed to a realist attitude toward only ineliminable notions faces no problem from the history of science, but then he has no position to defend. In order to count as any sort of realism, a position must entail *something* positive about Fresnel's ether despite its "eliminability".

4. Fresnel and Maxwell: the Ether, the Field and "Structural Realism"

The above remarks notwithstanding, there is in much that I agree with in the positions of Hardin and Rosenberg and of Kitcher. Both accounts point to important features of the relationship between Fresnel's ether theory and its successors. However neither account gets this relationship quite right because each seeks to defend a stronger version of realism than is, I believe, really defensible.

It is, so the realist wants to claim, vastly improbable that a theory should score the sort of extensive empirical success scored by presently accepted theories in the "mature sciences" and yet not have *somehow* "latched on to" how things are "underneath" the empirical phenomena. To defend this intuitive claim against cases of theory-change, the realist needs to show that, from the point of view of the later theory, the fundamental claims of the earlier theory (in so far as they played integral roles in that theory's empirical success) were—though false—nonetheless in some clear sense "approximately correct". He needs to show that, from the point of view of the later theory, we can still *explain* the success enjoyed by the earlier one.

A natural assumption is that such an explanation requires a demonstration *either* that the parts of the earlier theory rejected by the later one were redundant *or* that no real "rejection" was involved (but only a "re-description"). However, in this particular historical case at least, the most straightforward and least revisionary account of the explanation of the success of the earlier theory provided by its successor fits neither of those patterns.

For convenience (and temporarily) freeze the history of science at the point where the "mature" (non-medium-based) version of Maxwell's theory had been accepted. From that vantage point, there is an easy explanation of the success of Fresnel's elastic-ether theory of light—one which requires no Whiggish "reinterpretation" of Fresnel's thought. From the later point of view, Fresnel clearly misidentified the *nature* of light, but his theory nonetheless accurately described not just light's observable effects but also its *structure*. There is no elastic-solid ether of the kind Fresnel's theory (problematically but nonetheless importantly) involved; but there is an electromagnetic field. The field is not underpinned by a mechanical ether and in no clear sense "approximates" it. Similarly there are no "light waves" in Fresnel's sense, since these were supposed to consist of motions of material ether-particles. Nonetheless disturbances in Maxwell's field do obey *formally* similar (in fact, and unusually, mathematically identical) laws to some of those obeyed by the "materially" entirely different elastic disturbances in a mechanical medium.

Unless—surely very much in the spirit of *anti*-realism—we think of these theoretical notions as characterised by their observable effects, then we have to allow that Fresnel's most basic ontological claim that the vibrations making up light are vibrations of real material ether particles subject to elastic restoring forces was entirely wrong. A displacement current in a *sui generis* electromagnetic field and a mechanical vibration transmitted from particle to particle are more like "chalk and cheese" than are real chalk and cheese. But if Fresnel was as wrong as he could have been about *what* oscillates, he was right, not just about the optical phenomena, but right also that those phenomena depend on the oscillations of something or other at right angles to the light. His theory was more than empirically adequate, but less than true; instead it was *structurally correct*. There is an important "carry-over" from Fresnel to Maxwell, one at a "higher" level than the merely empirical, but it is a carry over of *structure* rather than content. Both Fresnel's and Maxwell's theories make the passage of light consist of wave forms transmitted from place to place, forms obeying the same mathematics. Hence, although the periodic changes which the two theories postulate are ontologically of radically different sorts—in one material particles change position, in the other field strengths change—there is nonetheless a structural, mathematical continuity between the two theories.

All this is reflected in the fact that if you perform the following meta-level operation on Fresnel's theory you "turn it into" a genuine sub-theory of Maxwell's:

> Go through Fresnel's theory and, wherever he talks about a molecule of the ether's being forced away from its equilibrium position, replace that talk by talk of a forced change in the electromagnetic field strength.

(Another way to put this is that if you go through Fresnel's theory and replace the notion of a molecule's being forced from its equilibrium position by a theory-neutral term such as "optical disturbance" ; and then *reinterpret* "optical disturbance" as "forced vibration of the electromagnetic field" then what you get is a sub-theory of Maxwell.)

Nothing "Whiggish" is being perpetrated here: I do not assert—indeed I explicitly deny—that this is what Fresnel's *theory* "really" amounted to "all along". What you get as a result of this process is *not* Fresnel's theory but a structurally identical *facsimile* of it. But it's the fact that this facsimile is entailed by the later theory that explains why, from the vantage point of the later theory, the empirical predictive success of Fresnel's theory was no lucky accident.

This account, in terms of the "ontological" falsity, but structural correctness of Fresnel's theory, might appear insufficiently different from those of Hardin and Rosenberg and of Philip Kitcher to justify the fuss. But let's think through what their claims that Fresnel was "really" talking about the field or waves in it "all along" really mean. If those claims mean just that the entity that "really" plays the causal role in producing a given range of phenomena that Fresnel attributed to a highly attenuated elastic medium is—*according to the science of the later nineteenth century*—the electromagnetic field, then it is of course no more than the truth. (Just in the same sense that Aristotle was—according to twentieth century science—referring to bodies moving along geodesics in spacetime when he talked about bodies seeking their natural place.) But the conditional character of such judgments is, of course, crucial. The judgments are always theory-dependent. If we ask in turn what is the "real" referent of Maxwell's term "electromagnetic field", the question can again only be answered within the context of the theories accepted at some given time. The answer according to *currently* accepted theories is a quantum field carrying probability waves—a notion radically different from anything envisaged by Maxwell himself, perhaps even more radically different from it (if we can make sense of such comparisons) than Maxwell's own notion of field is from Fresnel's notion of the elastic ether. (It seems to be a historical accident —of no more than conventional significance—that science happens to continue to use the word "field" whereas, *on the whole*, it has dropped the term "ether".) Since the "continuity of causal role" to which Hardin and Rosenberg appeal now extends beyond Maxwell, presumably they must allow that what Fresnel was *really* "really" referring to "all along" was the quantum field carrying probability waves. *Or rather* —since this theory may of course itself eventually be replaced—that we don't know what the real "real referent" of Fresnel's notion the "ether" is. What we do know is that there was a certain structural or syntactic continuity between the *theories* of Fresnel and Maxwell (and again—though this time involving the "correspondence principle"—between Maxwell's and the quantum theory of "the" field).

It was exactly this point that Poincaré had in mind when he said that both Fresnel's notion of an elastic vibration and Maxwell's notion of displacement current are

> merely names of the images we substituted for the real objects which Nature will hide for ever from our eyes (1905, 162)

Poincaré insisted that adopting this view

> cannot be said [to amount to] reducing physical theories to simple practical recipes [i.e. to instrumentalism]; [Fresnel's] equations express relations, and if the equations remain true [better; "are preserved in the later theory"], it because the relations preserve their reality ....The true relations ....are the only reality we can attain. (*ibid*)

Poincaré did not think of himself as proposing a *restriction* of a stronger view about theories and their relation to the world, but rather as pointing out that, *in view of the fact that in science we can never "get outside" of our theories but only view reality through those theories we currently accept*, this structural version is the only view (the only version of "realism") that makes any sense.

If the switch from Fresnel to Maxwell is typical (and I have given no reason in this paper to think it is), then—against the currently fashionable "pessimistic induction"—there are (inductive) grounds for optimism, optimism that science is progressing towards a correct account of the universe, but that progress is at the structural, rather than the "ontological" level. If Poincaré is correct, then any feeling that this is less optimism than we could reasonably expect is based on a surely mistaken (but easily adopted) view that we can somehow have direct access to the furniture of the universe, unmediated by our theories.

References

Buchwald, J.Z. (1989), *The Rise of the Wave Theory of Light* . Chicago and London: University of Chicago Press.

Fresnel, A.J. (1818), "Memoire Couronné sur la Diffraction" (page references to the reprinted version in Fresnel *Oeuvres Complètes*, I, Paris 1865).

_ _ _ _ _ _ _. (1822), *De la Lumière*. (page references to the reprinted version in Fresnel *Oeuvres Complètes*, I, Paris 1865).

Hamilton, W.R. (1833), "Essay on the Theory of Systems of Rays", *Transactions of the Royal Irish Academy 17*, 1833, 1.

Hardin, C.L. and Rosenberg, A. (1982), "In Defence of Convergent Realism", *Philosophy of Science 49:* 604-615

Kitcher, P. (1993), *The Advancement of Science*. Oxford and New York: Oxford University Press.

Laudan, L. (1981), "A Confutation of Convergent Realism", *Philosophy of Science 48*: 19-49.

_ _ _ _ _ _. (1984), "Realism without the Real", *Philosophy of Science 51*, 156-62.

Lloyd, H. (1833), "On the Phenomena exhibited by Light in its passage along the axes of Biaxial Crystals", *Transactions of the Royal Irish Academy 17*, 145.

Poincaré, H. ([1905] 1952), *Science and Hypothesis*. Originally published as *Science et Hypothèse* (Paris:). New York: Dover.

Whittaker, E.T. (1951), *A History of the Theories of Aether and Electricity. The Classical Theories*. London: Thomas Nelson.

Worrall, J. (1989), "Fresnel, Poisson and the White Spot: the Role of Successful Predictions in the Acceptance of Scientific Theories" in D.Gooding *et al* (eds): *The Uses of Experiment*. Cambridge: Cambridge University Press.

Part IX

**QUANTUM MECHANICS:
DECOHERENCE AND RELATED MATTERS**

Making Sense of Approximate Decoherence[1]

Guido Bacciagaluppi and Meir Hemmo

University of Cambridge

1. Introduction

One of the main goals of no-collapse interpretations of quantum mechanics is to show how results of measurements are objectified, or, in a more poignant terminology, how in a measurement situation *facts* arise. One needs to show how in their dynamical evolution the measured system and, perhaps more importantly, the measuring apparatus acquire, or at least appear to acquire, the definite properties that are otherwise explained by the collapse of the state. This can be seen as a special case of the problem whether quantum mechanics alone can account for the classical behaviour of macroscopic systems; in particular, the permanent definiteness of at least those properties of macroscopic systems that are directly observable, such as the position of the pointer of a measuring apparatus. In other words, whether quantum mechanics can account for macrofacts of the familiar kind.

The decoherence approach (Zurek 1981, 1982, 1993; Joos and Zeh 1985) addresses precisely this more general question of how macrofacts arise. We describe this in section (2). In this approach, a macroscopic system is considered in its interaction with a very large or infinite environment. It is shown that, due to this interaction, correlations arise between the observable of interest (the 'pointer position') and some observable in the environment (the 'record observable'). The state is interpreted as an effective mixture of the macroscopically distinguishable terms. Zurek (1993) among others justifies this by adopting an Everettian stance in the interpretation of the formalism.

In this paper we address the problem of approximate decoherence: that is, as we explain in section (3), the problem that exactly stable correlations arise only in the infinite-time limit. At finite times these correlations are only approximate. But this apparently means that the macrofacts we are familiar with are only approximately defined, and this is conceptually puzzling. To take a very casual example, does it make sense to say that a *cat* is only approximately dead or alive?

In another paper we have analysed the problem of imperfect measurements in the modal interpretation of quantum mechanics (Bacciagaluppi and Hemmo 1994). There is a deep analogy between the two problems, and we apply the results of that analysis to the problem of approximate decoherence. We argue (in section (4)) that if from a certain time onwards the mutually correlated observables of system and environment are sufficiently close (in a well-defined sense) to the exact pointer position and record observable, then this will explain the apparent stability of the desired correlations. We then analyse quantitatively, in section (5), whether and how fast this closeness arises, and we conclude that, indeed (if one accepts the interpretation of decoherence given by its proponents), approximate decoherence is enough to explain why a cat always appears to be dead or alive.

2. Decoherence And Facts

The decoherence approach (see (Zurek 1993) and references therein) considers a macroscopic system not in isolation but as an open quantum system in interaction with an external, ideally infinite, environment. More precisely, it describes how in these situations the phase relations in the quantum state of the macroscopic system are irretrievably lost, because they get distributed over the degrees of freedom of the environment (loss of quantum coherence). It is argued that the interaction Hamiltonian depends explicitly on some distinguished macroscopic observable of the system, typically its position. In the models studied, the state vector of system and environment has the form

$$|\Psi(t)\rangle = \sum_i \lambda_i |\varphi_i\rangle \otimes |E_i(t)\rangle, \tag{1}$$

where the vectors $|\varphi_i\rangle$ are the eigenstates of the macroscopic observable on which the interaction Hamiltonian depends (from now on we shall refer to it as the 'pointer position', but it could be also, say, the position of a Brownian particle). It is shown that typically, in the limit of infinite time, the states $|E_i(t)\rangle$ of the environment approach orthogonality, and $|\Psi(t)\rangle$ tends to

$$|\Psi\rangle = \sum_i \lambda_i |\varphi_i\rangle \otimes |E_i\rangle. \tag{2}$$

To obtain an irreversible behaviour, it is essential that the environment be idealised as an infinite system. However, as Zurek (1993) points out, a sufficiently large environment, with very large recurrence times, will yield irreversibility for all practical purposes.

In the infinite-time limit, the reduced density operator for the system will be

$$\rho = \sum_i |\lambda_i|^2 |\varphi_i\rangle\langle\varphi_i|, \tag{3}$$

that is, it will be diagonal in the pointer position basis. At finite times, the off-diagonal terms will be proportional to the scalar products between different vectors $|E_i(t)\rangle$. There are various estimates of how fast the off-diagonal terms approach zero. The fall-off is exponential, and for any environment that is realistic for actual experiments, the rates are ridiculously fast. Joos and Zeh (1985) consider, for instance, a dust particle (say, of radius 10^{-3} cm) that interacts with different kinds of environment. If the environment is air, the off-diagonal terms between position states that are, say, 10^{-4} cm apart converge to zero like $e^{-10^{36}t}$ (!). Even in perfect vacuum, if the particle is just exposed to thermal radiation (at room temperature), convergence is like $e^{-10^{19}t}$.

But even if the off-diagonal terms are exactly zero, this does not yet mean that the cat is dead or alive. The conventional rule for assigning properties to quantum mechanical systems states that an observable has a definite value if and only if the quantum state is an eigenstate of that observable. But the state $|\Psi\rangle$ of the combined system is pure, so that the mixture (3) is improper (d'Espagnat 1976, 86-87). This means we cannot apply the ignorance interpretation and assign a definite value to the pointer position. How do facts arise then?

Decoherence yields stable correlations between the pointer position and a record observable in the environment. Zurek (1993) suggests that we interpret the record observable as a (time-dependent) 'memory', more or less in the sense of Everett, so that the pointer position has a particular value *relative* to a particular record in the environment. This move provides a solution to the problem of improper mixtures. In this paper we shall accept this view uncritically. On this account, macrofacts (the permanent definiteness of pointer position) are stable correlations between pointer position and record observable. Given this notion of a macrofact, decoherence seems to show how macrofacts arise. However, independently of whether this notion is adequate or not, *approximate* decoherence presents a challenge to the claim that stable correlations between pointer position and record observable arise at all.[2]

3. Are Facts Wishful Thinking?

We shall now turn to the so-called problem of approximate decoherence. The problem is that at any finite time the process of decoherence is not over yet. For finite t the reduced density operator $\rho(t)$ of the macroscopic system will have (small) off-diagonal terms in the pointer basis. But in order to interpret a density operator as a mixture of states $|\varphi_i\rangle$, it is necessary that the density operator be exactly diagonal in the basis $\{|\varphi_i\rangle\}$. Remember that the aim of decoherence is to derive stable correlations between the pointer position and some record observable in the environment. At finite times, the state of the system is given not by (2), but by (1):

$$|\Psi(t)\rangle = \sum_i \lambda_i |\varphi_i\rangle \otimes |E_i(t)\rangle,$$

where the $|E_i(t)\rangle$ are not orthogonal. But if the $|E_i(t)\rangle$ are not orthogonal, they simply do not define any environment observable, and there are absolutely no 'records' before $t = \infty$!

However, there are environment observables that at finite times are correlated with some observable of the system. These are given by the polar decomposition[3] of the state $|\Psi(t)\rangle$:

$$|\Psi(t)\rangle = \sum_i \tilde{\lambda}_i(t) |\tilde{\varphi}_i(t)\rangle \otimes |\tilde{E}_i(t)\rangle. \tag{4}$$

In fact, for each t, $\{|\tilde{E}_i(t)\rangle\}$ is an orthonormal set, and as such defines an observable. The environment and system observables thus defined will converge towards the 'true' record observable and pointer position, respectively. However, they will generally be incompatible with the corresponding limit observables. And further, since they explicitly depend on time, there is still no sense in which before $t = \infty$ we have *stable* records of anything at all; that is, no facts arise!

In passing, we wish to point out that the formulation in terms of the polar decomposition theorem brings out the analogy between the problem of approximate decoherence and the so-called problem of imperfect measurements. The latter was raised by Albert and Loewer (1990) (see also (Elby 1993)) as a criticism of the modal interpretation of quantum mechanics in the variants advocated by Kochen (1985), Healey (1989) and Dieks (1989). In this interpretation, facts are secured by having an interpretation rule for assigning values to observables that is more liberal than the conventional rule. This more liberal rule associates facts also with the basis vectors appearing in the polar decomposition of the quantum state, with certain probabilities. This might appear to solve the measurement problem, because the final state in an ideal measurement, as described by von Neumann, has the form of a polar decomposition

$$|\Phi\rangle = \sum_i \lambda_i |\varphi_i\rangle \otimes |\psi_i\rangle, \tag{5}$$

with respect to the eigenstates of the measured observable and of the pointer observable of the measuring apparatus. Albert and Loewer point out that, in reality, measurements are *imperfect* and that the actual state of system and apparatus will be a different vector $|\tilde{\Phi}\rangle$ with polar decomposition

$$|\tilde{\Phi}\rangle = \sum_i \tilde{\lambda}_i |\tilde{\varphi}_i\rangle \otimes |\tilde{\psi}_i\rangle. \tag{6}$$

No matter how close the two vectors may be, in general, the observables associated with the two polar decompositions (5) and (6) will be incompatible and, as stressed by Albert and Loewer, can even be maximally incompatible.

Notice that approximate decoherence arises because of finite-time effects (like in (Elby 1993)), not because of an imperfection in the Hamiltonian (like in (Albert and Loewer 1993)). For a discussion of this distinction, see (Bacciagaluppi and Hemmo 1994). Still, if we had the determination of Albert and Loewer, we could attack the decoherence approach with the same vehemence as they have attacked the modal interpretation. Indeed, it seems inadmissible to disregard the fact that the state of the macroscopic system and its environment at the time t is $|\Psi(t)\rangle$, even if this state is extremely close to $|\Psi\rangle$. For it is *precisely* this move that introduces the desired macrofacts. As long as the actual state is $|\Psi(t)\rangle$, it is not clear that facts arise *at all*, and if any facts should arise, they would certainly be the wrong ones. If the decoherence approach only shows that cats are approximately alive or dead, we are back to square one!

4. Of Facts Close And Far Away

If one's instinctive reaction to this challenge is that it simply cannot be right, then one's intuition is probably that the issue is quantitative. The 'approximate' facts will surely not be qualitatively different from the 'exact' ones; rather, the instability of the correlations will be so small as to pass unnoticed, and the 'approximate' facts will be indistinguishable from the 'exact' ones.[4] This problem has been examined in detail in the context of imperfect measurements, thanks to the determination with which Albert and Loewer have criticised the modal interpretation.

In (Bacciagaluppi and Hemmo 1994) we have argued that if the actual facts are close enough (in a suitable sense of the word) to the ideal ones, the two situations will be indistinguishable: definite pointer readings can be explained also if an observable sufficiently close to pointer position is definite. We can modify our argument to suit the case of approximate decoherence, because as a means of identifying facts, polar decompositions play a formally similar role in both approaches[5]. Namely, let us introduce the following *approximate fact:* from a certain time t the observables of system and environment that are singled out by the polar decomposition are *stably close* (in a sense defined below) to the exact pointer position and record observable. We argue now that this situation is indistinguishable from the fact consisting of a stable correlation between pointer position and record observable. What is *your* intuition about the difference between, on the one hand, a stable correlation between spin up in the z-direction and the corresponding record observable, and, on the other, correlations between spin up in some directions $\tilde{z}(t)$ and corresponding observables in the environment, where $\tilde{z}(t)$ is always close to z? Will the two situations appear very different, because there are no permanent correlations in the second case and the observables involved are incompatible with spin-z and the record observable? Or will they rather appear the same, because $\tilde{z}(t)$ is always close to z?

We believe that the second intuition just sketched is legitimate, and so we shall proceed from the assumption that correlations of observables *stably close* to the desired ones (which are approximate facts) are indistinguishable from *stable* correlations of the desired observables (which are facts). However, we have to make precise what we mean by 'close'. The polar decomposition theorem singles out *bases* in the Hilbert space of, say, the macroscopic system. The appropriate criterion for closeness of two bases is that the vectors in the bases be pairwise close in Hilbert-space norm (maybe after relabelling; see also (Dieks 1994)).

The bases singled out by the polar decomposition consist of the eigenvectors of the reduced density operators for system and environment. It is not difficult to show that these density operators will be close (in trace class), if $|\Psi\rangle$ and $|\Psi(t)\rangle$ are. However, even when the two density operators are close, one has to distinguish two cases.

(1) The eigenvalues are *far from degeneracy*: take

$$A := \begin{pmatrix} 1/3 & 0 \\ 0 & 2/3 \end{pmatrix} \text{ and } B := \begin{pmatrix} 1/3 & \varepsilon \\ \varepsilon & 2/3 \end{pmatrix}. \tag{7}$$

One can easily see both that A and B are close to order ε and that their eigenvectors are close. The example represents the general case: whenever the eigenvalues of A are far from degeneracy, any other B close to it will have eigenvectors that are close to those of A. This means that if the coefficients in the polar decomposition of $|\Psi\rangle$ are far from degeneracy, then as soon as $|\Psi(t)\rangle$ is close to $|\Psi\rangle$, the right facts will arise.

(2) The eigenvalues are *nearly degenerate*: take

$$A := \begin{pmatrix} 1/2 + \varepsilon & 0 \\ 0 & 1/2 - \varepsilon \end{pmatrix} \text{ and } B := \begin{pmatrix} 1/2 & \varepsilon \\ \varepsilon & 1/2 \end{pmatrix}. \tag{8}$$

We can easily see that they are close to order ε. However, their eigenvectors are *not close*. A and B are diagonal with respect to the bases

$$\left\{ \begin{pmatrix} 1 \\ 0 \end{pmatrix}, \begin{pmatrix} 0 \\ 1 \end{pmatrix} \right\} \text{ and } \left\{ \begin{pmatrix} 1/\sqrt{2} \\ 1/\sqrt{2} \end{pmatrix}, \begin{pmatrix} 1/\sqrt{2} \\ -1/\sqrt{2} \end{pmatrix} \right\}, \tag{9}$$

respectively. But these basis vectors are not close: in fact, they are as *far apart* as possible, and define *maximally incompatible* observables. So, it seems that if the coefficients in the polar decomposition of $|\Psi\rangle$ are nearly degenerate, then no matter how close $|\Psi(t)\rangle$ gets to $|\Psi\rangle$, the right facts will not arise!

5. The Cat Is In The Bag!

In this section, we give a precise analysis of the problem in a simplified model where the system has only two degrees of freedom. We use the results of the similar analysis we gave in (Bacciagaluppi and Hemmo 1994) of the problem of imperfect measurements. We shall see that this analysis can make sense of approximate decoherence.

Let us first look at the case where the polar decomposition of $|\Psi\rangle$ is, in fact, exactly degenerate. This case is simpler, and we shall use it as an example for the kind of analysis needed. We shall then give the final results for the important case, namely non-degeneracy, and in particular, near-degeneracy.

When the limit state $|\Psi\rangle$ has an exactly degenerate polar decomposition, the actual state has the form

$$|\Psi(t)\rangle = 1/\sqrt{2}\left(|\varphi_1\rangle \otimes |E_1(t)\rangle + |\varphi_2\rangle \otimes |E_2(t)\rangle\right). \tag{10}$$

Here, $\{|\varphi_1\rangle, |\varphi_2\rangle\}$ is the 'pointer position basis', while $|E_1(t)\rangle$ and $|E_2(t)\rangle$ are vectors in an infinite-dimensional Hilbert space that approach orthogonality. It is easily seen that, in the pointer position basis, the reduced density operator for the macroscopic system is

$$\rho(t) = \frac{1}{2}\begin{pmatrix} 1 & \alpha(t) \\ \overline{\alpha(t)} & 1 \end{pmatrix} \tag{11}$$

where $\alpha(t) := \langle E_1(t)|E_2(t)\rangle$ (estimates of 'rates of decoherence', as in (Joos and Zeh 1985), which give the fall-off of the off-diagonal terms, tell us in fact how fast the environment states approach orthogonality).

Let us write $\alpha(t) = |\alpha(t)|e^{i\omega(t)}$. The eigenvalues and eigenvectors of $\rho(t)$ are, respectively,

$$1/2(1 \pm |\alpha(t)|) \quad \text{and} \quad \frac{1}{\sqrt{2}}\begin{pmatrix} \pm e^{i\omega(t)} \\ 1 \end{pmatrix}. \tag{12}$$

These eigenvectors diagonalize the reduced density operator for the system, that is, they are the basis vectors $|\tilde{\varphi}_1(t)\rangle$ and $|\tilde{\varphi}_2(t)\rangle$ that are singled out by the polar decomposition of $\Psi(t)$. Now, in the limit $\alpha(t) \to 0$, the convergence of the $|\tilde{\varphi}_i(t)\rangle$ will depend crucially on the behaviour of $\omega(t)$, the relative phase of the vectors $|E_1(t)\rangle$ and $|E_2(t)\rangle$. If $\omega(t)$ does not converge, the eigenvectors $|\tilde{\varphi}_i(t)\rangle$ will not converge at all. But even if it does, the $|\tilde{\varphi}_i(t)\rangle$ will not converge towards the pointer basis vectors $|\varphi_i\rangle$! In fact, the scalar product between the $|\tilde{\varphi}_i(t)\rangle$ and the $|\varphi_j\rangle$ is

$$\left|\langle \varphi_i | \tilde{\varphi}_j(t)\rangle\right|^2 = 1/2, \tag{13}$$

for all pairs i, j and *for all times* t. The observables correlated with 'records' are always *maximally incompatible* with pointer position.

One can repeat the same analysis for a non-degenerate limit state (the details are in (Bacciagaluppi and Hemmo 1994)). $\rho(t)$ can always be written (*modulo* normalization factor) as

$$\rho(t) = \begin{pmatrix} r & \alpha(t) \\ \overline{\alpha}(t) & 1/r \end{pmatrix}, \tag{14}$$

with r positive. The size of $\delta := r - 1/r$ determines the degree of degeneracy (the smaller $|\delta|$ is, the more degenerate is the state). The scalar products corresponding to (13) turn out to be functions of $\kappa(t) := |\alpha(t)/\delta|$. When $\kappa(t) \ll 1$, these scalar products become

$$|\langle \varphi_1 | \tilde{\varphi}_2(t) \rangle|^2 \approx \kappa^2(t), \quad |\langle \varphi_1 | \tilde{\varphi}_1(t) \rangle|^2 \approx 1 - \kappa^2(t), \tag{15}$$

and analogously with subscripts 1 and 2 interchanged. So $\kappa(t)$ can be interpreted as the angle between $|\tilde{\varphi}_i(t)\rangle$ and $|\varphi_i\rangle$. And since $\alpha(t)$ falls off at a tremendous rate, the convergence of the eigenvectors is very fast.

What happens when the limit state is nearly degenerate? When δ is small, initially $\kappa(t) \gg 1$. But then one gets

$$|\langle \varphi_i | \tilde{\varphi}_j(t) \rangle|^2 \approx 1/2, \tag{16}$$

and the $|\tilde{\varphi}_i(t)\rangle$ are far from the $|\varphi_i\rangle$, just like in the degenerate case! However, δ is fixed, and as soon as $|\alpha(t)| \ll |\delta|$, we will get the same behaviour as in (15), and the eigenvectors will converge. So, provided the polar decomposition is not degenerate, the eigenvectors always converge to the pointer position basis: near-degeneracy induces only a *delay* in convergence.

How significant is this delay? The angle $\kappa(t)$ between the eigenvectors is equal to the degree of decoherence $|\alpha(t)|$ *divided* by the degree of degeneracy $|\delta|$. Since $|\alpha(t)|$ is an exponential $e^{-\gamma t}$, the delay is

$$T := \gamma^{-1} \log |\delta|^{-1}. \tag{17}$$

So, if we use again the values taken from (Joos and Zeh 1985) we get estimates for T that vary between $\log|\delta|^{-1} \times 10^{-11}$ s and $\log|\delta|^{-1} \times 10^{-28}$ s. And this is completely insignificant, unless δ is unreasonably small.

We grant that, in principle, δ could be ridiculously small, or even zero. In this case, pathological behaviour of the kind of (16) could arise even over long time scales[6]. But the assumption seems plausible that states with ridiculously small degeneracies δ are extremely improbable, both in nature and, indeed, in experiments. Of course, there are experiments that involve nearly degenerate states, for instance EPR experiments. But, at least to our knowledge, even in the most accurate EPR experiments, δ is small enough to produce high correlations between the outcomes of measurements on the two wings of the experiment, but it is not small enough to produce long delays in the convergence of the Schmidt states. With this proviso, we think we have, indeed, provided a solution to the problem of approximate decoherence, and thus believe that if one accepts an Everett-like view, decoherence can explain why cats always appear to be either dead or alive.

6. Decoherence and the Modal Interpretation

Our analysis of decoherence applies not only in the context of an Everett-like interpretation, but also in the context of the modal interpretation of Kochen, Healey and

Dieks. In fact, it was inspired by Albert and Loewer's criticism of the modal interpretation, as well as by a proposal of Dieks (1994) to combine the decoherence approach with the modal interpretation. A philosophical analysis of this 'modal decoherence' framework will be given elsewhere. The main idea of it is that the modal interpretation rules can be applied to the states resulting from the interaction between a system and its environment. Of course, this is not an extension of the modal interpretation, but rather an application of it to systems of the form of a macroscopic system and its environment.

As noticed before, in section (2), the decoherence approach needs a modification of the conventional interpretation of the quantum formalism in order to overcome the problem of improper mixtures and secure the existence of facts. The Everett interpretation (or some variant of it) is a way of bridging this interpretive gap. Dieks (1994) points out that the modal interpretation is equally capable of securing the facts needed by the decoherence approach. Furthermore, in the modal interpretation these facts are objective properties of system and environment existing here and now, rather than relative to an Everett branch. Decoherence, on the other hand, can help the modal interpretation solve the problem of imperfect measurements raised by Albert and Loewer. If the microsystem and the measuring apparatus are considered in isolation, the criticism is quite formidable. But if they are considered in interaction with an external environment, then the total state will approximately decohere with respect to the pointer position basis. The results of the previous section imply that, according to the modal interpretation rules, an observable with eigenvectors that are close to those of pointer position will become definite, irrespective of any imperfections in the interaction between microsystem and apparatus. And closeness of these eigenvectors to those of pointer position is guaranteed by the time scale of decoherence, which is much shorter than the time scale of actual measurements.

The modal interpretation and the decoherence approach thus complete each other. The modal interpretation gives you the existence of facts. And decoherence ensures that these facts correspond to cats that are dead or alive, rather than to weird superpositions that are incompatible with our definite experience.

Notes

[1] Our special thanks go to Harvey Brown, Dennis Dieks, Michael Redhead and, in particular, Jeremy Butterfield, whose criticism helped us remain coherent. We further wish to thank Joseph Berkowitz, Thomas Breuer, Rob Clifton, Michael Dickson, Andrew Elby, Gordon Fleming, Renata Grassi, Adrian Kent, Larry Landau, Klaas Landsman and Constantine Pagonis for a number of discussions and comments that contributed to the development of the ideas and content of this paper. We thank Constantine Pagonis also for his patience in teaching us how to use Microsoft Word. GB acknowledges financial support from the Arnold Gerstenberg Fund and the British Academy, MH from the Cambridge Overseas Trust, the Overseas Research Scheme and Anglo-Jewish Association.

[2] Notice that the decoherence approach also provides the Everett interpretation with a criterion for selecting *preferred bases*, namely those presenting stable correlations between pointer and record.

[3] The polar (biorthogonal, Schmidt) decomposition theorem states that for each vector $|\Phi\rangle$ in a product Hilbert space $H_1 \otimes H_2$, there are orthonormal bases $\{|\phi_i\rangle\}$ and $\{|\psi_j\rangle\}$

in the two Hilbert spaces, for which the expansion of $|\Phi\rangle$ has no cross terms, i.e. reduces to

$$|\Phi\rangle = \sum_i \lambda_i |\phi_i\rangle \otimes |\psi_i\rangle.$$

Unless some of the coefficients are equal (in absolute value), this representation is unique (up to phase factors). In the bases associated with the polar decomposition of the state the reduced density operator is clearly always diagonal.

[4] One might want to relax the requirement of diagonality of the density matrix. But this is mandatory for an ignorance interpretation. Also, if this requirement is relaxed, there will be many different bases with respect to which the state approximately decoheres.

[5] The relevance of the polar decomposition for the decoherence approach has been noted also by Albrecht (1992), who has an eye in particular for the consistent histories approach (we are indebted to Adrian Kent for this reference). See also (Gell-Mann and Hartle 1993) and (Paz and Zurek 1993) for the relation between the so-called 'Schmidt histories' and decoherence.

[6] Dieks (1994) has sketched an argument that is meant to overcome simultaneously the difficulties both with near-degeneracy and degeneracy.

References

Albert, D. and Loewer, B. (1990), "Wanted Dead or Alive: Two Attempts to Solve Schroedinger's Paradox", in *PSA 1990* volume 1, A. Fine, M. Forbes and L. Wessels (eds.). East Lansing, Mi: Philosophy of Science Association, pp. 277-285.

_____. (1993), "Non-Ideal Measurements", *Foundations of Physics Letters* 6: 297-305.

Albrecht, A. (1992), "Investigating Decoherence in a Simple System", *Physical Review* D 46: 5504-5520.

Bacciagaluppi, G. and Hemmo, M. (1994), "Modal Interpretations of Imperfect Measurements", submitted to *Foundations of Physics*.

Dieks, D. (1989), "Resolution of the Measurement Problem Through Decoherence of the Quantum State", *Physics Letters* A 142: 439-446.

_____. (1994), "Objectification, Measurement and Classical Limit According to the Modal Interpretation of Quantum Mechanics", in *Proceedings of the Symposium on the Foundations of Modern Physics*, P. Busch, P. Lahti, and P. Mittelstaedt, (eds.). Singapore: World Scientific.

Elby, A. (1993), "Why 'Modal' Interpretations of Quantum Mechanics Don't Solve the Measurement Problem", *Foundations of Physics Letters* 6: 5-19.

d'Espagnat, B. (1976), *Conceptual Foundations of Quantum Mechanics*, 2nd edition. Reading, Mass.: Benjamin.

Kochen, S. (1985), "A New Interpretation of Quantum Mechanics", in *Proceedings of the Symposium on the Foundations of Modern Physics,* Lahti, P. and Mittelstaedt, P., (eds.). Singapore: World Scientific.

Healey, R. (1989), *The Philosophy of Quantum mechanics: An Interactive Interpretation.* Cambridge: Cambridge University Press.

Gell-Mann, M. and Hartle, J. (1993), "Classical Equations for Quantum Systems", *Physical Review* D 47: 3345-3382.

Joos, E. and Zeh, H.D. (1985), "The Emergence of Classical Properties Through Interaction with the Environment", *Z. Physik* B 59: 223-243.

Paz, J. P. and Zurek, W. H. (1993), "Environment-Induced Decoherence, Classicality, and Consistency of Quantum Histories", *Physical Review* D 48: 2728-2738.

Zurek, W. H. (1981), "Pointer Basis of Quantum Apparatus: Into What Mixture Does the Wave Packet Collapse?", *Physical Review* D 24: 1516-1525.

_ _ _ _ _ _ _. (1982), "Environment-Induced Superselection Rules", *Physical Review* D 26: 1862-1880.

_ _ _ _ _ _ _. (1993), "Preferred States, Predictability, Classicality, and the Environment-Induced Decoherence", *Progress in Theoretical Physics* 89: 281-312.

The 'Decoherence' Approach to the Measurement Problem in Quantum Mechanics[1]

Andrew Elby

University of California—Berkeley

1. Introduction

The decoherence approach to the measurement problem invokes dissipative interactions between a measuring apparatus and its environment to explain, within the context of 'pure' quantum mechanics (QM), why such devices appear to possess definite pointer readings. By 'pure' QM, I mean Schrödinger evolution with no wavefunction collapse. Several classes of interpretations of pure QM can rely on decoherence. One class is the 'modal' interpretations of van Fraassen (1979), Dieks (1989), Healey (1989), and others. Another class is the relative-state and many-world interpretations.

I will argue that decoherence cannot help these interpretations address the general metaphysical challenges raised against them. But decoherence can help pick out a 'special' basis that determines which observables receive definite values. I'll explore to what extent decoherence rescues the modal (biorthogonal) basis-selection rule, and Zurek's (environmental interaction) basis-selection rule, from the basis degeneracy problem and the imperfect measurement problem. 'Basis degeneracy' occurs when a selection rule does not pick out a unique basis. The 'imperfect measurement' problem occurs when a selection rule, designed to choose the pointer-reading basis after an ideal measurement, chooses a basis not even close to the pointer-reading basis after a non-ideal measurement. Decoherence, we'll see, only partially alleviates these technical concerns. To set the stage, I must review one aspect of the measurement problem.

2. The Measurement Problem

Consider a spin-1/2 particle initially described by a superposition of eigenstates of S_z, the z-component of spin:

$$|\Phi\rangle = c_1|S_z=+\rangle + c_2|S_z=-\rangle.$$

Let $|R=+\rangle$ and $|R=-\rangle$ denote the 'up' and 'down' pointer-reading eigenstates of an apparatus that measures S_z. According to pure QM, if the apparatus ideally measures the particle, the combined system evolves into

$$|\varphi\rangle = c_1|S_z=+\rangle \otimes |R=+\rangle + c_2|S_z=-\rangle \otimes |R=-\rangle. \qquad (1)$$

Common sense insists that, after the measurement, the pointer reading is definite. But according to the eigenvector-eigenvalue rule, the pointer reading is definite only if the quantum state is an eigenstate of **R**, the pointer-reading operator. Since |φ> is not an **R**-eigenstate, the pointer reading is indefinite. Before discussing how decoherence fits into all this, I'll lay some groundwork by outlining the modal interpretation.

3. Modal Interpretations

Modal interpretations address the measurement problem by renouncing the eigenvector-eigenvalue rule: Some observables possess definite values even when the quantum state isn't an eigenstate of the corresponding operator. The quantum state itself determines which observables these are.

Most modal interpretations rely on the biorthogonal decomposition theorem. This theorem states that any state vector describing two subsystems can, for a certain choice of bases, be expanded in the simple 'biorthogonal' form $\sum_i c_i |A_i>\otimes|B_i>$, where the $\{|A_i>\}$ and $\{|B_i>\}$ vectors are orthonormal, and are therefore eigenstates of Hermitian operators **A** and **B** associated with subsystems 1 and 2, respectively. Modal interpreters assert that when $\sum_i c_i |A_i>\otimes|B_i>$ is the unique biorthogonal decomposition of the quantum state with respect to subsystem 1 and 2, then observables A and B both possess definite values. For instance, consider the particle/apparatus system described by eq. (1). According to modal interpretations, the particle has a definite z-component of spin, and the pointer has a definite reading. In this way, modal interpreters explain why ideal measurements have definite results.

4. Decoherence: Formalism and Basic Interpretation

In this section, I outline the formalism of decoherence. I then present the basic interpretive claim shared by almost all interpretations that rely on decoherence.

4.1. Formalism

Zurek (1991), Zeh (1993), and others use general plausibility arguments and worked examples to argue the following: The measuring apparatus undergoes a 'dissipative' interaction with its environment. This interaction quickly destroys the coherence between the two branches of the superposition in eq. (1). In this way, the environment picks out the pointer-reading basis.

To see what this means formally, let $|E_+>$ denote the state of the environment (i.e., the rest of the universe) after it interacts with a particle/apparatus system in state $|S_z=+>\otimes|R=+>$. Similarly for $|E_->$. When a particle/apparatus system described by eq. (1) interacts with its environment, the universe evolves into

$$|\Psi> = c_1|S_z=+>\otimes|R=+>\otimes|E_+> + c_2|S_z=->\otimes|R=->\otimes|E_->. \qquad (2)$$

As time passes, the environmental states corresponding to different pointer readings quickly approach orthogonality. Formally, as $t\to\infty$, $<E_+|E_->\to 0$. (For all practical purposes, this 'decoherence' takes less than a trillionth of a second.) In this limit, the reduced density operator describing the particle/apparatus system, found by tracing over the environmental degrees of freedom, is the mixture

$$\rho_m = |c_1|^2 |S_z=+><S_z=+||R=+><R=+| + |c_2|^2|S_z=-><S_z=-||R=-><R=-|. \qquad (3)$$

Put roughly, the environment 'damps out' the interference terms in the density operator $|\varphi\rangle\langle\varphi|$. I must stress that according to pure QM, eq. (3) describes the particle/apparatus system only because eq. (2) describes the universe, with $\langle E_+|E_-\rangle = 0$ in the infinite-time limit. As Albert and Loewer (1990) point out, however, eq. (2) does not describe most real-life (non-ideal) measurements. If an initially spin-up particle has nonzero probability of yielding a 'down' measurement outcome, and an initially spin-down particle has nonzero probability of yielding 'up,' then Schrödinger's equation implies that the post-measurement state is

$$|\varphi'\rangle = c_{11}|S_z=+\rangle \otimes |R=+\rangle + c_{12}|S_z=+\rangle \otimes |R=-\rangle + c_{21}|S_z=-\rangle \otimes |R=+\rangle + c_{22}|S_z=-\rangle \otimes |R=-\rangle, \tag{4}$$

This conclusion fails only if an initially spin-up particle, whenever it yields a down measurement result, always has its spin flipped into a down eigenstate; and vice versa.

As Elby (1993) shows, even if our equipment is flawless, imperfect measurements follow inevitably from wavefunction tails and from environmental fluctuations whose existence is implied by QM. For this reason, a Stern-Gerlach experiment cannot be made ideal, even in principle. Although c_{12} and c_{21} can be made unbelievably small, they cannot be eliminated.

Why is this important to decoherence-based interpretations? Because, after the particle/apparatus system interacts with its environment, the final state is given not by eq. (3), but by

$$|\Psi'\rangle = c_{11}|S_z=+\rangle \otimes |R=+\rangle \otimes |E_{++}\rangle + c_{12}|S_z=+\rangle \otimes |R=-\rangle \otimes |E_{+-}\rangle + c_{21}|S_z=-\rangle \otimes |R=+\rangle \otimes |E_{-+}\rangle + c_{22}|S_z=-\rangle \otimes |R=-\rangle \otimes |E_{--}\rangle. \tag{5}$$

As $t \to \infty$, these four environmental states approach mutual orthogonality. But at any finite time, they are not strictly orthogonal. For now, I'll examine how decoherence theorists interpret eq. (3). In section 6, we'll see whether these interpretations successfully carry over to non-ideal measurements.

4.2. Interpretation

Decoherence-based interpretations—and we'll see that several exist—almost all agree on the following:

When the state of the universe takes the form of eq. (3) with the environmental states (nearly) orthogonal, then the pointer reading, or some observable 'close' to the pointer reading, is 'definite'.

Many interpretations fit into this framework, due partly to the different senses in which an observable can be 'definite'. I now explore some of the possibilities.

5. Interpretations relying on decoherence

I'll carve up decoherence-based interpretations into different classes based on their answers to two crucial questions:

(A) Does 'definite' mean 'definite in the absolute, classical sense'?

(B) Does the eigenvector-eigenvalue rule hold?

5.1. Decoherence interpretation #1: 'Definite'='classically definite', but the eigenvector-eigenvalue rule fails

According to decoherence interpretation #1, an observable (e.g., the pointer reading) can possess a definite value even when the quantum state isn't an eigenstate of the corresponding operator. To pick out which observables become definite, we can rely on the form of the quantum state, as modal interpreters do; or we can invoke formal properties of the relevant interaction Hamiltonian (see section 6.4). Although these approaches differ in formal detail, they're both part of the same *program* of letting the interactions between subsystems determine which observables acquire definite values.

Therefore, decoherence interpretation #1 is just a modal interpretation, perhaps with a different basis-selection rule. Interestingly, because Healey and Dieks stick with the usual modal (biorthogonal) basis-selection rule, they must explicitly consider environmental interactions to 'rescue' their interpretations from the Albert-Loewer challenge. Modal theorists hope to show that, after entangling with its environment, the pointer acquires a definite reading. More on this below. For now, my point is that modal theorists can use the interaction between the particle/apparatus system and its environment to pick out a special basis.

5.2. Decoherence interpretation #2: 'Definite'='classically definite,' and the eigenvector-eigenvalue rule holds.

According to this interpretation, which is *prima facie* appealing to many physicists I've spoken with, we can assign an ignorance interpretation to the mixture describing the particle/apparatus system. But despite its intuitive appeal, decoherence interpretation #2 is inconsistent. Here's why:

By assumption, the eigenvector-eigenvalue rule holds. Therefore, the apparatus has a definite pointer reading only if the quantum state is an eigenstate of $\mathbf{R}$. But the quantum state, given by eq. (2) or by eq. (5), is not an eigenstate of $\mathbf{R}$.

The inconsistency of decoherence interpretation #2 illustrates D'Espagnat's (1976) point that within pure QM, we cannot assign an ignorance interpretation to an 'improper' mixture (i.e., a mixture obtained by tracing over degrees of freedom of a system with which the subsystem of interest is entangled).

5.3. Decoherence interpretation #3: Relative-state

According to this view, the pointer reading becomes definite not in some absolute sense, but relative to its branch of the superposition. Within each branch, the eigenvector-eigenvalue rule holds.

Before discussing whether decoherence solves the ontological problems associated with relative-state and many-world interpretations, I'll briefly discuss what these interpretations are supposed to mean. Both Zurek and Zeh, two of the most respected decoherence theorists, stress that their interpretations flesh out Everett's 'relative-state' interpretation, not deWitt's many-world interpretation. (See deWitt and Graham's 1973 anthology.) Although Zurek and Zeh (and Everett) never spell out the precise ontology of their interpretations, I'll try to reconstruct an argument that captures (or at least supports) their views.

According to deWitt, after a measurement, each branch of the relevant superposition lives in its own world. If these separate worlds are physically inaccessible to each other, then no interactions can occur between inhabitants of the different worlds, even in principle. Therefore, no 'interference' can occur between different branches of the superposition, even in principle. But Zurek and Zeh espouse pure QM, according to which Schrödinger's equation governs all state evolution, and hence all interference effects permitted by Schrödinger's equation are possible in principle. For this reason, Zurek and Zeh want the different branches of eq. (2) to inhabit different 'realities' that could in principle (though not in practice) interfere. This, along with the radical metaphysics of the many-world view, could partially explain why Zurek and Zeh ally themselves more with Everett than with deWitt.

Unfortunately, the ontology of the Everett-Zurek-Zeh view is unclear. To see why, consider a system in state

$$|\Psi\rangle = c_1|S_z=+\rangle \otimes |R=+\rangle \otimes |E_+\rangle + c_2|S_z=-\rangle \otimes |R=-\rangle \otimes |E_-\rangle.$$

We can 'see' interference between the two branches of the superposition by measuring $Q=S'\otimes R'\otimes E'$, where S' doesn't commute with S_z, R' doesn't commute with R, and E' doesn't commute with $E=|E_+\rangle\langle E_+|+|E_-\rangle\langle E_-|$. Although we cannot in practice measure Q, the quantum formalism does not rule out such measurements in principle. Because the 'up' and 'down' branches can interfere, those branches cannot be said to inhabit 'separate' physical realities. Therefore, what it means for an observable to become definite 'relative to its branch' is ambiguous.

Some physicists downplay the severity of this metaphysical problem. They argue as follows: Sure, when the two branches interfere, it becomes meaningless to assert that the pointer reading is definite relative to its branch. But most of the time, the up and down branches *don't* interfere. During these times, it's unproblematic to claim that the pointer reading is definite, relative to its branch.

This counterargument fails to resolve the ontological ambiguities raised above. When the two branches aren't interfering, do two 'copies' everything exist? If not, then in what sense are both measurement results actualized? If so, and if the two branches don't inhabit separate worlds (in deWitt's sense), then how do they co-exist in space and time?

Since I have nothing to add to the general arguments for and against relative-state and many-world interpretations, I won't press these questions any further. My point is this: First, the ontology of the relative-state (as opposed to many-world) framework adopted by some decoherence theorists is, at best, ambiguous. Second, decoherence cannot help us to address the metaphysical difficulties facing relative-state and many-world interpretations. If you think these interpretations make no sense, decoherence cannot change your mind.

5.4 Summary

In section 5, I sketched the three most popular decoherence-based interpretations. Other such interpretations, though logically possible, have not been developed to my knowledge. Decoherence cannot help the modal, relative-state, and many-world interpretations fend off general metaphysical criticisms. What decoherence *can* do is help these interpretations pick out a 'special' basis. In the modal view, this special basis determines which observables acquire definite values. In the relative-state and many-world view, this special basis determines how physical reality 'branches' (in some sense). I'll now explore to what extent decoherence can help these interpretations select the pointer-reading basis.

6. Decoherence and basis-selection rules

Any interpretation relying on a 'special' basis must specify formal rules that pick out the basis. Since we're trying to explain why measurements seem to result in definite pointer readings, a successful basis-selection rule must choose the pointer-reading basis, or something very 'close' to the pointer-reading basis, in almost all situations we want to call 'measurements.' (A observable R' is 'close' to the pointer-reading observable R if the **R**'-eigenstates are very nearly **R**-eigenstates, i.e., if $\langle R'=r_n|R=r_n\rangle \approx 1$ for all n.) With respect to basis selection, the interpretations discussed above potentially suffer from two major obstacles: The imperfect measurement problem, and the basis degeneracy problem. In this section, I'll define these problems and explain why they arise. Then, I'll explore to what extent decoherence can overcome these technical difficulties.

6.1. Basis degeneracy problem

This difficulty arises when a basis-selection rule doesn't always choose a unique basis. As an example, consider the usual 'modal' rule, also advocated by deWitt for many-world interpretations, of letting the biorthogonal decomposition pick out a special basis. If any two $|c_i|$'s are equal, then the quantum state has multiple biorthogonal decompositions. For instance, consider the particle/apparatus system in state

$$|\varphi\rangle = c_1|S_z=+\rangle \otimes |R=+\rangle + c_2|S_z=-\rangle \otimes |R=-\rangle.$$

If $c_1=c_2=2^{-1/2}$, then $|\varphi\rangle$ can be rewritten as

$$|\varphi\rangle = 2^{-1/2}[|S_x=+\rangle \otimes |R'=+\rangle + |S_x=-\rangle \otimes |R'=-\rangle],$$

where

$$|S_x=\pm\rangle = 2^{-1/2}[|S_z=+\rangle \pm |S_z=-\rangle]$$
$$|R'=\pm\rangle \equiv 2^{-1/2}[|R=+\rangle \pm |R=-\rangle].$$

Because of this degeneracy, nothing is 'special' about the pointer-reading basis, at least, not if we stick with the modal basis-selection rule. According to most modal theorists, if the biorthogonal decomposition isn't unique, then none of the relevant nondegenerate observables acquires a definite value. This is troublesome, at least in principle, because we want S_z-measurement of a particle initially in state $|S_x=+\rangle$ to yield a definite pointer reading.

6.2. Imperfect measurement problem

This problem arises when a basis-selection rule, designed to choose the pointer-reading basis after an ideal measurement, selects another basis when the measurement isn't ideal. As Albert and Loewer point out, the modal basis-selection rule suffers from this problem. Let me show why. Consider the spin-1/2 particle initially in state

$$|\Phi\rangle = c_1|S_z=+\rangle + c_2|S_z=-\rangle.$$

As mentioned in section 4.1, due to unavoidable wavefunction tails, the Stern-Gerlach experiment yields the following post-measurement state of the particle/apparatus system:

$$|\varphi'\rangle = c_{11}|S_z=+\rangle \otimes |R=+\rangle + c_{12}|S_z=+\rangle \otimes |R=-\rangle$$
$$+ c_{21}|S_z=-\rangle \otimes |R=+\rangle + c_{22}|S_z=-\rangle \otimes |R=-\rangle.$$

According to the biorthogonal decomposition theorem, a basis exists in terms of which $|\varphi'\rangle$ can be biorthogonally expanded:

$$|\varphi'\rangle = d_1|S'=+\rangle \otimes |R'=+\rangle + d_2|S'=-\rangle \otimes |R'=-\rangle,$$

where $|S'=\pm\rangle$ are eigenstates of some operator $\mathbf{S'}$ that doesn't commute with $\mathbf{S_z}$, and $|R'=\pm\rangle$ are eigenstates of some operator $\mathbf{R'}$ that doesn't commute with $\mathbf{R}$. The modal basis-selection rule implies that $\mathbf{R'}$, but not $\mathbf{R}$, has a definite value. Furthermore, by altering our state preparation of the particle, we can always choose c_1 and c_2 such that R' won't be 'close' to R, no matter how small c_{12} and c_{21} are. This is the imperfect measurement problem. Let's see to what extent decoherence can address the basis degeneracy and imperfect measurement problems.

6.3 Decoherence and the modal selection rule

In this section, I argue that decoherence cannot rescue modal interpretations from either the basis degeneracy problem or the imperfect measurement problem. First, consider ideal measurements. The post-measurement state of the universe is

$$|\Psi\rangle = c_1|S_z=+\rangle \otimes |R=+\rangle \otimes |E_+\rangle + c_2|S_z=-\rangle \otimes |R=-\rangle \otimes |E_-\rangle,$$

with $|E_+\rangle$ and $|E_-\rangle$ almost orthogonal. If $|c_1|=|c_2|$, then the biorthogonal decomposition of the apparatus with the particle/environment subsystem is not unique, and therefore the pointer reading does not acquire a definite value, according to the modal basis-selection rule. To escape this degeneracy, however, modal theorists can invoke the tri-decompositional uniqueness theorem recently proved by Elby and Bub (1994):

Tri-decompositional uniqueness theorem: Suppose $|\Psi\rangle$ can be 'tri-decomposed' as $|\Psi\rangle = \sum_i c_i |A_i\rangle \otimes |B_i\rangle \otimes |C_i\rangle$, where $\{|A_i\rangle\}$ is a linearly independent set of vectors, $\{|B_i\rangle\}$ is also a linearly independent set, and $\{|C_i\rangle\}$ is a non-collinear set of vectors. Then there exist no alternative linearly independent sets of vectors $\{|A_i'\rangle\}$ and $\{|B_i'\rangle\}$, and no alternative non-collinear set $\{|C_i'\rangle\}$, such that $|\Psi\rangle = \sum_i d_i |A_i'\rangle \otimes |B_i'\rangle \otimes |C_i'\rangle$. (Unless the new and old bases differ only by phase constants.)

Since $|S_z=+\rangle$ and $|S_z=-\rangle$ are linearly independent, as are $|R=+\rangle$ and $|R=-\rangle$, and since the $|E_\pm\rangle$ vectors are non-collinear, the theorem implies that no other apparatus operator $\mathbf{R'}$, and no other particle operator $\mathbf{S'}$, exist such that $|\Psi\rangle$ can be re-expanded as $d_1|S'=+\rangle \otimes |R'=+\rangle \otimes |E'_+\rangle + d_2|S'=-\rangle \otimes |R'=-\rangle \otimes |E'_-\rangle$. This is true even if $c_1=c_2$. In words, the 'biorthogonal tri-decomposition' of $|\Psi\rangle$ uniquely picks out the pointer-reading basis.

Unfortunately, this result applies only to ideal measurements. After a non-ideal measurement, the universe ends up in state

$$|\Psi'\rangle = c_{11}|S_z=+\rangle \otimes |R=+\rangle \otimes |E_{++}\rangle + c_{12}|S_z=+\rangle \otimes |R=-\rangle \otimes |E_{+-}\rangle$$
$$+ c_{21}|S_z=-\rangle \otimes |R=+\rangle \otimes |E_{-+}\rangle + c_{22}|S_z=-\rangle \otimes |R=-\rangle \otimes |E_{--}\rangle,$$

In general, $|\Psi'\rangle$ cannot be tri-decomposed; tri-decompositions, unlike biorthogonal decompositions, do not always exist. And when $|\Psi'\rangle$ is biorthogonally decomposed in terms of the apparatus and the particle/environment, the coefficients d_1 and d_2 could be equal, depending on the initial state of the particle. For this reason, decoherence cannot save modal theories from basis degeneracy.

Nor does decoherence save modal theorists from the imperfect measurement problem. Recall that, in state $|\Psi'\rangle$, the $|E_{ij}\rangle$ states are nearly (but not precisely) orthogonal. Therefore, the biorthogonal decomposition of the apparatus with the particle/environment does not pick out the pointer-reading basis.

In response, a modal theorist could emphasize that since the $|E_{ij}\rangle$ states are nearly orthogonal, a basis 'close' to the pointer reading basis gets chosen unless $|c_1|$ is very close to $|c_2|$. As Zurek shows, decoherence typically occurs over a 'half life' of about 10^{-40} seconds. Therefore, when we consider state $|\Psi'\rangle$ a thousandth of a second after measurement, an observable extremely close to the pointer reading gets picked out by the modal selection rule, unless $|c_1|$ differs from $|c_2|$ by less than one part in 10^{16} or so. In practice, we cannot fine tune our state-preparation devices this precisely. Consequently, an observable incredibly close to the pointer reading gets picked out in (almost) all practical situations.

Nonetheless, the modal basis-selection rule is adequate *only if* we can

(a) Not worry about the in-principle possibility of $|c_1|\approx|c_2|$, in which case the apparatus observable that becomes definite isn't even close to the pointer reading, and this state of affairs persists for a noticeable length of time; and

(b) Explain why we perceive the pointer reading to be definite, when an observable close to the pointer reading is definite.

The palatability of (a) depends on taste. But (b) is more problematic than it appears. Intuitively, we want to say that when someone looks at the pointer, she directly perceives the pointer's definite reading; and if an observable sufficiently close to R is definite, then the pointer reading 'looks' definite to the observer. But according to modal interpretations, the definite values picked out by the biorthogonal decomposition play no role in choosing which new observables become definite when a new subsystem interacts with the 'old' system. Therefore, when the observer looks at the pointer, she does *not* directly perceive the definite value associated with the pointer. Instead, the pointer interacts quantum mechanically with her brain. Whether the observer acquires a definite belief depends entirely on the resulting biorthogonal decomposition of her brain with the rest of the universe, *not* on the definite value possessed by the pointer. Therefore, even though an observable close to the pointer reading is definite, we cannot automatically conclude that the observer acquires a definite belief about the pointer reading. As shown in Elby (1994), a separate—and somewhat shaky—line of reasoning is required to show that the observer's brain biorthogonally decomposes as needed to be assigned a definite belief. Crucially, this line of reasoning relies upon assumptions neither implied nor suggested by considerations of decoherence. Instead, the legitimacy of these assumptions depends on delicate matters of fact about brain neurophysiology. Therefore, for modal interpretations, decoherence does not guarantee a solution to the imperfect measurement problem.

Let me elaborate briefly. As shown in Elby (1994), the modal interpretation assigns a definite belief to the observer only if (i) opposing beliefs correspond to brain states that are nearly (though not strictly) orthogonal in Hilbert space; and (ii) the set of all 'up' belief states corresponds to a nearly (though not strictly) closed subspace of Hilbert space. Decoherence does not imply (i) or (ii). Decoherence implies only that nearly-orthogonal brain states quickly decohere, as described in section 4.1. Therefore, decoherence does not solve the imperfect measurement problem.

(Interestingly, if the 'up' belief states form a *strictly* closed Hilbert subspace, or if 'up' and 'down' belief states are *strictly* orthogonal, then the observer does not acquire a definite belief. For this and other reasons, strong arguments against (i) and

(ii) exist. My point here is that decoherence supplies no counterarguments to these strong arguments.)

6.4. Decoherence and Zurek's basis-selection rule

Zurek offers an alternative to the modal basis-selection rule. Instead of relying on the form of the quantum state, he lets the apparatus/environment interaction Hamiltonian, H_{int}, pick out a basis. Here's how:

Let R' denote an arbitrary apparatus observable that doesn't commute with the pointer reading, R. Using 'toy' examples, along with general considerations, Zurek argues that H_{int} commutes with **R**, but does not commute with any **R'**. In rough terms, the interaction between the apparatus and its environment picks out the pointer-reading basis.

To see the physical motivation behind this selection rule, pretend that the apparatus's time evolution depends only on its interaction with the environment. In other words, 'turn off' the apparatus's internal Hamiltonian, H_{app}. In this pretend universe, if the apparatus begins in a pointer-reading eigenstate at time t=0, it remains in that eigenstate, because $[H_{int}, \mathbf{R}]=0$. In words, the apparatus/environment interaction leaves the pointer reading undisturbed. By contrast, the environment would knock the apparatus out of an **R'**-eigenstate.

Because H_{int} is tremendously complicated in all but the simplest examples, we don't yet know whether Zurek's basis-selection rule avoids the basis degeneracy problem. But based on the examples worked out so far, the prospects look promising.

What about the imperfect measurement problem? Apparently, Zurek's basis-selection rule cannot suffer from this difficulty, because the basis picked out by the apparatus/environment interaction in no way depends on the measurement interaction between the apparatus and the 'particle.' Formally, the special basis depends only on the apparatus/environment interaction Hamiltonian H_{int}, not on the particle/apparatus interaction Hamiltonian $H_{measurement}$. Therefore, it doesn't matter how imperfect $H_{measurement}$ is. Nonetheless, given our incomplete knowledge of neurophysiology, it's not clear whether the interaction Hamiltonian between an observer's brain and its environment picks out a definite-belief state.

Furthermore, Zurek does not specify when the relative-state 'branching' occurs, i.e., at what time the pointer reading acquires a definite value (relative to its branch). Since the environment interacts with the apparatus before, during, and after the measurement, it's not clear when the measurement ends, so to speak. To address this difficulty, we must look beyond H_{int}.

A crucial though little-appreciated consequence of both Zurek's basis-selection rule and the modal basis-section rule is this: A single particle in a heat bath acquires a definite value for position (or some observable very close to position). According to Zurek's rule, if we consider the particle to be an 'apparatus,' then it acquires a definite position, because the particle's position commutes with the interaction Hamiltonian between the particle and its thermal environment. And according to modal interpreters, the particle acquires a definite value for an observable very close to position. This is because the thermal interaction rapidly destroys the coherence between states corresponding to different particle position. Hence, the biorthogonal decomposition of the particle with its environment picks out an observable close to position.

Personal taste will dictate your willingness to accept the conclusion that microscopic particles acquire 'definite' positions via interacting with their environment. I'll just point out that, within a many-world or relative-state framework, this conclusion implies that a 'branching' occurs every time a particle interacts with a heat bath in a certain way. But within modal theories, this conclusion carries less ontological baggage, because no branching happens.

In summary, modal theories avoid the ontological problems associated with relative-state and many-world interpretations. But the modal (biorthogonal) basis-selection rule suffers from the basis degeneracy problem and the imperfect measurement problem. Since Zurek's basis-selection rule may partially sidestep these problems, I'd like to propose that modal theorists consider incorporating Zurek's selection rule. Perhaps they could combine Zurek's rule and the usual modal selection rule in some manner. I lack space here to explore this possibility.

7. Conclusion

Decoherence cannot help modal, relative-state, or many-world interpretations fend off general metaphysical criticisms. The value of decoherence lies in its ability to pick out a special basis. Unfortunately, decoherence solves the basis-degeneracy problem only in idealized cases. More important, decoherence helps modal theorists address the imperfect measurement problem only if we don't worry about the in-principle possibility of $|c_1|\approx|c_2|$, in which case the apparatus observable that becomes definite isn't even close to the pointer reading. Also, we must show that brains acquire definite beliefs about pointer readings. As discussed above, since an observer does not directly perceive the definite value associated with the pointer, showing that an observable close to the pointer reading becomes definite does *not* establish that an observer acquires a definite belief. Zurek's basis-selection rule might not assign the observer's brain a definite belief, either. But his rule probably gets rid of basis degeneracy, and unequivocally assigns definite positions to pointers. Perhaps by combining Zurek's rule with the modal basis-selection rule, modal theorists can sidestep the metaphysical problems associated with relative-state interpretations, and also the technical problems associated with biorthogonal decompositions.

Notes

[1] I'd like to thank David Albert, Guido Bacciagaluppi, Dennis Dieks, and Martin Jones for wonderful discussions and correspondence about these issues.

References

Albert, D. and Loewer, B. (1990), "Wanted Dead or Alive: Two Attempts to Solve Schrödinger's Paradox", in A. Fine, M. Forbes, and L. Wessels (eds.), *Proceedings of the 1990 Biennial Meeting of the Philosophy of Science Association, Volume 1*. East Lansing: Philosophy of Science Association, pp. 277-285.

D'Espagnat, B. (1976), *Conceptual Foundations of Quantum Mechanics*. Reading, Massachusetts: Addison-Wesley-Benjamin-Cummings.

B.S. deWitt and R. N. Graham, (eds.), The Many-Worlds Interpretation of Quantum Mechanics. Princeton: Princeton University Press.

Dieks, D. (1989), "Quantum Mechanics without the Projection Postulate and its Realistic Interpretation", *Foundations of Physics* 19: 1395-1423.

_____. (forthcoming), "Objectification, Measurement, and Classical Limit according to the Modal Interpretation of Quantum Mechanics", in P. Busch, P. Lahti, and P. Mittelstaedt (eds.), *Symposium on the Foundations of Modern Physics 1993*. Singapore: World Scientific.

Elby, A. (1993), "Why 'Modal' Interpretations of Quantum Mechanics Don't Solve the Measurement Problem", *Foundation of Physics Letters* 6: 5-19.

_____. and Bub, J. (forthcoming), "The Tri-orthogonal Uniqueness Theorem and its Relevance to the Interpretation of Quantum Mechanics", *Physical Review A..*

_____. (1994, unpublished), 'Modal Interpretations do not Explain Why we Acquire Definite Beliefs about Pointer Readings."

Healey, R.A. (1989), *The Philosophy of Quantum Mechanics: An Interactive Interpretation*. Cambridge: Cambridge University Press.

_____. (1993), "Why Nonideal Quantum Measurements Have Outcomes", *Foundations of Physics Letters* 6: 37-50.

van Fraassen, B. (1979), "Hidden Variables and the Modal Interpretation of Quantum Mechanics", *Synthese* 42: 155-165.

Joos, E. and Zeh, H.D. (1985), "The Emergence of Classical Properties Through Interaction with the Environment", *Zeitschrift fur Physik B* 59: 223-243.

Zurek, W.H. (1991), "Decoherence and the Transition from Quantum to Classical", *Physics Today* 46 no. 10: 36-44.

_____. (1993), "Negotiating the Tricky Border Between Quantum and Classical: Zurek Replies", *Physics Today* 46 no. 4: 84-90.

Wavefunction Tails in the Modal Interpretation[1]

Michael Dickson

University of Notre Dame

1. The Measurement Problem and the Modal Interpretation

Consider the general form of the state of a quantum system, $|\psi\rangle = \sum_i c_i |\alpha_i\rangle$ (eq. 1), where the $|\alpha_i\rangle$ are (non-degenerate) eigenvectors of an operator, A, corresponding to an observable, A. According to the standard interpretation of quantum mechanics (the precise details of which do not concern me here), A does not have a definite value on the system in the state $|\psi\rangle$. But now suppose we measure A by coupling the system to a measuring apparatus. The composite system is then $|\Psi\rangle = \sum_i c_i |\alpha_i\rangle \otimes |\varphi_i\rangle$ (eq. 2) where the $|\varphi_i\rangle$ are the indicator-states of the apparatus. Again, the standard interpretation tells us that the apparatus is not in a definite indicator-state. But how can this be? Measurements do have definite results. The apparatus is always in a definite state.

I have just reviewed one version of the measurement problem. The modal interpretation of quantum mechanics proposes a solution. It begins with a distinction between two types of state, the theoretical state and the physical state (different terms are used by different authors):

Theoretical State: The state assigned by quantum mechanics, used to calculate predictions.
Physical State: A specification of the values, if any, of observables on a system.

Armed with this distinction, the modal interpretation can allow that (2) is the correct theoretical state of the composite system, while nonetheless claiming that the apparatus is in a definite physical state. The modal interpretation thus denies one of the fundamental (though rarely mentioned) assumptions of the standard interpretation, namely, that the theoretical state and the physical state of a system are the same. To put it differently: Observables *can* have definite values on a system even when the system is not in an eigenstate of the relevant operator.

Of course, more needs to be said. In particular, any version of the modal interpretation must answer two questions. (i) Under what conditions can one attribute a definite physical state to a system? (ii) When those conditions are met, what attributions are allowed?

In the next section, I shall review one type of answer to these questions. It is the answer given by Kochen (1985), Dieks (1988, 1989a, 1989b), and Healey (1989) in their versions of the modal interpretation.[2] Henceforth I call this answer the KDH proposal. In Section 3, I shall show how inaccuracy is an intrinsic feature of any quantum mechanical measurement, and I shall review briefly why some authors have considered inaccuracy to be a stumbling block for the KDH proposal. In Section 4, I shall argue that it is not in fact a problem, and in Section 5 I shall suggest a possible extension of the argument. For those familiar with the dispute (and in particular, Healey's (1993a, 1993b) replies to the problem of inaccurate measurements) I note now that my argument does not depend on supposing that inaccurate measurements are to be modelled one way rather than another. The model that I use arises from the measurement process itself when described by the Schrödinger equation and a typical interaction Hamiltonian.

2. The KDH Proposal

The KDH proposal relies on the following theorem:[3]

Biorthonormal Decomposition Theorem: For any $|\Psi\rangle$ in the tensor-product Hilbert space $H_\alpha \otimes H_\beta$, it is possible to express $|\Psi\rangle$ as $|\Psi\rangle = \sum_i c_i |\alpha_i\rangle \otimes |\beta_i\rangle$, where $\{|\alpha_i\rangle\}$ and $\{|\beta_i\rangle\}$ are complete orthonormal sets. Furthermore, this biorthonormal form is unique, provided that $|c_i| = |c_j|$ implies $i = j$.

In Section 4, I shall need to refer to the following proof of this theorem.[4]

Proof. If $\{|\xi_i\rangle\}$ is another orthonormal basis for H_β, then $|\Psi\rangle$ can always be written as $|\Psi\rangle = \sum_{ij} d_{ij} |\alpha_i\rangle \otimes |\xi_j\rangle$. For each value of i, define $|\zeta_i\rangle = \sum_j d_{ij} |\xi_j\rangle$, so that $|\Psi\rangle$ becomes $|\Psi\rangle = \sum_i |\alpha_i\rangle \otimes |\zeta_i\rangle$. Of course, the $|\zeta_i\rangle$ are not orthogonal, or even normalized. First, normalize by defining $|\beta_i\rangle = |\zeta_i\rangle / |\langle \zeta_i | \zeta_i \rangle|^2 =_{df} |\zeta_i\rangle / c_i$ so that $|\Psi\rangle = \sum_i c_i |\alpha_i\rangle \otimes |\beta_i\rangle$.

Can we make the $|\beta_i\rangle$ mutually orthogonal as well? Yes. To see why, consider the (kernel) operator $K = \text{Tr}_{(\beta)} |\Psi\rangle\langle\Psi|$, the partial trace of $|\Psi\rangle\langle\Psi|$ in H_β. Because $|\Psi\rangle$ is normalized $|K|^2$ is clearly a bounded operator. And K is symmetric. Hence the following lemma applies:

Lemma (from the Hilbert-Schmidt Theorem—see note 3): If $|K|^2$ is a bounded symmetric kernel operator, then there exists a complete set of orthonormal eigenvectors, $\{|\alpha_i\rangle\}$, satisfying $|\alpha_i\rangle = \lambda_i K |\alpha_i\rangle$.

Taking the $|\alpha_i\rangle$ in the expansion of $|\Psi\rangle$ to be the eigenvectors of K, the $|\beta_i\rangle$ can be chosen to be mutually orthogonal. By the lemma, $|\alpha_i\rangle = \lambda_i K |\alpha_i\rangle$ implies $\langle \alpha_j | \lambda_i K |\alpha_i\rangle = \delta_{ij}$. But also (letting $\{|e_i\rangle\}$ be another basis of H_β, and leaving tensor products implicit):

$$\langle \alpha_j | \lambda_i K |\alpha_i\rangle = \text{Tr}_{(\beta)} \lambda_i \langle \alpha_j|\Psi\rangle\langle\Psi|\alpha_i\rangle = \sum_k \lambda_i \langle e_k|\langle\alpha_j|\Psi\rangle\langle\Psi|\alpha_i\rangle|e_k\rangle$$
$$= \sum_k \lambda_i \langle\Psi|\alpha_i\rangle|e_k\rangle\langle e_k|\langle\alpha_j|\Psi\rangle = \lambda_i \langle\Psi|\alpha_i\rangle\langle\alpha_j|\Psi\rangle.$$

Therefore, $\langle\Psi|\alpha_i\rangle\langle\alpha_j|\Psi\rangle = (1/\lambda_i)\delta_{ij}$. Now note that $\langle\Psi|\alpha_i\rangle = c_i |\beta_i\rangle$, so that $\langle\Psi|\alpha_i\rangle\langle\alpha_j|\Psi\rangle = \langle\beta_i|c_i^*c_j|\beta_j\rangle$ and therefore $c_i^*c_j \langle\beta_i|\beta_j\rangle = (1/\lambda_i)\delta_{ij}$, which proves the theorem (apart form uniqueness, which should be clear—but keep in mind that uniqueness fails when there is degeneracy among the $|c_i|$ or, equivalently, among the λ_i).

The KDH proposal is that whenever the theoretical state of a system has a unique biorthonormal decomposition, the subsystems corresponding to the spaces H_α and H_β have definite physical states given by the values corresponding to one of the $|\alpha_i\rangle|\beta_i\rangle$. These values are values for the observables A and B that correspond to the operators A and B whose eigenvectors are the $|\alpha_i\rangle$ and $|\beta_i\rangle$. The probability that the physical states are the ones corresponding to $|\alpha_i\rangle$ and $|\beta_i\rangle$ is $|c_i|^2$. (This last part of the proposal justifies the epithet 'modal'.) The KDH proposal is silent about the case where the biorthonormal decomposition is not unique, a case that I ignore here.[5]

If we suppose the $|\varphi_i\rangle$ in (2) to be orthogonal, then (2) is the biorthonormal decomposition of the composite system after measurement, and the proposal attributes a definite indicator-state to the apparatus. But what happens when the $|\varphi_i\rangle$ are not orthogonal? In that case, (2) is not the biorthonormal decomposition, and it might turn out that, according to the KDH proposal, the apparatus does not have a definite indicator-state. To see whether this particular disaster strikes the KDH proposal, we need to look at how non-ideal measurements arise in quantum mechanics.

3. Non-Ideal Measurements

One reason for inaccuracy in measurements is that experimenters cannot build a flawless apparatus. Hence detectors sometimes do not fire when they should, and sometimes fire when they should not. Such inaccuracies, though in practice unavoidable, are not theoretically fundamental. They arise from limitations of engineering, and are handled theoretically by more or less ad hoc (but not necessarily unjustified) modifications of the theoretical description of a measurement. They are not naturally accommodated by the quantum formalism, and no interpretation of quantum mechanics should be expected to deal with this type of inaccuracy in anything but an ad hoc way.

But there is a further type of inaccuracy that is intrinsic to quantum theory and arises straight from the quantum-mechanical description of measurement. This second type of inaccuracy is therefore fundamental to quantum theory, and an interpretation of quantum theory must be able to handle it. What follows is a brief description of this second type of inaccuracy.

A truly quantum-mechanical (non-relativistic) description of measurement begins with the Schrödinger equation and an appropriate interaction-Hamiltonian. To see how it works, consider an object system described by $|\Psi\rangle$, whose free Hamiltonian is H_O, and a measuring apparatus described by $|\varphi\rangle$, whose free Hamiltonian is H_M. Let the momentum operator for the apparatus be given by P, and let the measurement-interaction be described by an interaction-Hamiltonian, $H_I = \gamma f(t) A \otimes P$. Here γ is a constant characterizing the strength of the interaction and $f(t)$ is a scalar-valued function characterizing the time-dependence of the interaction. The interaction occurs over a finite time, and $f(t)$ is normalized, $\int_0^\tau f(t)dt = 1$, where τ is the duration of the interaction (which begins at $t = 0$).

The aim is to measure the observable A (corresponding to the operator A) through the coupling H_I, which results in a correlation between the eigenvectors of A and the position-states of the apparatus. We might suppose, for example, that the 'apparatus' is just the spatial part of the state of a particle whose spin is being measured by a standard Stern-Gerlach apparatus. (Note that the argument to follow does not depend on the particular model used here, and in particular that the P in H_I can be replaced by other operators.)

One generally assumes that the measurement is an 'impulsive' measurement—τ is very small and that γ is very large.[6] Hence one may ignore the evolution due to H_O and H_M during the measurement interaction, and write the Schrödinger equation as (setting $\hbar = 1$): $i(\partial|\Psi\rangle / \partial t = \gamma f(t) A \otimes P|\Psi\rangle$. The solution just after the measurement is

$$|\Psi(t)\rangle = \exp\left[-i\gamma \int_0^\tau f(t)dt \; A \otimes P\right]|\Psi(0)\rangle. \tag{3}$$

In a typical measurement, the general form of the initial state of the composite system is given by: $|\Psi(0)\rangle = \sum_i c_i |\alpha_i\rangle \otimes |\varphi_0\rangle$ (eq. 4), where the $|\alpha_i\rangle$ are, as before, the eigenvectors of A, and $|\varphi_0\rangle$ is the ready-state of the apparatus.

Given (4) and the normalization of $f(t)$, the solution (3) becomes

$$|\Psi(\tau)\rangle = \sum_i c_i |\alpha_i\rangle \exp[-i\gamma a_i P]|\varphi_0\rangle, \tag{5}$$

where a_i is the eigenvalue corresponding to $|\alpha_i\rangle$. Now note that $\exp[-i\gamma a_i P]$ is a translation operator. It shifts the position of the apparatus by γa_i. (In general, if instead of P we use an operator S in the definition of H_I, then $\exp[-i\gamma a_i S]$ shifts the state of the apparatus for the conjugate operator.) Indicate the shifted state $\exp[-i\gamma a_i P]|\varphi_0\rangle$ by $|\varphi_i\rangle$. Then the state of the composite system after measurement is: $|\Psi(\tau)\rangle = \sum_i c_i |\alpha_i\rangle \otimes |\varphi_i\rangle$ (eq. 6). If the $|\varphi_i\rangle$ are mutually orthogonal, then (6) is the biorthonormal decomposition of $|\Psi(\tau)\rangle$, and the KDH proposal implies that the apparatus has for its physical state a definite indicator-state. Moreover, this definite indicator-state is, under the KDH proposal, perfectly correlated with the physical state of the object system.

But are the $|\varphi_i\rangle$ mutually orthogonal? Hardly ever. For any i and j we have

$$\langle \varphi_i | \varphi_j \rangle = \langle \varphi_0 | \exp[-i\gamma(a_i - a_j)P] | \varphi_0 \rangle \tag{7}$$

which is hardly ever zero.[7] The exponential in (7) is inherited directly from the solution of the Schrödinger equation. It is difficult to suppose, therefore, that a quantum mechanical account of measurement will always, or even often, result in a biorthonormal form where the states of the apparatus are orthogonal.

Instead, those states will typically be 'highly localized'. Consider the position wavefunctions, $\varphi_i(x) = \langle x|\varphi_i\rangle$. If $\varphi_i(x)$ is concentrated in a region of space, with tails reaching to infinity, then $\varphi_i(x)$ is 'highly localized'. Let the region where $\varphi_i(x)$ is large be the 'localization region' of $|\varphi_i\rangle$. Because of their tails, the $|\varphi_i\rangle$ will have at least a tiny overlap, the size depending on the size of the tails, the width of the localization regions, and the amount of translation effected by the operators $\exp[-i\gamma a_i P]$. But no matter how small the overlap is, the $|\varphi_i\rangle$ are not orthogonal, and the biorthonormal decomposition of $|\Psi(\tau)\rangle$ is not (6).

For example, let H_α be two-dimensional; let the eigenvectors of A be $|\alpha_1\rangle$ and $|\alpha_2\rangle$; and let the initial wavefunction for the apparatus, $\varphi_0(x)$, be a highly localized wavepacket (with tails). It might represent, for example, the position of a spin-1/2 particle entering a Stern-Gerlach apparatus. The shifted wavefunctions, $\varphi_1(x)$ and $\varphi_2(x)$, are the outgoing states, representing the particle heading for the $|\alpha_1\rangle$ and $|\alpha_2\rangle$ detectors, respectively. Because $\varphi_1(x)$ and $\varphi_2(x)$ have non-zero overlap, the state

$|\Psi(\tau)\rangle = c_1|\alpha_1\rangle \otimes |\varphi_1\rangle + c_2|\alpha_2\rangle \otimes |\varphi_2\rangle$ is not the biorthonormal decomposition of the composite system after measurement.

One's immediate reaction is: 'Disaster for the KDH proposal'. After all, although they are not completely localized, $\varphi_1(x)$ and $\varphi_2(x)$ do represent the indicator-states of the apparatus. But according to the proposal, $\varphi_1(x)$ and $\varphi_2(x)$ do not correspond to the definite-valued states for the apparatus. What are the definite-valued states, and are they physically acceptable? Or are they, at least sometimes, the very kind of superposition that the Modal Interpretation seeks to avoid? Some authors, notably Elby (1993) and Albert and Loewer (1990, 1991, 1993), have suggested that in the case of inaccurate measurements, the KDH proposal does result in definite states that are unacceptable superpositions of indicator states. (See Hemmo (1993) for a discussion as well.) In the next section I shall suggest that detailed calculation vindicates the KDH proposal.

But first, it is important to see clearly what the potential difficulty is. The problem is not merely that the states attributed to the apparatus by the KDH proposal are superpositions. In the standard interpretation too, the indicator-states are superpositions, but they are highly localized superpositions. Such highly localized superpositions, even for macroscopic objects, are acceptable because they are effectively indistinguishable from completely localized states. (I realize that this claim raises its own set of questions about the interpretation of the wavefunction, but I cannot address those here.) Hence any interpretation that is committed only to highly localized superpositions is acceptable. But the potential problem with the KDH proposal is that it might, in the case of inaccurate measurements, attribute to the apparatus non-localized superpositions, perhaps even superpositions of macroscopically distinguishable states. I shall now address this problem.

4. Resolution of the Problem

Continue with the example of the two-level system from Section 3. To calculate the biorthonormal decomposition, we begin by finding the eigenvectors of the operator K, which in this case is $K = \text{Tr}_{(\beta)} |\Psi\rangle\langle\Psi| = \Sigma_i \langle e_i|\Psi\rangle\langle\Psi|e_i\rangle$ (eq. 9), where the $|e_i\rangle$ form an orthonormal basis in H_β. Thinking of the $|\varphi_i\rangle$ in their position-representation, it is easiest to take the $|e_i\rangle$ to be position (improper) eigenstates, denoted $|x\rangle$. The sum in (9) is therefore an integral, and we have $K = \int_{-\infty}^{+\infty} dx \langle x|\Psi\rangle\langle\Psi|x\rangle$, which is:

$$K = \int_{-\infty}^{+\infty} dx \left(c_1\varphi_1(x)|\alpha_1\rangle + c_2\varphi_2(x)|\alpha_2\rangle\right) \times \left(c_1^*\varphi_1^*(x)\langle\alpha_1| + c_2^*\varphi_2^*(x)\langle\alpha_2|\right)$$

Let $\varphi_{ij} =_{df} \int_{-\infty}^{+\infty} dx \, \varphi_i(x)\varphi_j^*(x)$. Because $\varphi_{ij} = \varphi_{ji}^*$, we may define $\eta = \varphi_{12} = \varphi_{21}^*$. Now, let us choose a representation for the $|\alpha_i\rangle$ where $|\alpha_1\rangle = \binom{1}{0}$ and $|\alpha_2\rangle = \binom{0}{1}$. Then (noting that $\varphi_{ii} = 1$)

$$K = \begin{pmatrix} |c_1|^2 & c_1 c_2^* \eta \\ c_2 c_1^* \eta^* & |c_2|^2 \end{pmatrix}$$

To calculate the eigenvalues of K, find the values of λ that satisfy $\det(\lambda K - I) = 0$ (where I is the identity operator on H_α). Calculation yields two values:

$$\lambda_1 = \frac{1 + \sqrt{1 - 4|c_1|^2|c_2|^2(1 - |\eta|^2)}}{2|c_1|^2|c_2|^2(1 - |\eta|^2)} \quad \lambda_2 = \frac{1 - \sqrt{1 - 4|c_1|^2|c_2|^2(1 - |\eta|^2)}}{2|c_1|^2|c_2|^2(1 - |\eta|^2)}$$

Call the corresponding eigenvectors $|\chi_1\rangle$ and $|\chi_2\rangle$. To find $|\chi_1\rangle$, define $|\chi_1\rangle = \begin{pmatrix} \mu \\ \nu \end{pmatrix}$ so that

$$\begin{pmatrix} \mu \\ \nu \end{pmatrix} = \lambda_1 \begin{pmatrix} |c_1|^2 & c_1 c_2^* \eta \\ c_2 c_1^* \eta^* & |c_2|^2 \end{pmatrix}.$$

Then $(\nu/\mu) = (\lambda_1 c_2 c_1^* \eta^*)/(1 - \lambda_1 |c_2|^2)$. As η gets small, ν/μ approaches zero, because the denominator approaches a constant and the numerator approaches zero. Now, the normalization condition $|\mu|^2 + |\nu|^2 = 1$ implies that $1 + (|\nu|^2/|\mu|^2) = (1/|\mu|^2)$, so that as ν/μ approaches 0, μ must approach 1.[8] In other words, as η gets small, $\begin{pmatrix} \mu \\ \nu \end{pmatrix}$ approaches $\begin{pmatrix} 1 \\ 0 \end{pmatrix}$, up to an overall phase. Similarly, $|\chi_2\rangle$ goes to $\begin{pmatrix} 0 \\ 1 \end{pmatrix}$, up to an overall phase.

From the proof given in Section 2, it is clear that the $|\chi_i\rangle$ are states of the object-system in the biorthonormal expansion, $|\Psi(\tau)\rangle = c_1|\chi_1\rangle \otimes |\pi_1\rangle + c_2|\chi_2\rangle \otimes |\pi_2\rangle$, where the $|\pi_i\rangle$ are the corresponding states of the apparatus. According to the KDH proposal, the apparatus is in the physical state corresponding to one of the $|\pi_i\rangle$. What are the $|\pi_i\rangle$, and are they acceptable indicator-states, in the sense of Section 3?

To find the $|\pi_i\rangle$, turn again to the proof of the biorthonormal decomposition theorem. From that proof, it is clear that $\sqrt{1/\lambda_1}\,|\pi_1\rangle = \langle\chi_1|\Psi(t)\rangle$. Hence,

$$\sqrt{1/\lambda_1}\,|\pi_1\rangle = (\mu \quad \nu)\left[c_1\begin{pmatrix}1\\0\end{pmatrix}|\varphi_1\rangle + c_2\begin{pmatrix}0\\1\end{pmatrix}|\varphi_2\rangle\right]$$

so that $|\pi_1\rangle = \sqrt{\lambda_1}\left[\mu c_1|\varphi_1\rangle + \nu c_2|\varphi_2\rangle\right]$. The case $i = 2$ is analogous.

In general, such an expression for $|\pi_1\rangle$ would be disastrous, for it is a superposition of two macroscopically distinct states of the apparatus. But, as I noted in Section 3, we have learned to live with highly localized superpositions. And, amazingly enough, the $|\pi_i\rangle$ are highly localized, provided that η is small. The reason is that when η is small μ is nearly 1 and ν is nearly 0. Hence the expression for $|\pi_1\rangle$ given above is nearly $|\varphi_1\rangle$. That is, the overlap between $|\pi_1\rangle$ and $|\varphi_1\rangle$ is nearly 1. The state $|\pi_1\rangle$ is an acceptable indicator-state that more or less accurately indicates the state of the measured system.

Hence the KDH proposal does not break down in the case of inaccurate measurements, provided we assume that φ_{ij} is small whenever $i \neq j$. (Keep in mind that I have considered only cases where the decomposition is unique. See notes 4 and 7.)

5. Approximate Decoherence

But is this assumption admissible? Yes. Otherwise, it does not make much sense to say that a 'measurement' has occurred, for there will be insufficient correlation between the states of the apparatus and the system.

To see this point in an example, consider again the case where the 'apparatus-state' is the spatial part of the state of a spin-1/2 particle being measured by a Stern-Gerlach apparatus. The ready-state, $\varphi_0(x)$, gives the position of the particle before it enters the magnetic field. (Note that even if $\varphi_0(x)$ were completely localized, it would immediately evolve into a state with tails.) This highly localized state with tails will be shifted by the interaction, and the result will be a superposition of two highly localized states with tails. Provided that the shift is large enough (*i.e.*, much larger than the localization width of the initial wavefunction), the overlap between the superposed states will be very tiny—it is the overlap of their tails only. And we must assume that the shift is large enough to effect this separation, lest the interaction be insufficient to get the correlation needed for measurement. Hence in this case, and generally, φ_{ij} must be small when $i \neq j$, and the KDH proposal is satisfactory.

But why consider only measurements? Suppose that we perform a pseudo-measurement, in which the interaction strength is too small to get a sufficient separation. In that case, φ_{ij} is not small for some $i \neq j$, and the definite states are physically unacceptable superpositions.

In the case of interactions between two microscopic systems, the result of such pseudo-measurements could clearly be non-localized superpositions. We might suppose, for example, that the magnetic field generated by the Stern-Gerlach device is almost homogeneous. Then the outgoing wavepackets would not be separated enough to distinguish them clearly, and the KDH proposal would assign a non-localized superposition to the particle. But in such microscopic cases we should not worry, for we already have to live with non-localized superpositions at the quantum level.

It is less clear what happens in the macroscopic case. On the one hand, it might turn out exactly as the microscopic case does. That is, the macroscopic object might, according to the KDH proposal, end up in a non-localized superposition, as would occur when φ_{ij} is not small for some $i \neq j$. On the other hand, there are reasons to suppose that the overlap between macroscopically distinct states is always small. One very nice argument has been suggested by DeWitt (1994), though work has been done by many under the heading 'decoherence theory'. (See, for example, Zurek (1993), and Gell-Mann and Hartle (1990), as well as several references in these.)

DeWitt's argument relies on three apparently weak assumptions. The first is that macroscopic objects interact with particles (of at least two different momenta) from the environment. The second is that these interactions are given by multiples of a delta-function in space, $g\delta(x - X)$, where x is the coordinate of the particle, X is the coordinate of the macroscopic object,[8] and g is the interaction-strength. The third is that the mass of the macroscopic object is much larger than the mass of the particles. DeWitt goes on to argue that the states of the macroscopic object 'decohere,' that is, that there is very little overlap between macroscopically distinct states.

We are still far from a proof that any interaction involving macroscopic systems will result in a state whose biorthonormal decomposition yields acceptable states, but DeWitt's argument suggests that such a proof might be found along the same lines as the proof in Section 4 that measurement interactions do result in such states.[10] In any case, it is clear that inaccurate measurements do not pose any problem for the KDH version of the modal interpretation. I do not mean to suggest that the modal interpretation (in any version) is without problems, but its problems are the topic of a different discussion.

Notes

[1]Thanks to James Cushing for invaluable comments and discussion. Thanks also to Guido Bacciagaluppi, Dennis Dieks, and Meir Hemmo for their helpful remarks.

Thanks to the University of Notre Dame and the Mellon Foundation for generous financial support while this work was being done.

[2]Other versions of the modal interpretation include those by van Fraassen (1979, 1981, 1991) and Bub (1992, 1993). Bub's version came under attack from Elby (1993), the charge being again its failure to deal with inaccurate measurements. But Elby himself seems to recognize that Bub's version is less damaged by the problem. My view is that it is not damaged at all, as Bub (1993) argues.

[3]Kochen (1985) bases his version on the polar decomposition theorem. (See Reed and Simon (1975, 196–198) for a statement and proof of the theorem.) Kochen's approach implies the approach via the biorthonormal decomposition theorem.

[4]I thank James Cushing for suggesting this version of the proof, an expansion of one given by Schrödinger (1935). A proof of the Hilbert-Schmidt theorem, which is the meat of this proof of the biorthonormal decomposition theorem, can be found, for example, in Cushing (1975).

[5]Degeneracy of the $|c_i|$ does cause problems for the KDH proposal as I have outlined it here. There are various ways to deal with the problem. One is that taken by Dieks (1989a), who suggests that the measure of actual states where degeneracy holds (or, the measure of time over which any actual physical system is in a degenerate state) is zero. Hence degeneracy is a not a problem. The second approach, found for example in Clifton (1994), is to extend the KDH proposal to cover cases of degeneracy. The central idea there is to take as definite-valued the projection onto the subspace spanned by all vectors in a degenerate set.

[6]Bohm (1951) gives a beautiful example of an impulsive measurement. Note that impulsive measurement is not the only kind of measurement. Aharonov et al. (1993) have shown the viability of adiabatic measurement, where τ is large and γ is small.

[7]Equation (7) can sometimes be zero. Consider it in the following (equivalent) form:

$$\langle\varphi_i|\varphi_j\rangle = \int_{-\infty}^{+\infty} dp\, \exp[-i\gamma(a_i - a_j)p]|\varphi_0(p)|^2.$$

where p are momentum eigenvalues. There are functions $\varphi_0(p)$ that make this integral zero, though they do not seem to be 'typical' wavefunctions. In any case when (7) is zero, then the biorthonormal decomposition gives exactly the correct result and nothing more need be said. I thank James Cushing (again!) for pointing out to me that (7) can sometimes be zero.

[8]There are some subtleties here. First, if one begins by calculating μ/ν rather than ν/μ, then the calculation shows that rates of convergence are important. To see why, note that as η becomes small, λ_1 approaches $1/|c_1|^2$. Further,

$$(\mu/\nu) = (\lambda_1 c_1 c_2^* \eta)/(1 - \lambda_1 |c_1|^2)$$

so that as $|\eta|$ becomes small both numerator and denominator approach zero. But a more detailed look shows that the denominator approaches zero as $|\eta|^2$ whereas the numerator approaches zero as $|\eta|$, so that μ/ν approaches infinity. (Normalization takes care of the rest.) The second subtlety is that if $|c_1|^2$ and $|c_2|^2$ are identical, then v/μ likewise approaches infinity. In that case, the argument fails. However, keep in mind that I have all along assumed no degeneracy, so that $1 - \lambda_1|c_2|^2$ is non-zero in the limit as η gets small.

I am grateful to Guido Bacciagaluppi and Meir Hemmo (1994a) for further suggesting to me that because rates of convergence are important, problems can arise when the coefficients are *nearly* degenerate. Dennis Dieks has suggested to me that the argument of his mentioned in note 4 could defuse this problem. Andrew Elby has also suggested to me that decoherence could save the day here. I shall simply ignore the case of near-degeneracy.

[9]But how is *X* defined? It is clearly not the coordinate of the center of mass. But I shall skip this detail here.

[10]In fact, Bacciagaluppi and Hemmo (1994) explore exactly this question. They claim that if interactions of a macroscopic object with its environment are taken into account, then the near-decoherence of the states of the macroscopic object translates into acceptable states for the macroscopic object under the KDH proposal. Dieks (1989b, forthcoming) has also suggested that decoherence theory and the modal interpretation might be usefully combined, though his suggestions are along somewhat a different line. *Cf.* Elby (1994) for another discussion.

References

Aharanov, Y., Anandan, J., and Vaidman, L. (1993), "Meaning of the Wave Function", *Physical Review A* 47:4616-4626.

Albert, D. and Loewer, B. (1990), "Wanted Dead or Alive: Two Attempts to Solve Schrödinger's Paradox", in A. Fine, M. Forbes, and E. Wessels, (eds.), *PSA 1990*, vol. I. East Lansing, MI: Philosophy of Science Association.

———. (1991), "Some Alleged Solutions To the Measurement Problem", *Synthese* 88:87-98.

———. (1993), "Non-Ideal Measurements", *Foundations of Physics Letters* 6:297-305.

Bacciagaluppi, G., and Hemmo, M. (1994a), "Modal Interpretation of Imperfect Measurements", University of Cambridge preprint.

———. (1994b), "Making Sense of Approximate Decoherence", this volume.

Bohm, D. (1951) *Quantum Theory.* New York: Prentice-Hall.

Bub, J. (1992), "Quantum Mechanics Without the Projection Postulate", *Foundations of Physics* 22:737-754.

———. (1993), "Measurement: It Ain't Over Till It's Over", *Foundations of Physics Letters* 6:21-35.

Clifton, R. (1994), "Independently Motivating the Dieks-Kochen Modal Interpretation of Quantum Mechanics", University of Western Ontario preprint.

Cushing, J.T. (1975), *Applied Analytical Mathematics for Physical Scientists*. New York: John Wiley & Sons.

DeWitt, B. (1994), "How Does the Classical World Emerge From the Wave Function?", in F. Mansouri and J.J. Scanio, (eds.), *Topics on Quantum Gravity and Beyond*. Singapore: World Scientific.

Dieks, D. (1988), "The Formalism of Quantum Theory: An Objective Description of Reality?", *Annalen der Physik* 7:174-190.

_____. (1989), "Quantum Mechanics Without the Projection Postulate and Its Realistic Interpretation', *Foundations of Physics* 19:1397-1423.

_____. (1989), "Resolution of the Measurement Problem Through Decoherence of the Quantum State" *Physics Letters A* 142:439-446.

_____. (forthcoming), "The Modal Interpretation of Quantum Mechanics, Measurements and Macroscopic Behavior", *Physical Review A*.

Elby, A. (1993), "Why 'Modal' Interpretations of Quantum Mechanics Don't Solve the Measurement Problem", *Foundations of Physics Letters* 6:5-19.

_____. (1994), 'The "Decoherence" Approach to the Measurement Problem in Quantum Mechanics", this volume.

Gell-Mann, M. and Hartle, J. (1990), "Quantum Mechanics in the Light of Quantum Cosmology", in W. Zurek, (ed.), *Complexity, Entropy, and the Physics of Information*. New York: Addison-Wesley.

Healey, R. (1989), *The Philosophy of Quantum Mechanics: An Interactive Interpretation*. Cambridge: Cambridge University Press.

Healey, R. (1993a), "Measurement and Quantum Indeterminateness", *Foundations of Physics Letters* 6:307-316.

_____. (1993b), "Why Error-Prone Quantum Measurements Have Outcomes", *Foundations of Physics Letters* 6:37-54.

Hemmo, M. (1993), "Review of R. Healey's The Philosophy of Quantum Mechanics: An Interactive Interpretation", *Foundations of Physics* 23:1137-1145.

Kochen, S. (1985), "A New Interpretation of Quantum Mechanics", in P. Lahti and P. Mittelstaedt, (eds.), *Symposium on the Foundations of Modern Physics*. Teaneck, NJ: World Scientific Publishing Co..

Reed, M. and Simon, B. (1979), *Methods of Modern Mathematical Physics*. San Diego, CA: Academic Press.

Schrödinger, E. (1935), "Discussion of Probability Relations Between Separated Systems", *Proceedings of the Cambridge Philosophical Society* 31:555-563.

van Fraassen, B. (1979), "Hidden Variables and the Modal Interpretation of Quantum Theory", *Synthese* 42:155-165.

van Fraassen, B. (1981), "A Modal Interpretation of Quantum Mechanics", in E. Beltrametti and B. van Fraassen, (eds.), *Current Issues in Quantum Logic*. New York: Plenum.

van Fraassen, B. (1991), *Quantum Mechanics: An Empiricist View*. Oxford: Clarendon Press.

Zurek, W. (1993), "Preferred States, Predictability, Classicality, and the Environment-Induced Decoherence", in J. Halliwell, (ed.), *The Physical Origins of Time Asymmetry*. Cambridge: Cambridge University Press.

Part X

GAMES, EXPLANATIONS, AUTHORITY, AND JUSTIFICATION

The Microeconomic Interpretation of Games[1]

Chantale LaCasse and Don Ross

University of Ottawa

It is clear that many philosophers of science are inclined to be sceptical about the scientific ambitions of economics. Sometimes this scepticism amounts to a forthright denial of economists' entitlement to be granted the epistemic authority normally claimed by scientists (for example, Dupré 1993). A more modest ambivalence may be signalled by a philosopher's refusal to treat the intuitions of economists as *data* for the philosophy of science, in the way that (e.g.) physicists' ontological and epistemological judgements are now routinely treated. The following paper is part of a much larger project that criticizes the basis for this scepticism, at least where microeconomics is concerned.[2] This general project is partly motivated by our belief that philosophers have often viewed microeconomics from a highly distorting set of perspectives. One source of these distortions is a sociological accident: the philosophy of economics, as a part of the philosophy of science, has been largely owned since its outset by Popperians and their intellectual descendants.[3] Economic methodologists, deferring to philosophers, have therefore adopted an explicitly Popperian conceptual vocabulary; and this in turn has interfered with the perception of economics by even *non*-Popperian philosophers of science. A second source of distortion, and the one that will be our concern in what follows, is that philosophers, as a result of their wider interests and preoccupations, have implicitly attributed ambitions to economics that microeconomics has been slowly but consistently abandoning for over a century. In particular, many philosophers seem to have been chiefly interested in microeconomics as a result of their expectation that it can underwrite their search for a rigorous normative theory of decision. This expectation, however, saddles microeconomics with a burden that it cannot carry; and the resulting disappointment has motivated *some* of the philosophical dissatisfaction with the discipline. The clearest instance of the above mistake is to be found in a recent survey of the literature on the economic conception of rationality by Hollis and Sugden (1993). Since their understanding of the relevant economic concepts is otherwise quite sophisticated, we shall take them as representative opponents and concentrate our arguments against theirs.

Hollis and Sugden express unequivocal disappointment with the microeconomic paradigm as it expresses itself in game theory. Their view is that game theory, despite (or perhaps because of) its being the most sophisticated "logic of practical reason ...[for]... anyone whose notion of rationality is instrumental and whose view of the social world is instrumentalist," is "beset with paradoxes" (Hollis and Sugden 1993,

32). These paradoxes consist in the fact that, according to game theory, pairs or groups of rational agents are sometimes incapable of reaching equilibrium outcomes that are plainly superior according to other criteria, such as moral ones. This leads them to conclude that "there is something amiss with the general aim of abstracting from all 'moral psychology' to a world of fully transparent agents with common knowledge of rationality and a synchromesh between preferences and choices" (Ibid); and this 'general aim,' is, in turn, the one that they attribute to the leading foundational work in microeconomics. Their diagnosis of this situation is that the rationality concept as incorporated in game theory is inadequate; and it should be repaired, they urge, through attention to "an empirical moral psychology which seeks behavioral regularities" (Ibid, 29). Any mainstream microeconomist would interpret this as equivalent to the claim that the central concept of their discipline needs profound reanalysis, and that microeconomics must thus be rebuilt on new foundations.

We insist on this interpretation with special vehemence, since we endorse (though we cannot here argue for) the view that game theory is the most plausible source of unifying foundations for all of microeconomics (Ross and LaCasse 1993). We will argue, however, that Hollis and Sugden's disenchantment rests on their having asked game theory to solve problems for which its role in microeconomic theory does not, and cannot be expected to, suit it. This tension expresses itself most convincingly in subtle but important differences between Hollis and Sugden's conception of what a game *is*, and the nature of the objects of analysis that are explicitly recognized by game theory. Before proceeding to our critique of Hollis and Sugden, therefore, we will present a brief analysis of the concept of a game as it occurs in microeconomic theory and applications.

We first provide a philosophically 'uninterpreted' representation of the concept of a game, in strategic ('normal') form. By 'uninterpreted,' we mean that it gives the analysis of the concept that is constant across textbooks, and that must be preserved under all philosophical interpretations *if* those interpretations are to be interpretations of games as microeconomics understands them. There should be nothing controversial or surprising about this representation, which we provide in order to fix aspects of our subsequent 'interpreted' characterization that are otherwise ambiguous. With respect to the formal conception of games, there is, after all, consensus; philosophical problems break out in the space opened by differences among informal interpretations.

A *game* in strategic form is a list
$$G = \{N, S, \pi(s)\}$$
where
(a) N is the *set of players*:
$$N = \{0,1,2,3,...,n\}$$
(b) S is the *strategy space* for the players:
$$S = X_{i=0}^{n} S_i$$
where S_i is the strategy set for agent *i*. A strategy is a vector of actions for *i*, and the strategy set is the set of all such vectors.
(c) $\pi(s)$ is a *vector of payoff functions*, one for each agent, excluding player 0. Each payoff function ranks the consequences, for the agent in question, of the strategies for all agents:
$$\pi(s) = (\pi_1(s),...,\pi_n(s))$$
$$\pi: S \to \Re^n.$$

We now introduce a barrage of concepts, in terms of which we will then give our 'interpreted' characterization of the objects defined above, as these figure in conventional microeconomic applications (including applications in evolutionary game theory). The players in a game are *Dennettian agents*. The qualifier 'dennettian' is in acknowledgement of Dennett's careful separation, over a large body of work, of the concept of agency, on the one hand, and the concepts of deliberation and consciousness, on the other (see Ross forthcoming). A Dennettian agent, then, is not necessarily presumed to be either deliberative or conscious. Henceforth, we will refer to Dennettian agents simply as 'agents.' An agent is included among the defined participants in a given game if her acts potentially influence future allocations. The point of saying 'potentially' is to include in the specification of a game anybody who could be involved, even if, in equilibrium, they have nothing to do.[4] An agent i is *rational* iff, given any acts by other agents, each act taken by i is such that it secures the consequence which is most highly ranked for i. This is a fanatically narrow, and strong, definition of rationality by comparison with most philosophical uses. However, for present purposes we insist on using the concept of game theoretic rationality (and of microeconomic rationality in general) as technical concepts. In all and only games involving incomplete information, one agent is indexed by 0. This agent is a generator of probabilistic influences on outcomes, and is known as *Nature*; technically, she is the unique agent in a game that is not rational. Nature's strategies are referred to as *states of nature*. An *outcome* is an allocation of resources determined by the acts of the agents. An agent i has *control* if a change in i's acts is sufficient to change the outcome. A *consequence* for i is the value for i of a function δ that maps outcomes onto the real numbers. δ may be determined using expected utility theory if i is a consumer, or by reference to fitness considerations if i is an agent in an evolutionary game, or by some other non-arbitrary method of ranking outcomes for agents. Given that outcomes are determined by the agents' acts, and given that these acts are specified by their strategies, it follows that the specification of δ together with strategies implies the existence of the vector of payoff functions π as defined above.

We now put this flurry of concepts to work. Consider an initial allocation of resources distributed among Dennettian agents. Then a *game* is a set of acts by one to n rational Dennettian agents, and, possibly, a non-rational Dennettian agent called 'Nature,' where at least one agent has control over the outcome of the set of acts, and where the Dennettian agents are potentially in conflict, in the sense that a Dennettian agent potentially ranks the consequences of each outcome differently from the other agents.

Let us now highlight a few aspects of this characterization that will be relevant to our subsequent discussion. First, it makes no reference, implicit or otherwise, to *beliefs* (or other cognitive states) of agents. In this respect, the foundations of game theory are consistent with the long line of efforts in microeconomics, beginning with Pareto and running through the work of Samuelson, Savage, Afriat and Varian (to mention only the highlights) that aimed at bleaching all substantive psychological hypotheses out of the fundamental axioms. Second, note that our characterization depicts games as *objects*, rather than as *deliberative processes*. This foreshadows the most important respect in which the decision theoretic understanding of games is apt to depart from the standard microeconomic conception. Our analysis is non-committal as to what *sort* of object a game is. Games are clearly among the objects of microeconomic study. But since the concept of a game as we have presented it is formalizable, and since even the concepts introduced in our philosophical interpretation *can* all be construed as strictly technical concepts, our picture to this point is consistent with Rosenberg's (1992) thesis that (the defensible portion of) economics studies mathematical objects. This is not the interpretation we favour, however; we prefer the hypothesis that the objects of (at least) microeconomics are 'Real Patterns' in the sense of Dennett (1991), that is, abstract *empirical* objects. For present purposes, however, this point need not be insisted upon. We need only claim that, as a matter of fact, microeconomists are chiefly concerned with equilibrium properties of games (though we would *explain* and *justify* this concern by appeal to the fact that it is as a result of the stability associated with equilibria that games emerge as Real Patterns). In its capacity as the study of the properties of games, we claim, microeconomics is a positive, and not a normative, science.

The polemical targets of this last claim are not exclusively philosophers. It is very common for authors of textbooks in game theory to gloss the content of the theory as though its aims were ultimately normative. Here are two representative quotations from standard sources:

> Game theory deals with the choices people may make, or, better, the choices they should make (in a sense to be specified), in the resulting equilibrium outcomes, and in some aspects of the communication and collusion which may occur among players in their attempts to improve their outcomes [Luce and Raiffa 1957, 4].

> Its [Game Theory's] aim is to investigate the manner in which rational people should interact when they have conflicting interests [Binmore 1992, vii].

It is not difficult to guess at why such interpretations seem natural, even to authorities on game theory. Our interest in game theoretic analysis often stems from our hope that we may gain some insight into what a rational agent *would* do in the case at hand; and we readily suppose that what a rational agent *would* do is what an hypothetical agent *ought to* do. Though, as we have just seen, this slide is sometimes made by economists, it is particularly inviting for philosophers, who are apt to be more conscious than economists of economics' historical roots in the clearly normative project in political and moral theory that began with Hobbes and was developed by Hume and Bentham. Furthermore, the concerns of the early neoclassical economists appear to be in direct descent from Hobbes'. Classical microeconomics is constantly concerned with "efficiency", in particular the Pareto efficiency, of equilibrium allocations. The results which are most celebrated by economists, and most well known outside the discipline, are precisely those which state that actions by agents who individually lack control over outcomes lead to Pareto optimal states. The new resource that game theory brings to microeconomics is a set of methods that permit us to systematically examine cases in which agents *have* explicit control over the ultimate results of interactions. This involves no shift in the traditional economic conception of rationality, but it does make it unsurprising that game-theoretic analysis does not in general lead to Pareto optimal outcomes. It *ensures* that microeconomics as enriched by game theory will yield what Hollis and Sugden call paradoxes, that is, situations where agents do not achieve socially desirable outcomes by using optimal strategies. In setting the ambitions for game theory that they do, Hollis and Sugden thus guarantee their own disappointment.

Let us begin with a quick reconstruction of the argument that motivates Hollis and Sugden's treatment of microeconomics as intended to serve normative aims. They begin from the premise that either a science is positive or it is normative. If a science is positive, it should be descriptively accurate, have explanatory power and be predictively successful. Microeconomics only appears to have these virtues if its axioms are emptied of empirical content, that is, if they are interpreted as vacuously claiming that agents are rational with respect to some unknown utility function. So it is not successful as a positive science. If it is successful at all, it must therefore be as a normative science.

We have not been careful in this sketch of the argument, because variations on it will be very familiar to philosophers. (For as sophisticated a version of it as has been given, see Rosenberg 1992.) For the sake of argument, let us suspend our disbelief in its third premise and in its conclusion: suppose that microeconomics *is* a normative science. This now serves as a premise for another argument of Hollis and Sugden, which we reconstruct a bit more carefully. We observe that microeconomics is concerned with what rational agents do, and with the consequences of these actions as distributed over interacting communities of agents. Presumably, the relevant normative ambition is to tell us what should happen in a world of *ideally* rational agents; and, presumably, what should happen in such a world should be *good* from the standpoint of reasonableness and elementary moral considerations. Therefore, the prescriptions of economics should lead to outcomes where agents, on the whole and on balance, fare well. 'Faring well' in the economic sense, involves at least Pareto optimality. So prescriptions that agents act rationally so as to reach a Pareto inferior state are paradoxical. But microeconomic theory, and game theory in particular, generate a host of such paradoxes, the Prisoner's Dilemma being the most famous example. Therefore, Hollis and Sugden conclude, microeconomics fails as a prescriptive theory of rationality because, according to its assumptions, "ideally *rational* agents are either paralysed, when *reasonable* people would not hesitate, or are *rationally required* to make choices which *reasonable* people would reject" (Ibid, 2). The economic concept of rationality thus fails to capture the properties of rationality that (should?) most concern us. We are therefore motivated to change it; and empirical moral psychology stands as the most promising source of enrichment.

Hollis and Sugden reveal much about their attitude toward the aims of game theory when they say that

> there are two schools of thought about its [game theory's] purpose. For any given game, one might expect the theory to answer the question, "What is it rational for the players to do?" (or "What is it rational for them to believe?"). Some game theorists try to answer these questions, but others merely look for *equilibria* [Ibid, 9].

The difference between our view and theirs is all implied in that 'merely'. Equilibria differ from non-equilibria by virtue (among other things) of a certain special kind of stability: once a community of strategically interacting agents find themselves using equilibrium strategies, no one agent would be motivated to change her plans were she given the opportunity to do so. It is therefore often very useful and informative to identify equilibria - useful and informative enough to justify the existence of a non-prescriptive science of microeconomics. Evidently, Hollis and Sugden disagree. We are not content to declare a stand-off at this point, however. Instead, we direct attention to another aspect of the last quotation: Hollis and Sugden's suggestion that game theory can be interpreted as telling agents what to do or as telling them what to *believe*. This, we will argue, is a mistake, and not an innocent one.

What sorts of beliefs might game theory be taken to prescribe? These can't be beliefs as they figure within expected utility theory. In one of the celebrated achievements of twentieth century microeconomics, Savage (1954) showed how the preferences of an agent can be expressed in functional form so as to uncover her subjective beliefs about the state of nature and her ranking of consequences. The economist offers no comment on, and holds no theory about, the sources or justifications of these beliefs. There can be no justification in this context for speaking of dictating beliefs to agents; nor, given the relevant conception of instrumental rationality that is at issue, can any meaning be attached to a distinction between 'rational' and 'irrational' beliefs. The beliefs to which Hollis and Sugden refer, then, must be beliefs about the plans of the other players in the game (excluding Nature). We thus interpret Hollis and Sugden as suggesting that game theory should tell one what to believe about what other players are going to do. Given these sorts of beliefs, the idea appears to go on, one can choose the strategy that maximizes expected utility. Thus they identify game theory as seeking to provide a "universal logic of practical reason" (Ibid, 32). Quoting

Hobbes, they say that deployment of this logic should be "the pace which leads to the benefit of mankind" (Ibid, 20). A game, for Hollis and Sugden, is simply a decision problem where the interaction of agents must be taken into account; that is, it is a decision problem in a strategic setting. Thus game theory is taken to derive its significance from the service that it can provide to decision theory, and if it cannot provide such service, this must be because it provides an *incorrect* logic of practical reason.

If game theory is to tell us what to do, it must do this by way of its notion of equilibrium. And if, in turn, this is equivalent to telling us what to believe, then it must be possible to interpret some aspect of Nash equilibrium as making reference to actions and/or to beliefs. Consider Hollis and Sugden's informal definition of Nash equilibrium:

> The core notion of equilibrium in game theory is *Nash equilibrium*. The strategies of two players are in Nash equilibrium if each is a *best reply* to the other - that is, if each player's strategy maximises his expected utility, given the other player's strategy [Ibid, 9].

This conforms to the standard definition of Nash equilibrium, with one caveat. A pair of strategies *are* a Nash equilibrium; they are not *in* Nash equilibrium. In light of Hollis and Sugden's general interpretation of game theory, we suspect that this is not a casual slip. Saying that two strategies are *in* Nash equilibrium suggests that they are in a certain sort of *state*, or in a relationship to each other that allows Nash equilibrium to obtain. This suggestion is confirmed by what Hollis and Sugden then go on to say:

> Such pairs of strategies are equilibria in the general sense that *players' beliefs are mutually consistent*. As a rational agent, each player maximises his expected utility, given *his beliefs about* what the other player will do. But in Nash equilibrium specifically, a player's action is also expected utility maximising in relation to what the other player *actually does* [Ibid; our emphases].

In summary, then, Hollis and Sugden's interpretation seems to go as follows. What it means to say that an agent is rational is that she maximizes expected utility given her beliefs. Her beliefs concern the plans of the other players. In equilibrium, these beliefs must be correct. Therefore, the equilibrium must prescribe both the correct actions and the correct beliefs.

In fact, however, a Nash equilibrium is *by definition* a vector of strategies which is such that agent i gets a higher payoff from playing her equilibrium strategy, *given the strategies* of all other agents, than she could earn by the use of any other strategy; and this is true for all such agents i. A Nash equilibrium is *not* a pair (actions, beliefs). A vector of strategies specifies more than just the acts taken in equilibrium, because it specifies all acts which could potentially have been taken for any play of the game. Describing this vector thus involves no direct reference to beliefs. One might imagine, however, that such reference, while not explicit in the definition of equilibrium, is implicit. Could beliefs be hiding in the concept of the payoff function? This function, as we have seen, links vectors of strategies to numbers which rank the consequences of outcomes for agents. When Nature is a participant in the game, any combination of strategies yields a lottery over outcomes, and the probabilities given in these lotteries are objects of belief; but they concern the states of nature, not the plans of the other rational agents, and they are shared by all of the rational agents. If players hold beliefs about each other, these could only be that each agent i believes that, with probability one, the other agents are using their equilibrium strategies. Speaking this way will cause no trouble, but only because it locates beliefs at a point where they can do no work in the theory. When the agents in a game play an equilibrium vector of strategies, each agent i believes that, with probability one, this is indeed what the players are doing. So we may multiply the agent's utility on the certain outcome by one and call it expected utility. Re-scaling units by a factor of one cannot be significant. But this is not the main problem. Recall that, on Hollis and Sugden's conception, the beliefs which figure in the equilibria of games are supposed to be the beliefs which the agent *should* hold about what the other agents will do. So now imagine an agent, poised to participate in a game with two Nash equilibria. What can game theory tell this agent with respect to her beliefs? That she *should* believe that with probability one the other agents will act as in the first equilibrium *and* that she should believe that with probability one they will act as in the second equilibrium? This obviously makes no sense.

We summarize this part of our criticism as follows. Either the statement of a Nash equilibrium does not specify beliefs at all, or the beliefs it specifies have as their contents expectations of events with probability one. In the latter case, in equilibrium, all players hold common and sure beliefs about what everyone else will do. These beliefs cannot involve agent i's expectations about what the other players might do *under any possible circumstances*; they can only represent i's beliefs about the other players *given* everyone's strategies. Thus if beliefs figure in Nash equilibria, then the contents of these beliefs cannot be independent of the strategies of the players. Since we can, therefore, ascribe them only once we have fixed the game, they *cannot* serve as useful prescriptions, nor can they themselves be prescribed on the basis of the resources of game theory. So either the identification of a Nash equilibrium does not specify beliefs in the first place, or it cannot prescribe the beliefs it specifies.

We thus agree with Hollis and Sugden that game theory is not the salvation of decision theory. But given the *correct* conception of a game, this is not an objection to the microeconomic conception of rationality that game theory incorporates. For Hollis and Sugden, games are *decisions* about how to act. By contrast, on our account games are simply sets of acts. The sets that most interest us are the sets in equilibrium, that is, the sets with the property that each agent's plan secures the best consequence for her given the acts of the other agents. This interest in equilibria is *not* motivated by a conviction that these sets are morally or practically best, but by the fact that they are stable, and that games, as Real Patterns in the world, tend to converge on them. In our notion of a game, there is no room for talk of agents' beliefs about the strategies of other agents, because according to us microeconomics is not concerned with the specific processes by which agents *arrive* at equilibria; and at equilibrium, all agents' strategies are given. Note that this is consistent with our deliberate specification of the agents in games as *Dennettian* agents, that is, as not necessarily capable of deliberation. Games converge on equilibria due, in every particular case, to some sort of selection; but many forms of selection, including natural selection and market selection, operate through invisible hands.

Hollis and Sugden might object that our criticism of their account rests on our having changed the subject. That is, they might protest that their target is the microeconomic project as it has actually been practised, and that they are not responsible for ensuring that their arguments also apply to every possible reconceptualization of that project. We would reply, however, that our conception is at least as consistent with the relevant intellectual history as is theirs. In games as we conceive them, the use of expected utility theory conforms exactly to the role envisaged by von Neumann and Morgenstern (1947), and developed by their successors. This is not true of the use of the expected utility concept required by Hollis and Sugden. Players in their games hold subjective beliefs about what other players will do, and this would force the game theorist to treat both states of Nature *and* the strategies of agents other than i as *events*, that is, as elements in the lotteries over which i maximizes her expected utility. This is explicitly inconsistent with Savage's restriction of the relevant concept of an event as being a state of Nature, because the strategies of other agents are objects of choice which themselves depend on the strategy choice of player i. Hollis and Sugden seem to recognize this, when they comment that

> Savage's axioms are not designed to apply to games against rational opponents; for Savage, 'events' are states of nature ... [I]t is far from clear that Savage's axioms can provide an interpretation of subjective probabilities within games [Ibid, 24].

However, because they take for granted that game theory *must* seek to describe and explain the decision problems of expected utility maximizers, they have no choice but to depart from Savage's intentions, and from microeconomists' practice.

We close with some reflections on interdisciplinary relations. Decision theory is a branch of inquiry over which philosophers can reasonably claim proprietary rights, but game theory is held in joint custody. We have agreed with Hollis and Sugden that game theory in particular, and microeconomics in general, will probably not be able to serve the purposes of philosophers in the way that they have often hoped; but it is philosophical arrogance to suppose that there must therefore be something wrong with the foundations of microeconomics. We do not charge Hollis and Sugden *personally* with arrogance. However, their *view* is arrogant, and in this it seems to us to resemble a good deal of the literature in the philosophy of economics. Some philosophers of science (e.g., Cartwright 1983) have recently come to recognize that philosophers have often distorted their views of various sciences by regarding them as though they were mainly servants of philosophical projects. The philosophy of economics can only be improved by this sort of modesty.

Notes

[1] Thanks to Paul Dumouchel, Maurice Lagueux, Paisley Livingstone, Robert Nadeau, and the other members of the Groupe de Recherche en Épistémologie Comparée at l'Université du Québec à Montréal, for their helpful criticism.

[2] For a full summary of the larger project, see Ross and LaCasse (1993).

[3] For example, most of the leading surveys of the philosophy of economics (e.g., Blaug 1980, Caldwell 1982), including very recent ones (Redman 1991) take as their central problem the question of whether economics has the structure of a successful Lakatosian research program.

[4] For example, consider a game involving a monopolist and a potential market entrant. The potential entrant is included in the game even if in equilibrium she does not enter the market.

References

Binmore, K. (1992), *Fun and Games*. Lexington: D.C. Heath.

Blaug, M. (1980), *The Methodology of Economics*. Cambridge: Cambridge University Press.

Caldwell, B. (1982), *Beyond Positivism: Economic Methodology in the Twentieth Century*. London: Unwin.

Cartwright, N. (1983), *How the Laws of Physics Lie*. Oxford: Oxford University Press.

Dennett, D. (1991), "Real Patterns", *The Journal of Philosophy* 88:27-51.

Dupré, J. (1993), *The Disorder of Things*. Cambridge: Harvard University Press.

Hausman, D. (1992), *The Inexact and Separate Science of Economics*. Cambridge: Cambridge University Press.

Hollis, M. and Sugden, R. (1993), "Rationality in Action", *Mind* 102:1-35.

Luce, R., and Raiffa, H. (1957), *Games and Decisions*. New York: Wiley and Sons.

Neumann, J. von, and Morgenstern, O. (1947), *Theory of Games and Economic Behaviour*. Princeton: Princeton University Press.

Redman, D. (1991), *Economics and the Philosophy of Science*. Oxford: Oxford University Press.

Rosenberg, A. (1992), *Economics: Mathematical Politics or Science of Diminishing Returns?* Chicago: University of Chicago Press.

Ross, D. (forthcoming), "Dennett's Conceptual Reform", *Behavior and Philosophy* 21.

Ross, D., and LaCasse, C. (1993), *Toward a New Philosophy of Positive Economics*: Working Paper #9324, Groupe de Recherche en Épistémologie Comparée, Université du Québec à Montréal.

Savage, L. (1954), *The Foundations of Statistics*. New York: Wiley and Sons.

Ceteris Paribus Laws and Psychological Explanations[1]

Charles Wallis

University of Rochester

1. Introduction

Two important components of Fodor's philosophy of cognitive science rely heavily upon ceteris paribus laws, thus requiring that Fodor forward an analysis of ceteris paribus laws. First, because Fodor has long held (1975, 1990) that *all* of the laws of the special sciences are ceteris paribus laws, he must demonstrate the scientific legitimacy of ceteris paribus laws. Second, ceteris paribus laws play a prominent role in Fodor's most recent (1990) theory of representational content, and solution to the "disjunction problem," a well-known problem for his approach. Fodor recently (1991a) offered an analysis of ceteris paribus laws for the special sciences (particularly cognitive science). I ask two questions with regard to Fodor's analysis: Does the analysis underwrite Fodor's theory of representation? Does the analysis provide adequate solutions to the problems for ceteris paribus laws as articulated by Hempel (1965, 1988) and Giere (1988)? I answer both questions negatively.

2. Nomic Covariance

Before turning to Fodor's analysis of ceteris paribus laws, let us review the philosophical difficulties leading Fodor to forward his analysis: Fodor is the principle spokesperson and architect of a tradition in the philosophy of cognitive science. This tradition holds that cognitive science will result in the formulation of universal, conditional "...intentional laws that connect, for example, states of believing that $P \& (P \rightarrow Q)$ to states of believing that Q." (Fodor 1990, 145) The tradition further asserts that such intentional laws are available within folk knowledge of human mentality or easily extrapolated therefrom:

> ...commonsense intentional psychology should be construed as a prototheory of the etiology of behavioral comports with the idea that commonsense intentional explanation/prediction is (at least implicitly) at least from time to time a species of explanation/prediction by law subsumption. (Ibid. 1991a, 19)

Within the Fodorian tradition, the main problem facing philosophers is generating a naturalized theory of representation underwriting the assignment of values to variables in covering laws like the one above. Though Fodor has modified his position over the years, he continues to play a significant role in developing one such theory of representation, nomic covariation. Nomic covariation assigns contents to states as follows:[2]

A state, S_b, represents Bs as Bs iff the system tokens a S_b when, only when, and because instances of B are present.

For example, S_b represents barns as barns according to nomic covariation iff the system tokens a S_b when, only when, and because of the presence of instances of barnhood.

However, nomic covariation fails to underwrite univocal content ascriptions in that ostensive error also determines content: Even in cases where one explicitly thinks about barns, one might occasionally mistake a barn facsimile for a barn. Intuitions urge that in such cases one has misrepresented the facsimile as a barn, but nomic covariation forces one to construe the ostensive error as a veridical representation of a disjunctive property. When a barn facsimile causes a S_b, the "only when" and "because" of covariationist definition dictates that S_b correctly represents the disjunctive property that one might express as "barn or barn facsimile," hence, the moniker "disjunction problem."

As a result of the disjunction problem, one also finds within the Fodorian tradition attempts to bolster nomic covariance by appealing to ideal perceptual conditions (e.x., Fodor 1990a). The idealized laws of mainstream science inspire and provide the tacit support for the nomic covariationist's idealization strategy: In normal cases of scientific idealization theorists abstract from various negligible parameters in real systems to formulate laws which, though not strictly true of any actual system, predict adequately and quantify a real relation in actual physical systems.

Two examples clarify the strategy behind mainstream scientific idealization: Real gases are not composed of Newtonian molecules; point masses entering into perfectly elastic collisions. Nor is heat the only energy source in gases. However, because other forces (like gravitational forces) prove negligible in many cases, one can idealize away from them to quantify the inverse relationship between the pressure and volume of a gas and the temperature and number of moles: **$PV=nRT$**.[4]

Real pendulums have variable frictional resistance in their arm pivot and air resistance against their arms and bobs. However, as frictional and air resistance decrease, a pendulum's periodicity comes to vary inversely with the square root of the arm length divided by gravitational acceleration. Hence, one idealizes from frictional and air resistance to quantify a real relationship between arm length and periodicity in a manner that proves predictively adequate:[3] **$T=2\pi(L/g)^{1/2}$**.

The covariationists' use of idealization in response to the disjunction problem professes to emulate noncontroversial cases of scientific idealization like the pendulum and ideal gas laws. Idealized nomic covariation defines representation as follows:

A state, S_b, represents Bs as Bs iff *under ideal circumstances* the system tokens S_b when, only when, and because of the presence of Bs.

According to the above approach, S_b represents barns as barns since, under ideal circumstances, the system tokens S_bs when, only when, and because of instances of barnhood. S_b does not represent *barn or barn facsimile* because barn-facsimilehood will not cause the system to token S_bs under ideal conditions. Such an idealization from error, admits the covariationist, breaks down. Nevertheless, the idealization allows psychological laws to capture a real relation (the representation relation) in actual systems. Moreover, since most cases approximate ideal circumstances, psychological laws utilizing idealized nomic covariation predict adequately.

Covariationist idealization divides tokenings of a state by a system into content-imbuing (ideal) and content-fixed (malfunction/atypical) classes. One legitimately idealizes away from error, claims the covariationist, because cases of error are coextensive with the cases of malfunction and/or atypical situations from which science legitimately idealizes: A pendulum in which the arm pivot is rusted tight does not swing. Such pendulums do not undermine the idealized pendulum law because circumstances prevent the pendulum from operating *qua* pendulum. The same holds for cognitive malfunctions: One's tokening of a S_b as a result of a hammer-blow does not introduce disjunctive content, as hammer-blows cause one to malfunction.

Of course, no malfunction occurs when one tokens a S_b as the result of a barn facsimile. The features prompting one to token a S_b for a real barn prompt one's facsimile tokenings—that is the whole point of facsimiles. Dismissing such cases as malfunctioning requires one also dismiss barn cases as malfunctions, making all cognition into malfunction. The covariationist must respond differently to cases like barn facsimiles: At pressures greater than atmospheric pressure, or at high temperatures, forces ignored by the idealized gas law become non-negligible. The predictive accuracy of the idealized gas law plummets. The gas does not malfunction. Yet, these cases do not count against the idealized gas law because circumstances are atypical. The case violates the assumptions of the idealization, thereby falling outside the intended application of the law. In like manner, claims the nomic covariationist, circumstances are non-ideal when the system tokens a S_b in response to a barn facsimile.

The covariationists' move looks suspiciously circular unless they specify a means of ruling out barn facsimile cases as atypical: One must have a reason for labeling barn facsimile cases as less than ideal, and the reason cannot be that barn facsimiles cause barn tokens (S_bs). In other words, one must define ideal perceptual conditions so that malfunctions and atypical circumstances prove coextensive with cases of error, and one cannot avail oneself of the notion of error, nor other intentional or semantic notions in formulating and motivating the definition. After all, given our environment, one might plausibly motivate an idealization about gases that holds up to atmospheric pressure. But, why call cases resulting from ostensively *normal* functioning that may well *commonly* occur in actual cognitive environments either malfunctions or atypical circumstances?! One intends a theory of representation to explain both cases of successful performance of a cognitive task using veridical representation *and* cases of unsuccessful performance traceable to misrepresentation. Unlike gas at high pressure, the barn facsimile case, a failure to identify barns as barns due to misrepresentation, is within the intended application of the (representational) theory.

Difficulties with disjunctive content and ideal perceptual conditions have lead Fodor to modify his definition of representation. Fodor now (1990) holds that the following constitute sufficient conditions for "X"s to represent Xs:

1 'Xs cause "X"s' is a [ceteris paribus] law.

2. Some "X"s are actually caused by Xs.

3. For all Y not=X, if Ys qua Ys actually cause "X"s, then Ys *causing* "X"s is asymmetrically dependent on Xs *causing* "X"s. (Fodor 1990, 121)

Fodor claims to avoid the problems with circularity plaguing idealized nomic covariation by appealing to ceteris paribus laws instead of ideal conditions.[5] However, switching from ideal conditions to ceteris paribus laws should not silence critics unless it actually solves problems for Fodor's theory of content while avoiding the liabilities of ideal conditions. Moreover, Fodor's appeal to ceteris paribus laws requires that he address the traditional problems with ceteris paribus clauses.

3. Traditional Problems with Ceteris Paribus Clauses

I have discussed idealization in mainstream science as if one finds no problems associated with the traditional view of laws as universally generalized conditionals in closed deductive systems, and the fact that scientific laws have ceteris paribus clauses.[6] There are a number of serious problems. These problems are of crucial importance to Fodor because he relies upon the notion of ceteris paribus lawhood in defining representation, and because he holds that *all* of the laws of the special sciences are ceteris paribus laws. If Fodor cannot find a way to understand ceteris paribus laws as scientifically legitimate, he stands to lose a great deal.

Hempel recognizes that specifying ceteris paribus clauses for laws like the ideal pendulum law proves more complicated than portrayed in the previous section. Hempel would ask one to consider a pendulum with an iron bob swinging over a magnet. Assuming the magnet exerts a strong force, the pendulum law proves radically false. Hempel's initial response is to claim that laws have implicit "provisos." Such laws have the following form (**P**=proviso, **I**=is an instance of the system, and **R**=has the lawlike relation):[7]

$$(x)((Px \ \& \ Ix) \rightarrow Rx)$$

Hempel found three problems with this approach: (1) Accepting the above notion of laws requires one accept that laws are essentially inexpressible, as one will not be able to list, much less anticipate, all of the provisos for a given law. The inability to list all provisos is problematic not only because scientific laws appear to be finite expressions, but also because the standard deductive notion of prediction and explanation depends upon laws forming closed deductive systems. One cannot have a closed deductive system if one cannot write down one's axioms. (2) One needs a mechanism for discriminating allowable from unallowable provisos to prevent laws from becoming empirically vacuous. Absent constraints on admissible provisos one might deal with each anomalous case by introducing another proviso. (3) The language of the theory, even the science of the time, may prove incapable of expressing provisos. Magnetic forces, for example, were neither expressible nor explainable within the classical mechanics of the pendulum law. This poses a problem both for understanding how the acceptance of idealizations could be a rational process, and in relation to Hempel's first problem (expressing all provisos within laws, i.e., having a closed, deductive system).

4. Fodor on Ceteris Paribus

The above problems are very serious. Some philosophers (ex. Giere 1988) argue that the above difficulties provide another reason to reject the notion of scientific laws and theories as embodied in logical empiricism. Fodor, in contrast, intends his analysis to render his appeal to ceteris paribus laws unproblematic. One ought ask three questions regarding the analysis: (A) Does the analysis provide a plausible rationale for accepting the ceteris paribus laws that Fodor needs for his definition of representational content, i.e., univocal laws relating properties to particular brain state types [e.x. $cp(\text{Barn} \rightarrow S_b)$]? (B) Does the analysis exclude the disjunctive laws that pose problems for Fodor's definition of representation [e.x. $cp((\text{Barn v Barn Facsimile}) \rightarrow S_b)$]? In order for Fodor's analysis of ceteris paribus lawhood to help in the dissolution of the disjunction problem, it must disallow disjunctive ceteris paribus laws. (C) Does the analysis provide adequate solutions to Hempel's problems? In other words, does it provide the basis of a thoroughgoing understanding ceteris paribus laws as scientifically legitimate?

Fodor (1991a) claims that ceteris paribus laws can have mere exceptions, like the rusted pendulum, as well as exceptions which are genuinely anomalous. In other words, Fodor claims that ceteris paribus laws might have exceptions not covered by any possible Hempelian proviso. In Fodor's terms, a ceteris paribus law, ex. $cp(A \rightarrow B)$, can have "Realizers" for which there are "Completers" (mere exceptions), and, in certain cases, realizers for which there are no completers (absolute exceptions); where Fodor defines completers as follows:

Let $A(R_i)$ be an event type in which A is realized by R_i. Let C be an arbitrary event type. Then C is a *Completer* relative to a realization of A by R_i if and only if:

i. $A(R_i)$ & C is (strictly) sufficient for B.
ii.   It's not the case that $A(R_i)$ alone is sufficient for B.
iii.   It's not the case that C alone is sufficient for B. (1991a, p.23)

Fodor allows absolute exceptions only when one finds "many *other* laws in the network" for which the instantiation (realizer) is not an absolute exception; where the relevant network for a law is the set of laws containing the antecedent of the law within their antecedent. (1991a, 28)

Does the above analysis underwrite the laws Fodor needs (A)? 'Barn $\rightarrow S_b$' seems a bad candidate for ceteris paribus lawhood. Fodor insists that an anomalous exception cannot violate most or all of the laws in the network. So, absent many other unviolated laws in the network, an absolute exception, E, to Barn $\rightarrow S_b$ refutes its claim to lawfulness. There are absolute exceptions (e.x. barns that do not cause an S_b). But, if E is an absolute exception to Barn $\rightarrow S_b$, one finds few, if any, laws having barnhood in their antecedent, and for which E is not an absolute exception.[8] First, as one finds no actual barns in the head, one finds no other psychological laws governing the system that have barnhood as their antecedent. When one tries cooking up other laws, these new laws seem derivative upon the Barn $\rightarrow S_b$ connection. Such derivative laws have absolute exceptions whenever the original law has them. Consider the following:

$cp($Barn $\rightarrow R_b)$ (report of barn)
$cp($Barn $\rightarrow M_b)$ (belief about barn stored in memory)

The above "laws" depend upon one's ability to first token a barn representation. After all, who reports seeing a barn when they do not seem to see a barn? Who remembers seeing a barn when they cannot recognize the barn as a barn?

The derivative nature of the above "laws" does not demonstrate the impossibility of a legitimate network of laws. Nevertheless, for Fodor's analysis to be plausible, he must demonstrate the plausibility of such networks, especially since unviolated legitimate networks look *prima facie* implausible. Without such a demonstration, Fodor's analysis of ceteris paribus clauses fails to support the psychological laws that Fodor needs for his theory of representation. I conclude that Fodor's analysis fails to provide a rationale for accepting the needed laws.

Does the analysis exclude the disjunctive laws Fodor wants to shed (B)? I claim it does not. A ceteris paribus conditional with a large disjunctive antecedent seems a preferable candidate for ceteris paribus lawhood because it has a wider domain of application: Just as one prefers the law of universal gravitation to a law quantifying gravitational attraction between two particular objects, one prefers the disjunctive law to the univocal law.

One might argue that ceteris paribus conditionals with large disjunctive antecedents, having more ways to satisfy their antecedents, seem more likely to have anomalous exceptions, making them worse candidates for lawhood.[9] Fortunately, any network including Barn $\rightarrow S_b$ and its relevant "many other laws" would also include laws like: cp((Barn v Barn Facsimile) $\rightarrow M_b$), cp((Barn v Barn Facsimile) $\rightarrow R_b$), and etc.. If Barn $\rightarrow S_b$ survives anomalies by appeal to a network of laws, the disjunction and alternative disjunct laws should also survive. The danger of refutation via absolute exception is no greater than for univocal laws. Fodor's analysis, therefore, fails to shed the unwanted disjunctive laws. The result is not the needed idealization away from error required by the disjunction problem. Appealing to trivial ceteris paribus laws makes lawful covariance ineffective against the disjunction problem in that it makes ceteris paribus laws relating each disjunct (and the disjunctive property) to the state equally legitimate. What one represents now depends upon the ceteris paribus law one cites.

I conclude that, in the worst case scenario, which seems *prima facie* likely, Fodor's analysis fails to underwrite the legitimacy of the ceteris paribus laws he requires. In the best case scenario, Fodor's laws become legitimate only if their unwanted rivals also become legitimate. In either case, Fodor's analysis fails to aid in the dissolution of the disjunction problem.

Fodor might accept the latter result, claiming that he needs only the relevant ceteris paribus law, and that the asymmetric dependence clause in his definition of representation resolves difficulties with error and the disjunction problem. Given that scenario, Fodor must rely upon his asymmetric dependence clause to rule out disjunctive content. I reject the move for two reasons: First, I have argued (1992 and 1993) that contemporary theories of vision in cognitive science do not exhibit an asymmetric dependence of the misrepresented upon the represented. Appeals to a nonexistent asymmetric dependence will not resolve the disjunction problem. Second, Fodor appeals to ceteris paribus laws to resolve the disjunction problem when discussing asymmetric dependence (1991, 302-3).[10] Appealing to asymmetric dependence to resolve the disjunction problem when discussing idealization is, therefore, circular.

Fodor's analysis fails to make needed laws legitimate while prohibiting unwanted laws. Does it address Hempel's problems (C)? Fodor's analysis would solve problems (1) and (3) by construing "ceteris paribus" as an operator attached to the law, not as an additional antecedent within the law. One could then write the law in finite form and, provided one has a semantics for the operator, understand the law as a universal conditional. I find many problems with this solution. To start, Fodor's analysis explicitly appeals to networks of laws, which must be, on Fodor's own supposition, constituted of ceteris paribus laws. (Fodor 1990, 157) Fodor's analysis proves circular, appealing to ceteris paribus laws to analyze ceteris paribus laws. In response, Fodor suggests that some networks can contain a mix of strict laws (exceptionless, mainstream scientific laws) and ceteris paribus laws, and that one can appeal to the notion of a strict law to recursively define a network. (1991a, 31-32) Even if Fodor calls upon the notion of a strict law, it fails to help him. Hempel, after all, denies the existence of strict laws. Hence, Fodor's solution to Hempel's problems presupposes a solution to Hempel's problems. Fodor retreats at this point to nomic necessity (1991a, 32-3), claiming that his theory of ceteris paribus laws

...takes the notion of nomological necessity per se for granted. So, I'm not trying to say what laws are; only how hedged ones differ from strict ones. (1991a, p22)

Fodor's retreat is circular yet again. As Fodor must realize, the most plausible analysis of nomological necessity (Lewis 1973, 1986) is in terms of laws. Appeals to nomic necessity only serve to mask an appeal to laws. Fodor might reject Lewis' notion of nomic necessity, choosing to define the notion himself. However, he would be wise to consult Wesley Salmon, (1976 and 1990) who convincingly argues that the notions of laws, modalities, and counterfactuals are so tightly intertwined that any plausible analysis of one will likely invoke at least one of the others.

Fodor's analysis also requires that one *count* the unviolated laws in the network. However, theories and laws have no unique characterization.[11] Equivalent characterizations can have variable numbers of laws sharing an antecedent. For instance, the Ramsey sentence of a theory is a single conjunctive law. Hence, whether an absolute exception undermines a ceteris paribus law, i.e., whether one finds sufficient numbers of unviolated laws, is underdetermined by Fodor's analysis.

Furthermore, Fodor fails to provide an adequate semantics for the ceteris paribus operator. Fodor seems to intend the ceteris paribus operator to be a modal operator, since he defines the truth conditions for the operator, in part, by reference to counterfactuals:

$cp\phi$ is true iff in every accessible world either ϕ holds, or $\exists C$ such that $(C \rightarrow \phi)$ holds, or ϕ does not hold, but many other laws in the network having ϕ's antecedent hold.

The definition itself is deviant, defining the truth of $cp\phi$, not merely in terms of the truth of ϕ in other accessible worlds, but also in terms of unspecified other sentences to which its antecedent can be conjoined, or with which it shares an antecedent. One would normally define the truth conditions for $cp\phi$ in terms of the truth of ϕ across accessible worlds.

In my mind the above deviation renders Fodor's semantics no semantics at all. Even allowing the deviation, Salmon (1990, 130) has offered two objections to attempting to understand laws in terms of modal operators, both of which pose unanswered problems for Fodor's analysis. Salmon first notes that the most popular account of modal semantics (Lewis 1973, 1986) requires the existence of a large number of possible worlds existing independent of the actual world. Such a semantics, notes Salmon, would appeal to the superempirical to provide content for empirical laws. Salmon also points out that one evaluates the accessibility or similarity between possible worlds in terms of laws. So, the semantics for the *cp* operator would appeal to laws, an ineffectual move in the context of Hempel's problems, and dubious on Salmonian (1976, 1990) grounds. Worse yet, Fodor offers no reason why ceteris paribus laws would not play a role in determining accessibility, thereby introducing another circularity.

Finally, Fodor lacks the relevant rules of inference by which he would augment the logic of the basic empiricist view. That is, "$cp(\text{Barn} \rightarrow S_b)$, Barn $\therefore S_b$" is not a deductive inference, as the truth of the premises fails to guarantee the truth of the conclusion. Fodor's suggestion of "$cp(\text{Barn} \rightarrow S_b)$, Barn $\therefore cpS_b$" (1990, 158) fails since Fodor defines truth conditions for *cp* only for conditionals.

I conclude that Fodor's analysis fails to solve Hempel's first and third problems. Additionally, Fodor's analysis provides no means to distinguish allowable from trivializing certeris paribus clauses (problem 2).[12] Suppose, for instance, that one has a law concerning heat with the following two consequences: First, it dictates that the warmer of two contiguous bodies warms the other. Second, it dictates the warmer of two bodies of equal weight warms a (colder) third body the most when contiguous. If one has one kilogram of green wood at 55°C, a kilogram of iron at 49°C, and a kilogram of water at 43°C, the first dictate ranks the objects from warmest to coldest; wood, iron, water. The second dictate ranks them water, wood, and iron.

It is consistent with Fodor's analysis that one keep one's law in the face of this anomaly by appealing to each object's specific heat. Water, having a greater heat capacity in relation to the temperature differences, has the greatest capacity to warm (the greatest heat energy). Hence, one can explain the second ranking. On the other hand, either iron or wood will warm water, as it has the lowest temperature, explaining the first ranking. One need not abandon one's law, one merely adds a ceteris paribus operator. Fodor's analysis allows such a ceteris paribus law, because (i) the temperature and the proviso against the interference of specific heat are strictly sufficient for the temperature rankings, (ii) it is not the case that the temperature alone is sufficient for the rankings (as specific heat may interfere), and (iii) specific heat absent temperature is not sufficient for the rankings.

Fodor's analysis allows one to continue to hold one's heat law. This, of course, is scientifically perverse. Once one realizes that different objects have different specific heats and hence different amounts of heat energy when at the same temperature—once one realizes that "warm" means different things for different objects—one has good reason to abandon the law, not merely to prefix it with a ceteris paribus operator. Fodor's analysis fails to distinguish trivializing from substantive provisos.

5. Conclusion

Fodor's analysis of ceteris paribus laws fails on three crucial counts. It fails to provide a rationale for accepting the ceteris paribus laws required by Fodor's theory of content. It fails to exclude laws that pose problems for Fodor's theory of content. Finally, it fails to solve the traditional problems associated with a thoroughgoing understanding of ceteris paribus clauses. As a result, Fodor's analysis proves ineffectual against the problems that have plagued his position since its inception.

Notes

[1] I am indebted to the Mellon foundation, National Endowment for the Humanities, and to colleagues at the University of Rochester and the University of Arizona.

[2] Cummins (1989) and Fodor (1988).

[3] P=pressure, V=volume, n=number of moles, R=gas constant, and T=temperature (kelvins). Van der Waal's equation, $(P + a/V^2)(V - b) = RT$, introduces experimentally determined constants; b, for available volume in the cylinder given the size of the gas molecules, and a for reduced pressure due to attraction between molecules. Nevertheless, no single law is accurate for the entire range of pressure and temperature for any material.

[4] T=period, L=arm length, and g=gravitational attraction ($9.80 m/s^2$).

[5] As detailed later (see note 9), Fodor's appeal to asymmetric dependence requires he accept Burgian disjunctive scenarios immune to considerations of asymmetric dependence. In these cases, Fodor appeals to ceteris paribus laws. Fodor asserts that his theory illustrates "...how ceteris paribus laws can do serious scientific business, since it captures the difference between the (substantive) claim that Fs cause Gs ceteris paribus, and the (empty) claim that Fs cause Gs except when they don't." (Fodor 1990, 152) "I misrepresent relative brightness because the conditions violate an unspecified ceteris paribus clause," is as uninformative as "except when they don't," as becomes obvious.

[6] Thanks to Giere for his suggestions here and elsewhere. Hempel (1965 and 1988) and Giere (1988) do not exhaust this debate.

[7] Magnets do not exhaust the missing provisos for the pendulum law. One must introduce provisos covering cases in which the pendulum platform oscillates, cases where the bob is half full of water, i.e., the center of gravity is no longer a fixed point mass, cases where the pendulum arm is elastic, and etc..

[8] Fodor is ambiguous regarding the make-up of networks. I interpret Fodor as holding that only laws governing the system can constitute the network. Fodor might allow that *any* law from any science can be part of a given network. Sociological laws about barns would then be relevant to $cp(\text{Barn} \rightarrow S_b)$, making for trivial progress: Psychologists could merely issue laws with photons as part of their antecedent. These laws could withstand widespread anomalies because of the success of physics.

[9] On many accounts, ex. Popper, science aims for laws that have a greater chance of falsification.

[10] In the (1991, 302-3) passage Fodor acknowledges that asymmetric dependence requires that he accept that a single belief could manifest itself in an error not asymmetrically dependent upon correct tokenings (pepsi for water). He attempts to stave off disjunctive content by denying the existence of a legitimate disjunctive ceteris paribus law. "...consider the following (slight) difference between me and my almost-Twin; he thinks "You never find water in Pepsi bottles" and I don't think this. So then there are likely to be environments...in which he would have 'water' thoughts where I would not. ... ...suffice it to say that you get some room to wiggle if you think of the denotation in terms of the nomic relations among properties...you might then argue that the Pepsi guy and the non-Pepsi guy are both subsumed by a water → 'water' law despite postulated differences between them. (remember these are ceteris paribus laws,...."

[11] Lewis (1973) and Salmon (1990) acknowledge that nonunique characterizations pose a problem for possible worlds semantics. Note too that Fodor's extremely lax notion of lawhood would likely license an countable infinitude of laws, making it rather tough to evaluate anomalies.

[12] Churchland 1979, 23-4.

References

Churchland, P. (1979), *Scientific Realism and the Plasticity of Mind*. Cambridge, England: Cambridge University Press.

Cummins, R. (1989), *Meaning and Mental Representation*. Cambridge, MA: MIT Press.

Fodor, J. (1974), "Special Sciences," in *Synthese*. 28: 97-115.

_ _ _ _ . (1988), *Psychosemantics*. Cambridge, MA: MIT Press.

_ _ _ _ . (1990), "Theory of Content, II: The Theory," in *A Theory of Content and Other Essays*. Cambridge, MA: MIT Press.

_ _ _ _ . (1990a), "Psychosemantics," in W. Lycan (ed.) *Mind and Cognition*. Cambridge, MA: Blackwell.

_ _ _ _ . (1991), "Replies," in G Rey and B Loewer (eds.) *Meaning in Mind: Fodor and his Critics*. Cambridge, MA: Blackwell.

_ _ _ _ . (1991a), "Hedged Laws and Psychological Explanation," in *Mind*. 100: 19-34.

Giere, R. (1988), "Laws, Theories, and Generalizations," in A. Grunbaum and W. Salmon, eds., *The Limits of Deductivism*. Berkeley: University of California Press, pp. 37-46.

Hempel, C. (1965), *Aspects of Scientific Explanation*. New York: The Free Press.

Hempel, C. (1988), "Provisos," in A. Grunbaum and W. Salmon, eds., *The Limits of Deductivism*. Berkeley: University of California Press, pp. 19-36.

Lewis, D. (1973), *Counterfactuals*. Cambridge: Harvard University Press.

_ _ _ _ _. (1986), *On the Plurality of Worlds*. New York: Basil Blackwell.

Salmon, W. (1990), *Four Decades of Scientific Explanation*. Minneapolis, University of Minnesota Press.

_ _ _ _ _ _. (1976), "Foreword," in Reichenbach, H. (1954) *Nomological Statements and Admissible Operations*. Amsterdam: North-Holland Publishing Company.

Wallis, C. (1993), "T II: Judgement Day," unpublished manuscript.

_ _ _ _ _. (1992), "Asymmetric Dependence and Mental Representation," in *Psycoloquy* (3) 70.

The Epistemic Authority of Expertise

Robert Pierson

York University

All of us defer to the authority of experts. Living in a world of increasing specialization we find that deferral to experts is integral to even the most mundane habits of our ordinary life. I regularly complain about mistakes in meteorologist's forecasts, yet I still carry an umbrella when rain is predicted, even when a glance outside reveals that there is not a cloud in the sky. If I develop a rash on my arm and the doctor says that it is nothing more than an allergic reaction, I do not ask her to list all the evidence and reasoning that led her to that conclusion, but trust her judgment and get on with my day. Now, while we defer to experts as a matter of course, we often do so with a measure of resentment and fear.

If it were possible, I suspect that most of us would rather govern what we do according to our own good reasons rather than expert authority. But the demands of everyday life require us to make many more decisions and hold many more opinions than we could ever base on personally examined reasons. So we defer to the opinions of others, in hopes of learning from their experience and thereby supplementing the limited scope of our own. But to what extent is this a rational thing to do? When is it more rational to defer to the authority of experts rather than thinking for oneself?[1]

While there has been and continues to be much work done on the psychology of expertise, the use of experts in courtrooms, and expert systems in computer science and artificial intelligence, by comparison little work has been done on the politics of expertise, less again on the history and sociology of expertise, and with the exception of the debate between John Hardwig (1985, 1991) and Steve Fuller (1986), virtually nothing has been written in the past decade on the epistemology of expertise.

In this paper, I suggest a way of resolving Hardwig's and Fuller's epistemological debate, and thereby answering the question of the division of cognitive labour between the expert and the layperson. (I assume that we regularly and quite unproblematically engage those whom we take to be experts, and so I do not offer *criteria* for picking out experts, but simply recommend how one ought to epistemically relate to another *one takes to be an expert*. If my position is correct, however, it will be essential to define "expert" and the range of possible and actual expertise.) This paper comes in two parts. The first, presents Hardwig's argument, Fuller's corresponding criticism, and what amounts to Hardwig's rejoinder. The second, presents three lines of thinking that are confused in Hardwig and Fuller's debate and explain how they may be reconciled to better determine the question at hand.

Hardwig (1985) argues that the ordinary individual, or "layperson," holds more beliefs than she could reasonably be expected to have the relevant evidence for. Most people can do little more than trust: that the sun and not the earth is the center of our solar system, that human development is the result of billions of years of evolution; that premature ventricular contraction is the most common form of irregular heart beat; that we do not have effective long term disposal techniques for the radioactive tritium in our nuclear reactors. The things that we believe without possessing the relevant evidence for the truth of those beliefs are virtually infinite, and we, of course, are finite. Although we could acquire the relevant evidence for some of the beliefs that we hold in this way, we surely could not do the same for *all* such beliefs. Hardwig explains that we "believe too much; there is too much relevant evidence (much of it available only after extensive, specialized training); [our] intellect is too small and life too short." (Hardwig 1985, 335)

In most cases, however, the layperson knows of some other individual, an "expert," who through specialized training has acquired the evidence needed for rationally assessing these kinds of beliefs. Hardwig says we have a choice to make. From the fact that the layperson holds more beliefs than she could reasonably be expected to have the relevant evidence for, and from the fact just mentioned, that there are experts who *have* the relevant evidence, we must choose either (a) that most of the layperson's beliefs are irrationally held, or (b) that it is rational for the layperson to hold a belief if someone she recognizes as an expert has the evidence needed for rationally holding it. Hardwig chooses (b) because it is the only way, he says, to save our intuition that most of our beliefs are rationally held. From this choice, it follows, he concludes, "that the layperson is epistemically dependent on the authority of experts for all but the beliefs on which she herself is an expert." (Hardwig 1985, 336) Or in other words, it is generally more rational to defer to the authority of the relevant expert than it is to think for oneself.

Fuller (1986) disagrees, criticizing Hardwig's failure to address "the crucial normative question" most apropos to the philosopher of science: "to what extent is expertise relevant to the sorts of goals which normally cause us to seek knowledge?" (Fuller 1986, 279) Fuller complains that Hardwig simply presumes that the epistemic goals of the layperson and the expert are sufficiently similar so that the layperson may unproblematically rely on the knowledge gathered by experts when engaged in her own inquiries. But this establishes only that *if* expertise is relevant, that we should defer to the expert's authority. But *when is* expertise relevant, Fuller asks? In response, he raises a number of considerations designed to show that "even in a knowledge-intensive society there may be rational grounds for 'thinking for oneself' and rejecting a general policy of deferring to the authority of experts." (Fuller 1986, 279)

The reason for this asymmetry between the goals the expert and the layperson is plain, Fuller writes. Expert's are oriented toward the systematic isolation and manipulation of their disciplines' defining set of variables as a 'closed system,' while laypeople are not. As a result, there can be "no direct extrapolation from the closed systems in which [an expert] proposes and tests hypotheses to the complexities of the 'undisciplined' environment which prompts the layperson to act." (Fuller 1986, 271) It is simply irrational for the layperson to trust expert testimony, Fuller insists. It is unlikely, moreover, that this situation will ever improve, he tells us, as experts are driven largely by internal problems that rarely, and only then, accidentally coincide with problems of general public concern. In matters of belief or action, the layperson is the expert's epistemic peer, and as a result, the two ought to negotiate as equals "the relevance of the latter's research to the former's interest." (Fuller 1986, 282) Hardwig's account is wrong to insist upon deference to expert authority rather than thinking for oneself, because, in point of fact, Fuller concludes, the layperson must first think for herself to determine the *relevance* of expert opinion.[2]

In his most recent article, Hardwig once more argues against epistemic individualists like Fuller. His thesis is that "modern knowers cannot be independent and self-reliant, not even in their own fields of specialization." (Hardwig 1991, 693) It is a fact that in most disciplines if you do not trust, you cannot know. In an important sense, "trust is often epistemologically even more basic than empirical data or logical arguments: the data and the argument are available only through trust." (Hardwig 1991, 694) Hardwig tells us that there is a trend away from single-author papers in scientific journals toward an ever-increasing number of authors per paper. Contemporary science is cooperative not simply because scientists build upon the work of those who have preceded them, but also in the sense that research is now often done in teams and, in fact, by bigger and bigger teams. It is the deferral to another's expertise or testimony, Hardwig explains, that allows the bits of evidence gathered by many different researchers to be gathered into a unified whole that can justify the experiment's conclusion. By trusting each other's testimony, individual researchers are united into a team which now possess what no individual member of the team possesses: "sufficient evidence to justify their mutual conclusion." (Hardwig 1991, 697)

So what does this mean for Fuller's general policy of thinking for oneself? Clearly Hardwig has shown that an individual engaged in some scientific practice is often too limited by either time or ability to think for oneself. On some occasions, at least, it *is* more rational to defer to expert authority than it is to think for oneself. Although neither Fuller nor Hardwig have established their entire case, I nevertheless believe they raise significant considerations which point us toward a resolution of the rational division of cognitive labour. It seems to me that three distinguishable things are getting mixed up in their discussions. One is the claim that we believe more than we could reasonably be expected to have the relevant evidence for, therefore we must trust others. Call this line of thinking A. Another is that expertise is closed system-oriented, with the implication that criticism can only be intra-systematic. Call this line of thinking B. And another is cost-benefit analysis, that when one decides what to do, one should act so as to maximize utility or information. Call this line of thinking C.

Hardwig's position is designed to support A, but mistakenly he thinks that this can be done only in conjunction with B. He takes no account of C. Fuller, on the other hand, entirely misses Hardwig's point about A and mistakenly reads him as supporting only B, and so in response argues for C. In the remainder of this paper I will attempt to unravel these confusions, and thereby suggest a way in which these three lines of thinking can be reconciled so as to resolve the issue at hand.

A: Hardwig's principle insight was to see that we hold more beliefs than we could reasonably be expected to have the relevant evidence for, therefore, we must depend upon others. Mistakenly, however, Hardwig restricted the "other" that we could rationally depend upon to that of the testimony of experts. He assumed that line of thinking A requires support from line of thinking B, but such a move, Anthony Quinton explains, "utterly fails to recognize the extent to which we are cognitively members of one another." (Quinton 1982, 67) My personal knowledge, what I have discovered on my own or learned from experts and stored in my memory, together with what I have inferred from this, "constitutes a quantitatively minute fragment of the whole range of what I claim to know." (Quinton 1982, 67)

One thing that obscures this fact is the way in which Hardwig, Fuller and others use the word "testimony" to describe our cognitive relationship with others. It misleadingly over-intellectualizes the relationship in question. Our cognitive relationship with others is profoundly and necessarily more thorough-going than reliance on the "testimony" of another, let alone reliance on the "testimony" of an "expert," suggests. Quinton says that this kind of mistake has much to do with the model of knowledge implicit in epistemology and courtroom procedure. In such procedures we tend to presume that there are clear, distinct, atomic facts to be discovered, which speak for themselves. When considered this way, testimony becomes a small, and relatively esoteric, part of human cognition. These associations tend to obscure its ubiquity in cognitive affairs. Even the basic instruments of communication and criticism that are necessary for any degree of conscious epistemic dependence or independence are acquired only out of a deeply authoritative tradition. As Quinton points out, "...external authority is the original source not only of items of information ... but also of language, logic and method, the indispensable means for the formulation and critical assessment of our beliefs." (Quinton 1982, 70) When this fact is recognized, the problem of justifying the acceptance of the authority of experts moves to an entirely new level.

I've directed this criticism at Hardwig because he is at least aware of the inescapably social dimension of knowledge, albeit not of the same order that Quinton describes. The same, though, should be said twice over for Fuller, as he manifestly neglects this dimension, favouring instead the enlightenment ideal of epistemic individualism.[3] I will now turn the discussion to the issue Fuller sees to be at the heart of his dispute with Hardwig - the evaluation of expertise.

B: As you will recall, Fuller argues against Hardwig, "that in matters of belief or action, the layperson is rationally obliged to negotiate as the expert's epistemic peer the relevance of the latter's research to the former's interest." In large measure, I agree with this: the layperson ought not to accept uncritically expert knowledge as a programme for personal action. She must, in all such cases, assess whether her benefit in acting on the basis of a particular expertise is worth the cost of doing so. This is what I call line of thinking C, and I will take it up in the next section. In this section, however, I'll show that it is almost always irrational for a layperson, no matter what her interests are, to evaluate the knowledge claims of an expert. So long as the claim refers only to an expertises' defining set of variables, and is not intended to advise the layperson, then evaluation can only be intra-systematic. This is what I call line of thinking B.

Hardwig's claim that the layperson is intellectually inferior to the expert in matters on which the expert is expert, is true only when 1) the expert's discipline is closed-system oriented and 2) the expert's knowledge claims are grounded within that discipline.[4] So long as the expert is concerned only with manipulating and controlling her discipline's defining set of variables, there is no rational room for lay evaluation of the corresponding knowledge claims. The reason for this is plain. The layperson lacks the training and competence of the expert. And, in lacking these qualities, she will likely be unable to understand the expert's reasons, or, even if she does understand them, she may not be able to appreciate why they are good reasons. Michael Polanyi (1977) makes this point very convincingly with regard to the physical sciences:

> The popular conception of science says that science is a collection of observable facts that anybody can verify for himself. We have seen that this is not true in the case of expert knowledge, like that needed ... in the physical sciences. In the first place, for instance, a layman cannot possible get hold of the equipment for testing a statement of fact in astronomy or in chemistry. Even supposing that he could somehow get the use of an observatory or a chemical

laboratory, he would not know how to use the instruments he found there and might very possibly damage them beyond repair before he had ever made a single observation; and if he should succeed in carrying out an observation to check upon a statement of science and found a result that contradicted it, he could rightly assume that he had made a mistake, as students do in a laboratory when they are learning to use its equipment. (Polanyi 1977, 184-185)

The means and grounds for determining the truth or falsity of expert claims are not available to the layperson. With regard to expert claim p, layperson B must defer to the relevant expert authority because layperson B: 1) cannot perform the inquiry that could provide evidence for p, 2) is not competent, and perhaps could never become competent, to perform that inquiry, 3) may be unable to evaluate the importance of the evidence provided by the expert's inquiry, and 4) may even be unable to understand the evidence and how it supports p.

Fuller thinks that in all matters the layperson is the expert's epistemic peer. He is mistaken. Not only does the layperson lack the where-with-all to evaluate expert knowledge claims, but the latter's research is always relevant to the former's interest of believing the true and not now believing the false. And to consider any other lay interests in this context is to confuse line of thinking B with that of C.

One may well find this restriction to be unrealistic. How is the layperson to know whether or not the expert is working intra-systematically? Strictly speaking, of course, they can't. A determination a layperson can make with respect to this, however, is that the expert is not now making a claim intended to be binding on "me," the layperson. If this is so, then the layperson ought to think for oneself. Otherwise, the layperson ought to defer to the expert, trusting that the expert both knows her discipline and is not now being dishonest.

An obvious rejoinder to this line of thinking is that it depends upon agreement among experts, an occasion that is seldom if ever the case. What is the layperson to do in cases of divided expert opinion? If expert C claims p with respect to x, and expert D claims not-p with respect to x, to whose authority should the layperson defer? It seems that in such cases she is rationally obliged to think for herself. Keith Lehrer (1977) disagrees, arguing instead for a method of ranking the various experts. Although I suspect this method is as good as any for making this kind of discrimination, I am sure that becoming an expert yourself in the discipline where you have discovered divided opinion would be only moderately more difficult . Lehrer's statistical machinations will not do for our purposes, but nor do I believe that anything like it is required. Even if a layperson ends up appealing to a lesser rather than a greater expert, the lesser expert's opinion will still be better than their own. A layperson defers to the authority of experts, not because in so doing one is guaranteed the truth-of-the-matter, but because one lacks the means to determine the issue oneself. In deferring to experts, one is not deferring simply to particular knowledge claims, but to a process for making those claims. I will now turn to a discussion of Fuller's principle insight to our question at hand.

C: Now while there are occasions where lay interests have no bearing on expert knowledge claims, and so deferral to experts is rational, Fuller has persuasively shown, to the contrary, that there are also occasions where lay interests are highly relevant to expertise, and so thinking for oneself is rational. The times where this obtains, however, are much more circumscribed than Fuller believes. In view of thinking B, I restrict such occasions to: 1) when expert knowledge is extrapolated into a programme for personal action,[5] or 2) the expertise is not closed-system oriented but layperson oriented. I will begin with restriction 1).

We learned above that when an expert is concerned only with claims that result from the control and manipulation of her discipline's defining set of variables, then there is no rational room for extra-systematic or lay evaluation of those claims. However, when an expert extrapolates her knowledge claims into *a programme for personal or lay action*, she is making extra-systematic claims and so is open to extra-systematic or lay evaluation of those claims. Consider again the example drawn from medicine. When a doctor diagnoses a patient as having premature ventricular contractions, she is operating within the boundaries of her discipline - she is doing cardiology - and so there is no rational room for the lay dissent. However, when the doctor recommends that the patient visit a cardiologist for care of her irregular heartbeat, she is stepping beyond her discipline of cardiology to "advise" that person about treatment. In such cases, she is no longer simply controlling and manipulating her discipline's defining set of variables, but is recommending changes to the layperson's virtually unlimited set of variables. There would be many scenarios on which it would be rational for the patient to reject such "advice." It may be, for instance, that she is averse to spending the amount of time and money required to act on the doctor's diagnosis unless she is sure that it will have beneficial consequences. Although she realizes that she has an unhealthy heart that requires treatment, she may have other demands on her time and money which she considers to be more pressing.

The governance of a particular life *always* involves many more relevant variables than any one expert can control, manipulate or account for. A cardiologist's claim with respect to the heart is open to question only by another who is sufficiently trained in cardiological matters, and, by definition, this excludes the layperson. This is what I call line of thinking B. However, any claim with respect to how "I" ought to govern "my" life can only be rationally determined, assuming I am relatively sane, by "me." It may be that my heart needs professional attention, but surely the decision as to whether it will get it should be mine, for only I can assess whether the benefit in having my heart attended to is worth the cost of not fulfilling any one of my other priorities. My priorities are mine, they are not variables within the control of, or even accessible to, experts.

Similarly, some "expertise" are by nature layperson oriented, that is, they are in the specific business of "advising" or "consulting" laypeople. The job of such experts is to advice clients on a narrow selection of variables as prescribed by their expertise, (ie. strike or contract negotiators, insurance or investment brokers, accountants, therapists, etc.). To be rational in such cases, the "advised" ought to then determine whether her benefits of following the expert's advise outweigh the costs of doing so. If they do, then she should proceed with the recommended action, and if they do not, then not. Inasmuch as the expert is unable to control or manipulate the layperson's desires, reasons or environment, there is nothing rationally binding in her advice. In all such cases, it is more rational to think for oneself.

Conclusion

With so little work done in the epistemology of expertise, Hardwig and Fuller must surely be considered pioneers. But like many other pioneers, their sorting of the new territory has tended to extremes of interpretation. Hardwig grants sole epistemic authority to experts, while Fuller, to the contrary, grants it to laypeople. Drawing upon these two thinkers, my work has been to suggest a more complex relationship between experts and laypeople.

Expertise is of two dominant sorts, I concluded: closed-system oriented and layperson oriented. The first sort of expertise is concerned primarily with controlling and manipulating a discipline's defining set of variables as a closed or relatively closed system. The second sort of expertise is simply in the business of "advising" clients. When expert claims are the result of the first sort, then there is no rational room for lay evaluation of those claims, and so the layperson must defer to the experts. However, when experts either extrapolate from their closed-systems to produce programmes for personal or lay action, or if experts are of the second sort, then the layperson is rationally obliged to think for herself, which amounts to nothing more than determining whether the *benefit* of following the expert's advice is worth the *cost* of doing so.[6]

Notes

[1] I restrict this paper to a consideration of the epistemic authority of experts. The epistemic point of view is characterized by the twin goals of now believing the true and not now believing the false. On this definition, therefore, the expert has epistemic authority over the layperson when the former is better situated to sort the true from the false. Although there are important and related issues in the epistemic authority of non-experts as well as broader issues of the epistemic authority of testimony, such is beyond the scope of this paper.

[2] The idea behind Fuller's argument lies at the center of the prevailing model of what it means to be a rational person. A model that first came to the center of philosophical attention with Descartes' methodological doubt, and that is formatively captured by Kant's remark that one of the three basic rules for avoiding error in thinking is to "think for oneself."

[3] At first glance, it may appear that I am mistaken in taking this complaint to be a criticism of Fuller, for Fuller specifically takes issue with traditional philosophy of science's neglect of the social dimension of knowledge. Upon closer consideration, however, one can see that Fuller's talk of "social epistemology" focuses only on "how the *products* of our cognitive pursuits are affected by changing the social relations in which the knowledge *producers* stand to one another." (Fuller 1986, 3) He shows no familiarity with the more fundamental type of social dependence Quinton describes as underwriting such practices.

[4] Following Fuller, I take an expertise to be closed-system oriented if it restricts i) word usage, ii) borrowings permitted from other disciplines, and iii) appropriate contexts of justification and discovery. Generally, an expertise that is closed-system oriented controls its own academic department, programme of research, and historical lineage, in fact, the word "discipline" can be used as a substitute for "expertise." And in this sense, the word "expert" can be used interchangeably with the word used to name a practitioner of a discipline, such as, "economist," "physician," "meteorologist," "geologist."

[5] I am concerned in this paper only with action vis-á-vis oneself - how "I" as an autonomous agent should act. I suspect that expert "advice" would be more binding on individuals who represent various organizations, since in those cases all or most of the relevant variables are available for public and expert evaluation.

[6] Of course, there remains an interesting and very important question as to whether the layperson performs her cost-benefit analysis rationally, but this is not within the scope of this paper. My task has simply been to mark out the epistemological territory as it regards the "right" or "rational" relation between the expert and the layperson.

References

Fuller, S. (1986), *Social Epistemology*. Bloomington and Indianapolis: Indiana University Press.

Hardwig, J. (1985), "Epistemic Dependence", *The Journal of Philosophy* Volume LXXXII, No. 7: 335-349.

_____. (1991), "The Role of Trust in Knowledge", *The Journal of Philosophy* Volume LXXXVIII, No. 12: 693-708.

Lehrer, K. (1977), "Social Information", *Monist* LX, No. 4: 473-487.

Quinton, A. (1982), *Thoughts and Thinkers*. London: Duckworth.

Polanyi, M. (1977), *Meaning*. Chicago: University Press.

Circular Justifications[1]

Harold I. Brown

Northern Illinois University

1. Introduction

It is common practice to reject a justification out-of-hand if the argument that yields the justification is circular. Here are two recent examples. Siegel, criticizing Giere's (1988) proposal for a naturalistic philosophy of science, lists a number of questions about the justification relation between evidence and theory. Siegel then maintains that any attempt to provide a *scientific* account of this relation must fail:

> any answer to the question of the relationship between evidence and a justified theory, if arrived at scientifically, would depend upon exactly the same relationship between it and the evidence for it as it recommends for the relationship between any justified theory and the evidence for it. Because these general questions about the epistemology of science cannot be answered naturalistically without begging the question, they cannot be so pursued. (1989, 369.)

In a similar vein, Barnes and Bloor reject any attempt to justify a system of deductive logic:

> The basic point is that justifications of deduction themselves presuppose deduction. They are circular because they appeal to the very principles of inference that are in question. In this respect the justification of deduction is in the same predicament as the justifications of induction which tacitly make inductive moves by appealing to the fact that induction 'works'. Our two basic modes of reasoning are in an equally hopeless state with regard to their rational justification. (Barnes and Bloor 1982, 41-42.)

Perhaps the most famous example of a charge of circularity is Hume's contention that any attempt to use experience to support the principle that the future resembles the past fails because the attempt assumes the truth of the very principle to be established.

One convenient feature of the sweeping rejection of circular justifications is that the objector is freed from any further need to consider the details of the argument in question. But there are reasons for asking if such quick rejections are appropriate. An examination of recent literature on circular arguments indicates that it is far from clear how circularity is to be defined and, relative to a given definition, whether all circular arguments are *ipso facto* defective. Discussions are further complicated by a lack of general agreement on whether the term "circular" is to be understood as a descriptive or an evaluative term. For example, Wilson (1988) uses "circular" as a descriptive term and distinguishes between arguments which are circular and those which are viciously circular. But Sorenson (1991) uses "circular" as an evaluative term and declines to label an argument circular unless it is defective. Since Sorensen holds that some arguments of the form "p, therefore p" are legitimate, he maintains that arguments of this form need not be circular. Other writers are less clear about the descriptive and evaluative uses of their term.

In this paper I will distinguish between "circular" and "viciously circular," using the former as a descriptive term and the latter as evaluative. I will describe as "circular" those arguments that, in some sense, assume what they seek to justify. The real work then consists in specifying the relevant sense. Once this is done we can inquire whether a particular circular justification is in fact defective and if so, exactly where the defect lies. I will use "viciously circular" to label those justifications that are both circular and defective. One advantage of this terminology is that it leaves open such questions as whether all circular arguments are defective, whether all defective circular arguments share the same defect, and whether a particular characteristic is a defect *per se*, or only in some contexts.

There is one further complication that we should note at the outset. A familiar dialectic recurs throughout discussions of circularity. Suppose we begin with the hypothesis that all arguments which have a particular property are *ipso facto* defective, but we then discover this defect among arguments that are considered paradigmatically correct. Clearly something has to give but, as is often the case when an hypothesis yields an undesired conclusion, it is rarely clear exactly what is to be altered. Perhaps we have mislocated the defect and should reject our hypothesis; perhaps there are other considerations which indicate that the property in question is only a defect in certain contexts; perhaps we should reconsider some standardly accepted arguments; and so forth.

This dialectic can be illustrated by considering Biro's (1977, 1984) discussion of arguments that "beg the question." Biro maintains that begging the question is a specific epistemic defect: "an argument begs the question when one of its premises cannot be known without knowing the conclusion" (1984, 242). This is a plausible suggestion and we should note that the proposal both specifies a type of circularity and explains why a justification that depends on a circular argument of this kind is defective: the argument provides no new reasons for accepting the conclusion. But accepting Biro's diagnosis will require major revisions in standard logic. Obvious examples include simplification and conjunction. Moreover, as Woods and Walton (1975, 111-112) note, disjunctive syllogism will also have to go at least in truth-functional logics. For if the truth-value of "p or q" depends only on the truth-values of "p" and "q," and we already know that "p" is false, we have to know that "q" is true in order to know that "p or q" is true. Advocates of intensional and relevance logics will be happy with this outcome (see Jacquette 1993 for a recent discussion), but others will be more inclined to reject or modify Biro's account of circularity. Sanford (1971, 1981), for example, suggests that the defect in Biro's proposal lies in the attempt to treat an epistemic feature as a property of an argument *per se*, irrespective of context. Instead, Sanford maintains that the application of Biro's criterion depends on who we are talking to and what we are trying to accomplish. If someone has reasons for believing "p or q" but no reason for believing either of the disjuncts, then coming to know not-p, and thus q, may involve substantive new knowledge. (E.g., I may be aware that I have arrived at a weekend but be hung-over and uncertain if it is Saturday or Sunday.) Thus Sanford concludes that "Some arguments of the disjunctive syllogism form beg the question. Some do not," (1971, 198). Moreover, whether an argument exhibits this particular defect does not depend just on its logical form, but on features of the individuals using the argument and the context in which the argument is being used.

Sanford's pragmatic approach will set the context for formulating the thesis of this paper. I shall argue that whether a circular argument is viciously circular depends on a variety of specific contextual features. In particular, I shall argue that philosophers should be wary of rejecting justifications as viciously circular without examining the details of the argument that yields the justification even when there is a clear sense in which the argument is circular. Since this thesis is of a kind that can best be defended by providing examples, I will examine some arguments that exhibit a specific kind of circularity, but which are not *ipso facto* illegitimate.

2. An Observational Example

Suppose we develop an observational procedure to test an empirical hypothesis, H. The procedure will make use of some instruments, and some body of accepted science will be required for the design of these instruments and for analyzing their output. Suppose, in addition, that the required scientific background includes H. We have here a clear sense in which the proposed "test" of H is circular; indeed, this is one of the situations that is commonly included under the label "theory- laden observation." It is also commonly held that a procedure of this sort cannot provide empirical support for H, and that any attempt to extract such support from the procedure will be viciously circular. Moreover, one explanation of the source of this vicious circularity is at least implicit in the literature. The idea is that evidence supporting an hypothesis only comes from observational procedures that *test* that hypothesis, but one necessary condition for a legitimate test is that it be at least possible that the test yield a result that is unfavorable to that hypothesis. But, the argument goes, an observational procedure cannot possibly yield a result that is unfavorable to an hypothesis that is assumed in the design of that procedure and in the analysis of the resulting data. However, an example will show that this particular form of circularity is not always vicious.

While I shall describe an actual case from contemporary science, it is sufficient for present purposes that we consider its logical structure, quite independently of whether the observation has ever actually been carried out. (See Zensus and Pearson 1987, for a review of the actual situation as of late 1986 and Brown 1993 for further discussion.) Consider, then, a case in which earth-based telescopes are being used to measure the change in the angle between two distant celestial objects (call them A and B) at intervals of several years. Given these measurements, along with a body of accepted results, it is possible to determine the linear velocity at which, say, B is moving away from A. The following steps are required. First, B's red shift is determined; call this S. Given S and the formula for the Doppler effect, B's recession velocity can be determined. Now Newtonian physics and relativity theory yield different formulas for the Doppler effect. Since it is currently believed that relativity is correct, the calculation is naturally done using the relativistic formula:

$$\beta = (S^2 - 1)/(S^2 + 1),$$

where β is the recession velocity, taking the velocity of light in vacuo as one. At this point relativity plays an essential role in the procedure. Note especially that using the relativistic formula, β can never be equal to or greater than one. If someone were to offer the results of the procedure thus far as empirical support for relativity we would have an example of just the sort of viciously circular theory-ladenness that has concerned many philosophers; of course, no competent scientist would make such a proposal.

Next, the Hubble constant is used to calculate B's distance from the earth on the basis of its recession velocity. Once this distance has been determined, it can be multiplied by the measured change in angle between A and B, and the result can be divided by the known time between the two measurements to yield the B's tangential velocity relative to A. Let us assume that the angle has increased over time. There is nothing in this procedure that blocks the occurrence of a tangential velocity substantially greater than the velocity of light, even though the relativistic limitation on such velocities was assumed, indeed, played an essential role in the calculation.

In the present context, the moral of this story is that the essential use of an hypothesis in the interpretation of a set of observations does not automatically prevent an empirical outcome that challenges that hypothesis. Such challenges may be impossible in some cases, but this must be shown by *detailed examination* of the specific case in question. It would be a mistake to stop the analysis once the central role of relativity in the data analysis is noticed and to conclude that continued analysis could not possibly yield a result that contradicts relativity.

An important corollary to our moral is that if the procedure just described were proposed as a test of relativity, the temptation to reject the legitimacy of the test because it assumes the truth of relativity would be hasty. Moreover, if such a test were to yield a result in conformity with relativity, this would seem to be a fair confirmation. This view of what counts as a fair empirical test can be described as "quasi-Popperian" since it requires the possibility of disconfirmation for a fair test, but allows for genuine confirmation as well.

3. An Example from Logic

The idea of a fair test can be adapted to cases from logic, although the test will not be empirical. Let us ask exactly why Barnes and Bloor hold that a deductive justification of deduction must be circular. One possible reading is that if we use the very logic whose justification we are examining as the basis for this examination, then it is a foregone conclusion that we will find this logic to be justified. But an example will show that this general thesis is not correct.

Suppose we are evaluating a proposed system of logic, L, using a metalanguage, L^*, and there is a one-one correspondence between the argument forms permitted in L and those permitted in L^*. In other words, the same logic is embodied in the object language and the metalanguage. In particular, L and L^* both permit arguments of the following form:

$A: p$ or q
$\therefore p$.

If using the same argument forms in the object language and the metalanguage guarantees a vicious circularity, then it should be impossible to use A to challenge the validity of A. But consider the following metalinguistic argument: let "p" be "A is invalid" and let "q" be any true proposition. Our argument is:

A is invalid or q,
$\therefore A$ is invalid.

The premise is true and we have a challenge to the validity of A whether we consider the conclusion to be true or false. If the conclusion is true, the validity of A has been challenged; if the conclusion is false, we have an instance of A with a true premise and a false conclusion. However one may respond to this result, one point is clear: the fact that our argument proceeded on the assumption that A is a valid form did not prevent the argument from yielding a challenge to the validity of A. We can specify a clear sense in which the argument is circular: it makes use of the very argument form whose validity is being assessed. And we can also specify a clear sense in which the circularity is not vicious: it possible to construct an argument of the form in question that yields a challenge to the validity of that form.

Actually, this result should not be especially surprising. Logics that license invalid arguments are not sufficiently restrictive, thus they will yield a larger set of conclusions from a given set of premises than we could derive from those premises if we limited ourselves to valid arguments. This point can be underlined by considering another case in which we use a logic that is not sufficiently restrictive. Suppose that, unknown to us, L is inconsistent and that, again, we are studying L in a metalanguage that uses L as its logic. Since anything can be deduced using an inconsistent logic, we will surely be able to deduce the proposition, "L is inconsistent" in our metalanguage. Given that inconsistency is a logical defect, we have arrived at a challenge to L in spite of having assumed L to arrive at our result.

4. Foundations

Let us consider one further sense in which it may be claimed that a justification is viciously circular. Suppose we make essential use of a proposition, p, in a justificatory argument. Suppose, further, that the justificatory status of p is open to question and that our argument is aimed at increasing p's justification. Again we have a clear example of a circular argument and there is an historically important line of thought that will lead us toreject such attempts at justification. Here the idea is that it is not legitimate to use a proposition in a justification unless that proposition has itself been *fully justified*. If we accept this claim, then our admittedly circular justification may be described in either of two ways: If the justification is legitimate it is pointless, since p has already received the maximal justification possible. If the justification is not legitimate, we may mark this fact by describing the circular argument on which the justification is based as "vicious."

It will be recognized that this line of argument expresses one of the themes involved in foundational epistemology. In particular, it captures the thesis that each proposition in the edifice of knowledge must itself be fully justified before it can legitimately be used in the justification of any other proposition. It is not implausible that this is the kind of vicious circularity that is being referred to in the two quotations at the beginning of this paper. On this reading, Siegel is arguing for the necessity of a non-naturalistic foundational epistemology if science is to be justified, and he believes that such an epistemology can be developed. Barnes and Bloor are arguing for the impossibility of such an epistemology and thus that no philosophical justification of logic is possible. It is striking that, if this is the correct interpretation of the quoted remarks, we are dealing with a fundamental disagreement that is itself based on a prior agreement that only foundational justifications are cognitively legitimate.

This is not the place to enter into a general discussion of foundationalism, but it is worth noting that our discussion in Section 2 provides a reason for believing that the foundationalist impulse can be successfully resisted by showing that there are at least some cases in which a kind of naturalistic bootstrapping can yield genuine empirical tests that increase the confirmation of empirical claims that are assumed by those tests.[2]

5. Induction

Now let us turn briefly to Hume's blanket rejection of any attempt to use induction to justify induction. Given the discussion in this paper, we can agree with Hume that any attempt to inductively justify induction will be circular, but still ask whether the circularity is vicious. In particular, let us ask if "the principle of induction" is capable of a fair empirical test. But our answer to this question will depend on exactly how that principle is formulated. For example, if we read the claim that the future will resemble the past as the claim that every empirical pattern discovered up until a specific point in time will be found to hold after that point in time, then the principle is certainly capable of empirical evaluation. Indeed, it is known to be false since there are many examples of patterns identified in past experience have not been successfully projected into the future. A defensible principle of induction will have to be rather more limited in scope.

Consider one possible candidate: that there are recurring patterns in nature. This principle would license a continued search for such patterns even as the projectibility of various proposed patterns was empirically refuted. Is this principle amenable to empirical support? Given that we have some reasonable candidates for specific patterns that do recur, I suggest that any attempt to use these patterns as empirical evidence for the principle is problematic, but for reasons that have nothing to do with circularity. The principle in question is a pure existential and it is notoriously unclear what counts as empirical evidence *against* a pure existential. Thus if we agree that a fair empirical test requires that it at least be possible that the outcome pose a challenge to the hypothesis being tested, then any attempt to test the proposed principle will depend on a solution to a difficult problem in confirmation theory. But the issue is, once again, not one of circularity.

It may now be suggested that I have totally missed the point. The point is that we sometimes discover a specific pattern that has recurred in the past and we would like to be able to argue that this pattern will continue to occur in the future. But any such argument requires a premise that justifies projection of past experience into the future. It is the justification of this premise that is in question. Let me make two points in response. First, the claim that we need a justified principle that licenses projections of past experience into the future *before* we can rationally make any such projections is itself open to question. I submit that this is a piece of lurking foundationalism and I have suggested a quasi-Popperian alternative in Section 2: specific projections do not require prior justification as long as we are prepared to treat them as claims that are subject to empirical challenge.

Second, I agree that our empirical beliefs would be more secure if we could establish a principle that licensed (some) projections, but I urge again that we must have a specific candidate before us if we are to assess its epistemic status. I have considered two candidates; one of these we can reject as false on empirical grounds; the second may well be true, and attempts to provide an empirical justification for that principle are indeed problematic. But the problem has nothing to do with circularity which is, after all, the topic of this paper.

6. Vicious Circularity: An Example

I have mainly been concerned in this paper to challenge claims of vicious circularity, but this does not mean that I think all such claims are mistaken. Let me underline this point by considering an example of an argument that does embody an unacceptable circularity. The example I shall take is from Galileo's *Dialogue Concerning the*

Two Chief World Systems. A major aim of the *Dialogue* is to show that standard Aristotelian arguments against the motion of the earth fail to prove anything at all. This project leads Galileo to a careful analysis of many of these arguments and in some cases Galileo concludes that his opponents' arguments are question begging. Galileo's discussion of the tower argument provides an important example of this strategy (1967, 139-141). Let us examine this argument.

On the Aristotelian account of falling objects, a stone dropped from the top of a tower will move on a straight line towards the center of the universe, which necessarily coincides with the center of the earth. But, Aristotlians argued, if the earth is engaged in a daily rotation from west to east, the tower will be moving to the east as the stone falls, and the stone will land to the west of the tower. Since the stone lands at the foot of the tower, we have a straightforward empirical refutation of the view that the earth rotates. The following is a reconstruction of the argument. Since my aim is to illustrate a point about circularity I have not followed the literal discussion in Galileo's text, although I do not think that I have misrepresented anything in the text.[3]

A1. If a stone is dropped from the top of a tower then the stone moves on a straight line to the center of the earth.

A2. If the stone moves on a straight line to the center of the earth and the tower rotates around the center of the earth as the stone falls, then the stone will not land at the base of the tower.

A3. A stone is dropped from the top of a tower.

A4. The tower rotates around the center of the earth as the stone falls.

Thus, the stone will not land at the base of the tower.

Since the argument is valid and the conclusion is false, at least one of the premises must be rejected. The second and third premises can be accepted, so the decision focuses on premises A1 and A4. Simplicio, the Aristotelian spokesman in the *Dialogue*, assumes that the first premises is true, and thus concludes that A4 is to be rejected. But the truth of premise A1 is exactly what is at issue in the discussion, and it is the assumption that A1 is true that constitutes the vicious circularity. Note especially that this circularity is not a result of having used the Aristotelian premise in the argument—after all, the anti- Aristotelian fourth premise also occurs in the argument and the outcome of the argument plus the observational test of the conclusion is, as we have seen, that either A1 or A4 is false. The unacceptable circularity occurs when Simplicio simply rejects the possibility that A1 should be rejected, thereby literally assuming the conclusion that he seeks to support. Galileo would be equally guilty of begging the question were he to assume the truth of A4 and reject A1. But Galileo does not do this. His only claim is that the argument fails to prove anything about the rotation of the earth; and this is clearly correct. The proper conclusion is that either the first or the fourth premise is false (perhaps both are false). No further conclusion can legitimately be drawn on the basis of this argument without assuming the truth of one of the two views that are being debated.

7. Conclusion

The major conclusion that I wish to draw from this discussion is that philosophers have often been too quick to reject specific arguments and even entire research projects on the grounds of circularity. I have argued that before an admittedly circular argument is rejected, the exact problem generated by the circularity must be pin-pointed. In some cases, we have seen, a blatant circularity need not generate any troublesome problems. In other cases, circularities may indicate limitations on the proper use of an argument or there may be some other specific reason why an argument involves an unacceptable circularity. But such conclusions require detailed study of the particular argument in question.

Notes

[1] I want to thank C. A. Hooker, Tomis Kapitan and Harvey Siegel for comments on earlier drafts of this paper.

[2] See Hooker 1987 for a general analysis of circularities in scientific theories and Brown forthcoming for a response to the general challenge to naturalistic epistemology.

[3] See Finocchiaro (1980, 192-194 and 277-282) for a more literal reconstruction of the argument and a different diagnosis as to why the circularity renders the argument worthless.

References

Barnes, B. and Bloor, D. (1982), "Relativism, Rationalism and the Sociology of Science", in M. Hollis and S. Lukes, (eds.), *Rationality and Relativism*. Cambridge: MIT Press, pp. 21-47.

Biro, J I. (1977), "Rescuing 'Begging the Question'", *Metaphilosophy* 8: 257-271.

_____. (1984), "Knowability, Believability and Begging the Question: A Reply to Sanford", *Metaphilosophy* 15: 239-247.

Brown, H I. (1993), "A Theory-Laden Observation *Can* Test the Theory", *British Journal for the Philosophy of Science* 44: 555-559.

_____. (forthcoming), "Psychology, Naturalized Epistemology and Rationality, in R. Kitchener and W. O'Donohue, (eds.), *Psychology and Philosophy: Interdisciplinary Problems and Responses*. Allyn and Bacon.

Finocchiaro, M. (1980), *Galileo and the Art of Reasoning*. Dordrecht: Reidel.

Galileo (1967), *Dialogue Concerning the Two Chief World Systems*, S. Drake, (trans.). Berkeley: University of California Press.

Giere, R.N. (1988), *Explaining Science*. Chicago: University of Chicago Press.

Hooker, C.A. (1987), "On Global Theories", in *A Realistic Theory of Science*. Albany, NY: SUNY Press, pp. 109-138.

Jaquette, D. (1993), "Logical Dimensions of Question-Begging Arguments", *American Philosophical Quarterly* 30: 317-327.

Sanford, D. (1971), "Begging the Question", *Analysis* 32: 197-199.

_____. (1981), "Superfluous Information, Epistemic Conditions of Inference, and Begging the Question", *Metaphilosophy* 2: 145-158.

Siegel, H. (1989), "Philosophy of Science Naturalized? Some Problems with Giere's Naturalism", *Studies in History and Philosophy of Science* 20: 365-375.

Sorensen, R. (1991), "'P, Therefore, P' Without Circularity", *Journal of Philosophy* 88: 245-266.

Wilson, K. (1988), "Circular Arguments", *Metaphilosophy* 19: 38-52.

Woods, J. and Walton, D. (1975), *"Petitio Principii"*, *Synthese* 31: 107-127.

Zensus, J.A. and Pearson, T.J. (eds.) (1987), *Superluminal Radio Sources*. Cambridge: Cambridge University Press.

Part XI

PHILOSOPHY OF PSYCHOLOGY AND PERCEPTION

Is Cognitive Neuropsychology Possible?[1]

Jeffrey Bub

University of Maryland

The aim of contemporary cognitive neuropsychology is to articulate the functional architecture underlying normal human cognitive abilities, on the basis of patterns of performance over a variety of cognitive tasks involving subjects with varying degrees of brain-damage. An example of a contemporary functional architecture is the following model for the recognition and production of spoken and written words:

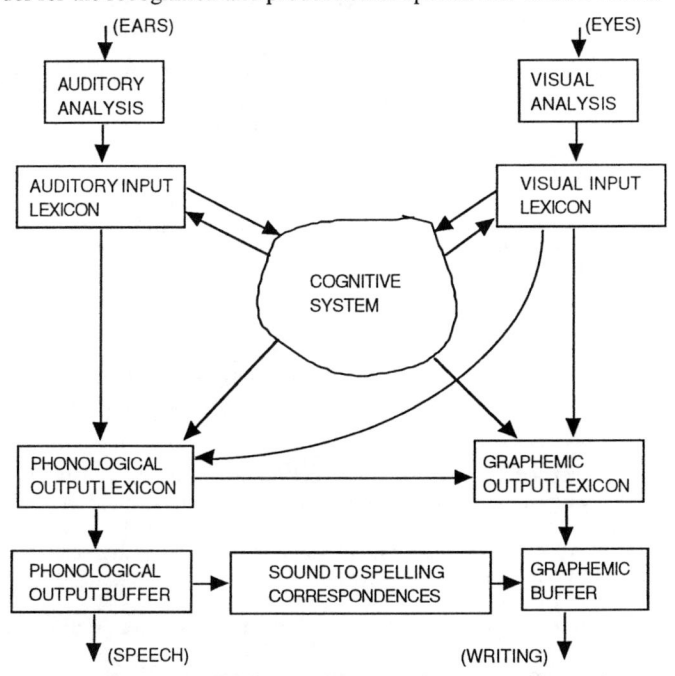

Figure 1. Functional architecture of the recognition and production of written and spoken words. From Shallice (1988), P. 139.

The modular processing components in this 'box-and-arrow' functional architecture represent functionally independent subsystems that perform computational transformations on the representations that originate as input along the connecting links, providing a theory of cognition at the algorithmic and representational level (not the implementational or 'hardware' level). The diagram incorporates the 'dual-route' model of reading, the existence of functionally separable mechanisms for converting lexical and sublexical units into a pronunciation: a lexical route from the visual analysis system through the visual input lexicon to the phonological output lexicon (either through the semantic system, or bypassing the semantic system), in parallel with a non-lexical route that directly converts graphemes to phonemes (Patterson, Coltheart, and Marshall 1985). The dual-route architecture accounts for features of normal and impaired reading performance, such as the ability to read with or without comprehension, the ability to read regularly spelled words (like 'hat') as well as irregularly spelled words (like 'yacht') and word-like nonsense words (like 'fip,' 'munt'), different forms of dyslexia, etc.

In contemporary cognitive neuropsychology, rival theories of the mental processes underlying cognitive abilities are either modular functional architectures that provide computationally explicit, information-processing accounts of how the mental representations required to perform these cognitive tasks are produced, or parallel distributed processing (PDP) networks, or mixed architectures in which some or all of the processing components of the classical architectures are PDP networks. Observed patterns of performance of different cognitive tasks under varying experimental conditions, including various forms of brain damage, place constraints on the possible functional architectures that sustain the cognitive abilities and disabilities reflected in this performance. The general methodological problem for cognitive neuropsychology is how evidence, in the form of patterns of cognitive deficits, bears on theory, in the form of rival functional architectures.

Throughout the history of the subject, questions have been raised, beginning in the 1880's with Freud (1891) and Bergson (1896), as to whether the methods of neuropsychology are adequate to its goals. The question has been re-opened in a new and sophisticated form by Clark Glymour (forthcoming), who formulates a discovery problem for cognitive neuropsychology, in the sense of formal learning theory, concerning the existence of a reliable methodology. It appears that the discovery problem is insoluble: granted certain apparently plausible assumptions about the form of neuropsychological theories and the nature of the available evidence, a reliable methodology does not exist!

The following discussion is divided into two sections. §1 is an exposition of Glymour's formulation of the discovery problem and his argument for a radical underdetermination of theory by data in cognitive neuropsychology. In §2, I argue for a reformulation of the discovery problem in terms of an alternative characterization of relevant evidence in neuropsychology.

1. Glymour's Formulation of the Discovery Problem for Cognitive Neuropsychology

Glymour points out that a modular functional architecture - a system of processing centers or modules and connecting links - can be understood as a directed graph: a set of vertices or nodes (corresponding to the processing modules), together with a set of directed edges between pairs of nodes (corresponding to the connecting links in the architecture).

A number of simplifying assumptions are introduced for these graphs (appropriate, because the aim is to prove a negative result: that the discovery problem defined below is insoluble, even under certain simplifying constraints for the class of possible theories and the type of evidence considered). It is assumed that all normal subjects have the same graph (the same architecture), and that the graphs are acyclic (no feedback loops). The graph of an abnormal subject is represented, in the first instance, as a subgraph of the normal graph (obtained by removing nodes and/or edges from the normal graph, subject to the constraint that if a node is removed, then all edges into or out of that node are removed as well). This assumption is modified later for the case of partially lesioned functional architectures. It is assumed that every subgraph of the normal graph will eventually occur among abnormal subjects (i.e., that any evidentially relevant pattern of performance is eventually observed).

Glymour associates the normal performance of a given cognitive task with an input-output pair, or more generally a set of relevant input nodes coupled with an output node, and assumes that the cognitive processes activated in the performance of the task - the functioning of the internal nodes - will produce either a 1 (normal response) or a 0 (abnormal response) at the output node, following an appropriate assignment of values to the relevant input nodes. So the input and output nodes have a different status to the internal nodes. The internal nodes represent information-processing modules. The input and output nodes represent task variables, either stimuli presented to the subject or measures of behavioral response.

Finally, it is assumed that *all* paths or routes between a relevant input node and an output node, associated with a given cognitive task in the normal graph, are required to be intact in an abnormal graph for the output node to take the value 1 on inputs for which the normal subject outputs 1 for that task. Glymour points out that one might alternatively assume that if *any* route between a relevant input node and an output node is intact in an abnormal graph, then the output node will take the value 1 on inputs for which the normal subject outputs 1 for the task. I shall refer to these alternative assumptions as the 'conjunctive' and 'disjunctive' multiple route assumptions, respectively. Glymour remarks that the nature of the discovery problem is not essentially altered by which of these assumptions is adopted. The standard interpretation of multiple processing routes in cognitive neuropsychology conforms to the disjunctive assumption, and I shall consider the consequences of this assumption in §2.

The discovery problem for cognitive neuropsychology is now formulated as the problem of selecting the true normal graph (or correctly identifying some structural feature of the true normal graph), with respect to a set of possible alternative normal graphs, on the basis of performance data from normal subjects, who instantiate the true normal graph, and brain-damaged subjects, who instantiate subgraphs or other modifications of the true normal graph. If we represent the normal performance of a cognitive task by a relevant set of input nodes and an output node - Glymour calls such an ordered pair a *capacity* - then each normal subject can be associated with the normal set of capacities determined by the true normal graph, and each brain-damaged subject with a characteristic subset of the normal set, determined by the precise nature of the damage to the true normal graph. Glymour refers to the characteristic sets of capacities associated with normal and brain-damaged subjects as *profiles*. Assuming that every abnormal graph is eventually instantiated among brain-damaged subjects, so that every abnormal profile eventually occurs in the temporal sequence of available performance data, the discovery problem has a solution if there is a procedure that always selects the true normal graph (or correctly identifies some structural feature of the true normal graph) after some finite accumulation of profiles, for each possible graph in the set of alternatives, and for each possible ordering of the normal and abnormal profiles associated with that graph.

Evidently, the discovery problem will be insoluble if it can be shown that the true normal graph, in a given set of alternative normal graphs, is underdetermined by the normal and abnormal profiles that can be generated from the set of alternative normal graphs. Glymour's analysis suggests that this might well be the case for alternative normal graphs that are possible on the basis of certain current assumptions in cognitive neuropsychology.

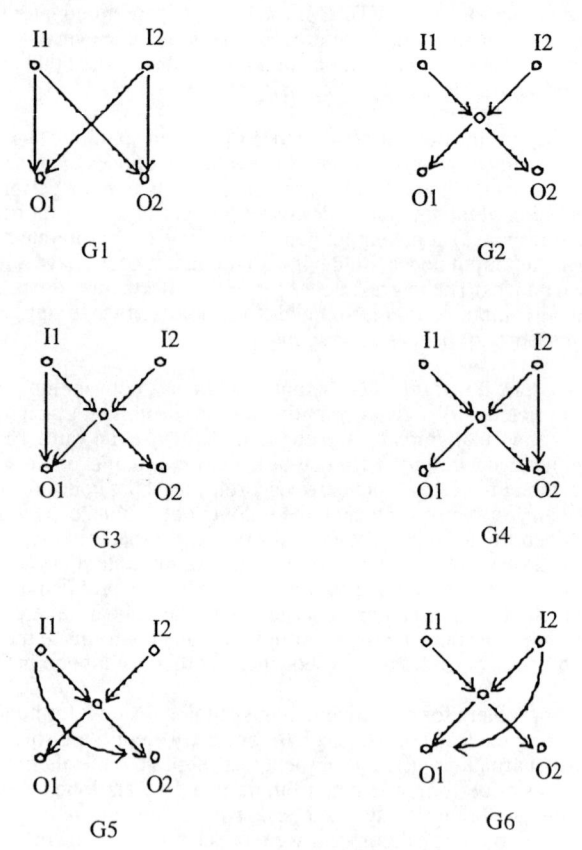

Figure 2. The six graphs of Glymour's initial discovery problem (I1, I2 are input nodes; O1, O2 are output nodes).

To start with, Glymour considers an elementary discovery problem defined by the following six alternative graphs, each with two input nodes and two output nodes. The first graph, G_1, contains no internal variables: each input node is directly connected to each output node by an edge, generating the normal profile: {<1, 1>, <1, 2>, <2, 1>, <2, 2>} (where the first member of each ordered pair designates the single relevant input node, and the second member designates the output node). The other five graphs each contain one internal node in addition to the two input nodes and two output nodes, and differ from each other and the first graph by the edges connecting the nodes. All six graphs have the same normal profile: {<1,1>, <1,2>, <2,1>, <2,2>}. To generate the abnormal profiles associated with these graphs, Glymour applies the conjunctive multiple route assumption. In $G_1, ..., G_6$, each capacity is defined by one relevant input node and an output node, and there is either a single route between the input and output node, or two routes: a direct route and an indirect route via the internal node. Assume, initially, that all abnormal profiles are generated from abnormal graphs that are subgraphs of the normal graph. In the case of G_1, each of the four edges can be removed separately, or any pair of edges, or any triple of edges, or all four edges, generating fifteen possible subgraphs, i.e., fifteen possible abnormal profiles representing the fifteen proper subsets of the normal profile (including the null profile). For G_2, all of these abnormal profiles are possible, except the four abnormal profiles that lack just one of the normal capacities, and the two abnormal profiles {<1, 1>, <2, 2>} and {<1, 2>, <2, 1>}. So, nine abnormal profiles can be generated from subgraphs of G_2. Adopting the conjunctive multiple route assumption, Glymour shows that the same nine abnormal profiles can be generated from each of the remaining four graphs, G_3, G_4, G_5, G_6, together with one additional abnormal profile that is different in each case:

	<1, 1>	<1, 2>	<2, 1>	<2, 2>
N				
P_1				X
P_2			X	
P_3		X		
P_4	X			
P_5			X	X
P_6		X		X
P_7	X			X
P_8		X	X	
P_9	X		X	
P_{10}	X	X		
P_{11}		X	X	X
P_{12}	X		X	X
P_{13}	X	X		X
P_{14}	X	X	X	
P_{15}	X	X	X	X

G_1: $P_1, ..., P_{15}$
G_2: $P_5, P_6, P_9, ..., P_{15}$
G_3: $P_4, P_5, P_6, P_9, ..., P_{15}$
G_4: $P_1, P_5, P_6, P_9, ..., P_{15}$
G_5: $P_3, P_5, P_6, P_9, ..., P_{15}$
G_6: $P_2, P_5, P_6, P_9, ..., P_{15}$

Figure 3. Table listing all possible normal (N) and abnormal ($P_1 ..., P_{15}$) profiles for graphs $G_1 ..., G_6$ ('X' indicates missing capacity in the abnormal profile), followed by a listing of abnormal profiles corresponding to $G_1, ..., G_6$, respectively.

The discovery problem for this restricted set of alternative graphs is soluble. Glymour formulates the appropriate discovery rule as follows: *Conjecture any normal graph whose set of normal and abnormal profiles includes all of the profiles seen in the* data, and which has no proper subset of profiles, associated with one of the graphs, that also includes all of the profiles seen in the data.

Now, Glymour points out that the discovery problem becomes insoluble if we introduce further complexity into the six graphs. If we replace any internal node by a subgraph, with the same edges directed into and out of the subgraph as were directed into and out of the node, and there is a route through the subgraph linking any edge into the subgraph with any edge out of the subgraph, then the resulting graph has the same normal profile and the same set of abnormal profiles as the original graph, and so is indistinguishable from the original graph on the basis of any performance data. (Glymour refers to this replacement procedure as 'pinching,' because in replacing a node by two nodes connected by an edge, we in effect 'pinch together' edges directed into the original node, or edges out, or both.)

So far, abnormal graphs have been considered as resulting from the normal graph by simply deleting edges and/or nodes. But brain damage is more plausibly viewed as a matter of degree, and not an all-or-nothing affair. One might assume that different cognitive tasks that involve the activation of the same module place different computational demands on the module. A module may be partially lesioned in such a way as to support the normal performance of some tasks, while allowing only impaired performance of other, more computationally demanding, tasks. More generally, if the modules are PDP networks, an abnormal graph would not be a subgraph of the normal graph, but a modified normal graph, in which internal nodes on paths connecting the input and output nodes of a set of normal capacities support designated subsets of these capacities, depending on the nature and extent of the lesions to the nodes. The different designated subsets of capacities would be defined for each node by a partial ordering relation on the capacities. One might interpret the order relation as characterizing which capacities place more demands than others on the computational resources of the node, or as reflecting the organization of the PDP network constituting the node.

These assumptions about the ordering of capacities with respect to the computational demands they place on internal nodes, in conjunction with the possibility of partial lesions to the graph elements, introduce a further type of complexity into the graphs. As Glymour shows, if we apply these assumptions to the original set of six graphs, some of the graphs become empirically indistinguishable, for certain orderings of the capacities supported by the internal node. Glymour concludes that, without independent arguments for the comparative processing demands of different cognitive capacities, the enterprise of identifying modular structure from patterns of deficits is hopeless.

2. Reformulation of the Discovery Problem

Consider, now, the disjunctive multiple route assumption, that if *any* route between a relevant input node and an output node is intact in a graph, then the output node will take the value 1 when the cognitive process for the capacity is activated. Assume that abnormal graphs are simply subgraphs of the normal graph.

Graphs G_1 and G_2 are characterized by the same normal and abnormal profiles as before. But now, for both G_3 and G_4 we can generate the three abnormal profiles {<1, 1>, <2, 2>}, {<1, 1>, <2, 2>, <1, 2>}, {<1, 1>, <2, 2>, <2, 1>} from appropriate subgraphs, in addition to the nine abnormal profiles characteristic of G_2. Similarly, for both G_5 and G_6 we can generate the three abnormal profiles {<1, 2>, <2, 1>}, {<1, 2>, <2, 1>, <1, 1>}, {<1, 2>, <2, 1>, <2, 2>} from appropriate subgraphs, in addition to the nine abnormal profiles of G_2. So, on the disjunctive multiple route assumption, G_3 and G_4 are empirically indistinguishable (but distinguishable from G_1 or G_2 or G_5 or G_6), and G_5 and G_6 are empirically indistinguishable (but distinguishable from G_1 or G_2 or G_3 or G_4), and the discovery problem is already insoluble for the original six graphs, without the introduction of additional complexity by 'pinching,' or deriving abnormal graphs by partially lesioning a normal graph in such a way as to selectively affect capacities.

G_1: $P_1, ..., P_{15}$
G_2: $P_5, P_6, P_9, ..., P_{15}$
G_3: $P_2, P_3, P_5, P_6, P_8, P_9, ..., P_{15}$
G_4: $P_2, P_3, P_5, P_6, P_8, P_9, ..., P_{15}$
G_5: $P_1, P_4, P_5, P_6, P_7, P_9, ..., P_{15}$
G_6: $P_1, P_4, P_5, P_6, P_7, P_9, ..., P_{15}$

Figure 4. Abnormal profiles for $G_1, ..., G_6$ on the disjunctive multiple route assumption.

Now, one might suppose that Glymour chose the conjunctive multiple route assumption, because the disjunctive assumption simply makes the discovery problem more difficult - indeed, underdetermination appears to arise immediately. But a similar empirical indistinguishability occurs with the conjunctive assumption, even without assuming the possibility of partial lesions, and not only through 'pinching.'

Consider the subgraph of G_2 - call it G_2' - obtained from G_2 by deleting the input node 2, the output node 2, and the edges joining these nodes to the internal node. G_2' is empirically indistinguishable from the subgraph of G_3 - call it G_3' - obtained from G_3 by deleting the same nodes and edges. Both the open triangle G_2' and the closed triangle G_3' are characterized by the same normal capacity <1, 1>, and by the same abnormal capacity, the null capacity.

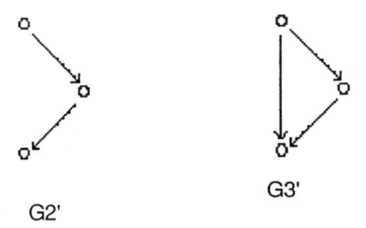

Figure 5. The subgraphs G_2' and G_3'.

The graph G_2 is empirically indistinguishable from the graph $G_{2,p}$, obtained from G_2 by pinching so that the single internal node, n, of G_2 is replaced by two internal nodes, n_1 and n_2, joined by an edge. But both these graphs are also empirically indistinguishable (on the conjunctive assumption, not the disjunctive assumption) from the graph $G_{2,p'}$, obtained from $G_{2,p}$ by 'bridging,' adding an edge from the first internal node n_1 to one of the output nodes. Notice that the bridge converts an open triangle to a closed triangle.

It might appear that I have simply identified an additional topological feature of the graphs as a source of underdetermination. But now the underdetermination between $G_{2,p}$ and $G_{2,p'}$ on the conjunctive multiple route assumption should be cause for suspicion: Surely cognitive neuropsychologists can distinguish graphs like $G_{2,p}$ and $G_{2,p'}$ on the basis of task performance data? The alternative between an open and a closed triangle is the alternative between a single-route subarchitecture and a dual-route subarchitecture, and cognitive neuropsychologists routinely distinguish such cases on the *disjunctive* multiple route assumption (as in the single-route versus dual-route models of reading) on the basis of dissociations in performance (assuming all-or-nothing lesions), or double dissociations (to avoid the possibility of so-called 'resource artefacts' if the lesions are partial).

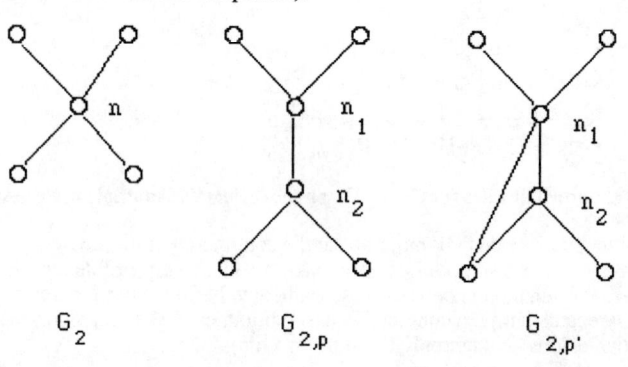

Figure 6. Graphs derived from G_2 illustrating pinching and bridging.

For example, if we assume all-or-nothing lesions, then either of the dissociations in reading performance presented by a pure surface dyslexic (who can read nonsense words like 'fip,' but not orthographically irregular words like 'yacht' (Marshall and Newcombe 1973)) or a pure phonological dyslexic (who can read words like 'yacht' but not nonsense words like 'fip' ((Beauvois and Derouesné 1981), (Funnell 1983)) will exclude a single-route model of reading (a single route for converting lexical and sublexical units into a pronunciation, representing an open triangle) in favor of a dual-route model (functionally separate lexical and nonlexical routes, representing a closed triangle). If we assume partial lesions, then the inference is more complicated, because either dissociation could (separately) reflect the fact that one of the reading tasks (reading orthographically irregular words, or reading pronounceable nonsense words) simply requires more of the processing resources of some of the modules activated in performing the task than the other - in effect, that both reading tasks involve the same route, but one of the reading tasks is simply more difficult than the other. The 'double dissociation' between the two kinds of dyslexia is generally taken to exclude such a 'resource artefact,' on the basis of certain assumptions (e.g., monotonicity) about the functioning of the modules (Shallice 1988, 232 ff), (Bub forthcoming).

Of course, cognitive neuropsychologists might simply be mistaken about the methodological significance of dissociations in distinguishing between such alternatives. I would argue, however, that the apparent underdetermination between cases like the open and closed triangles, or more generally between bridged and unbridged structures, for the ideal case of all-or-nothing lesions, is an artefact of the conjunctive multiple route assumption and Glymour's purely 'syntactic' approach to the discovery problem for cognitive neuropsychology, in which only the topology of the graphs representing functional architectures is taken as relevant, and evidence, in the form of task performance data, is presented in terms of inputs and outputs to the graphs.

I propose that in formulating the discovery problem we consider both the topological and the functional structure of the graphs, i.e., the functional or 'semantic' aspect of the nodes and edges. Task performance data would then be represented theoretically in terms of paths through the graphs, depending on the functional significance of the nodes and edges (in effect, via different 'correspondence rules' for each graph separately) - and not purely 'syntactically' in terms of input-output pairs.

To illustrate: Suppose we have a possibly infinite set of graphs, like Glymour's six graphs, with more internal structure and more input and output nodes, representing possible functional architectures for cognition, where the input and output nodes represent processing components like the visual analysis system, auditory analysis system, various output buffers for different modalities (speech, writing), etc., as in contemporary functional architectures for the recognition and production of spoken and written words. (Recall that on Glymour's analysis the input and output nodes represent task variables, either stimuli presented to the subject or measures of behavioral response.)

Now, each graph - as a functional architecture - is not merely a topological structure of nodes and edges. Particular nodes represent particular processing components. I assume that, for each task and for each graph, there exists a path through the graph corresponding to the task, depending on what the nodes and edges are supposed to do functionally (i.e., I assume the set of graphs is 'adequate' for the tasks considered in this sense). I allow that two tasks might correspond to the same path in some graphs, and to different paths in other graphs, but I do not assume, as Glymour does, that *all* paths between an input node and an output node must be intact for the performance of a task associated with a pair of input-output nodes. I think of a task as associated with a path in a graph.

I shall distinguish a *data profile* (d-profile) from a *theoretical* or *graph profile* (g-profile). A d-profile for a particular patient is the total set of cognitive capacities and incapacities presented by the patient - what the patient can and can't do, relative to normals, for all relevant cognitive tasks (where what counts as a 'relevant' task depends on the set of alternative graphs, and the way in which d-profiles are related to these graphs, i.e., relevance is defined by the discovery problem). For the moment, understand this in an all-or-nothing sense: The patient either performs a task normally, or abnormally.

Each graph is associated with a set of possible paths through the graph (representing the possible information flows from the various input nodes to the various output nodes). A g-profile for a given graph is a topologically possible set of intact and lesioned paths, where (initially) lesioning is understood as all-or-nothing (i.e., a path is either intact, or lesioned if some node or edge of the path is removed so that information flow along the path is impossible).

I assume that for every g-profile in the actual (true) graph, there exists a d-profile, and that we eventually see all these d-profiles (i.e., assume, in effect, that we can design tasks and find patients to instantiate all these d-profiles).

I define the 'ideal' discovery problem for cognitive neuropsychology as the discovery problem specified by the set of graphs with all-or-nothing assumptions for lesions, with evidence accumulating as d-profiles (again, in the all-or-nothing sense). It can be shown (Bub forthcoming) that this discovery problem has a solution, subject to the following constraint on 'pinching': If a given processing component n consists of two subcomponents, n_1 and n_2, where the representation computed by n_1 is passed as input to n_2, then either (i) this preliminary representation (the output of n_1) is available to the system independently of further processing by n_2 (which requires a second path branching out of n_1), or (ii) it is not. In the first case, we can distinguish the graphs via tasks that activate this alternative path, without activating the path through n_2. In the second case, n_1 and n_2 can be regarded as two internal stages of n that can't be revealed by evidence consisting of d-profiles. So pinching will be allowed for the ideal discovery problem only if the pinch is accompanied by an extra edge or 'bridge' (to satisfy (i)).

In effect, I grant Glymour's point about the underdetermination introduced by 'pure pinching.' Pinching without bridging does lead to underdetermination. But this only means that we can't tell whether modules are assembled (from more basic modules) or not via neuropsychological data, unless the output representations from these submodules are independently available (via bridges) to the system.

Neuropsychologists do not, of course, receive data as d-profiles. Rather, they design experiments to elicit data as dissociations in performance: single dissociations (task x abnormal, task y normal), or multiple dissociations (tasks x_1, x_2, ..., abnormal, tasks y_1, y_2, ..., normal). Other relevant evidence consists of double dissociations (task x normal, task y abnormal on a patient or patient-group P_1; task x abnormal, task y normal on a patient or patient-group P_2), and the nature of the errors in abnormal task performance. Dissociations function as relevant evidence for the ideal discovery problem. Double dissociations and error-types are relevant to the discovery problem for partial lesions.

Dissociations place constraints on g-profiles. For example, under the all-or-nothing assumption for functional lesions of the ideal discovery problem, a single dissociation between two tasks, x and y, requires two distinct paths: in each graph there must be a path for task x that is separate from the path for task y. As we get more and more dissociations and multiple dissociations for a particular patient, the data constitutes a partial d-profile for the patient that approximates more and more closely to the full d-profile for the patient. If the discovery problem is soluble for evidence accumulating as d-profiles, it is soluble for evidence accumulating as dissociations.

The non-ideal discovery problem involves the consideration of partial lesions and the assumptions underlying the logic of double dissociation experiments, and the methodological role of error-types. See (Bub forthcoming) for a discussion. For cognitive neuropsychology to claim validity as a science, it is not necessary that these assumptions are always applicable. It is sufficient that they can be justified for a nontrivial class of inductive problems concerning reading, writing, memory, object recognition, and speech production and comprehension, and this is at least arguably the case for much of current research in cognitive neuropsychology.

Note

[1]This material is based upon work supported by the National Science Foundation under Grants SBE-9012399 and SBE-9122696. I want to thank Clark Glymour and Dan Bub for helpful discussions.

References

Bergson, H. (1986), *Matiere et Memoire*. Paris: Alcan.

Beauvois, M.-F. and Derouesné, J. (1981), "Lexical or Orthographic Agraphia", *Brain* 104: 21-49.

_____. (1985), in K.E. Patterson, J.C. Marshall, and M. Coltheart (eds.), *Surface Dyslexia: Neuropsychological and Cognitive Studies of Phonological Reading*. London: Erlbaum.

Bub, J. (forthcoming), "Testing Models of Cognition Through the Analysis of Brain-Damaged Performance", *British Journal for Philosophy of Science*.

Freud, S. (1891), *Zur Auffassung der Aphasien*. Vienna: Deuticke.

Funnell, E. (1983), "Phonological Processing in Reading: New Evidence from Acquired Dyslexia", *British Journal of Psychology* 74: 159-180.

Glymour, C. (forthcoming), "On the Methods of Cognitive Neuropsychology," *British Journal for Philosophy of Science*.

Marshall, J.C. and Newcombe, F. (1973), "Patterns of Paralexia: A Psycholinguistic Approach", *Journal of Psycholinguistic Research* 2: 175-199.

Patterson, K., Coltheart, M. and Marshall, J. (eds.) (1985), *Surface Dyslexia*. London: Erlbaum.

Shallice, T. (1988), *From Neuropsychology to Mental Structure*. Cambridge: Cambridge University.

The Scope of Psychology

Keith Butler

Washington University

Psychology is a scientific discipline whose universe of discourse is the mind. What could be so controversial about that? Plenty; and (almost) all of the controversy involves questions about the *scope* of mind. The traditional view on this matter is most closely associated with Descartes. Though Descartes supposed that the mind is essentially nonmaterial, this dualism is not a part of the legacy Descartes left to the modern student of mind. What modern cognitive science has inherited from Descartes is the view that the mind is a private place, separated (in a number of respects) from the body and the environment that lies outside the physical boundaries on an individual. There is, however, a rising tide within the philosophy of psychology that seeks to push the boundries of the mind outside the skin (or at least the cranium). Tyler Burge has become famous for arguing that facts about one's linguistic and physical environments are partly constitutive of mental states. But new and very different anti-individualist arguments are beginning to emerge. John Haugeland, in a recent paper drawing on a number of sources (Haugeland 1993), has argued that the Cartesian divisions between mind, body, and environment melt away under the lights of modern systems analysis.[1]

The aim of this paper is to stand up for Descartes. Obviously, however, in the space of a single paper it will be impossible to establish the truth of internalism. I hope only to contain the damage Haugeland has done to the Cartesian legacy. I will argue that there is no reason to think that the universe of discourse for psychology, namely, the mind, should involve aspects of the environment; he has fastened on concerns that can plausibly be taken to lie outside the scope of psychology.

1. The Scope of Psychology

Sciences are sciences *of* something. Chemistry is the science of the basic elements and their interactions, biology is the study of living things, etc. If one wants to know about stars one studies astronomy, if one wants to know about rocks one studies geology. But what is it that one wants to know about when one studies psychology? That is the question Haugeland is raising. It is very facile to simply say that the domain of psychology is the mind; the issue in this paper can arise only because it is not perfectly clear what a mind is. Nor, for that matter, is it perfectly clear what behavior is.[2]

Nevertheless, the domain of psychology has something to do with mind and behavior, and the relation between them.

At first blush, it would seem, the psychologist wants to know *why* and *how* we behave. Let me illustrate with a simple example: What would it be to provide a psychological explanation of Joe Carter's home run in the final game of the 1993 World Series? There are, of course, several answers. One explanation would appeal to Carter's desire to win the game, and his belief that swinging at the ball has some probability of helping to fulfill that goal. Another sort of explanation would appeal to the cognition needed to make just the swing he made—a bat-swinging algorithm, as it might be. These are *psychological* explanations because they are confined to Joe Carter's contribution to the production of the home run. More specifically, they describe mental activity on the part of Joe Carter, mental activity that, *as a matter of fact*, resulted in a home run that won the World Series for Toronto. Indeed, the *same* psychological explanations might apply to Joe Carter even if the ball never made it out of the park, even if he never his the ball at all. The success of his efforts depends, at least in part, on factors quite independent of his psychology. It is not Joe Carter's fault if a gust of wind kicks up and keeps the ball in the park; all he can do is carry out the cognition and hope the outside world does its part.

None of this, of course, is an argument for internalism. But it does illustrate what seems to be a central fact about the science of psychology: Its topic of concern is what goes on *inside* the subject. Now, because psychology is a science, it is an empirical matter whether this preconception about the domain of psychology is accurate. Indeed, the *nature* of the beast may be very different from our preconceptions of it (at least if (Churchland 1981) has anything to say about it); but there does not *seem* to be any problem finding the *location* of the beast—it's on the inside! Thus, the task facing Haugeland is to convince us that psychological explanations might *better* be viewed as involving selected factors on the outside.

2. Mind Embodied and Embedded

Haugeland's (1993) goal is to expose an "intimacy" in the relation between mind, body, and environment. By this he means "...a kind of *commingling* or *integralness* of mind, body and world...[that] undermine[s] their very distinctness" (Haugeland 1993, 2). The challenge, as Haugeland sees it, is to make this position intelligible; for once it is understood, its plausibility will be manifest.

Haugeland's case is ingenious, and I cannot hope to do it justice in this short paper. But I would like to convey some of the more provocative points. He begins with a seductive example owing to (Simon 1969/82) involving an ant who travels across an irregular beach, the ever-changing surface of which is shaped by the random action of waves and wind. The ant's trajectory across the beach is a complex sequence of turns and weaves—much too complex to be the product of simple ant-cognition. It is, in one clear sense, the beach's complexity that is responsible for the complexity in the ant's path. Perhaps, says Simon in a move echoed by Haugeland, the same can be said for people.

There is a lesson in this realization; or maybe two:

On the one hand, one might heave a sigh of scientific relief: understanding people as behaving systems is going to be easier than we thought, because so much of the apparent complexity in their behavior is due to factors external to them, and hence external to our problem. On the other hand, one might see the problem itself as transformed: since the relevant complexity in the observed behavior depends on so much more than the behaving system itself, the investigation cannot be restricted to that system alone, but must extend to some larger structure of which it is only a fraction. (Haugeland 1993, 4)

Haugeland, clearly, is pushing for the latter option; and maybe he's got a point. He develops his case with the help of some insights from Simon's version of systems analysis. In particular, he wants to mark this distinction: interactions *between* components in a system vs. interactions *within* a component in a system. Interactions between systems are "narrow bandwidth" interactions; they are effected through a sort of bottleneck, such that what happens inside the component is well-insulated from what happens outside the component, except at the limited focus of the interface. Electrical components of a TV, for example, have narrow bandwidth interfaces. Interactions within a component, in contrast, are "high bandwidth" interactions; here the interface is not bottlenecked in any appreciable way. A random partitioning of the guts of a TV, one that doesn't respect the usual component divisions, is high bandwidth; and for that reason, it is not a scientifically interesting division. Indeed, the interesting divisions need not be corporeal or spatial. Social structures, particularly those that are organized around information flow, can be divided into components by appealing to intensity of interaction, regardless of spatial propinquity. Telemarketers working from the same room, for example, may never communicate with each other; the more interesting interfaces involve their bosses and clients.

Armed with these analytical insights, Haugeland invites us to consider several examples of manifestly intelligent behavior that involve high bandwidth interfacing between the behaver and its environment. The first, of course, is the ant and its interface with the beach. Had the ant navigated by means of an internal map of the beach, then it would enjoy only narrow bandwidth interfacing with the beach (through the soles of its feet). But the fact of the matter is that the ant has no such internal map, and participates in a free exchange of information with the beach as its trek progresses; "...the ant and beach must be regarded more as an integrated unit than as a pair of distinct components" (Haugeland 1993, 12).

A related example comes out of the work of (Brooks 1991), who has built an insect-like robot that navigates the MIT AI laboratory without the aid of a symbolic cognitive architecture. Instead, it has "layers" of largely independent behavioral units connecting sensation to action. Several layers are 'superimposed' and coordinated so that they don't interfere with each other in ways that damage performance. The most basic layers involve object avoidance, while others involve, e.g., picking up soda cans around the lab. According to the criteria for system decomposition, the most salient divisions are not between behaver and environment, but between the different layers. There is, in other words, high bandwidth interaction within layers, including the environment (which Brooks calls "its own best model"), and only narrow bandwidth interfacing between layers. As with the ant, it is not at all clear why the object of psychological inquiry is the behaver *per se* when the most interesting divisions do not include any such concept.

Haugeland also cites, with approval, Gibson's (e.g., 1979) ecological approach to psychology, and Dreyfus' (1972/79/92) persistent criticisms of classically oriented AI. Gibson's talk of "affordances" is well-suited to Haugeland's assertion that the behaver(perceiver)/environment interface is high bandwidth. Perceptual systems include not just the perceiver, but the perceived, and the relation between them is one of mere "pick-up". Affordances *unify* the perceiver and the perceived by essentially involving both. Dreyfus' well-known complaints about classical symbol systems do for mind/body interaction what Gibson has done for mind/environment interaction. The 'intensity of interaction' criterion calls into question the separateness of mind and body for cases Dreyfus presents. Performing any complex task even reasonably well requires a vast amount of information exchange between the brain and the body. There is very little sense, Haugeland thinks, to the notion of an instruction guiding complex behavior. There is instead a high bandwidth interface between brain and body something like that of ant and beach.

Lest we think that all of the above is plausible only in the case of simple organisms, or at least simple tasks, Haugeland asks us to consider knowing the way to San Jose as an example that (at least for those who live there) hits quite close to home. A signature feature of intelligence, one almost entirely lacking in the cases presented above, is the ability to deal with what is absent. San Jose (for most of us) is absent, and knowing the way to San Jose involves the capacity to deal with what is absent. It is a central commitment of Cartesian cognitivism that such a capacity must involve internal representations of what is not before us; how else can we hope to direct ourselves coherently? Haugeland's suggestion, building on lessons drawn from Simon, Brooks, Gibson, and Dreyfus, is that the world takes care of much of the task. Knowing the way to San Jose is knowing what road to take, and, as it were, letting the road do the rest. The relation between a driver and the road taken is strikingly like that of the ant and the beach. High bandwidth interaction such as this is the mark of *no* (relevant) distinction; the road, no less than the driver, is an essential part of an intelligent system.

3. A Cartesian Reply

At first blush, it looks as though Haugeland has done the impossible. He has made it intelligible, even plausible, to suppose that the seat of intelligence, the mind, abides in the public world. If the universe of discourse for psychology is the mind, then the science of psychology must expand its domain beyond current boundaries. Psychologists must fundamentally reconceive their task, and explore the myriad ways in which what is inside the head interfaces with what is outside the head. Or not.

Haugeland's brilliant case depends (by Haugeland's own admission) on shifting the sort of problem it is the business of psychology to address. In a passage quoted above, Haugeland invites us to reappraise the domain of psychology on the grounds that there is an interesting problem to be addressed that is not currently being addressed by psychologists. The brain is still dishing-up output, but, sometimes at least, funny things happen on the way to the world; the complexity of behavior owes much to the complexity of the environment, and its relation to the organism. It is hard to see, however, why that would motivate a shift in the focus of psychology. Consider the ant and the beach. Haugeland asserts that "...what we want to understand in the first place is the ant's *path*" (Haugeland 1993, 11). But this is debatable. An ant psychologist may have a passing interest in the complexity of the beach, and even how that influences the ant's path, but what she would really want to understand, I wager, is the *ant's* contribution to that path. And what a human psychologist is interested in is not the close coupling between the driver and the road, but what the driver brings to that coupling. It is quite uncertain, therefore, that Haugeland has given us compelling reasons to expand the universe of discourse for psychology. That there are other interesting questions to ask, even, perhaps, more interesting questions, does not eliminate questions of paradigmatically psychological interest.

But, Haugeland will be quick to point out, if we grant that the ant and the beach, or the driver and the road, enjoy high bandwidth interfacing, the onus is on us to articulate the criteria according to which we can make a (nonarbitrary) distinction between mind and world. There are at least two (related) avenues we might explore preliminarily, both of which suggest that the relevant bandwidth of an interface is, at best, insufficient to delineate component boundries in a system in general, and the scope of psychology in particular. The first can be introduced in the context of Brooks' layers of behavioral units. Even on the assumption that Brooks' subsumption architecture reveals a kernal of truth about the organization of behavior, it does not do the work Haugeland needs of it.[3] Any intelligent creature is involved in many behaviors, simple and complex, extending over many environments. But while the environments may come and go, the organism (at least ideally) does not. One clear point of contrast between the inner and the outer, then, is that the inner remains (relatively) stable across significant changes in the outer. The inner/outer dichotomy is just not the illusory and arbitrary dichotomy that Haugeland makes it out to be.

This is all the more striking when we consider the case of the driver who does a good deal more than drive to San Jose. She may also hold down a job, go to dinner with friends, play tennis, study financial portfolios and any number of other things. Each of these may in some way involve high bandwidth interfaces with the relevant environments, but it is only *she* who enjoys high bandwidth interfaces with all the relevant environments. The point is that the object of psychological inquiry is determined by more than a single behavioral episode. A large part of AI, both classical and connectionist, concerns the exploration of models that generalize beyond a limited behavioral domain; we want to figure out how a single system could do so many different things. High bandwidth interfacing may be *necessary* for intracomponent interaction, but there is no reason to expect that it is *sufficient*.

The recognition of the intersection of different behaviors dovetails with a second point that tells against Haugeland's attempt to broaden the scope of psychology. Minds are typically viewed as loci of control. The insides of the ant, Brooks' robot, and the driver control their respective behaviors in ways that the relevant environments do not. Organisms do not determine the nature of their actions independently of the environment (a brick wall is, after all, a *brick wall*); but they do determine their actions in a way that the environment does not, namely, by deciding or choosing or at least being motivated to take a particular course of action. And while it may be difficult to articulate just what the relevant differences are between the inner and the outer (what makes a choice a choice?), there can be no doubt that they are real. Again, the presence of a high bandwidth interface may well be a necessary feature of intrasystemic interaction; but it seems manifestly insufficient.

The critical remarks have so far dealt only with Haugeland's attempt to stretch the domain of psychology to include the environment. Part of his case, however, is directed at undermining the distinction between mind and body; given the high bandwidth interface between body and world, this is tantamount to undermining the distinction between mind and world. He aims to show not just that the mind is intimately related to the brain, but that the mental, insofar as it is involved with the brain, is just as intimately involved with the body. Borrowing points broached initially by Dreyfus, Haugeland tries to expose the comminglingness of brain and body essentially by appealing to the messiness of the interface between brain and motor neurons; to draw a sharp distinction here would be like disentangling a bowl of spaghetti (not Haugeland's analogy). Neural projections from the brain reach out to a bewildering array of motor neurons in our muscles; and these are in constant informational flux as a result of their high bandwidth interface with the ever-changing environment. There is simply no way that "simple instructions" or "well-defined, repeatable messages" could govern the richly textured interplay between mind, body and world. In Haugeland's eyes, we can no longer sustain a picture of "...two relatively independent separable components—a rational mind and a physical body, meeting at an interface—but rather a closely-knit unity" (Haugeland 1993, 22).

But recent work in the neural organization of motor control belies Haugeland's worries. (Gallistel 1980, esp. 275-280) has reported substantial evidence of a hierarchical organization in motor control. This is significant because it has the consequence that an initial motor command need not specify the details of its implementation in motor behavior (Ghez, 1985; Ghez and Fahn, 1985). The reason for this is just the same reason an army General need not stipulate how each troop is to behave when he hands down the order to assault. The General tells the Colonels, the Colonels pass the word on down the chain of command; the specific details of the implementation for the command are left to the individual troops. The same is true in the neural hierarchy that governs moter behavior. Through a series of (sometimes nested) servomechanisms, feedback information from the environment is accomodated by lower-level control centers (analogous to Majors and Captains in the military chain of command).[4] The top level of the neural hierarchy sets out the instruction; the lower levels fight it out with the environment to determine specific details of implementation. The same organizational structure, moreover, can be extended to cover rather more sophisticated behaviors of the type normally taken to issue from beliefs and desires (Butler, 1992).

But now we can see how the spaghetti-like interface between brain and body might admit of a structured over-lay. Haugeland is sceptical that any set of instructions "from the top", as it were, can adequately account for the behavior that results from the high bandwidth interface between brain and body. But our recognition of the motor control hierarchy allows us to make sense of central control in a way that Haugeland cannot. The presence of a high bandwidth interface between body and world has no relevance to the intelligibility of well-defined, repeatable instructions encoded in the brain. Servomechanisms in a hierarchically structured control system insulate the brain, and the mind, from the neuromuscular spaghetti, and hence from the outside world. A hierarchy of servomechanisms, in other words, creates a *narrow* bandwidth interface between brain and muscles. In light of this, we can appreciate why it is that a high bandwidth interface between body and world does not draw the mind out into the world.[5]

We can see now why a psychological explanation of Joe Carter's home run should not involve the peculiarities of his musculature and the near-indeterminacies of the environment. An explanation must provide an understanding, and to do so, it must abstract from details. High bandwidth interfaces simply contain too many details to be intelligible; they prevent us from finding what we're looking for in a psychological explanation. Joe Carter's beliefs and desires, even his algorithms and computations, are protected from his body and world by a hierarchical nexus of servomechanisms. Haugeland is quite right that high bandwidth interfaces characterize many of the divisions we've heretofore endorsed uncritically, but our own assessment affords us those same divisions upon further review.

4. Closing Remarks

Haugeland's project can be seen as the antithesis of eliminative materialism. Rather than eschew a privelaged conception of mind as a misguided postulate of an outmoded theory, he endeavors to find it everywhere—in our muscles, in a beach, and even on the road to San Jose. In fact, he would like to find intelligence in our tools, our institutions, and our culture. We might call this a *ubiquitous* materialism. Intelligent cooperative acitivity is everywhere around us. For example,

...the structure of an institution is implemented in the high bandwidth intelligent interactions among individuals, as well as between individuals and their paraphernalia. Furthermore, the expertise of those individuals could not be what it is apart from their participation in that structure. Consequently, the intelligence of each is itself intelligible only in terms of their higher unity. (Haugeland 1993, 32)

Intelligence is here portrayed as a sort of emergent property, the bases of which, in part, are the intelligences of individuals. But this cannot be right; when collections of things interact to give rise to what we might timidly call an "emergent" property, it is rarely if ever the case that the property that emerges is anything like the properties that it emerges from. Individual water molecules are not wet, but a big-enough collection of them is; individual neurons are not smart, but a big-enough collection of them is. Surely the interaction of individual intelligences, if it gives rise to any further property, does not give rise to more intelligence; such a supposition is, if nothing else, historically untenable.

In this quotation we see a glimpse of why Haugeland's project can seem so wrongheaded: He has tried to turn Cartesian psychology into a kind of anthropology or sociology or ecology; and it just won't fit. There already are sciences whose topic of inquiry is the interpersonal and environmental. More importantly, none of the concerns Haugeland has raised diminished (in substance or significance) the problems inside the head that psychology traditionally takes on. There is no room for an expanded psychology, no motivation for it, and no need for it. No one will deny that environmental information can be of great help in unlocking the secrets of the mind (witness the ecological factors that inform Marr's (1982) theory of vision), but there is enough going on inside for psychologists to worry about.

It is testimony to Haugeland's considerable philosophical skill, however, that he could bring such a bizzare thesis to the brink of plausibility. But we must not allow ourselves to be caught up in the excitement that such novelty and cunning inspire. As we assess the future course of psychology, we must proceed cautiously, keep our feet to the ground, and not lose our minds.

Notes

[1] (van Gelder 1993) has tried to preserve the cartesian conception of mind by turning only *cognition* over to the scientists; *mind* remains safely insulated from the inexorable forward march of science. But van Gelder's attempt to have it both ways (a sort of externalist internalism) is less compelling than Haugeland's bold thesis in that it is more difficult for him to escape the charge of semantical quibbling. Nonetheless, it is a challenging and interesting response to Haugeland's view, but one that is quite distinct from the points to follow in this paper.

[2] There are some rather radical analyses of behavior in the literature (e.g., Dretske 1988; Hornsby 1986). Moverover, most behavior of psychological interest cannot even be described without making psychological commitments. There are many ways to describe the movements one goes through in performing any action, and it is an old point that behavior is only explained under a particular description.

[3] It is quite natural to complain that Brooks' subsumption architecture is too simple to deal with the complexities of human existence, and so is an inappropriate model of

human cognition (see, e.g., Kirsh 1991). Though I think this is true, I do not begrudge Haugeland's attempt to make his case in increments. It would be bad enough, from my point of view, if Haugeland were right about ants and robots. I am therefore duty-bound to challenge his points even at this level.

[4]Servomechanisms are control structures that compare sensory signals to a stored representation (efference copy) and measure the difference between them (there are well-specified dimensions along which differences can be measured). Detected differences result in an error signal being sent to muscle groups whose activity will lessen the amount of difference between the sensory signal and the efference copy. This is the principle means of navigation in power-steering systems and guided missles. See (Merton 1973) for a readable introduction to servomechanisms.

[5]Massive environmental anomalies, of course, can corrupt the smooth flow of information between the environment and lower-level servomechanisms, and call for reassessments from above. But this is just what we would expect from a mind that is *largely* but not totally insulated from its environment; it presents no reason whatsoever to wash away the distinctions that separate mind, body, and world.

References

Brooks, R. (1991), "Intelligence Without Representation", *Artificial Intelligence* 47: 139-159.

Butler, K. (1992), "The Physiology of Desire", *Journal of Mind and Behavior* 13: 69-88.

Churchland, P. (1981), "Eliminative Materialism and the Propositional Attitudes", *Journal of Philosophy* 78: 67-90.

Dretske, F. (1988), *Explaining Behavior*. Cambridge, MA: MIT Press.

Dreyfus, H. (1972, 1979, 1992), *What Computers Can't Do*. New York, NY: Harper and Row (first two editions only). *What Computers Still Can't Do*. Cambridge, MA: MIT Press (third edition).

Ghez, C. (1985), "Voluntary Movement", in E. Kandel and J. Schwartz, (eds.), *Principles of Neural Science* . New York, NY: Elsevier, pp. 487-500.

Ghez, C. and Fahn, S. (1985) "The Cerebellum", in E. Kandel and J. Schwartz, (eds.), *Principles of Neural Science*. New York, NY: Elsevier, pp. 487-500.

Gibson, J. (1979), *The Ecological Approach to Visual Perception*. Boston, MA: Houghton Mifflin.

Haugeland, J. (1993), "Mind Embodied and Embedded", *Proceedings of Mind and Cognition: An International Symposium, May 28-30, 1993*. Taipei, Taiwan: Academia Sinica.

Hornsby, J. (1986), "Physicalist Thinking and Conceptions of Behavior", in P. Pettit and J. McDowell (eds.), *Subject, Thought, and Context*. Oxford: Clarendon Press, pp. 95-115.

Kirsh, D. (1991), "Today the Earwig, Tomorrow Man?', *Artificial Intelligence* 47: 161-184.

Marr, D. (1982), *Vision*. San Francisco, CA: Freeman.

Merton, P. (1973), "How Do We Control the Contractions of Our Muscles?", *Scientific American*, May: 30-37.

Simon, H. (1969, 1982), *The Sciences of the Artificial*. Cambridge, MA: MIT Press.

van Gelder, T. (1993), "The Distinction Between Mind and Cognition", *Proceedings of Mind and Cognition: An International Symposium, May 28-30, 1993*. Taipei, Taiwan: Academia Sinica.

Perception and Proper Explanatory Width

Mark Rollins

Washington University

1. Introduction

Marr's theory of vision (1982) is often said to exemplify wide psychology. The claim rests primarily on Marr's appeal to a high level theory of computational functions and secondarily on the sort of representations he posits for computational processes (Burge 1986, 28). I agree that Marr's theory embodies an exemplary form of wide psychology; what is exemplary about it is his appeal to perceptual tasks. But I shall argue that the result of invoking task considerations is that we need not adhere to Marr's own restricted construal of proper width. Indeed, the larger conclusion I want to draw is that there is no single conception of width that has a special place in explanation.

> That Marr *has* a conception of proper explanatory width for vision science can be brought out by considering two points:
>
> (1) First, on Marr's account, there is a distinctive (syntactical and algorithmic) level at which primitive computational processes are located. This follows from the fact that Marr's approach requires that we distinguish clearly the processes that implement computational goals from the biological processes that serve other noncomputational ends. It is this primitive level of computational processing that marks the boundary between the biological and the computational (cf. Pylyshyn 1986, lll).
>
> (2) Second, computational processes are identified with reference to perceptual tasks faced by the perceiver-agent. In the most general terms, the assumption is that the identification of computational processes takes its cue from a description of tasks set for the agent described in more-or-less familiar terms. These tasks are then decomposed into simpler and simpler tasks, until the level of mechanisms is reached. This view is also held by Dennett and Lycan, as one version of teleological functionalism. This version, *homuncular functionalism*, accepts the need for explanations in intentional terms, e.g. in terms of beliefs and desires, which require an agent or homunculus whose representations they are and who has access to their content. But the account is said to 'discharge the homunculus' in the end by virtue of the decomposition of the representa-

tional functions into nonrepresentational ones. As a model for explanation, this decomposition corresponds to a move from the intentional stance to the design stance. (Dennett p. 28). There are, however, two aspects of this view that need to be distinguished: the decomposiblity claim; and the focus on a certain level of decription of intentionally characterized tasks as the starting point of the decomposition. It is possible to accept one without accepting the other.

I want to reject both of the assumptions in Marr's notion of proper explanatory width. I deny that there can be a single best description of our functional organization which depends on the tasks we perform being all of a piece, i.e. describable in similarly general terms. What Marr's methodology assumes is not just that computational processes that serve ends like object recognition should be isolated from noncomputational supporting architecture; it also implies that the processes should be isolated *by locating them together at a certain level of abstract description.* That is, Marr's second tier explanations in algorithmic terms are thought to be *united* by a vocabulary with a certain degree of abstraction, which sets them apart from physical level descriptions. But I shall argue that, while it may always be possible to distinguish computational processes from some less abstractly described mechanism that implements them, the identification of computational processes should not be united by a common level of abstractness of their own.

Second, I shall argue against the assumption that the starting point of explanation should always be at the level of tasks set for the agent or whole person in intentional terms which are equally abstract in each case. It is open to the homuncular functionalist to argue on teleological grounds that what has primary value for thinkers and perceivers are ultimately true beliefs; thus, the production of true perceptual beliefs (e.g. about the identity of perceptual objects) could be set as the primary sort of goal posited in computational theory. But that, I think, would be a mistake (cf. Clark 1989, 35).

I should point out that Marr's own theory of vision does not actually take the production of true perceptual beliefs or other propositional attitudes as its primary computational goal, despite Marr's emphasis on the veridicality of perception (as that which needs to be explained). Rather, he posits tasks like object recognition. Still, the *recognition* that is required for a Marr-style task like object recognition is an intentional notion that implies the intelligent apprehension and use of representational content. What makes Marr's perceptual goals the target of my objection, as much as the production of propositional attitudes, is the fact that Marr tends to sort out the goals of computational theory by focussing on the highest level tasks of which the visual system is capable; tasks normally ascribed to the whole system or perceiving subject. My claim will simply be that this is not always appropriate. Perceptual tasks are much more multifarious than Marr's account would allow: computational goals are not all defined in similarily general or familiar terms.

The methodological point then is that perceptual task analyses of a more refined sort are required; and they open the door to descriptions of computational functions that are not all of a piece. The descriptions need not be confined to the same level of abstraction. *Perceptual teleology* (as we might call the identification of perceptual goals and tasks) thus begins to look like a multilayered affair. The need for guidance from the top down is not the issue. Rather, the issue concerns what the appropriate identification of the 'top' should be. That it is not always identified in the same sort of terms is one primary point I want to try to make.

A further clarification that needs to be made concerns Marr's treatment of *subpersonal* tasks. Marr's identification of computational goals is certainly not limited to the last stage of visual processing. He describes the ends that processes serve at every stage along the way. For example, edge detection is a computational goal, even though detecting edges is not really what I, the perceiver, do. Thus, the objection to his approach is not that he fails to provide for subpersonal perceptual tasks. Rather, the point is that such tasks seem to be largely *derivative* from a more primary level of description of high level systemic tasks; whereas I believe that sometimes they should not be construed in that way. As a computational goal, edge detection owes its place in Marr's theory to the fact that objects have edges; hence, it serves and is subordinated to the still higher, more intentional goal of object recognition. That may be feasible for edge detection, but I shall argue that not every subpersonal computational goal should be understood in this derivative sense. Less than fully intentional perceptual states are sometimes ends-in-themselves; they are narrower computational goals which are advantageous to perceivers on their own.

I shall argue in two ways: First, I cite evidence to suggest that perceptual structures and operations can serve multiple functions at different levels; and these are neither subordinate to a single sort of task nor merely alternative ways of describing a single function. Second, I argue that linking proper explanatory width to a task analysis of this sort provides the basis for a genuinely different model of explanation than that associated with classical computational theory and with the representational theory of mind. Further, it makes sense of an interesting claim that has sometimes been made, viz. that Marr's perceptual psychology is wide, but not in the same respects as propositional attitude psychology. Thus, the appeal to task analysis supports the conclusion that there are different degrees of proper explanatory width.

2. Multiple Teleologies

I begin by pointing out that the oft-noted *multiple realizability* of psychological functions corresponds to another relation between functions and structures that is less often discussed. According to the principle of multiple realizability, the relation of functions to structures is one-to-many. However, it is also possible for that relation to be *many-to-one*. One and the same physical state or process type can instantiate more than one functional state or process. In that case, we have the opposite side of multiple realizability, viz. *multiple teleology* (henceforth, MT); a single structure performing several functions.

In some animals, for instance, ear flaps can be used to flick flies and fan faces, as well as to catch sound. Likewise, neural structures sometimes have double functions. Evidence suggests that there are distinctive types of brain cells implicated in vision, which divide the task of vision into specialized functions. On the broad scale, there are parvocellular and magnocellular pathways originating in the retina and extending to visual cortex, which are dedicated to object recognition (the 'what' system) and spatial location (the 'where' system), respectively. Nonetheless, sometimes there is crossover: for instance, when spatial properties are used for object recognition. In that case, a single neural subsystem or pathway can play two different roles.

This example shows that MT applies to basic functions and processes, as well as to structures. It is the capacity to represent spatial properties, as much as the neural structures that support that capacity, that takes on a double role. In such cases, a certain causal power contributes to the overall behavior of the system in different ways, depending on the conditions under which it is employed and the task at hand. The implication is that a distinction can be made between the *occurance* of a function and the *deployment* of it in combination with other functions.

The first point of significance regarding this sort of MT is that it casts doubt on the hegemony of propositional attitude level explanation and on the autonomy of levels.

For one thing, the multiple computational ends that are served by a particular structure or function may not all have to be specified in the same sort of terms. For instance, a contour might be represented in the visual system as part of the process of forming a conscious perceptual belief. But it need not be. Contour representation may aid the unconscious avoidance of the surface of an unidentified object. In that case, the explanation of the second task may need to go no further than surface location, while the attribution of the conscious perceptual belief will require something more.

Moreover, because MT focuses on single structures with multiple functions (or processes that serve multiple ends), it reveals a sense of task diversity that is not simply due to the fact that a task can be described in various ways. That is, it shows how task diversity is not the same as task description relativity. What MT suggests is that different level descriptions are of *alternative* functions, which is not what the difference between terms, e.g. *waste eliminator* and *urea excreter*, suggests. The alternative functions might served at different times; but the relation is not necessarily temporal. (A face can be fanned and a fly can be flicked at one and the same time, by the same motion of a single structure.) What matters is that the two functions are not ordered; one does not implement the other. Thus the fact that there can be diferent kinds of explanations need not rest simply on the appropriateness of different ways of speaking for various *explanatory practices*, e.g. folk psychology versus cognitive science. It can rest on the evidence for MT, which implies the need for multi-leveled explanations entirely within the cognitive science domain.

What is especially important about MT, however, is that it weakens considerably a claim that has been made in defense of the Marr's methodology. The claim is that only those processes that help to realize computational goals expressed in terms of intentional, contentful, representational functions are *really* computational and cognitive. Thus, Fodor has argued that the 'contents' produced by lower level processes and the functions they perform are too *shallow* to explain thinking, reasoning, or perceptual categorization (1983, 94). Or, Pylyshyn has said, treating the processes that serve less exalted functions as part of the computational architecture threatens the principled distinction between computation and mere biophysical change (1986, 111).

However, one consequence of identifying cases in which two ends of different sorts are served by the same underlying process, is that the argument that only one of the ends is genuinely computational begins to look like a mere prejudice. The fact that two ends are served by what is identified clearly as a computational process (e.g., edge detection, described in algorithmic terms), where one of the ends is a standard computational goal (e.g. object recognition) implies prima face that the underlying process implements a computational function in the other case, as well, even though the end may not be described in fully intentional language. The two ends or tasks are on a par in two respects: They can be realized or performed by using the same ability, and both can be non-derived, final goals. If either function is the cause of behavior that is learned, that can be improved, and that admits of explanation in terms of some principles of content use, then the place of the nonintentionally described goal in computational teleology would seem hard to deny. What MT does, in effect, is bring out the basis for treating functions at different levels of explanation as computational; indeed, it establishes a presumption in their favor.

3. Evidence for MT and Why It Matters

It is important to make clear that MT is not merely the theoretical flip side the multiple realizability of functions. It needs to be shown that MT has both empirical warrant and philosophical significance. In theoretical terms, the result of MT is what might be called *token functionalism*: Any type of physical state, structure, or process will always instantiate some sort of function; but not necessarily always a function of the same sort. The structure-function relation, as much as the function-structure relation, will be type-token. A structure can be a *multiple realizer* of functions; or one function can implement several others. I shall discuss in a moment some important philosophical aspects of MT so construed. But first I want to point out several different concrete illustrations of how it might obtain.

(1) The most straightforward form of MT is found when a certain causal power (of the information carrying sort) interacts with more than one constellation of other causal powers. A good example comes from research on the primate visual system. Based on anatomical, physiological, and behavioral evidence, David Van Essen and his colleagues have identified thirty two distinct cortical areas associated with visual processing in the macaque monkey (1992). Among those visual areas, *three hundred and five* interconnected pathways have been found. What this anatomical cross talk shows is that nearly every function in early cortical analysis can contribute to the performance of two different higher order tasks. To that extent, MT is actually quite far-reaching in scope. Moreover, as Van Essen, et al, argue, the intertwining of processing streams requires a variety of information processing "strategies" to account for efficient information use. Such strategies are themselves unlikely to be comprehensible in terms of a single level of explanation.

(2) One strategy involves the *multiplexing* of information: While cells are functionally specialized, they can still carry more than one type of information. Although the interconnectedness of processing streams noted above is explained partly in terms of such multiplexing, from the point of view of MT, the ideas are really different. One concerns multiple routes of information flow, the other concerns multiple information carrying functions. The possibility of multiple functions for specialized cells raises the problem of *control*, a responsibility often assigned to an internal attentional mechanism. In terms of MT, that control amounts to an allocation of functional resources to different tasks.

(3) One way to understand multiple functionality is in terms of short term *modulations* in functions. (Again, the ideas are distinct: Multiple functions might be simultaneously present as unactivated potentials, rather than as abilities acquired quickly and then lost). Van Essen et al describe feedback pathways in the monkey's brain, for instance, which contribute to computational flexibility through regulating inputs and switching targets to which the results are transmitted (input and output gating). Such control processes, they speculate, have a pervasive importance and "allow the brain to reorganize its computational structure adaptively, on a rapid (~100ms) time scale, for optimal utilization of the incoming data and available neural resources" (1992, 422). Here, the actual routing of information along the multiple interconnected pathways noted in (1) above is explained by a modulation of functional connections, due presumably to experience and task demands.

Thus I think it can be shown that MT has a strong empirical grounding. Still, the hard question remains: How are we to understand the phenomenon? For one thing, it immediately presents a conceptual puzzle. It might be thought that the extension of MT to basic functions is foreclosed by the generally *holistic* character of functional role theory; that is, by the identification of functions as causal roles within an interconnected complex. If the identity of a function depends on its place in such a complex, any change in the causal chain would seem to reconstitute the causal role in question. And in order to account for MT, alternative higher order ends that are realized will have to be vested in some difference in cause and effect. In that case, we might think, there simply can not be a *single* basic function that contributes to two different higher order functions.

Of course, one way to respond to this puzzle is to note that the network construal of the identity of functions does not by itself require that a role be identified in terms of *every* causal connection it happens to have. The taxonomy of roles that underlie cognition can, for instance, ignore causal relations that are *computationally irrelevant*, given a particular theory of computation. Add to that the possibility of overlapping causal relations which appear in the identification of different roles, and the basis for MT begins to emerge. Thus information about shapes, for instance, can be the output of a certain process and provide input to two higher level processes. The latter share shape representation as part of the input/output relations that define them; but shape representation itself need not be specified with reference to either one of them in particular. It only requires a role in providing input to some such process. In that sense, a specialized role like shape representation can be located in any constellation that requires the inclusion of information about shapes.

However, a more profound and problematic version of multiple functionality can occur when a previously specialized function actually *changes* the contribution it makes. For instance, shape representation processes might supply input to further processes which do not employ the input *qua* information about shape. In Dennett's words, "the specialists are also recruited as *generalists*, to contribute to functions in which their special talents play no discernible role" 1992, 271).

The theoretical problem then is that the functions or processes in question somehow must both retain and give up their content-specific (i.e. special information carrying) roles. Dennett claims that we have no adequate model for such cases, in part because such "multiple, superimposed functionality" is very difficult to understand from the perspective of "reverse engineering."

Aside from the need for multiple levels of explanation (which is obvious in this case), that reverse engineering is the wrong approach to explanation is precisely the point for which I have been arguing. (Reverse engineering is essentially Marr's methodology expressed in Dennett's terms of the intentional, design, and physical stance.) In effect, the possibility of MT shows that Marr's approach must be on the wrong track. Thus I think the case against it has been made.

4. Degrees of Explanatory Width

Something of the larger philosophical significance of MT can be brought out by considering its bearing on the notion of degrees of explanatory width. It has been argued, by Patricia Kitcher and others, that Marr's theory of vision actually implies a level of explanation that falls somewhere *between* the design stance and the intentional stance described by Dennett (Kitcher 1988, 15; Sterelny 1990, 90). Thus, although Marr's theory is a version of wide psychology, it is wide in a different sense than intentional psychology of the sort that is inspired by folk wisdom and common sense. The distinction is supported in part by the fact that, on Marr's account, concepts — which seem essential for belief formation — are not required for perception. That makes sense: A perceiver can distinguish one object from another without having concepts for the objects involved. And Marr's account of perceptual recognition stops short of matching 3D representations with concepts drawn from memory. Thus, prima facie, the sort of perceptual content that falls within the scope of Marr's theory is not conceptual content and stops short of the content of thought and belief.

While the proponents of this interpretation do not put the point quite this way, the implication is that Marr's computational theory is wide, yet not wide *to the degree* that full-blown intentional psychology is. Thus, the argument suggests that in general there can be varying degrees of explanatory width. The interesting possibility that this opens up is that still further distinctions might be made in the sense or degree in which wide psychology is required for an account of perception. The question then is how the purported intermediacy of Marr's preferred degree of explanatory width should be understood. Two possible conceptions have to do with (a) causal sequences in stage-related processes and (b) the semantic properties of representations that are invoked in explanation. But neither of these conceptions is adequate.

First, as Kitcher describes it, any theory of belief must be "grounded in" a theory of perception, in the sense that perceptual information is relevant to the intentional objects of belief; the latter (even when identified by imperceptible features) are derived from the former (1988, 15-16). This might be taken to mean that belief production, as a later stage in information processing, is *causally dependent* on prior perceptual functions like object recognition.

However, the direction of a causal relation is clearly not be enough to warrant the claim that explanations in terms of the one are wider (or wide in a different sense than) explanations in terms of the other. The identification of subpersonal computational goals suggests that, in Marr's stage-dependent model of vision, the computational theory for the effects at later stages is no wider than it is for the causes at earlier stages. Edges, after all, belong to the world as much as objects do.

The second possibility is that the general division between narrow and wide explanation simply coincides with the absence or presence of certain *semantic properties*, notably reference or truth, which depend on a relation between the representational system and something else. Explanations are either wide or narrow, one might say, depending on whether those types of semantic property are invoked or not. They determine wide content; it is thus on the width of representational content that the width of explanations in terms of content depend. The claim that there can be different senses of width could then be put in similar terms. Explanatory width may vary, we might say, depending on precisely *which* semantic properties are invoked. Without concepts there are no propositional attitudes, hence no propositions to which to ascribe truth values. Still, the production of perceptual representations is supposed to proceed under a reliability or veridicality constraint, and the representations are supposed to refer to whatever sort of objects are their typical initiating causes. Marr's account is thus wide in requiring reliability and reference, we might argue, but not so wide as to require truth conditions (or the rationality that is assumed in belief attribution). That would leave explanatory width to be determined by the particular mix of representational properties on which explanation is supposed to depend.

The problem is that, as a basis for understanding explanatory width, the appeal to semantic properties has things backwards. To be sure, the question of what is required for the explanation of perception is partly a question of what sorts of representations, and thus what sorts of semantic properties, are required for perceptual processing. But to make the model of explanation *parasitic* on the theory of content, i.e. to derive an understanding of explanatory width from an account of semantic properties, is to leave the important questions unanswered. The semantic account does not tell us why a particular level of abstractness is appropriate for behavior of a certain sort; it does not tell us why generalizations of one particular kind or another are required of the performance of any given task; why only representations of a certain sort should be explanatorally adequate. The only way to answer these questions is to analyze the nature of tasks in a domain, e.g. in perception.

Thus, we should turn the relation between explanation and content around. Instead of assuming that the width of explanations depends on the width of the contents they ascribe, we can take the width of content ascribed in psychological explanations to depend on the width of explanation, which is established on some other grounds. The natural proposal then is to simply develop an account of levels of explanation directly in terms of relations among goals or tasks. Such an account can be grounded on the fact that complex tasks can be decomposed into simpler ones; for that suggests that complex task descriptions are more general or abstract than simple task descriptions. In effect, complex tasks supervene on simpler ones; explanations of higher-order tasks do not reduce to lower-order ones. That is because a complex task need not always be decomposed into the same subtasks. For instance, object recognition can be broken down into edge detection and shape representation; but it is no more wedded to one specific set of subtasks than it is to a set of algorithms. In that sense, tasks can vary in the level of abstractness of their description, i.e. in their relative independence from any particular analysis in terms of subtasks. Thus what sort of task is identified as the starting point of an analysis (the computational goal in Marr's sense) will determine the scope of the explanation required. Identifying the representational properties that are required to perform the task will then be a secondary matter.

The advantage of this way of thinking is precisely that it allows more flexibility in the construal of proper explanatory width. Although in principle degrees of width might be described in terms of various combinations of semantic properties, in fact, the possible combinations tend to be limited so as to reinforce a fundamental narrow-wide opposition. To break out of that opposition and move beyond the current ideological stalement, it is essential to emphasize the primary importance of task analyses on Marr's account.

This task-analytic approach is entirely consistent with Marr's conception of wide psychological explanation: Although representational content is required for perceptual tasks, as Burge notes, for Marr, it is the environment that "sets the task" (Burge, 1986, 43). Yet, in application, Marr's own conception of proper explanatory width tends to focus too exclusively on tasks at a certain level of abstractness. Thus my argument for the *need* for such an account differs from Marr's. It rests on the evidence for multiple functions and tasks for perceptual processes introduced in the first section; on the evidence for what I have termed MT. This argument ultimately works against Marr's own way of putting his model into play. What recent evidence for short-term perceptual plasiticity suggests is that we have the capacity to strategically reallocate resources in the performance of various tasks. To account for this capacity, a new and more task-relativistic model of proper explanatory width is now required. The question of what sort of explanation perceptual psychology should employ, narrow or wide, is the wrong question to ask.

References

Burge, T. (1986), "Individualism and Psychology," *Philosophical Review* 15:1, 3-45.

Clark, A. (1989), *Microcognition*. Cambridge: MIT.

Dennett, D. (1986), *The Intentional Stance*. Cambridge: MIT.

_ _ _ _ _ _. (1992). *Consciousness Explained*. Boston: Little Brown.

Fodor, J. (1983). *The Modularity of Mind*. Cambridge: MIT.

Kitcher, P. (1988), "Marr's Computational Theory of Vision," *Philosophy of Science* 55: 1-24.

Marr, D. (1982), *Vision*. San Francisco: Freeman.

Pylyshyn, Z. (1986), *Computation and Cognition*. Cambridge: MIT.

Sterelny, K. (1990), *The Representational Theory of Mind*. Oxford: Blackwell.

Van Essen, D., Anderson, C.H., and Felleman, D.J. (1992), "Information Processing in the Primate Visual System: An Integrated Systems Perspective," *Science* 255: 419-423.

Is Seeing Believing?[1]

David Hilbert

California Institute of Technology

1. Introduction

The precise nature of the connection between perceptual experience and empirical knowledge is one of the traditional problems of philosophy. That much of our knowledge is somehow based on our perceptual experience has been a common element of many otherwise disparate views as to how beliefs about empirical matters are justified. Although this relationship is now often conceived to be causal rather than logical, it is still rarely doubted that there is some kind of intimate connection between experience and justified belief. Any view of this kind leads quite naturally to the idea that the function of perceptual experience in our epistemic economy is to lead to knowledge of those matters of fact that cause or are represented by that experience. Although I do not wish to deny that this is one epistemic role that experience can play, I will argue below that there are others as well. I will also argue that perceptual experience, as traditionally conceived, is not a necessary component in the acquisition of knowledge by means of the senses. The relation between sensory experience and perceptual knowledge is both more complicated and less intimate than many philosophers assume.

It may not generally be the case that truth is stranger than fiction, but it is certainly so with regard to the possibilities for things going wrong with the processes of visual perception. Although it is traditional for philosophers to appeal to hypothetical rather than actual examples in motivating and justifying their conclusions, my impoverished imagination and some doubts that my project is entirely one of conceptual analysis leads me to draw my examples from the empirical literature. Accordingly I would like to start with a few summaries of reported cases in order to motivate my discussion and help suggest some plausible hypotheses about the relation between perceptual experience and perceptual belief. Although there are serious problems in interpreting the apparently bizarre patterns of behavior that some of the subjects in these cases exhibit, in my initial description I will take the published reports more or less at face value. I will later consider just how plausible is the description of the more troubling of these cases. These case histories will serve, I hope, to generate a sense of puzzlement about how perception works. Attempting to clear up these puzzles regarding abnormal cases will lead to some conclusions that, hopefully, apply to more normal cases as well.

2. Some Case Studies

2.1 Visual knowledge without visual experience

Case 1

Some cases of cerebral achromatopsia (color blindness due to brain damage) display a preserved ability to locate boundaries between areas on the basis of color information without any awareness of what property it is that defines the boundary. M.S. becomes ill with a virus and suffers neurological damage as a result. One striking finding is apparent complete color blindness. All peripheral components of the color vision system are, however, demonstrably intact. M.S. denies seeing color, cannot sort objects by color, cannot name the colors of objects shown to him, and in most circumstances behaves as if he were completely color blind. In some circumstances, however, M.S. is able to recognize figures whose shapes are specified only by color boundaries. When asked what differentiates figure from ground M.S. is unable to provide an answer. Visually, the figure looks exactly the same as the ground, according to M.S.[2]

Case 2

The patient suffered a stroke that resulted in extensive bilateral damage to several parts of the visual cortex among other areas.[3] In his own words, he had been, "Sitting, eating breakfast...looked up and everything went black" (31). His condition was (mis?)diagnosed as blindness due to bilateral cerebral infarction (stroke). He was re-examined two years later and at that time denied any recovery in his visual abilities. He had, however, been living independently and was able to cook for himself, type, and walk without assistance in unfamiliar surroundings. He had never availed himself of any services for the blind and had failed to apply for any of the benefits or services for the visually disabled available to him through the Veterans Administration.

Visual examination of the patient revealed that he had small (30 degrees) spared areas in the upper right quadrant of both visual fields. Within this spared area his visual acuity was no worse than that of a moderately nearsighted individual. The patient was able to read words, although inaccurately and with difficulty, and was able to recognize objects and faces presented visually. What makes this case more interesting than a simple case of earlier misdiagnosis is the patient's resolute refusal to admit that he was able to see. When asked how he was able to successfully perform these visually mediated tasks he denied any awareness of visual perception and made responses like "I feel it," "I feel like something is there," "it clicks," or "I feel it in my mind" (33). I should also mention that with the exception of some mild memory and language impairment the patient showed no evidence of confusion or dementia. His independent life-style and failure to apply for benefits on the basis of his claimed blindness would also seem to rule out a mercenary motive for his denial of visual experience.

2.2 Visual experience without visual knowledge

Case 3

B.M. has had the entire right half of her cortex removed to control very severe and otherwise untreatable epileptic seizures. In spite of this radical surgery, most of B.M.'s cognitive and perceptual abilities appear to be normal. Her IQ is normal and she performs within the normal range on a number of other tests. She is partially blind to stimuli in the left visual field, but in the spared visual field most tests of visual function put her within the normal range. In particular, her spatial acuity is normal, as is her color vision. There is one striking area, however, in which B.M.'s visual abilities are severely impaired. She has almost no ability to identify or recognize faces, a condition known as prosopagnosia. This impairment does not just involve remembering what name to associate with a face or whether she has seen a face before but extends to judgments as to whether simultaneously presented photographs are photographs of the same face or not. If the photographs are identical she can detect this identity but if there is a difference, say in expression, her ability to detect identity falls to near chance levels. In addition, familiarity does not seem to help, since B.M. fails to recognize photographs of her mother's face, with whom she lives, or even photographs of her own face.

In spite of her severe impairment on tasks involving facial identity it is clear that the source of B.M.'s difficulty is not a general problem with identity or an inability to detect the features of faces. B.M. is capable of producing names in response to verbal cues and of describing familiar people when prompted with their name; she can even recognize individuals on the basis of visual information. Her ability to classify faces by gender and age is nearly normal, as is her ability to make judgments about emotional states on the basis of facial expression. B.M.'s visual impairment is remarkably specific to facial identity and coexists with a relatively normal ability to visually mine facial features for information about age, gender, and emotional state.[4]

3. Some preliminaries

I propose to explore questions about the relation between perceptual experience and perceptually obtained knowledge or belief. Although I have no detailed account of the nature of perceptual experience, I can roughly sketch the general nature of perceptual experience and how it differs from perceptual belief. Perceptual experience is imagistic, rich and detailed, and its components are modality specific. By modality specific I mean that the elements of perceptual experience can be differentiated by the sensory modality that produces them. The visual experience of a square is different from the tactile experience of a square. The other three characteristics are closely connected, and some idea of what I have in mind may be made clear by considering the difference between the visual experience of a square object and the belief that that object is square. The visual experience of a square, as Berkeley pointed out long ago, must be of some color, and this color must be either homogeneous or differentiated. I can, by contrast, merely believe of an object that it is square without also believing it to be of some particular color. This difference is not confined to the secondary qualities. The visual experience of a square must represent it as standing in some more or less definite spatial relation to the other objects in my field of view; this again is not true of my belief. Although this crude characterization is clearly inadequate in several respects, I hope that the characteristics of visual experience will become somewhat clearer in what follows.

In addition to the general characteristics of perceptual experience just mentioned, it is commonly held that there is a limited set of qualities that are present in the experiences associated with each sense. For vision, my main object of concern in this paper, it is traditional to maintain that our experience includes only color and shape. Whatever knowledge we manage to obtain visually of other properties is somehow derived from, causally or logically, our experience of this limited set of sensible qualities. Although this traditional view has waxed and waned in popularity over the years, and relatively few contemporary philosophers would be willing to endorse it without serious qualification, it is this traditional conception of visual experience that I will be investigating in this paper. My justification for proceeding in this somewhat arbitrary way is twofold. First, I suspect there is more sympathy for this conception of visual experience than the published literature might indicate. Second, and more importantly, even those who do not subscribe to the traditional conception of visual experience typically wish to enrich, not impoverish, its properties. The puzzles I discuss below are puzzles even for such enriched conceptions of the content of visual experience. I will return to this point below.

4. Some Puzzles

4.1 Is seeing necessary for believing?

It is, of course, not true that it is necessary to see an object in order to have a belief about it. I can come to believe that the carpet in the seminar room is brown by being told by someone I trust that the carpet is brown. What would be more interesting to learn is that I can come to believe (accurately?) that the carpet is brown when the only source of information I have is the operation of my eyes and the rest of my visual system, without the visual experience that normally is involved in seeing color. In such a case, if this is a genuine possibility, vision is operating directly to give me beliefs about color without the mediation of any visual experience. Case 2 above is such a case. The patient denies any visual experience but at the same time acquires (more or less) accurate beliefs about the objects around him, beliefs whose only possible source is the operation of vision.[5]

There are reasons to be a little skeptical that Case 2 is really correctly described. The patient in Case 2 has mild language difficulties and his descriptions of his experience are somewhat ambiguous. Although he insists that he does not see in the way he did before his strokes, he does talk of feeling that something is there. Given his degraded visual abilities it may be that he still has visual experience but that it is different from the kind of visual experience he had before his strokes. Since the patient is also described as being somewhat aphasic, i.e., suffering from language difficulties, and suffering from fairly severe attentional deficits, a certain amount of skepticism about the case is in order. Case 1, although less general and consequently less striking, is partly for the same reason more convincing. M.S. is demonstrably completely color blind by almost every test imaginable. Moreover, M.S., like most achromatopsic subjects, describes his visual experience as lacking color. In addition, M.S. is not alone in his apparent combination of total loss of color vision with the ability to discriminate some boundaries that are defined solely in terms of color difference. There are now a handful of known cases like M.S., and they all share with M.S. the ability to recognize boundaries that separate areas that are experientially indiscriminable. These cases would seem to show that in this limited domain it is possible to have beliefs that are based on visual information without the relevant visual experience.

4.1 Is seeing sufficient for believing?

As we have seen, there is some reason to suppose that visual experience is not a prerequisite for obtaining accurate beliefs on the basis of visual information. It may seem like old news to be told that visual experience is also not sufficient for the formation of beliefs based on visual information. I can, for example, look at a begonia in good light, with my glasses on, and still fail to form the belief that there is a begonia before me. It is a familiar point that we can see an object and that object can present some distinctive appearance to us and we can still fail to recognize that object. I fail to recognize begonias even when visual experience is indicative of the presence of a begonia because I don't know what begonias look like. This sort of failure of visual experience to lead to the appropriate belief, although instructive, is not my concern here. These kinds of failures throw little light on the causal structures that produces visual belief.

What is more puzzling and more instructive than these familiar failures to recognize the things that we see are the kinds of pathology exemplified in Case 3. Here we have people who, although they appear to perform normally on most visual tasks, are strikingly deficient in one limited domain. B.M. is not lacking in experience of faces; she sees them everyday and is just as impaired in recognizing the most familiar faces as she is in recognizing new faces. She has no general difficulty in recognizing the things that she sees, since her visual abilities are nearly normal with regard to other categories of objects. And most important she does not lack visual experience of faces and their parts. Her ability to classify faces by gender, age, and emotional expression strongly suggests not only that she sees faces and their parts but that she is also visually aware of the details of these parts and their relation to one another. There is no evidence that her visual experience of faces differs in any marked way from that of an unimpaired person. None of the ordinary explanations as to how a person can have visual experience of an object and still fail to form the belief that normally accompanies that experience seem to apply.

4.3 Does seeing have anything to do with believing?

My discussion so far suggests, I think, a certain kind of model of how vision works. This model is incomplete, but consideration of it and the ways in which it is incomplete will prove instructive in arriving at a more adequate picture of visual perception. By way of motivating this model I will make a brief digression into artificial vision.

Computer scientists have recently devoted some attention to the task of developing artificial systems for recognizing faces. Some of these systems are reasonably successful in classifying faces as familiar or unfamiliar. That is to say, you train the system on a restricted set of faces and then present it with a broader range of faces. The system, if everything goes well, will produce one kind of output for the faces from the training set and a different kind of output for other faces. One characteristic of successful systems of this sort is that they ruthlessly discard information that is irrelevant to their task. If the algorithm does not require information about facial blemishes like moles or warts, the system will never extract these features from the image. In this regard, they are very different from human vision, since visual experience contains a large amount of information about features of the world that are irrelevant to the perceptual task at hand. Another interesting feature of these systems is that the features of the image they rely on in successfully performing their task are often quite different from those that seem salient to human beings. Although the presence or absence of moles may make a difference to those image properties the system does rely on, it may be the case that the presence or absence of facial moles as such is never represented anywhere in the course of processing. Face recognizers do not have as one of their goals the parsing of the face into individual features and may rely on much more abstract image properties instead. In fact, success in computational vision would come more easily than it does if researchers could rely on our intuitions about which features of the image are important for particular tasks and which are not.

Consideration of the cases presented above suggests that in some ways human vision is similar to these artificial vision systems. Prosopagnosia is plausibly thought to be the result of damage to brain systems that are devoted to extracting information about facial identity and nothing else. That is why the impairment can be so selective. Moreover, there is no reason to suppose that such a special purpose system would make any more use of those features of faces that are salient in visual experience than do artificial vision systems. Case 3 would seem to provide some positive evidence for this claim. In Case 1 we see that the outputs of visual processing can remain intact even if brain damage interferes with the representation in experience of some of the image features that contribute to generating those outputs. The part of the visual system that is concerned with detecting material boundaries has no concern with identifying the difference between the two sides of the boundary. Thus brain damage can produce the result that a boundary is detected in spite of the fact that there is no information in visual experience about *how* the boundary was detected.

These considerations suggest a model of human vision consisting of a collection of special purpose mechanisms that are individually concerned with detecting certain features of the visual world and collectively providing us with our perceptual beliefs. If this is true, then there is a puzzle about what role visual experience plays in vision. As we have seen, such dedicated mechanisms need not require anything resembling visual experience in order to carry out their assigned tasks. We can recognize faces, but why do we see eyes, noses, and warts as well? We can detect object boundaries and material changes, but what is the point of filling in these regions with color, as we normally do? Enriching the range of properties that may be represented in visual experience will not help. Adding spatial boundaries that are not color boundareas or adding facial identity to visual experience leaves the problem untouched. There is a still a puzzle about the presence in visual experience of the traditional elements of color and shape that appear, on the current model, irrelevant to producing our visual beliefs. Visual experience provides us with our rich phenomenology, but on this model it seems disconnected from the main business of vision, which is telling us what the world is like. It may be pleasant to be aware of some of the features that underlie the outputs of visual processing, but this awareness can be misleading. Morevoer, it seems to contribute nothign practical to our well-being. The question at hand is then, "Can we, with our rich phenomenology, do anything that could not be done by someone completely lacking visual experience, like the patient in Case 2?"

5. Some Answers

5.1 A few cautionary remarks

We should not be overly hasty in jumping to sweeping conclusions about the irrelevance of visual experience to normal functioning on the basis of cases like these. The case description does not tell us that the patient suffers no deficiency in the formation and assessment of perceptual beliefs. Although his abilities are surprising, there may be other abilities that he lacks. Some deficiencies in visual functioning are surprisingly hard to detect on the basis of casual observation of behavior. Color blindness, for example, is a defect of visual function that is only clearly revealed in rather special circumstances. Although color blindness has presumably been present in the human population throughout recorded history the first clear recognition of this common disorder occurs in the seventeenth century and the nature of the deficit was not well characterized until the early nineteenth century.

5.2 A limited proposal

As a partial step towards restoring visual experience to some of its former glory I will argue that there is at least one way in which it contributes to visual belief. One difference between those of us with visual experience and those real and imagined people that lack visual experience in whole or part is that we possess, on the basis of our visual experience, information about how we came to have our visual beliefs. The patient in Case 2 just knows, or as he puts it "feels things in his mind." He is ignorant of how he came to know or believe the propositions he believes as a result of the operation of his visual system. Similarly, in Case 1, M.S. has no information about what difference identifies the boundaries he believes are present. Having visual experience allows us to know that some of our beliefs are visually derived; having color experience allows us to know that some of our beliefs are based on color differences. In these cases the visual experience does not provide the evidence on which the belief is based, nor does it figure in the causal production of those beliefs. Nevertheless, it provides information about those beliefs.

Since the experience of *seeing* a person to my left is different from the experience of *hearing* a person to my left, this difference can tell me that my belief that there is a person to my left is based on or derived from visual as opposed to auditory evidence. Similarly, since there is a difference between the experience of seeing a *color* boundary and seeing a *texture* boundary, this difference can allow me to know that my belief that there is a boundary present is based on color as opposed to texture differences. This kind of information may seem of little practical value but a little reflection shows otherwise. Our senses are far from perfect and their reliability varies with the circumstances of perception. Not only does the reliability of the senses vary with external conditions, but the different senses vary in different ways. In a dim light vision is very unreliable for many tasks while audition is unaffected. In a noisy environment audition is severely impacted while vision may be perfectly reliable. Even within a single sense there is similar variation in reliability. Under some illuminants color differences are unreliable guides to finding edges, while texture differences are quite conclusive. Under other viewing conditions, texture is unreliable while color is trustworthy. Thus, having information about the sources of our belief can be very helpful in determining how trustworthy our beliefs are in various circumstances. If I am walking down a dark street and entertain the belief that there is a person lurking in the bushes, it is of great practical value to know whether this belief is based on hearing him breath, as opposed to visually discerning a dim outline. People without perceptual experience would be impaired in their ability to adjust the confidence with which they hold their perceptual beliefs to the circumstances of perception.

There is some empirical evidence to support this contention. Studies have been done of achromatopsics who have spared ability to detect color boundaries in which they are also asked to provide confidence ratings about their judgments of the presence or absence of a boundary. Their confidence in their judgments did not change with variation in task difficulty and performance, while normal subjects did vary their confidence appropriately. Lacking color experience, subjects like M.S. are not able to tell that their judgments are based on color information and are not able to accurately assess the reliability of such judgments as the conditions relevant to color perception vary. Thus although color experience may not directly produce the relevant beliefs it is implicated indirectly, in the assessment of the reliability of such beliefs.

It is not my claim that such a mechanism for informing perceivers of the origins of perceptual belief necessarily must operate through the medium of perceptual experience. Rather, I am claiming that it is a fact about human beings that we do obtain this information in this way. There may be other possible sentient beings who obtain this information using other mechanisms; nothing I have said here in any way bears on this possibility. Unless this possibility is ruled out, it remains an open question as to why visual experience plays this role in us. Such possibilities seem to suggest that the role of perceptual experience in reliability assessment is not the whole story about the epistemic function of experience.

5.3 Some vague conclusions

I do not think that the role of perceptual experience is limited to enabling us to assess the reliability of our perceptual beliefs, but I will only wave my hand in the direction of some other possibilities. One obvious problem with the conception of human perception as a collection of special purpose mechanisms is that human beings are not specialized animals. The most distinctive feature of human beings is the degree of plasticity in our behavior and the range of different environments in which we can function. Special purpose mechanisms do not adapt with any degree of ease to new tasks. Our ability to learn new discriminations and novel systems of perceptual categories is impressive, and I suspect that something like visual experience is implicated in this ability. There may, after all, be something right about the philosophical tradition that perceptual belief requires perceptual experience, although the exact nature of the connection may not be what is traditionally supposed.

Notes

[1] An earlier version of this paper was read to the Philosophy Department of UC-Riverside and I am grateful for their comments. I would also like to thank Frank Arntzenius, Alex Byrne, Fiona Cowie, Carl Hoefer, Marc Lange, Nigel Thomas, and Jim Woodward for helpful comments and conversations.

[2] M.S.'s color vision was first characterized in Mollon et al. (1989). Further details can be found in Heywood et al. (1991).

[3] The description of this case is drawn from Hartmann et al. (1991).

[4] The description of this case is drawn from Sergent and Villemure (1989).

[5] It might be objected that Case 2 merely shows that *awareness* of visual experience is not causally necessary for visual belief. I am not sure whether this separation of experience and awareness is well motivated but in any event it mereley shifts the location of my puzzle. The question now becomes one of the connection between *conscious* experience and belief. See McGinn (1991), Chap. 4 for a discussion of these issues.

References

Heywood, C.A., Cowey, A., Newcombe, F. (1991), "Chromatic discrimination in a cortically colour blind observer," *Eur. J. Neuroscience* 3, pp. 802-812.

McGinn, C. (1991), *The Problem of Consciousness*, Oxford: Basil Blackwell.

Mollon, J.D., Newcombe, F., Polden, P.G., and Ratcliff, G. (1980), "On the presence of three cone mechanisms in a case of total achromatopsia." In G. Verriest (ed.), *Colour Vision Deficiencies V.* Bristol: Adam Hilger, pp. 130-135.

Roeltgen, D.P., Loverso, F. L. (1991), "Denial of visual perception," *Brain and Cognition* 16, pp. 29-40.

Sergent, J., and Villemure, J. (1989), "Prosopagnosia in a right hemispherectomized patient," *Brain* 112, pp. 975-995.

Simplicity, Cognition and Adaptation: Some Remarks on Marr's Theory of Vision[1]

Daniel Gilman

The Pennsylvania State University

1. Introduction

David Marr's theory of vision has been a source of both fascination and confusion for many in cognitive science. There has, of course, been substantial technical interest in the particulars of Marr's model. But beyond this we have seen his work cited in debates about the nature of mental representation and computation, the structure of cognition, the role of theoretical knowledge in perception, how and indeed whether one ought to apply results from cognitive science to antecedent questions in epistemology. Marr's theory is cited in such debates because it is widely seen as a successful, rigorous account of a large scale cognitive/perceptual process (if not as *the* successful such account). But it has been suggested that Marr's theory might be mistaken as a whole; that Marr might have been wrong not just about one or several of the sub-mechanisms of vision, and not just in offering an incomplete account, but wrong in his basic approach to the study of vision. I am especially interested in Patricia Kitcher's (1988) claim that Marr's theory may be wrong by virtue of its being an "optimizing" theory.[2] I am going to argue that this criticism is largely misplaced, that Marr's theory is not purely an optimizing theory. I am going to suggest, further, that establishing this prompts interesting questions for the cognitive and brain sciences; notably, questions having to do with inter-level—and inter-disciplinary—constraints on inquiry and questions about the role of simplicity as an aesthetic constraint on techniques, theories, models and explanations in the cognitive and brain sciences. Whereas simplicity has often been taken as an extra-empirical virtue in the sciences (see, for example, Chandrasekhar 1987 and Schoemaker 1991) here we are pushed to consider contexts in which its utility is an empirical question.

2. Biology, Cognition and Engineering

Kitcher is concerned about whether Marr's approach is appropriate to the study of biological organisms. She suggests that mechanisms meeting optimal design criteria are liable to be too good to be true; genuine adaptations work *well enough* but tend to bear little resemblance to the sorts of things we would see as elegant engineering solutions to the problems posed by the environments in which organisms evolve. Kitcher provides a sort of panda's thumb[3] argument against Marr. She underscores the importance of seeing the visual system as an evolved functional system, or adaptation, in a biological organism. Organisms, on this line, possess mechanisms jury-rigged from available phenotypic properties of ancestors;[4] consequently such mechanisms are liable to appear inelegant from a theoretical design standpoint. Just as the panda's thumb—an enlarged radial sesamoid—is "from a design point of view . . . pretty klutzy" (Kitcher 1988, 22) so we might expect other evolved devices to be somehow awkward. Thus, when models of such devices are developed top-down, supposing optimal design, their authors work from an assumption that might very well turn out to be empirically incorrect. Marr and his followers, she asserts, have worked in just such a top-down fashion, with just this supposition of optimality.

> If they are wrong in this very substantial assumption . . . then the unified theory of vision that Marr dreamt of will be impossible. The passages from global Computational theories to local theories, and to algorithms, and to biological hardware will not be at all smooth, or perhaps even possible. (1988, 23)[5]

Kitcher is worried about what she sees as Marr's top-down optimizing approach to—or "functional decomposition" of—the task of vision. She certainly does not think that functional decomposition of the task of grasping would have led to anything like an accurate account of the panda's grasping ability. In that case, she says,

> Only a very loose task description is part of the story, the pseudothumb enables the panda to hold bamboo shoots. Most of the details of the capacity are provided by the 'implementation' level. (1988, 22)

What's difficult to see here is how vision science might proceed in an analogous fashion. A description of the visual task as "loose" as that of the grasping task provided above would be very sparse indeed.[6] Could vision science be developed *entirely* in a bottom-up fashion if this were the extent of the higher level theoretical understanding guiding our biological enquiries? Consider this: one examines, say, the anatomy and physiology of cells in area 17 of the cortex guided by the understanding that *somehow* these cells aid in the process of discovering the form and location of objects in the world. But *at what scale*? Single neurons? Sub cellular structures? Metabolic or electro-chemical events within neurons? Along neural-membranes? At synapse? Between clusters of neurons? If clusters, how large and organized according to what principles? Are there *in principle* answers to these questions? Even if we pick an important level of focus, what's to guide our inquiries such that they'll likely be helpful to those working on related brain centers? More difficult still, in what terms might we develop a rigorous account of functional interaction between vision and other cognitive mechanisms? And without any sense of the decomposition of visual tasks, how do we isolate anatomical elements that, for example, serve some sort of information processing task in vision (supposing some do) from those that provide for some sort of support function? How do we know when we're done identifying centers of visual processing? Of course historical enquiry *might* give us a leg up on such questions. But when we're looking at little bits of cortex with extremely complex functional interconnections the fossil record is liable to be rather less help than it might be with the anatomical ancestors of grasping devices. This is not to disparage the utility of an evolutionary component to a multi-level, interdisciplinary approach to vision. That genetic or historical research might provide insight into, e.g., developmental questions, or the environmental factors shaping level one requirements of the system, is hardly controversial. But the notion that such a complex, emergent functional system might be unpacked from *just* these sorts of investigation seems seriously problematic.

3. Metatheory, Theory and Practice in Vision Science

Part of what's so attractive about Marr's theory of vision is that it comes with an explicit discussion of a broad range of metatheoretical issues. In addition to articulating the computational theory itself, we find Marr making numerous comments where he looks to be pushing a top-down approach. He says, for example, that "Although algorithms and mechanisms are empirically more accessible, it is the top level, the level of computational theory, which is critically important from an information processing point of view." (Marr 1982, 27) Indeed, emphasis on understanding the computational problem—the what and the why—*first* can be found throughout both the first and last chapters of *Vision*.

Given the emphasis on computational theory, it is not surprising that many see Marr as offering a top-down *method* for problem solving in cognitive science. Perhaps it is not too big a step to see the top-down approach as a commitment to optimization. Hence we see Kitcher's worry about the empirical adequacy of Marr's computational models, especially given Marr's discussion of Barlow's first dogma.[7] (Marr 1982, 27) But it is useful to observe that Marr's work is not developed from a strict top-down perspective *in fact*. It is not simply that Marr undercuts the exclusivity of this perspective in his metatheoretical remarks even as he appears to establish it.[8] More important is that we find Marr *developing* his computational theory with an eye towards psychophysical and even neurophysiological and neuroanatomical data. It is important because it reflects a commitment to taking seriously the complex nature of the visual system in particular and the mind in general. Suppose we look at some of the details of Marr's theory to see how this is so.

Kitcher discusses several components of Marr's model. Among them are spatial frequency filters, which play a crucial role in helping to isolate features in the retinal image which correspond to genuine object boundaries in the world. To take the adaptationist stance, the problem is that the early physiological mechanisms of vision are good at making fine light/dark discriminations in the retinal image and the visual system needs to identify which light/dark image boundaries correspond to distinct physical phenomena. The system does that by blurring the image at different scales and registering the intensity changes detectable at each scale. It then looks for places where boundaries are coincident in location and orientation for several contiguous scales. An assumption built into the system is that each such location is likely to correspond to a single physical phenomenon.

Marr suggests that two ideas are crucial to the process of detecting the relevant intensity changes. First, since intensity changes occur at very different scales in the image, efficient detection is greatly facilitated by the presence of several operators. Second, we note that a sudden intensity change will produce a peak or a trough in the first derivative of the function describing image intensity and that it will produce what Marr calls a "zero-crossing" in its second derivative; that is, the value of the second derivative will cross from positive to negative (or negative to positive) at the point where a light/dark boundary is located. (1982, 54) One wants, then, a differential operator that can be tuned to different scales. Marr and Hildreth (1980) proposed the filter $\nabla^2 G$; ∇^2 is the Laplacian operator $(\delta^2/\delta x^2 + \delta^2/\delta y^2)$ and G is the two-dimensional Gaussian distribution: $G(x,y) = e^{-(x^2+y^2)/(2\pi\sigma^2)}$. What constrains this choice of operator? What constrains the number of the filters?

Note, first, that the idea of tuned channels in early vision is not original to Marr; Marr cites Campbell and Robson's (1968) adaptation experiments as demonstrating the existence of multiple orientation and spatial frequency selective "channels" in early vision and as giving rise to substantial investigation of the psycho-physical properties of such channels. Note, too, his reference to the quantitative model of the structure of such channels in humans developed in Wilson and Giese (1977) and Wilson and Bergen (1979). (Marr 1982, 62) These early models suggest four channels. In concert with Marr's discussion, the spatial receptive fields of these filters all approximate the shape of a "DOG" (the difference of two Gaussian distributions) filter. Marr favorably reports that "Essentially all of the psychophysical data on the detection of spatial patterns below 16 cycles per degree at constant threshold can be explained by this model, together with the hypothesis that the detection process is based on a form of spatial probability summation in the channels." (1982, 62)

Though Marr questions whether there isn't a smaller channel than the N channel, the smallest allowed by Wilson and Bergen, we notice that his doubts are inspired, at least in part, by analysis of human acuity and resolution. (1982, 63) Indeed, on the basis of "oblique masking" experiments Wilson later revised his model to include six classes of spatial frequency mechanisms. (See, e.g., Wilson, McFarlane and Phillips 1983) More important, perhaps, is the general observation that the computational model is developed with an eye towards psycho-physical data; that is, with an interest in fitting human performance. If this seems an intuitively obvious goal of any computational model of human perceptual faculties, it nonetheless offers a distinct departure from the elegance suggested by Kitcher's analogy of the Panda's thumb; consider the engineering project of building not an idealized grasping device *tout court*, but a device which grasps only as well as the Panda does.

Marr is not merely sensitive to psycho-physical data on human performance and the models derived from these. He also cites development of a quantitative model of the neurological elements he holds responsible for the filters. Beginning with Kuffler's (1953) description of a circularly symmetric receptive field and antagonistic surround for retinal ganglion cells, we proceed to Rodieck and Stone's (1965) notion that the center/surround antagonism is explained by a central excitatory region superimposed upon a larger inhibitory dome covering the entire receptive field. In examining ganglion cells in the cat, Enroth-Cugel and Robson (1966) then suggest that the two domes are Gaussian—hence the DOG mechanisms—and that they may be classified either X cells or Y cells according to their temporal response properties. (Marr 1982, 64) Marr then compares the predicted responses of X cells, supposing they run $\nabla^2 G$, with measured responses of X cells in the retina and LGN.[9] (See Marr 1982, 64-65 and Marr and Ullman 1981) Of course the electro-physiological studies are not sufficient to fix the array of filters described by Wilson and Marr, nor are they sufficient to fix the role of these filters in Marr's larger model. But they do provide interesting "bottom-up" constraints.[10]

A great body of electro-physiological research on response properties of neurons in the visual system has followed Hubel and Wiessel's pioneering work on edge-detection mechanisms in the cortex. I would suggest that this research has had substantial effects on many disparate elements of Marr's theory of vision. Beyond the $\nabla^2 G$ mechanisms cited above I'll mention the mechanism for detecting oriented zero-crossing segments (see Marr and Hildreth 1980 and Marr 1982, 66) and, indeed, the representational primitives of, e.g., edge segments and boundaries (which we find represented in the Primal sketch and exploited by the various sub-mechanisms leading to the 2 1/2-D sketch and the 3-D model). We are reminded, here, of neural mechanisms tuned to edges of specific orientation, to terminating edges and so forth in the striate cortex. In general, the move to integrate psycho-physical results with electrophysiological ones can be seen to have broad influence in vision research. (see, e.g., Phillips and Wilson 1984)

A thorough account of psycho-physical constraints on human performance is well beyond the scope of this essay, as is an account of possible neurological correlates. Readers will find examples of both throughout Marr's work. I'll later raise the question of whether the amount or quality of real interdisciplinarity we see here is ideal. Notice, for now, that it permeates Marr's work, and that it is substantial in its importance for the development of his theory and for our perspective on this theory with regard to optimization.

In citing Marr's and others' use of physiological data I am not seeking to endorse any particular results as the last word on some neurological mechanism nor am I endorsing all the particulars of Marr's computational model. Certain mechanisms discussed in Marr's treatment of vision (e.g., spatial frequency filters, the larger process of stereopsis, etc.) are supported by such a diversity of evidence—evidence from very different experimental paradigms—that their *existence* seems beyond question. But I am not making such a claim for all of the sub-mechanisms posited by Marr. And I am especially concerned to avoid an overly strong commitment to the particulars of our quantitative models of these mechanisms; I've already mentioned a case where we've seen revision of such a model since the publication of *Vision* and, apart from the historical fact that more substantial revisions have been made elsewhere, I want to allow that any of the details are *in principle* revisable independent of our commitment to Marr's overall theory of vision or approach to cognitive science.[11] I am rather concerned to show the *fact* of such inter-level, and inter-disciplinary cross-fertilization. And I want to consider how that fact bears on Marr's notion of what it means to have an understanding of a complex information processing system, on how inter-level constraints develop in practice and about how such constraints ought to develop.

4. Marr and Optimization

Optimization—roughly, the science of optimizing assets or function—is a complex mathematical field with diverse applications in physics, engineering, computer science, cybernetics, economics, manufacturing and biology. For certain problems solutions are straightforward and familiar. In practice, one typically is concerned with maximizing (or minimizing) the value of some numerical quantity that corresponds to, e.g., units of production, weight, temperature, etc. The quantity being maximized (or minimized) typically consists of a function of several variables subject to a set of constraints. We might make three general observations at the outset: first, not all optimization problems admit unique solutions; second, not all optimization problems are clearly soluble; and third, simply framing an optimization problem often is not so much a straightforward technical task as a complex pragmatic, and even philosophical, problem of how to characterize the function in question, and how to identify the relevant variables and constraints.

There's a sense in which the charge of optimization leveled against Marr looks to follow reasonably from some of his discussion of computational theories, and there's a sense in which it looks to be a simple mistake. In his famous discussion of the cash register, Marr claims that a critical feature of the operation described (addition) is that it is uniquely determined by the constraints of the problem. (1982, 23) And in discussing stereopsis he suggests that the "critical step in formulating the computational theory of stereopsis is the discovery of additional constraints on the process that are imposed naturally and that limit the result sufficiently to allow a unique solution." (1982, 104) Perhaps there is an occasional tendency to push rough constraints to an extreme.[12] But the larger theory of vision he proposes is not nearly so well constrained and neither, typically, are the sub-processes he discusses.

Many real world optimization problems lack a unique solution. My concern though is with the kinds of constraints we see developing. Frequently these seem to suggest that the problems being posed by Marr are not optimization problems at all. Real world environmental constraints on vision are liable to be in many cases stochastic and in others simply indeterminate. Bottom-up constraints posed by the stuff of the system itself are also likely to be incompletely specified. I want to suggest that much of the content of Marr's theory depends neither on optimization of function for a simplified world that ignores those features nor—in a pure top-down sense—as an unconstrained optimization problem.

It may be helpful to consider Herbert Simon's notions of "satisficing" or "bounded rationality" and heuristic search. (See, e.g., Simon 1957) Simon suggests that economic agents faced with optimizing problems that—practically if not theoretically—are insoluble nonetheless arrive at acceptable decisions. They do so, he says, by heuristically guided search for decisions that are good enough, or satisfactory, given the functional demands that gave rise to the original optimizing problems; they neither determine nor seek to implement an optimal solution. I want to suggest that Marr typically offers intuitively guided search for solutions that satisfice for the functional problems encountered in developing a theory of the visual system; it's not supposed that the system is an optimal solution to some more precise version of the problem: discover what's out there in the world and where it is given the information in the light stimulus available; and it's not supposed that our model of the system should gloss deviations from optimality found in the system. Neither do we find evidence of a definite method for generating components of the model.

Consider again the filter $\nabla^2 G$. This may be as good a candidate for the optimization charge as any of the mechanisms in Marr's account; indeed he defends it as such. (See Marr and Hildreth 1980) His claim hinges on the fact that rough, general physical considerations prompt two conflicting localization requirements; first, in order to reduce the range of scales at which intensity changes take place one wants the filter to be smooth and limited in the frequency domain; second, because of the typical character of visual stimuli and because of the way these stimuli typically are related to their physical causes, the filter should be smooth and localized in the spatial domain. Following Leipnik (1960) Marr argues that only one distribution—namely, our Gaussian—optimizes this relation. (Marr and Hildreth 1980, 191)

But recall the discussion of section III, and the complex of psychophysical and electro-physiological considerations constraining our notion of function here. And note that the argument in the preceding paragraph does not complete the form of a filter; it does not specify the number or sizes of filters; and it does not establish the set of filters as determining optimal results in the raw primal sketch. Consider too that the filter is appropriate given a certain sort of gray-level image as stimulus—a fact fixed not just by external physical constraints but by a certain idealization of the physiology of our system of sensory transduction. Completing our picture of these mechanisms is something of a pragmatic juggling act, involving an assumption about linear variation that is only approximately true in smoothed images, (Marr and Hildreth, 193-194) comparison with obvious alternative filters and fit with extant psychophysical data.[13] Apart from the fact that fit with the empirical data palliates the problem of optimization to the extent that we find it, we have, in the end, a series of mechanisms that do not look to be optimal in Kitcher's sense. More important than the claim *that* the mechanisms are not thus optimal is that we begin to see the complexity of the task of *characterizing* these local functions.

Elsewhere in the theory we repeatedly fail to find corresponding optimization arguments. The initial image (the I(x,y) array) is simply an idealization of the graded outputs of the sheet of photoreceptors. The various representations (e.g., the primitives of the R.P.S. or the 2 1/2-D sketch) are not claimed to be optimal. Throughout, the treatment of sub-mechanisms, and conjectured sub-mechanisms, suggests satificing more than optimizing (consider stereopsis, shape from shading, occlusion, etc.); the dominant considerations seem to have more to do with what is tractable, and with fit to data, where we have such, than with unconstrained optimization. In general, particular mechanisms, embedded assumptions, etc.—and the representations they generate—are supposed to tend to be useful for the larger process of vision. But it is not supposed that their function—considered locally—is ideal, just as it is not supposed that the larger process of vision always delivers veridical, or even clear or complete, representations of the world scanned by the organism.

Kitcher is rightly concerned about neat *a priori* solutions to complex empirical problems. But it is likely that what concerns her is not genuine optimization. Her admonition attending the tale of the panda's thumb looks to be a worry about more-or-less elegant solutions to loosely posed problems. The "optimal" solution might be any of an enormous number of engineered devices that recognize, perhaps, some general physical and task-imposed constraints but which address very few of the constraints or variables that might have been important in a particular evolutionary situation. So perhaps we're concerned that an optimal solution to the grasping problem unconstrained will be inappropriate to the real world pressures that shaped the device in question. But it seems that we're also being cautioned against a broader class of solutions; that is, highly efficient—but not necessarily optimal—solutions to engineering problems posed under limited constraints. It is not clear how we are to identify an aesthetically inappropriate class of conjectured models *prior* to particular empirical evidence.[14] If there is a sort of "representativeness heuristic" bearing on aspects of model selection *in practice*, we might well suspect its application to vary critically across training networks, laboratories and even individual researchers.

On the one hand, our problems typically are insufficiently constrained for us to consider optimized solutions. Even with a simple grasping device it's not entirely clear how to constitute the design problem as an unconstrained optimization problem, let alone solve it. On the other hand, to the extent that we develop relevant constraints and variables and solve well-defined problems of detail there seems to be less force to the objection against optimization. At the extreme, we might see Wilson, McFarlane and Phillips as having faced a sort of optimization problem once they had their data sets. But surely their curve-fitting techniques are not inappropriate to biological explanation *in principle*.

Adaptationist strategies in general look more like cases of satisficing than optimization. The relevant adaptationist supposition is—I take it—that naturally occurring environments are harsh judges of functionality so we ought to expect evolved devices to be relatively efficient, robust, etc. That's not to say that they are optimal according to some framing of the optimality problem. It seems that the typical assumption is, again, more one of functionality or satisficing. Indeed, William Wimsatt (1986) has suggested that all adaptations meet the definitive criteria for heuristic search procedures.[14] We notice,

> First, ... that adaptations, even when functioning properly, do not guarantee survival and production of offspring. Second, they are nevertheless cost-effective ways of contributing to this end. Third, any adaptation has systematically specifiable conditions under which its employment will actually decrease the fitness of the organism... Fourth, these adaptations serve to transform a complex computational problem about the environment into a simpler problem, the answer to which is usually a reliable guide to the answer to the complex problem.(Wimsatt, 1986, 298)

Of course not all traits are the end-products of evolutionary satisficing (or other bounded search procedures) and, depending on the recognized constraints, there is likely to be a broad range of useful solutions to a given functional demand. But adaptations proper do offer some satisfactory degree of functionality, a degree that is of fitness advantage. And we need keep in mind that we are not, in studying cognition, exclusively concerned with understanding adaptations since many interesting cognitive capacities may not be such in any direct sense.

5. Conclusion

I hope that sections III and IV were sufficient to show that some of Kitcher's worries about Marr's program may have been misplaced, in part because Marr does not exemplify a purely top-down methodology, in part because his theory is not everywhere an optimizing theory and in part because optimization need not always been seen as damaging.

Marr and others have complained that AI is sometimes over-hasty in declaring a program equivalent to a theory. (See, e.g., Marr 1982, 347 and McDermott 1976) We might see an analogous problem more generally in cognitive science; that is, a tendency to make the particulars of one's own disciplinary strengths, or one's own modeling techniques, into methodological constraints for the whole of cognitive science.[16] This may be part of what we see in Marr's rhetorical push for top-down problem solving in cognitive science. I've been concerned to show that this rhetorical emphasis is significantly at odds with what I take to be one of the central strengths of Marr's approach; namely, its thoroughgoing interdisciplinarity. What's important is not simply that inquiry at one level can inform inquiry at another, but that we see the beginnings of systematic account of how different levels of explanation can constrain each other. The twin commitments, to the rigor of an information processing account on the one hand, and to the complexity of functional decomposition (and composition) that are involved in unpacking such an account on the other, are what look to be especially powerful. And there is, of course, the substantial success at demonstrating how such functions may be realized. Perhaps what's most interesting about the optimization question is that it points to the very difficult issue of how one characterizes a function and its constraints, *across* levels of inquiry, both for the system as a whole and for its components. Making such constraints explicit, and investigating their consequences, looks to be central to the joint progress of the cognitive and brain sciences.

By defending Marr's interest in inter-level constraints, I am not asserting that he—or others in the larger computational research program—placed no more than ideal stress on simplicity, or on optimizing considerations, nor that they achieved anything like an ideal degree of emphasis in attending to the various levels in practice. In fact, I suspect an extraordinary utility to a deeper understanding of both neurological and environmental constraints on cognitive processes. Moreover, I would suggest that we are liable to see longstanding discontinuities in the levels themselves; that is, I see no good reason to suppose that mature cognitive and brain sciences will offer comprehensive, continuous causal models for all the phenomena they seek to explain. Indeed, I suspect that part of what's important about developing the levels of explanation simultaneously is not just that one cannot predict from where insights will arrive, but that one may inevitably need to consult multiple levels of explanation to develop useful accounts of certain cognitive and brain processes. Where such discontinuities may prove to be intractable remains to be seen. Furthering our understanding of how our various inquiries into cognition can constrain each other should help to delimit such discontinuities, just as it serves to make our understanding of cognition more (optimally) coherent.

Notes

[1] Any early version of this material was presented at the 9th International Congress of Logic, Methodology and Philosophy of Science. Thanks to those in attendance, and to Mark Detweiler and Ron McClamrock, for their helpful comments.

[2] Daniel Dennett has advanced similar claims in Dennett (1991)—a fact perhaps surprising given his longstanding support of optimizing assumptions in cognitive science. I should point out that both Kitcher and Dennett present their criticisms as representing plausible and serious difficulties for Marr's program and not as claims that the program is certainly wrong.

[3] Following S.J. Gould's (1980) discussion of development of the panda's thumb as an adequate but surely inelegant grasping device.

[4] Through, of course, processes of mutation that operate on the underlying genotype; and, as is well familiar, the map between genetic specification and development of phenotypic traits is not at all simple, and neither is that between selection pressures and propagation of such traits.

[5] Gould's own use of the argument is a bit different. He's concerned, I think, not to show that we ought to expect a universal dearth of elegant design in evolved functional systems but that we will find inelegant mechanisms to be ubiquitous.

[6] Marr's comment that seeing is the ability "to know what is where by looking" (1982, 3) might have the right level of detail.

[7] "A description of the activity of a single nerve cell which is transmitted to and influences other nerve cells, and of a nerve cell's response to such influences from other cells, is a complete enough description for functional understanding of the nervous system." (Barlow 1972, 380).

[8] Marr also observes of "the methodology or style of this type of approach" that "there is no real recipe for this type of research—even though I have sometimes suggested that there is." (1982, 331) He says that he's "laid particular stress on the level of computational theory, not because I regard it as inherently more important than the other two levels . . . but because it is a level of explanation that has not be previously recognized and acted upon." (330)

[9] Measured by Rodieck, R. and Stone, J. (1965) and Dreher, B. and Sanderson, K. (1973).

[10] Notice that the electro-physiology precedes publication of Wilson's model and Marr's larger theory, though one might see at least the later measurements as simultaneous with early attempts to model these psychological mechanisms.

[11] At the same time, there may be an over-strong tendency to discount the early models. Substantial revision may be not so much matter of reconceiving a phenomenon outright as one of tuning our representation of it.

[12] For example, he reasonably suggests that in "real life time is often of the essence" and then suggests that for analysis of motion the relevant systems are "likely to use the earliest representations that they possibly can." (19892, 105)

[13] Again, recall section 4's discussion of the various considerations involved in the design of this model.

[14] Interestingly, it looks as though we're being cautioned against accounts that are overly elegant, not against the gerrymandered or baroque. There may be a suspicion, amongst some biologists, against certain sorts of formal elegance that can have extra-empirical appeal in the physical sciences. But even if this is true, it's not clear *how* we might elevate such a habit to a methodological standard and we ought to be chary about attempting to do so. We need to keep in mind, too, that whether our models are optimizing models, and if not, that the question of to what extent they depict function as deviating from optimality, is partly an artifact of how we choose to characterize the function in the first place.

[15] I do not mean to suggest that Wimsatt's own view, here, is an adaptationist one.

[16] For an example of this on a grand scale, readers might consider the debate about whether the particulars of various connectionist modeling architectures constitute THE structure of cognition.

References

Barlow, H.B. (1972), "Single Units and Sensation: A Neuron Doctrine for Perceptual Psychology", *Perception* 1: 371-394.

Campbell, F. and Robson, J. (1968), "Application of Fourier Analysis to the Visibility of Gratings", *Journal of Physiology* (London) 197: 551-566.

Chandrasekhar, S. (1987), *Truth and Beauty, Aesthetics and Motivations in Science*. Chicago: University of Chicago Press.

Dennett, D. (1991), "Cognitive Science as Reverse Engineering: Several Meanings of 'Top-Down' and 'Bottom-Up'", presented August 10, 1991 at the 9th International Congress of Logic, Methodology and Philosophy of Science

Dreher, B. and Sanderson, K. (1973), "Receptive Field Analysis: Responses to Moving Visual Contours by Single Lateral Geniculate Neurons in the Cat", *Journal of Physiology* (London) 234: 95-118.

Enroth-Cugel, C. and Robson, J. (1966), "The Contrast Sensitivity of Retinal Ganglion Cells of the Cat", *Journal of Physiology* (London) 187: 517-522.

Gould, S.J. (1980), *The Panda's Thumb, More Reflections in Natural History*. New York: Norton.

Kitcher, P. (1988), "Marr's Computational Theory of Vision", *Philosophy of Science* 55: 1-24.

Kuffler, S.W. (1953), "Discharge Patterns and Functional Organization of Mammalian Retina", *Journal of Neurophysiology* 16: 37-68.

Leipnik, R. (1960), "The Extended Entropy Uncertainty Principle", *Inf. Control* 3: 18-25.

Marr, D. (1982),*Vision*, New York: Freeman.

Marr, D. and Hildreth, E. (1980), "Theory of Edge Detection", *Proceedings of the Royal Society of London* B207: 187-217.

Marr, D. and Ullman, S. (1981), "Directional Selectivity and its Use in Early Visual Processing", *Proceedings of the Royal Society of London* B211: 151-180.

McDermott, D. (1976), "Artificial Intelligence Meets Natural Stupidity", *SIGART Newsletter*, 57: 4-9.

Phillips, G. and Wilson, H. (1984), "Orientation Bandwidths of Spatial Mechanisms Measured by Masking", *Journal of the Optical Society of America* 1:2: 226-232.

Rodieck, R. and Stone, J. (1965), "Analysis of Receptive Fields of Cat Retinal Ganglion Cells", *Journal of Neurophysiology* 28: 833-849.

Schoemaker, P. (1991), "The Quest for Optimality: A Positive Heuristic of Science?" *Behavioral and Brain Sciences* 14: 205-245.

Simon, H. (1957), *Models of Man*. New York: Wiley.

Wilson, H.R., McFarlane, D. and Phillips, G. (1983), "Spatial Frequency Tuning of Orientation Selective Units Estimated by Oblique Masking", *Vision Research* 23:9: 873-892.

Wilson, H.R. and Bergen, J. (1979), "A Four Mechanism Model for Spatial Vision", *Vision Research* 19: 19-32.

Wilson, H.R. and Giese, S. (1977), "Threshold Visibility of Frequency Gradient Patterns", *Vision Research* 17: 1177-1190.

Wimsatt, W. (1986), "Heuristics and the Study of Human Behavior", in D. Fiske and R. Shweder, eds., *Metatheory in Social Science: Pluralisms and Subjectivities*. Chicago: The University of Chicago Press.